Marine Molecular Biotechnology

Subseries of Progress in Molecular and Subcellular Biology
Series Editor: Werner E. G. Müller

Progress in Molecular and Subcellular Biology **43**
Series Editors: W.E.G. Müller (Managing Editor)
Ph. Jeanteur, Y. Kuchino, A. Macieira-Coelho, R. E. Rhoads

Volumes Published in the Series

Progress in Molecular and Subcellular Biology

Volume 27
Signaling Pathways for Translation: Stress, Calcium, and Rapamycin
R.E. Rhoads (Ed.)

Volume 28
Small Stress Proteins
A.-P. Arrigo and W.E.G. Müller (Eds.)

Volume 29
Protein Degradation in Health and Disease
M. Reboud-Ravaux (Ed.)

Volume 30
Biology of Aging
A. Macieira-Coelho

Volume 31
Regulation of Alternative Splicing
Ph. Jeanteur (Ed.)

Volume 32
Guidance Cues in the Developing Brain
I. Kostovic (Ed.)

Volume 33
Silicon Biomineralization
W.E.G. Müller (Ed.)

Volume 34
Invertebrate Cytokines and the Phylogeny of Immunity
A. Beschin and W.E.G. Müller (Eds.)

Volume 35
RNA Trafficking and Nuclear Structure
Dynamics
Ph. Jeanteur (Ed.)

Volume 36
Viruses and Apoptosis
C. Alonso (Ed.)

Volume 38
Epigenetics and Chromatin
Ph. Jeanteur (Ed.)

Volume 40
Developmental Biology of Neoplastic Growth
A. Macieira-Coelho (Ed.)

Volume 41
Molecular Basis of Symbiosis
J. Overmann (Ed.)

Subseries: Marine Molecular Biotechnology

Volume 37
Sponges (Porifera)
W.E.G. Müller (Ed.)

Volume 39
Echinodermata
V. Matranga (Ed.)

Volume 42
Antifouling Compounds
N. Fusetani and A.S. Clare (Eds.)

Volume 43
Molluscs
G. Cimino and M. Gavagnin (Eds.)

Guido Cimino Margherita Gavagnin (Eds.)

Molluscs

From Chemo-ecological Study to Biotechnological Application

With 105 Figures, 9 in Color, and 18 Tables

Springer

Professor Dr. Guido Cimino
Istituto di Chimica Biomolecolare
Consiglio Nazionale delle Ricerche
Via Campi Flegrei, 34
80078 Pozzuoli (Naples)
Italy

E-Mail: gcimino@icmib.na.cnr.it

Professor Dr. Margherita Gavagnin
Istituto di Chimica Biomolecolare
Consiglio Nazionale delle Ricerche
Via Campi Flegrei, 34
80078 Pozzuoli (Naples)
Italy

E-Mail: mgavagnin@icmib.na.cnr.it

ISSN 1611-6119
ISBN-10 3-540-30879-2 Springer-Verlag Berlin Heidelberg New York
ISBN-13 978-3-540-30879-9

Library of Congress Control Number: 2005937052

Springer-Verlag is a part of Springer Science + Business Media
springer.com

Typesetting: SPI Publisher Services, Pondicherry
Cover illustration: nudibranch *Jorunna funebris* by courtesy of Dr. E. Mollo
Cover design: design & production GmbH, Heidelberg, Germany

Printed on acid free paper 39/3150 YK 5 4 3 2 1 0

Preface

The first volume of "Marine Molecular Biotechnology" – a subseries of "Progress in Molecular and Subcellular Biology" - selected a very stimulating topic: "Sponges (Porifera)". The book proves that these animals are only apparently simple. All chapters discover new scenarios with implications for evolution, associated microbiology, biodiversity, sustainable exploitation and, of course, good science. This success prompted the editors to continue this series selecting other topics. Professors Müller and Schröder suggested "Molluscs" and we were generously invited to join them in this exciting adventure.

Analogously to sessile organisms, slow moving marine invertebrates are apparently without defence against both attacks from predators and infections from micro-organisms even though they can select the best habitat for their success in survival. Molluscs, and in particular gastropods, fall in this category. They are generally protected by the shell and, sometimes, also by toxins. Surprisingly, the venomous compounds from some shelled molluscs can aid people to overcome the terrible pains of terminal diseases. An example are the venoms of some *Conus* molluscs which possess analgesic properties fifty times as stronger than that of morphine. Other molluscs, the opisthobranchs, are only partially protected by the shell. They were successful in their survival by constructing a very effective arsenal of chemical weapons either sequestered from the organisms upon which they feed or biosynthesized by themselves.

During the 70's, many outstanding scientists (Prof. J. Faulkner, Prof. W. Fenical and Prof. P. Scheuer) attracted the attention of the scientific community with their exciting pioneering studies on opisthobranchs. Since then, many groups have worked on this topic. The studies have moved slightly from chemical ecology, to advanced biochemistry and applied biotechnology. Many intriguing molecules have been isolated from molluscs and some of them are now in an advanced clinical phase. Three of the five PharmaMar compounds, at present tested in human clinical trials, were detected by studying marine molluscs.

The volume "Molluscs" offers to readers an almost complete coverage of the most stimulating topics related to molluscs, with the contributions of many authoritative scientists active in this field. Organisms from all seas are treated with the exception of those recently reviewed from the Mediterranean Sea.

An explicative guide could be useful to the reader to navigate through the volume. After an ecological introduction in the first chapter (Avíla), toxins from bivalves and prosobranchs are extensively discussed in the following three chapters (Uemura, Fattorusso and Marì). Darias reports a comprehensive overview of the bioactive molecules from pulmonate gastropods. The subsequent chapters deal exhaustively with molluscs from distinct geographical areas, i.e. Antarctica, South Africa and South America

(Davies-Coleman), Australia and New Zealand (Garson), India, China and Egypt (Wahidullah), and Japan (Miyamoto). Some relevant specific topics are reported by Kamiya (bioactive proteins), Matsunaga (trisoxazole macrolides), and Proksch (alkaloids). The two following chapters describe biosynthetic studies on molluscs from the West coast of North America (Andersen) and from Mediterranean littorals (Fontana) and introduce one of the most intriguing topics exhibited by opisthobranchs: the ability to construct *de novo* their bioactive compounds. At present, outstanding groups in the world are very active in the synthesis of molecules isolated from molluscs. However, this interesting topic is only partially treated here. The synthesis of peptides and depsipetides (Spinella) has been selected due to the very promising antitumor activity of these molecules. Finally, some potent anticancer agents in clinical trials are described in the last chapter (Cuevas).

"Molluscs" is dedicated to Prof. Kenneth L. Rinehart, Prof. Guido Sodano and Prof. Salvatore De Stefano.

The outstanding scientific activity of K.L. Rinehart is mentioned in Fernàndez's foreword.

Here, we want to remember that the first work (1979) of our group and many other studies on opisthobranchs were carried out thanks to the valuable contribution of our colleagues and friends Guido and Salvatore.

Guido Cimino and Margherita Gavagnin
Istituto di Chimica Biomolecolare (CNR) – Pozzuoli (Naples)

Prof. S. De Stefano and Prof. G. Sodano

Acknowledgements. We are deeply grateful to Mr. Raffaele Turco for his precious help in the editing work of this book.

Preface by the Series Editor

Life originated in the oceans and has evolved there over a much longer time than on land, so the diversity of life in marine habitats is far greater than its terrestrial counterpart. Oceans cover nearly 70% of earth's surface and provide more than 90% of habitats for the planet's life forms. The first living organisms appeared in the sea more than 3500 million years ago and evolutionary development has equipped many marine organisms with the appropriate mechanisms to survive in a hostile milieu in terms of extreme temperatures, changes in salinity and pressure, as well as overcoming the effects of mutation, or bacterial and viral pathogens. The diversity in species is extraordinarily rich not only in coral reefs but also in other almost undisturbed natural marine habitats. Marine organisms have developed exquisitely complex biological mechanisms showing cross-phylum activity with terrestrial biota. In terms of evolution and biodiversity, the sea appears to be superior to the terrestrial ecosystem and marine species comprise approximately half of the total biodiversity, thus offering a vast source from which to discover useful therapeutics.

Several marine organisms are sessile and soft bodied. The question thus arises: how do these delicate-looking simple sea creatures protect themselves from predators and pathogens in the marine environment? While answering this interesting ecological question, researchers found that marine organisms have chemical defensive weapons (secondary metabolites) for their protection. Outstanding taxa that are extremely rich in those bioactive secondary metabolites are the mollusks. Intensive evolutionary pressure from competitors, that threaten by overgrowth, poisoning, infection, or predation, has armed these organisms with an arsenal of potent chemical defense agents. They have developed the ability to synthesize such chemical weapons or to obtain them from marine microorganisms. Those compounds help them to deter predators, keep competitors at bay, or paralyze their prey.

Investigations in the field of chemical ecology have revealed that the secondary metabolites not only play various roles in the metabolism of the producer but also in their strategies in the given environment. The diversity of secondary metabolites produced by marine organisms has been highlighted in several reviews and now comprehensively in this monograph. They range from derivatives of amino acids and nucleosides, macrolides, porphyrins, terpenoids, to aliphatic cyclic peroxides and sterols. There is ample evidence documenting the role of these metabolites in chemical defense against predators and epibionts. The studies on marine chemical ecology in mollusks cover three different aspects. Firstly, the diversity of chemical compounds produced by different organisms; secondly, the potential functions of these metabolites in nature: and finally, the strategies for their use for human benefit.

It is the merit of one of the most efficient experts working in the field of marine natural products, Prof. Guido Cimino (Napoli), to have called together prominent colleagues working in the field of natural products from mollusca to highlight and push forward research on bioactive secondary compounds from these animals. Guido Cimino is a pioneer who succeeded in establishing that various patterns in the evolution of chemical defense exist, including detoxification, modification, and sequestration of metabolites and the positioning of those in places where they will effectively repel predators. I am sure that this monograph will be a platform for future successful developments in this field.

Werner E.G. Müller
University of Mainz

Foreword 1

It is an honour for me to accept Professor Guido Cimino's invitation to write a foreword to the volume "Molluscs" of the "Marine Molecular Biotechnology" series, edited by Professor Werner E. G. Müller.

Mankind has always been very dependent on the sea, but the discovery of a new source of medicines in the organisms living in the oceans has opened up an enormously interesting new frontier. We founded PharmaMar in 1986 to explore this new frontier. Today, I am even more convinced of the potential of marine organisms as a source of medicines, since the company has five marine anticancer compounds undergoing clinical trials, with more than 4000 cancer patients treated so far. It is relevant in the context of this book that three of these molecules have been isolated from molluscs or derived from those present in molluscs, to which this volume is dedicated.

I would also like to express my recognition to the scientists working in marine organic chemistry who contributed to the discovery of those antitumour molecules, which are derived from molluscs that are in clinical trials: Professor Paul J. Scheuer for the discovery of Kahalalide F from the sacoglossa *Elysia rufescens*; Professor Kenneth L. Rinehart for the discovery of Spisulosine (ES-285) from the lamellibranch *Mactromeris polynima*; Professor George R. Pettit for the discovery of the first Dolastatin from the anaspidea *Dolabella auricularia*; and Professor Guido Cimino for the discovery of Jorumycin from the nudibranch *Jorunna funebris*, from which the PM-104 (Zalypsis®) analog is derived. And, for the treatment of chronic pain, the conotoxin Prialt, which was discovered by Dr. Baldomero Oliveira and his colleagues from the neogastropoda *Conus magus*, also deserves consideration.

I shall also take this opportunity to say a few words about Ken Rinehart, who passed away a few months ago. It goes without saying that Ken Rinehart was one of the most productive scientists researching marine organic chemistry, and a point of reference that we will all sorely miss in the future. I regret that he did not live to see ecteinascidin-743 (ET-743), which was discovered by his group, commercialised for the treatment of certain cancers, such as ovarian cancer or sarcomas. When these new treatments become available, I hope in the near future, they will represent a legacy from Ken to the scientific community

Ken Rinehart was for many years a member of the PharmaMar Board of Directors. He was also the person who selected the name *PharmaMar* for our company. Throughout the years, he served on many scientific committees where strategic decisions were made, and participated in several PharmaMar scuba diving expeditions. We very much enjoyed having him so involved with our company. Ken will always be with us.

José Maria Fernàndez Sousa-Faro
PharmaMar – Madrid

Prof. K.L. Rinehart

Foreword 2

Molluscs are the largest of all marine invertebrate groups, consisting of gastropods, bivalves, scaphopods, cephalopods, aplacophorans, monoplacophorans, and polyplacophorans, many of which have been widely used as food by humans. On the other hand, shells of gastropods and bivalves have been used for making tools and ornaments.

Molluscs have been overlooked as biotechnological resources, except for Tyrian purple (or royal purple), a brilliant dye derived from gastropods of the superfamily Muricacea used in the eastern Mediterranean and in China. Perhaps it represents the earliest documented application of marine biotechnology. However, recent progress in marine biotechnological research has shown that molluscs are potential resources for biomedical and other biomaterials as partly described in this book.

Gastropods and bivalves sequester a variety of chemicals from food organisms; bivalves often accumulate toxins from phytoplanktons referred to as harmful algae (HABs) and cause food poisoning not only in humans but also in marine mammals, which pose serious problems to food safety as well as to marine environments. Therefore, HABs and shellfish poisonings are an important area in marine biotechnology. Opisthobranchs are a group of gastropods that are lacking in the shell for physical defence, and have instead developed chemical defences. They sequester defensive substances such as toxins, antifeedants, and allelochemicals from their food, e.g., seaweeds, sponges, coelenterates, bryozoans, and tunicates, which results in significant regional variations in their defensive substances. The recognition mechanism of defensive chemicals by nudibranchs may be applicable to many areas, especially to drug delivery systems. The chemical defence of opisthobranchs is a good model for understanding chemically mediated interactions of marine organisms. A variety of unusual peptides isolated from herbivorous opisthobranchs are powerful anticancer agents; several of them are currently under human clinical trials. These peptides are actually of algal origin, mostly cyanobacteria (blue-green algae). Fortunately, most of these peptides can be supplied by chemical synthesis, differing from the case of most marine natural product drug candidates.

Defensive compounds are also synthesized by gastropods. Particularly interesting are polygodial and polypropionates, the former of which is synthesized from mevalonates by nudibranchs of the genus *Dendrodoris*. It is a wonder of nature that this powerful antifeedant is used for the same purpose by terrestrial plants. Polypropionates, which are a rare class of marine natural products, are contained in both pulmonates and sacoglossans, the latter of which contain active chloroplasts sequestered from green algae. Again, mechanisms of this sort of symbiosis and chemical recognition are interesting subjects.

Proteins and peptides of molluscs have not been well explored. As found in many animal species, a variety of antimicrobial peptides (AMPs) and proteins are reported from bivalves and gastropods. They are involved in innate immunity and potential antimicrobial agents. Cone snails contain numerous numbers of small peptides tabbed as conotoxins possessing various pharmacological activities, most of which have enormous therapeutic potential. In fact, •-conotoxin MVIIA, an N-type Ca^{2+} channel blocker, has recently been approved as an analgesic in the USA. Biopolymers such as glue proteins produced by bivalves, especially mussels, have potential for biotechnological applications.

Cephalopods are unique among molluscs; they can swim fast and use ink for defence. Perhaps this property prevents them from having interesting chemicals for their defence. Biotechnological investigation of cephalopods is very limited.

There is no doubt that molluscs are an important biotechnological resource as briefly mentioned above. Obviously, we need to exploit them more deeply from application-oriented viewpoints.

Nobuhiro Fusetani
Hokkaido University

Contents

Molluscan Natural Products as Biological Models: Chemical Ecology, Histology, and Laboratory Culture

C. Avila

Abstract. The utility of some natural products from molluscs has been known for centuries. However, only recently have modern technologies and advances in the fields of chemistry, chemical ecology, anatomy, histology, and laboratory culture allowed the exploitation of new, unprecedented applications of natural products. Recent studies have dealt with (a) the role that these compounds have in the sea in protecting the animals (e.g., chemical defense), or in mediating their intraspecific communication (e.g., pheromones), (b) the geographical differences in similar or related species (and the implications of this in chemical ecology and phylogeny), and (c) the localization of these metabolites in molluscan tissues (by means of the most modern technologies), among others. The methodology for the laboratory culture of some species has also been established, thus offering new insights into this interesting field. Further applications of all these challenging studies are currently being developed.

1.1 Introduction

Molluscs include more than 100,000 species living in marine, freshwater, and terrestrial habitats (Barnes et al. 1988; Hickman et al. 2002; Brusca and Brusca 2003). The main groups and their general trends are reported in Table 1.1. It is interesting to note that about 98% of them are either gastropods or bivalves, while the remaining groups are not so abundant (Table 1.1). Gastropods and bivalves have been the main groups studied for natural products so far, while the other groups have been scarcely studied. Gastropods, the largest group, include the typical marine snails (Prosobranchs), sea hares and sea slugs (Opisthobranchs), as well as terrestrial snails and slugs (Pulmonates). As reported in Table 1.1, most groups either possess a shell, and therefore have a mechanical defense against predators, or they possess effective swimming mechanisms to escape from predators (e.g. cephalopods). Soft-bodied molluscs have always been the favorite choice for natural product chemists because the chances of finding defensive chemicals are expected to be higher. However, only

Conxita Avila
Centre d'Estudis Avançats de Blanes (CEAB-CSIC), C/Accés a la Cala Sant Francesc 14
17300 Blanes (Girona), Catalunya, Spain

Progress in Molecular and Subcellular Biology
Subseries Marine Molecular Biotechnology
G. Cimino, M. Gavagnin (Eds.): Molluscs

Table 1.1. Main groups in the phylum Mollusca

	Caudofoveata	Solenogastres	Polyplacophora	Monoplacophora	Gastropoda	Cephalopoda	Bivalvia	Scaphopoda
Radula	yes/no	yes/no	yes	yes	yes	yes	no	yes
Foot	no	grooved	large, muscular	weak, circular, flat	large, crawling	modified into tentacles	conic, burrowing	burrowing
Shell	no (calcareous scales)	no (calcareous scales)	eight plates	univalve	coiled or reduced	chambered or reduced	bivalve	tubular
Gills	yes	no	multiple	yes	yes	yes	laminar	no
Feeding	deposit-feeders	carnivores (cnidarians)	algal grazers	deposit-feeders	all types	carnivores	deposit-feeders, filter-feeders	deposit-feeders
Habitat	marine	marine	marine	marine	marine, freshwater, terrestrial	marine	marine, freshwater	marine
Species number	70	180	650	10	80,000	1,000	20,000	350
Natural products studied	no	no	no	no	yes	yes	yes	no

seldom has it been ecologically proved that the chemicals described really are useful for defending the molluscs themselves in their own habitat.

From all these different groups, probably even less than 400 species have been studied for natural products (Fig. 1.1), since many species from different geographical areas have been repeatedly studied over the years. Although the number of chemical compounds studied and the number of published papers have constantly increased for almost two decades (Faulkner 2002 and references therein; Blunt et al. 2005 and references therein; Fig. 1.1), this means that a maximum of about 0.4% of molluscs have been chemically analyzed and therefore there is still a lot to know about them. The natural products found in molluscs, however, are of high complexity and structural diversity (Pietra 2001). Also, if we look at the analyzed species in the different groups, we realize that only 0.25% of the studies deal with cephalopods, 6.4% with prosobranchs, 7.6% with bivalves, 14.2% with pulmonates, and finally, as the most studied group, 71.6% with opisthobranchs (Faulkner 2002 and previous reports). Both nudibranchs and sacoglossans, two of the most studied groups, belong to the opisthobranch gastropods.

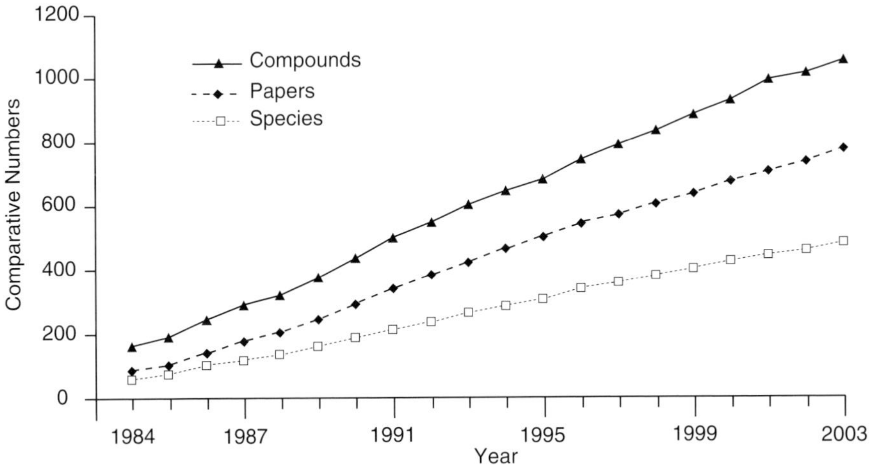

Fig. 1.1. Comparative numbers of papers, species, and compounds mentioned in the successive reports published by John D. Faulkner from 1984 to 2002 and by John W. Blunt et al. from 2003 to 2005 (Faulkner 2002 and previous reports mentioned therein; Blunt et al. 2005 and previous reports)

The use of molluscs in traditional medicine goes back to the times of Dioscorides and Pliny the Elder, as reported by Caprotti (1977) and Herbert et al. (2003). Many marine molluscs are currently used in traditional medicine in Africa, China, the Philippines, and Korea (Herbert et al. 2003 and references therein). Currently, several compounds from molluscs are in a preclinical or clinical phase for their use in the pharmaceutical industry, such as dolastatin 10, ziconotide, aplyronine A, and kahalalide F (Proksch et al. 2002). In fact, the ziconotide from *Conus magus* has already been synthesized, thus avoiding the problem of collecting (Proksch et al. 2002).

A wide range of compounds have recently been described in molluscs. However, it is not our purpose here to review all the recent findings on mollusc chemicals (see Harper et al. 2001; Faulkner 2002 and previous reports; Blunt et al. 2005), but as an example, some studies will be mentioned. Some sea hare metabolites show antimicrobial and/or antitumor activity (e.g., Faulkner and Stallard 1973; Rinehart et al. 1981; Ichida and Higa 1986; Pettit et al. 1987, 1990; Yamazaki 1993; see Avila 1995 and references therein; Melo et al. 2000). Recently, an anti-HIV protein, bursatellanin-P, was isolated from the purple secretion of an opisthobranch (Rajaganapathi et al. 2002). A prosobranch and a sacoglossan opisthobranch also possess anti-HIV factors (Orlando et al. 1996; Hamann et al. 1996) and some chemicals from sea hares provide interesting immunological results that could include the induction of apoptosis (Iijima et al. 2003 and references therein). Dolastatin 10, from the opisthobranch *Dolabella auricularia*, has been used successfully for the treatment of human prostate cancer (Turner et al. 1998). Kahalalide F from the sacoglossan *Elysia rufescens* and ES-285 from the bivalve *Spisula polynyma* are important anticancer agents currently under pharmaceutical development (Jimeno 2002). Another example is the potential biomedical use of chitosan obtained from gastropods and bivalves by Zentz et al. (2001) due to its biological characteristics. Bivalves also provide proteins (molluscan shell proteins; MSP) which may have interesting biological functions (e.g., Sarashina and Endo 2001) and antitumoral compounds of pharmacological interest (e.g., Takaya et al. 1998). Fontana et al. (2000) described a new antitumor alkaloid from a nudibranch which also shows antimicrobial activity. Osteogenetic activity has been reported for extracts of bivalves (Mouries et al. 2002 and references therein). Furthermore, oxidative stress markers have been studied by Cavas et al. (2005) in two sacoglossans (Opisthobranchia). Abalones (*Haliotis discus*) provide compounds with interesting immunological properties (Yoneda et al. 2000). And, finally, *Conus* species are a continuous source of interesting pharmacologically and neurologically active compounds, with more than 50,000 different conotoxins (e.g. Yang et al. 2000; Olivera and Cruz 2001) and at least 80 filed patents (Kohn 2005).

Several reviews have covered the topic of chemical ecology in molluscs (Karuso 1987; Faulkner 1988, 1992; Cimino and Sodano 1989; Pawlik 1993; Avila 1995; Cimino and Ghiselin 1998; Williams and Walker 1999; Amsler et al. 2001a; Stachowicz 2001). In general, they covered different aspects, such as chemical defense in shell-less molluscs, molluscan venoms, shellfish poisons from microalgae, and chemical cues for settlement and metamorphosis. Some reviews dealt with natural products from some mollusc groups, such as porostome nudibranchs (Gavagnin et al. 2001), dorids and sacoglossans (Cimino et al. 1999; Cimino and Ghiselin 1998, 1999), or gastropods in general (Cimino and Ghiselin 2001), incorporating the evolutionary perspective in their analysis. In the most studied group, the opisthobranchs, Faulkner and Ghiselin (1983) have already discussed the importance of the acquisition of defensive chemicals during evolution, thus allowing reduction of the shell. This may have many ecological implications, such as the advantage of searching for new food sources, the exploitation of new habitats, and the development of mantle glands or structures, among others.

In this chapter, we intend to focus on natural products from molluscs and their possible use as biological models in three main topics: chemical ecology, histology, and laboratory culture. In these fields, the development of new biotechnological tools plays an essential role.

1.2 Chemical Ecology of Molluscs

Chemical ecology examines the roles of naturally occurring compounds in plant and animal interactions (Paul 1992). Natural products are important in the interactions between organisms (either intra- or interspecific) and with the environment. Although we are still far from understanding the processes involved in chemical signaling in the sea, advances have been noteworthy (Zimmer and Butman 2000). One of the goals now is to determine the mechanisms by which chemicals with ecological activity contribute to structuring the communities. There is increasing evidence that chemical signals are important for mediating ecological interactions in the sea, at many different levels: metamorphic inducers, chemical defenses, pheromones, chemical cues, etc. (see Zimmer and Butman 2000 for a review).

Interesting studies have dealt with prosobranchs, showing that kairomones and pheromones regulate reproductive behavior (Moomjian et al. 2003), as well as feeding behavior (Rittschof et al. 2002). In fact, alarm pheromones were reported long ago for prosobranchs (Atema and Stenzler 1977 and references therein) and opisthobranchs (e.g., Cimino et al. 1991a). Also, prostaglandins and related eicosanoids are related to egg production in pulmonates and spawning in bivalves and they are involved in neurophysiology (Stanley-Samuelson 1994 and references therein). In opisthobranchs, prostaglandins have also been described for nudibranchs, displaying different roles, such as defensive compounds and cerata contraction (e.g., Cimino et al. 1991b; Di Marzo et al. 1991).

As the most studied factor, chemical defense stands as the main topic in molluscan chemical ecology, although it is by no means the only one. The importance of correctly demonstrated in situ activity against co-occurring predators has been the subject of repeated debate (Faulkner 1992; Avila 1995). However, many compounds are still assumed to have a defensive role without reliable ecological experiments. Until now, the ecological activity of these compounds has been evaluated in situ, against co-occurring predators, for only a few species (Avila and Paul 1997; Johnson and Willows 1999; Avila et al. 2000; Iken et al. 2002; Rogers et al. 2002). Some interesting results include those of Becerro et al. (2001) for *Elysia rufescens* and its algal prey, where kahalalide F protects both the sacoglossan and the alga from fish predators in the field at natural concentrations. Also in studies dealing with Antarctic species, co-occurring predators have been shown to be deterred by natural compounds of the nudibranchs *Austrodoris kerguelenensis* and *Bathydoris hodgsoni* (Avila et al. 2000; Iken et al. 2002). The methodological difficulties in carrying out in situ experiments or using co-occurring predators are probably responsible for the scarcity of the information available.

Furthermore, the effects of natural compounds may be different in different kinds of predators and habitats, although there are very few studies on this. For example, this has been shown in algal compounds (Paul 1992; Steinberg 1992; Cronin et al. 1997), where there are also differences at a geographical level. Marine invertebrates from Antarctica have been studied in recent years, yielding interesting results, although so far there have been fewer studied than Mediterranean and tropical invertebrates (McClintock and Baker 1997; Pietra 2001; Amsler et al. 2001b). Several studies of Antarctic molluscs have revealed the potential interest of their natural products (e.g., Davies-Coleman and Faulkner 1991; McClintock et al. 1994; Bryan et al. 1995; Avila et al. 2000; Iken et al. 2002). Further chemical studies have provided more details on the structure of the natural compounds and their possible origin (Gavagnin et al. 2003a,b and references therein). In fact, in nudibranchs, some compounds are amazingly similar in both temperate and polar regions. Polygodial from *Dendrodoris* species (e.g., Cimino et al. 1985; Avila et al. 1991) and hodgsonal from the Antarctic *Bathydoris hodgsoni* (Iken et al. 1998; Avila et al. 2000) are highly active sesquiterpene aldehydes, effective against potential predators (Fig. 1.2). Similarly, diacylglycerids found in the Antarctic nudibranch *Austrodoris kerguelenensis* (e.g., Gavagnin et al. 1995, 1999a,b, 2003a; Iken et al. 2002) are structurally related to those previously isolated from the Mediterranean nudibranch *Doris verrucosa* (e.g., Cimino et al. 1988; Avila et al. 1990; Gavagnin et al. 1997; Fig. 1.2).

A review of the natural products of opisthobranch molluscs was published ten years ago (Avila 1995). Since then, many papers have dealt with opisthobranchs and described interesting new chemistry, although not all of them can be mentioned here. The geographic variation of natural products in *Asteronotus cespitosus* (Fahey and Garson 2002) and in *Cadlina luteomarginata* (Kubanek et al. 2000) has provided new insights into the field. Kubanek et al. (2000) suggested that, in some nudibranchs, de novo biosynthesis may be modulated by habitat-specific external factors, thus working only when dietary compounds are not available. The authors suggested this represents an intermediate stage in the evolution of nudibranch chemical defenses between the probable primitive chemical sequestration from diet and the more evolved processes of de novo biosynthesis. The fact that some nudibranchs may only biosynthesize when dietary compounds are not available is an open question to be tested in other species. Among nudibranchs, only *C. luteomarginata* and *Dendrodoris grandiflora* have been reported to possess both dietary sequestered compounds and biosynthetic chemicals (Cimino et al. 1985; Avila et al. 1991; Kubanek et al. 2000).

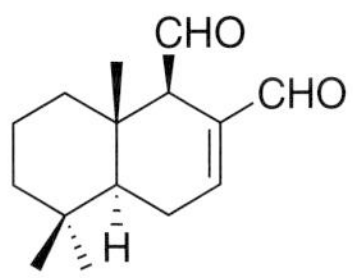

polygodial

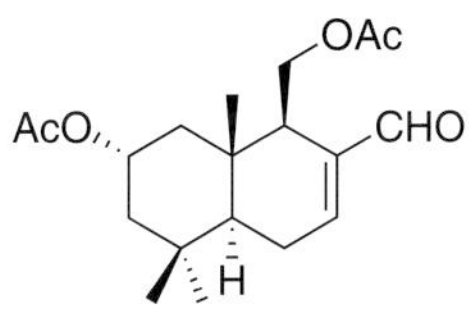

hodgsonal

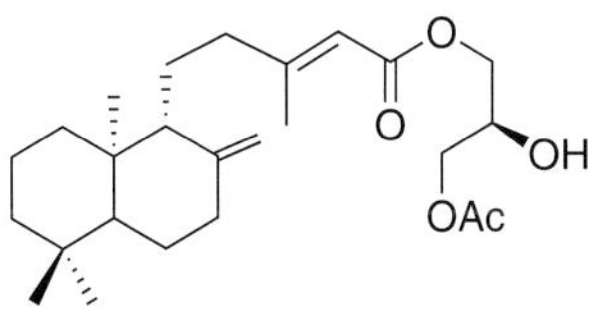

verrucosin-5

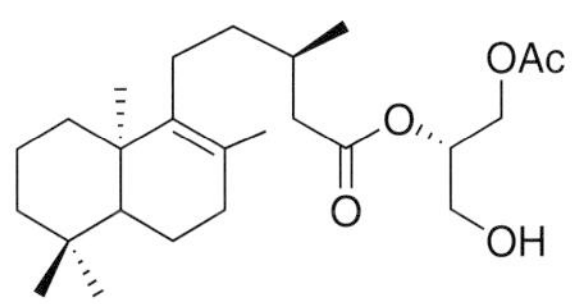

2′-(3′-acetoxy)-glyceryl-labda-8-en-15-oate

Fig. 1.2. Chemical similarities in secondary metabolites of nudibranchs from the Mediterranean and from Antarctica. Structures of polygodial from Mediterranean *Dendrodoris limbata* and hodgsonal from Antarctic *Bathydoris hodgsoni*, verrucosin-5 from Mediterranean *Doris verrucosa* and *ent*-labdane diacylglycerol from Antarctic *Austrodoris kerguelenensis*

Dietary chemicals are selected by a still unknown mechanism. Faulkner (1992) proposed two different mechanisms by which the selection of chemicals could be made, but this has not been studied in detail as far as we know. In this sense, the sea hare *Stylocheilus striatus* (previously *S. longicauda*) accumulates very different metabolites when offered in artificial diets (Pennings and Paul 1993). Also, Fontana et al. (1994) carried out some laboratory experiments showing that a chromodoridid species was able to accumulate selected compounds in its mantle glands, taken from a sponge not usually preyed on in the field. These experiments could support the idea that the initial role of accumulation structures was that of excretion or autoprotection from dietary chemicals, thus evolving later into a defensive mechanism.

Classically, sea hares are supposed to be protected by dietary chemicals obtained from algae (Faulkner 1992; Paul et al. 2001). Recently, new experiments with relevant ecological information have been provided by Rogers et al. (2002 and references therein), some of them studying different diets and their effect on predation. Surprisingly, although some sea hares were relatively unpalatable compared to squid, some dietary compounds seemed to be ineffective as a defense against some reef fish (Rogers et al. 2002). Pennings et al. (2001) reported, however, that more than diet, fish species may influence the results of predation experiments. Since only a few species have been studied so far, it seems that more experiments are needed to ascertain whether natural products are used in chemical defense in other sea hare species and with different kinds of predators.

Regarding the origin of the compounds, the number of biosynthetic compounds with respect to those obtained from the diet continues to increase from previous reports (Garson 1993; Cimino and Sodano 1994; Avila 1995; Cimino et al. 2001, 2004). The sesquiterpene aldehydes of the nudibranch *Acanthodoris nanaimoensis* are another example of de novo biosynthesis (Graziani and Andersen 1996). Further studies on biosynthesis include Fontana et al. (1999, 2003), Jansen and de Groot (2004), and others reviewed by Garson (2001). In fact, de novo biosynthesis has been demonstrated not only in different opisthobranch groups, but also in pulmonates (e.g. Garson et al. 1994).

Very little information exists about chemicals related to environmental interactions, mainly changes in phospholipids and other lipids related to temperature. Some examples are changes in the phospholipids of a bivalve (Sanina and Kostetsky 2002), lipids and fatty acids from *Clione limacina* from both polar areas (Kattner et al. 1998), and a recent comparison among fatty acids from Antarctic and Mediterranean gastropods (Avila et al. 2004). Chemotaxis, a mechanism allowing a predator to find its prey by detecting its chemical cues, is another interesting topic that has been scarcely addressed in molluscs (Castiello et al. 1978, 1980; Rogers and Paul 1991; Avila 1998b; Stachowicz 2001; Iken et al. 2002). The difficulties in carrying out these kinds of studies may be responsible for the limited research in this area.

Chemical cues for settlement and metamorphosis in molluscs have been studied mainly for commercially interesting species (see below). Very few studies address the induction of metamorphosis by using possible chemical cues present in the natural environment (Hadfield and Scheuer 1985; Pawlik 1992a,b; Avila 1995, 1998a; Tamburri et al. 1996; Lambert et al. 1997; Krug and Manzi 1999; Hadfield and Paul 2001; Stachowicz 2001). Indeed, in very few cases has the natural chemical cue been identified for marine invertebrates, and actually, in molluscs, this is true only for one species (the bivalve *Pecten maximus*; Yvin et al. 1985; Cochard et al. 1989). Other studies have only partially purified the natural chemical inducers in some bivalves, prosobranchs, and opisthobranchs (Pawlik 1992a,b and references therein; Hadfield and Scheuer 1985; Tamburri et al. 1996 and references therein; Lambert et al. 1997; Krug and Manzi 1999; Hadfield and Paul 2001).

1.3 Histology: from Tissues to Cell Location

In recent years, an effort has been made to locate chemicals within the animal's body (e.g., García-Gómez et al. 1990; Avila 1995; Avila and Durfort 1996; Avila and Paul 1997; Cimino et al. 1999, Kubanek et al. 2000; Thoms et al. 2003). Location is important not only in order to know where the compounds are stored, transformed, or biosynthesized, but also to ascertain their ecological role. However, the relationship between the location of a natural product in molluscs and its ecological significance has only been addressed in a few cases (Cronin et al. 1995; Avila and Paul 1997).

Shelled molluscs are usually assumed to be mechanically protected by their shell, and therefore, assuming a scarce presence of chemical defenses, few efforts have been made to study their natural products. However, many of these molluscs also produce defensive secretions. Bryan et al. (1997) reviewed secretions from molluscs which may have a defensive role. Among them, *Calliostoma canaliculatum* is a prosobranch that secretes a yellow exudate from the mantle, containing chemical defenses which have a negative effect on sympatric sea stars (Bryan et al. 1997). Pulmonates and opisthobranchs also produce interesting secretions (Faulkner 1992; Avila 1995; Cimino and Ghiselin 2001).

Some species possess special structures to store these chemicals in the external part of the body. The mantle dermal formations (MDFs) of Chromodorididae, for example, are now recognized as important key characters in the evolution of opisthobranchs (Wägele 2004). But, in fact, we now know that not only chromodoridids, but also other groups of opisthobranchs (at least one sacoglossan and one polycerid), possess MDFs (Wägele 2004). This renders invalid all the previous assumptions about mantle glands being an exclusive characteristic of the Chromodorididae. Furthermore, our recent studies show that more opisthobranch groups possess structures similar to MDFs than initially thought (Wägele, Ballesteros and Avila, unpublished data). In fact, even the more typical MDFs can show many differences at the histological and cytological levels. We are working to clarify the situation of defensive glands among the different opisthobranch groups (Wägele, Ballesteros and Avila, unpublished data). As an example of this, two different kinds of MDFs in chromodoridids, one with a weak muscular layer (*Hypselodoris villafranca*) and another with a strong muscular layer (*Risbecia tryoni*), are shown in Fig. 1.3.

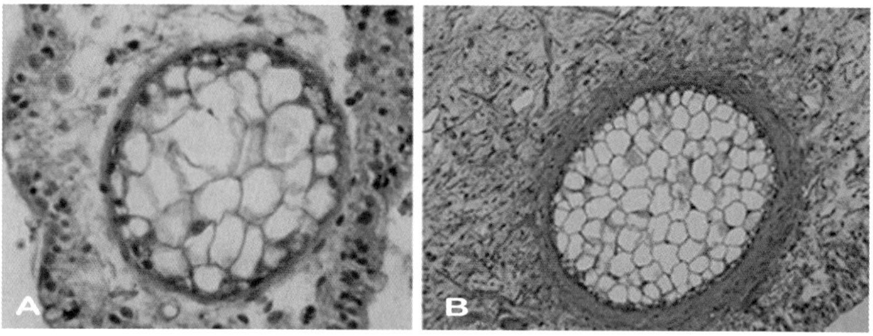

Fig. 1.3. Histological sections of mantle dermal formations (MDFs) from two nudibranchs: (**A**) *Hypselodoris villafranca* (0.2 mm diam.); (**B**) *Risbecia tryoni* (0.5 mm diam.)

The location of natural products at the cellular level has so far been highly problematic. We have used immunocytological methods, trying to create antibodies for longifolin, with, as yet, no positive results (Avila, Preisfeld and Wägele, unpublished data). Another new methodology which could provide good results is the use of oligonucleotid aptamers (Famulok 1999). Any of these methods, if successful, will make it possible to detect known compounds at the histological and cytological levels. Simultaneously, we will be able to locate where the biosynthesis takes place, if the compound is de novo biosynthesized, or alternatively, the transportation pathway from the digestive gland to the accumulating structures, if the compound is from dietary origin.

Some metabolites are located in the egg masses. The regulation of reproductive phenomena has proven to be complex. Remarkably, egg-laying in gastropods (*Lymnaea stagnalis*) and cephalopods (*Sepia officinalis*) may be regulated by complex interactions between peptides and monoamines (Li et al. 1992; Zatylny et al. 2000). Sea hare dolabellanins located in the reproductive organs and eggs seem to be involved in antimicrobial protection for the egg masses (Iijima et al. 2003). In many cases, the location of chemicals in the egg ribbons has an unknown function so far (Thoms et al. 2003). Benkendorff et al. (2000 and references therein) reported the antimicrobial activity of the compounds found in the egg masses of the prosobranch *Dicathais orbita*. Furthermore, they showed that changes in antimicrobial activity correspond to the egg developmental pattern and suggested that this protects embryos from autotoxicity (Benkendorff et al. 2000). In a recent study, Benkendorff et al. (2005) demonstrated that unsaturated fatty acids are responsible for the bacteriolytic activity of extracts from some mollusc egg masses.

Another interesting field within this area is the possibility that symbionts may be related to the presence of some chemicals. In fact, there is increasing evidence in marine invertebrates that symbionts may play an important role in producing bioactive chemicals or precursors within the hosts (Proksch et al. 2002; Constantino et al. 2004). Perhaps in the future molluscs too will be a good source of microorganisms which may produce interesting compounds. As a matter of fact, symbiotic bacteria have been described in *Dendrodoris* molluscs (Klussmann-Kolb and Brodie 1999) and they seem to be somehow involved in the processes of reproduction and egg-laying. These bacteria do not seem to have any relation with the molluscan natural products, since as far as we know, in *Dendrodoris* species, the described bacteria are not located where the natural products are found (Avila et al. 1991). Moreover, in *Tylodina perversa*, sequestration of diet chemicals is not related to the presence of sponge symbionts, which are not present in the molluscs (Thoms et al. 2003). However, it has recently been suggested that symbionts may have a role in the synthesis of compounds involved in cellulose digestion in wood-boring bivalves (Xu and Distel 2004 and references therein). Also, bacteria associated with the larvae of cultured bivalves have been shown to produce antibiotics (Fdhila et al. 2003). New symbionts are continuously being isolated from bivalves and sea slugs (Ishikura et al. 2004 and references therein), although they are not related to marine natural products as far as we know. Finally, it is worth mentioning that endosymbiosis in photosynthetic slugs is well documented (Green et al. 2000 and references therein), although again there seems to be no relation to the presence of secondary metabolites. A very peculiar case is that of *Melibe pilosa*, in which sterols are produced by zooxanthellae (Whiters et al. 1982).

1.4 Laboratory Culture: Producing Bioactive Compounds

Many molluscs can be cultured in the laboratory. The process, however, is not easy because many molluscs lay small eggs from which veliger larvae emerge. These larvae may undergo metamorphosis inside the egg capsule and crawl out as an adult, or may metamorphose outside the capsule. If they come out from the capsule as larvae, they may develop without feeding stages or they may need to feed on phytoplancton for days, weeks, or even a few months. Then, when ready, they metamorphose into adults.

Most species require an inductor of the metamorphosis, usually the adult prey species (but sometimes intermediate diets or other factors), which triggers the metamorphic sequence. Furthermore, metamorphosis is not a homogenous phenomenon and there is variation not only between different egg masses, but also within a single egg mass (e.g., Avila 1998a and references therein). It has been suggested that this variability in metamorphosis is important in the ecological strategies of the species (Krug 2001).

All these aspects make laboratory culturing highly complicated (e.g., Rupp et al. 2004). However, detailed methods and new techniques for growing larvae and inducing metamorphosis have been described, particularly for species with commercial interest, many of which are cultured in the field for the aquaculture industry. There are many examples of methods and details for culturing molluscs described in the literature, including algal foods, induction of metamorphosis (both natural and artificial), culturing conditions, etc., for prosobranchs (e.g., Aldana-Aranda and Patiño-Suárez 1998; Shie and Liu 1999; Gallardo and Sánchez 2001; Bahamondes-Rojas and Bretos 2002; Amos and Purcell 2003; Guzman del Proo et al. 2003; Daume et al. 2004), opisthobranchs (e.g. Kriegstein 1977; Hubbard 1988; Hansen and Ockelmann 1991; Gibson 1995; Plaut et al. 1995; Avila et al. 1997; Avila 1998a), cephalopods (e.g., Hanlon et al. 1997; Vidal et al. 2002; Anil 2003; Koueta and Boucaud-Camou 2003; Navarro and Villanueva 2003; Vaz-Pires et al. 2004), and bivalves (e.g., De la Roche et al. 2002; Parnell 2002; Navarte and Pascual 2003; Lora-Vilchis et al. 2004; Pernet et al. 2004; Rupp et al. 2004; Zhuang et al. 2004). Obviously, it should be considered that in some cases cultured animals may show differences in their chemicals depending on the diet offered, as reported for bivalves and cephalopods (Douglas et al. 1997; Delaporte et al. 2003; Navarro and Villanueva 2003).

The use of these culturing techniques could be applied to species producing interesting natural products in order to increase the material available for chemical studies. Unfortunately, it is easy to see that many of the previously mentioned studies have been carried out with mollusc groups that are not necessarily interesting from the natural products point of view. Culturing the species producing bioactive natural compounds has also the added benefit of protecting natural populations from intensive exploitation (Yan 2004). Interestingly enough, the use of antimicrobial peptides obtained from different organisms (including bivalves) has recently been suggested as a tool for improving the culture of commercial species (Bachère 2003 and references therein).

As a matter of fact, we have recently been culturing some nudibranch species in order to see when certain biosynthetic compounds are produced. As an example, in *Doris verrucosa*, verrucosins do not seem to be present during the early larval stages, while xylosil-MTA is present only during the first days of development (Avila and Gavagnin, unpublished data). In contrast, young *Dendrodoris limbata* obtained in the laboratory, about 1 mm long, contained only 7-deacetoxyolepupuane (Avila 1993); and this is relevant in the pivotal role suggested for this compound in the biosynthetic pathway of this species (Fontana et al. 1999).

Another interesting possibility would be developing cultures of cells or tissues involved in the production of compounds, as suggested for porifera (Rinkevich 1999). This has been reported to be a very difficult task for marine invertebrates in general (Rinkevich 1999; Sennett 2001) and for *Conus* snails in particular (Fainzilber 2004). However, it has been done successfully with neurons from pulmonates and opisthobrachs (Cohan et al. 2003 and references therein), muscle cells from bivalves (Plotnikov et al. 2003 and references therein), mantle cells from prosobranchs (Sud et al. 2001; Poncet et al. 2002, 2003 and references therein), hemocytes from bivalves (Serpentini et al. 2002), decapsulated embryos of pulmonates (Kuang et al. 2002), and digestive-gland cells in prosobranchs and bivalves (Carefoot et al. 2000 and references therein), as well as in other previous studies reviewed by Rinkevich (1999) and Mulcahy (2000). Cardiac cells and gill cells of bivalves are actually a tool for studying the toxic effects of marine pollutants (Domart-Coulon et al. 2000; Pennec et al. 2002 and references therein). In all these studies, molluscan cell lines have been useful for neurobiology, toxicity, immunology, pharmacology, physiology, biochemistry, and other subjects. However, as far as we know, cell culture is still unexplored for natural products production in molluscs.

Of course, synthesis and genetic engineering are further chemical tools which may be used to produce natural products, out of the scope of our review, and which so far have only been seldom used for mollusc products (e.g., Pietra 2001; Sennett 2001).

1.5 Conclusions

In recent studies, many natural products have been reported to be used for defense by molluscs, although only a few have demonstrated their effectiveness against co-occurring predators. Similarly, few studies locate the natural products in the body of the mollusc, thus providing information on the possible origin and role of the compounds. The scope of studying mollusc natural products has to expand beyond the description of new metabolites and their role in chemical defense and has to approach other aspects, such as the different roles for chemical, histological, and cytological approaches, evolutionary chemical ecology, culturing species, and perhaps, cells. Under the aforementioned perspective, molluscs with de novo biosynthesis are foreseen as "producers" of bioactive substances, or perhaps as biotransformers of dietary compounds. Further development of the earlier mentioned new biotechnological tools, as well as others to come, is necessary for success. Obviously, a multidisciplinary approach for studying all these topics is necessary. Furthermore, organisms are whole systems and not isolated parts (Greene 2005), and thus, our approach should be open to considering all aspects together. Descriptive natural history, biology, and ecology are fundamental to applied biology, and chemical interactions are only a part of the "cake".

The observation of nature provides amazing surprises. In fact, molluscs may induce the biosynthesis of natural products in other organisms. For example, in the gorgonian *Pseudopterogorgia elisabethae*, the production by biosynthesis of the diterpene pseudopterosins is increased by predation of the mollusc *Cyphoma gibbosum* (Thornton and Kerr 2002), in a way that resembles that of insect–plant interactions. Perhaps in the near future, molluscs will be also used to increase the biosynthesis of interesting natural compounds in prey organisms.

A recent review has looked at the diversity of natural products (Pietra 2001). Several aspects of the levels at which this can be considered are discussed there. Conservation is an important factor when talking about biodiversity, whether considering this in terms of natural products, genes, cells, species numbers, or ecosystems. For example, the risks derived from cone snail harvesting have received attention recently in prestigious journals (Duda et al. 2004 and references therein), which have pointed out that there is an increasing awareness of the importance of conservation. A tentative list of "endangered natural products" (Pietra 2001) includes several molluscs: the opisthobranchs *Hexabranchus sanguineus*, *Volvatella* sp., *Philinopsis speciosa* and *Haminoea cymbalum*, and the pulmonate *Peronia peronii*. In fact, many more species should be added to this list. To reduce this impact, some solutions are suggested: recent studies have shown that it is possible to release cultured animals in the field to allow natural restocking for some prosobranchs (Purcell et al. 2004 and references therein) and bivalves (Lafrance et al. 2003 and references therein). Even with partial remediation for particular species, efforts should be maximized in order to study nature without destroying it, not only because it is the source of still unknown natural compounds, but because our own survival depends on taking good care of our planet.

Acknowledgments. Thanks are due to S. Taboada and M. Ballesteros for their help in obtaining some bibliographic references. Financial support from the ECOQUIM projects (REN2003-00545,

REN2002-12006-E ANT, CGL2004-03356/ANT) and the collaboration program CNR-CSIC (2003IT0003) are also acknowledged.

References

Aldana-Aranda D, Patiño-Suárez V (1998) Overview of diets used in larviculture of three Caribbean conchs: queen conch *Strombus gigas*, milk conch *Strombus costatus* and fighting conch *Strombus pugilis*. Aquaculture 167:163–178

Amos MJ, Purcell SW (2003) Evaluation of strategies for intermediate culture of *Trochus niloticus* (Gastropoda) in sea cages for restocking. Aquaculture 218:235–249

Amsler CD, Iken KB, McClintock JB, Baker BJ (2001a) Secondary metabolites from Antarctic marine organisms and their ecological implications. In: McClintock JB, Baker BJ (eds) Marine chemical ecology. CRC Press, Boca Raton, pp. 267–300

Amsler CD, McClintock JB, Baker BJ (2001b) Secondary metabolites as mediators of trophic interactions among Antarctic marine organisms. Am Zool 41:17–26

Anil MK (2003) Laboratory rearing and breeding of spineless cuttlefish *Sepiella inermis* Orbigny (Mollusca: Cephalopoda). J Aquacult Trop 18:35–44

Atema J, Stenzler D (1977) Alarm substance of the marine mud snail, *Nassarius obsoletus*: biological characterization and possible evolution. J Chem Ecol 3:173–187

Avila C (1993) Sustancias naturales de moluscos opistobranquios: estudio de su estructura, origen y función en ecosistemas bentónicos. PhD thesis, University of Barcelona, Barcelona

Avila C (1995) Natural products of opisthobranch molluscs: a biological review. Oceanogr Mar Biol Annu Rev 33:487–559

Avila C (1998a) Competence and metamorphosis in the long-term planktotrophic larvae of the nudibranch mollusc *Hermissenda crassicornis* (Eschscholtz, 1831). J Exp Mar Biol Ecol 231:81–117

Avila C (1998b) Chemotaxis in the nudibranch *Hermissenda crassicornis*: does ingestive conditioning influence its behaviour in a Y-maze? J Molluscan Stud 64:215–222

Avila C, Durfort M (1996) Histology of epithelia and mantle glands of selected species of Doridacean mollusks with chemical defensive strategies. Veliger 39:148–163

Avila C, Paul VJ (1997) Chemical ecology of the nudibranch *Glossodoris pallida*: is the location of diet-derived metabolites important for defense? Mar Ecol Progr Ser 150: 171–180

Avila C, Ballesteros M, Cimino G, Crispino A, Gavagnin M, Sodano G (1990) Biosynthetic origin and anatomical distribution of the main secondary metabolites in the nudibranch mollusc *Doris verrucosa*. Comp Biochem Physiol 97B:363–368

Avila C, Cimino G, Crispino A, Spinella A (1991) Drimane sesquiterpenoids in Mediterranean *Dendrodoris* nudibranchs: anatomical distribution and biological role. Experientia 47:306–310

Avila C, Grenier S, Tamse CT, Kuzirian AM (1997) Biological factors affecting larval growth in the nudibranch mollusc *Hermissenda crassicornis* (Eschscholtz, 1831). J Exp Mar Biol Ecol 218:243–262

Avila C, Iken K, Fontana A, Cimino G (2000) Chemical ecology of the Antarctic nudibranch *Bathydoris hodgsoni* Eliot, 1907: defensive role and origin of its natural products. J Exp Mar Biol Ecol 252:27–44

Avila C, Fontana A, Esposito M, Ciavatta ML, Cimino G (2004) Fatty acids of Antarctic gastropods: distribution and comparison with Mediterranean species. Iberus 22:33–44

Bachère E (2003) Anti-infectious immune effectors in marine invertebrates: potential tools for disease control in larviculture. Aquaculture 227:427–438

Bahamondes-Rojas I, Bretos M (2002) Induction of spawning and early development in *Fissurella picta* (Mollusca: Archaeogastropoda) from Southern Chile. J Shellfish Res 21:185–191
Barnes RSK, Calow P, Olive PJW (1988) The Invertebrates, a new synthesis. Blackwell, Oxford
Becerro M, Goetz G, Paul VJ, Scheuer PJ (2001) Chemical defenses of the sacoglossan mollusc *Elysia rufescens* and its host alga *Bryopsis* sp. J Chem Ecol 27:2287–2299
Benkendorff K, Bremner JB, Davis AR (2000) Tyrian purple precursors in the egg masses of the Australian muricid, *Dicathais orbita*: a possible defensive role. J Chem Ecol 26:1037–1050
Benkendorff K, Davis AR, Rogers CN, Bremner JB (2005) Free fatty acids and sterols in the benthic spawn of aquatic molluscs, and their associated antimicrobial properties. J Exp Mar Biol Ecol 316:29–44
Blunt JW, Copp BR, Munro MHG, Northcote PT, Prinsep MR (2005) Marine natural products. Nat Prod Rep 22:15–61
Brusca RC, Brusca GJ (2003) Invertebrates. Sinauer Associates, Sunderland, MA
Bryan PJ, Yoshida WY, McClintock JB, Baker BJ (1995) Ecological role for pteroenone, a novel antifeedant from the conspicuous antarctic pteropod *Clione antarctica* (Gymnosomata: Gastropoda). Mar Biol 122:271–277
Bryan PJ, McClintock JB, Hamann M (1997) Behavioral and chemical defenses of marine prosobranch gastropod *Calliostoma canaliculatum* in response to sympatric seastars. J Chem Ecol 23:645–658
Caprotti E (1977) Molluschi e medicina nel Io secolo dC (studi di malacologia prerinascimentale, II). Conchiglie 13:137–144
Carefoot TH, Taylor BE, Land S (2000) Use of isolated digestive-gland cells in the study of biochemical and physiological processes in gastropod molluscs. Comp Biochem Physiol 125A:497–502
Castiello D, Cimino G, De Rosa S, De Stefano S, D'Izzo G, Sodano G (1978) Studies on the chemistry of the relationship between the opisthobranch *Peltodoris atromaculata* and the sponge *Petrosia ficiformis*. Colloq Int CNRS 291:413–416
Castiello D, Cimino G, De Rosa S, De Stefano S, Sodano G (1980) High molecular weight polyacetylenes from the nudibranch *Peltodoris atromaculata* and the sponge *Petrosia ficiformis*. Tetrahedron Lett 21:5047–5050
Cavas L, Yurdakov K, Yokes B (2005) Antioxidant status of *Lobiger serradifalci* and *Oxynoe olivacea* (Opisthobranchia, Mollusca). J Exp Mar Biol Ecol 314:227–235
Cimino G, Ghiselin MT (1998) Chemical defense and evolution in the sacoglossa (Mollusca: Gastropoda: Opisthobranchia). Chemoecology 8:51–60
Cimino G, Ghiselin MT (1999) Chemical defense and evolutionary trends in biosynthetic capacity among dorid nudibranchs (Mollusca: Gastropoda: Opisthobranchia). Chemoecology 9:187–207
Cimino G, Ghiselin MT (2001) Marine natural products chemistry as an evolutionary narrative. In: McClintock JB, Baker BJ (eds.) Marine chemical ecology. CRC, Boca Raton, pp. 115–154
Cimino G, Sodano G (1989) The chemical ecology of Mediterranean opisthobranchs. Chem Scripta 29:389–394
Cimino G, Sodano G (1994) Transfer of sponge secondary metabolites to predators. In: Van Soest RWM, Van Kempen TMG, Braekman JC (eds.) Sponges in time and space. Balkema, Rotterdam, pp. 459–472
Cimino G, De Rosa S, De Stefano S, Morrone R, Sodano G (1985) The chemical defense of nudibranch molluscs. Structure, biosynthetic origin and defensive properties of terpenoids from the dorid nudibranch *Dendrodoris grandiflora*. Tetrahedron 41:1093–1100
Cimino G, Gavagnin M, Sodano G, Puliti R, Mattia CA, Mazzarella L (1988) Verrucosin-A and -B, ichthyotoxic diterpenoic acid glycerides with a new carbon skeleton from the dorid nudibranch *Doris verrucosa*. Tetrahedron 44:2301–2310

Cimino G, Passeggio A, Sodano G, Spinella A, Villani G (1991a) Alarm pheromones from the Mediterranean opisthobranch *Haminoea navicula*. Experientia 47:61–63

Cimino G, Crispino A, Di Marzo V, Sodano G, Spinella A, Villani G (1991b) A marine molluscs provides the first example of in vivo storage of prostaglandins: prostaglandin-1,15-lactones. Experientia 47:56–60

Cimino G, Fontana A, Gavagnin M (1999) Marine opisthobranch molluscs: chemistry and ecology in sacoglossans and dorids. Curr Org Chem 3:327–372

Cimino G, Ciavatta ML, Fontana A, Gavagnin M (2001) Metabolites of marine opisthobranchs: chemistry and biological activity. In: Tringali C (ed.) Bioactive compounds from natural sources. Taylor & Francis, London, pp. 578–637

Cimino G, Fontana A, Cutignano A, Gavagnin M (2004) Biosynthesis in opisthobranch molluscs: general outline in the light of recent use of stable isotopes. Phytochem Rev 3:285–307.

Cochard JC, Chevolot L, Yvin JC, Chevolot-Magueur AM (1989) Induction de la metamorphose de la coquille Saint Jaques *Pecten maximus* L. par des derives de la tyrosine extraits de l'algue *Delesseria sanguinea* Lamouroux ou syntetiques. Haliotis 19:129–154

Cohan CS, Karnes JL, Zhou FQ (2003) Culturing neurons from the snail *Helisoma*. Methods Cell Biol 71:157–170

Constantino V, Fattorusso E, Menna M., Taglialatela-Scafati O (2004) Chemical diversity of bioactive marine natural products: an illustrative case study. Curr Med Chem 11:1671–1692

Cronin G, Hay ME, Fenical W, Lindquist N (1995) Distribution, density, and sequestration of host chemical defenses by the specialist nudibranch *Tritonia hamnerorum* found at high densities on the sea fan *Gorgonia ventalina*. Mar Ecol Progr Ser 119:177–189

Cronin G, Paul VJ, Hay ME, Fenical W (1997) Are tropical herbivores more resistant than temperate herbivores to seaweed chemical defenses? Diterpenoid metabolites from *Dictyota acutiloba* as feeding deterrents for tropical versus temperate fishes and urchins. J Chem Ecol 23:289–302

Daume S, Huchette S, Ryan S, Day RW (2004) Nursery culture of *Haliotis rubra*: the effect of cultured algae and larval density on settlement and juvenile production. Aquaculture 236:221–239

Davies-Coleman MT, Faulkner DJ (1991) New diterpenoic acid glycerides from the Antarctic nudibranch *Austrodoris kerguelensis*. Tetrahedron 47:9743–9750

De la Roche JP, Marín B, Freites L, Vélez A (2002) Embryonic development and larval and post-larval growth of the tropical scallop *Nodipecten* (=*Lyropecten*) *nodosus* (L. 1758) (Mollusca: Pectinidae). Aquacult Res 33:819–827

Delaporte M, Soudant P, Moal J, Lambert C, Quéré C, Miner P, Choquet G, Paillard C, Samain JF (2003) Effect of a mono-specific algal diet on immune functions in two bivalve species, *Crassostrea gigas* and *Ruditapes philippinarum*. J Exp Zool 206:3053–3064

Di Marzo V, Cimino G, Crispino A, Minardi C, Sodano G, Spinella A (1991) A novel multifunctional metabolic pathway in a marine mollusc leads to unprecedent prostaglandin derivatives (prostaglandin-1,15-lactones). Biochem J 273:593–600

Domart-Coulon I, Auzoux-Bordenave S, Doumenc D, Khalanski M (2000) Cytotoxicity assessment of antibiofouling compounds and by-products in marine bivalve cell cultures. Toxicol In Vitro 14:245–251

Douglas DJ, Kenchington ER, Bird CJ, Pocklington R, Bradford B, Silvert W (1997) Accumulation of domoic acid by the sea scallop (*Placopecten magellanicus*) fed cultured cells of toxic *Pseudo-nitzschia multiseries*. Can J Fish Aquat Sci 54:907–913

Duda TF, Bingham JP, Livett BG, Kohn AJ, Raybaudi-Massilia G, Schultz JR, Down J, Sandall D, Sweedler JV (2004) How much at risk are cone snails? Science 303:955

Fahey SJ, Garson MJ (2002) Geographic variation of natural products of tropical nudibranch *Asteronotus cespitosus*. J Chem Ecol 28:1773–1785

Fainzilber M (2004) How much at risk are cone snails? Science 303:955–956

Famulok M (1999) Oligonucleotide aptamers that recognize small molecules. Curr Opin Struct Biol 9:324–329
Faulkner DJ (1988) Feeding deterrents in mollucsc. In: Fautin DG (ed.) Biomedical importance of marine organisms. Mem Calif Acad Sci 13:29–36
Faulkner DJ (1992) Chemical defenses of marine molluscs. In: Paul VJ (ed.) Ecological roles of marine natural products. Comstock, Ithaca, pp. 119–163
Faulkner DJ (2002) Marine natural products. Nat Prod Rep 19:1–48
Faulkner DJ, Ghiselin MT (1983) Chemical defense and evolutionary ecology of dorid nudibranchs and some other opisthobranch gastropods. Mar Ecol Progr Ser 13:295–301
Faulkner DJ, Stallard MO (1973) 7-Chloro-3,7-dimethyl-1,4,6-tribromo-1-octen-3-ol. A novel monoterpene alcohol from *Aplysia californica*. Tetrahedron Lett 14:1171–1174
Fdhila F, Vázquez V, Sánchez JL, Riguera R (2003) DD-Diketopiperazines: antibiotics active against *Vibrio anguillarum* isolated from marine bacteria associated with cultures of *Pecten maximus*. J Nat Prod 66:1299–1301
Fontana A, Giménez F, Marín A, Mollo E, Cimino G (1994) Transfer of secondary metabolites from the sponges *Dysidea fragilis* and *Pleraplysilla spinifera* to the mantle dermal formations (MDFs) of the nudibranch *Hypselodoris webbi*. Experientia 50:510–516
Fontana A, Ciavatta ML, Miyamoto T, Spinella A, Cimino G (1999) Biosynthesis of drimane terpenoids in dorid molluscs: pivotal role of 7-deacetoxyolepupuane in two species of *Dendrodoris* nudibranchs. Tetrahedron 55:5937–5946
Fontana A, Cavaliere P, Wahidulla S, Naik CG, Cimino G (2000) A new antitumor isoquinoline alkaloid from the marine nudibranch *Jorunna funebris*. Tetrahedron 56:7305–7308
Fontana A, Tramice A, Cutignano A, d'Ippolito G, Gavagnin M, Cimino G (2003) Terpene biosynthesis in the nudibranch *Doriopsilla areolata*. J Org Chem 68:2405–2409
Gallardo CS, Sánchez KA (2001) Induction of metamorphosis and its effect on the growth and survival of postmetamorphic juveniles of *Chorus giganteus* (Gastropoda: Muricidae). Aquaculture 201:241–250
García-Gómez JC, Cimino G, Medina A (1990) Studies on the defensive behaviour of *Hypselodoris* species (Gastropoda: Nudibranchia): ultrastructure and chemical analysis of mantle dermal formations (MDFs). Mar Biol 106:245–250
Garson MJ (1993) The biosynthesis of marine natural products. Chem Rev 93:1699–1733
Garson MJ (2001) Ecological perspectives on marine natural product biosynthesis. In: McClintock JB, Baker BJ (eds.) Marine chemical ecology. CRC, Boca Raton, pp. 71–114
Garson MJ, Jones DD, Small CJ, Liang J, Clardy J (1994) Biosynthetic studies on polypropionates: a stereochemical model for siphonarins A and B from the pulmonate limpet *Siphonaria zelandica*. Tetrahedron Lett 35:6921–6924
Gavagnin M, Trivellone E, Castelluccio F, Cimino G, Cattaneo-Vietti R (1995) Glyceryl ester of a new halimane diterpenoic acid from the skin of the Antarctic nudibranch *Austrodoris kerguelenensis*. Tetrahedron Lett 36:7319–7322
Gavagnin M, Ungur N, Castelluccio F, Cimino G (1997) Novel verrucosins from the skin of the Mediterranean nudibranch *Doris verrucosa*. Tetrahedron 53:1491–1504
Gavagnin M, De Napoli A, Castelluccio F, Cimino G (1999a) Austrodorin-A and -B: first tricyclic diterpenoid 2′-monoglyceryl esters from an Antarctic nudibranch. Tetrahedron Lett 40:8471–8475
Gavagnin M, De Napoli A, Cimino G, Iken K, Avila C, García FJ (1999b) Absolute stereochemistry of diterpenoid diacylglycerols from the Antarctic nudibranch *Austrodoris kerguelenensis*. Tetrahedron Assym 10:2647–2650
Gavagnin M, Mollo E, Calado G, Fahey S, Ghiselin M, Ortea J, Cimino G (2001) Chemical studies of porostome nudibranchs: comparative and ecological aspects. Chemoecology 11:131–136
Gavagnin M, Carbone M, Mollo E, Cimino G (2003a) Further chemical studies on the Antarctic nudibranch *Austrodoris kerguelenensis*: new terpenoid acylglycerols and revision of the previous stereochemistry. Tetrahedron 59:5579–5583

Gavagnin M, Carbone M, Mollo E, Cimino G (2003b) Austrodoral and austrodoric acid: nor-sesquiterpenes with a new carbon skeleton from the Antarctic nudibranch *Austrodoris kerguelenensis*. Tetrahedron Lett 44:1495–1498

Gibson G (1995) Why be choosy? Temporal changes in larval sensitivity to several naturally occurring metamorphic inducers in the opisthobranch *Haminaea callidegenita*. J Exp Mar Biol Ecol 194:9–24

Graziani EI, Andersen RJ (1996) Investigations of sesquiterpenoid biosynthesis by the dorid nudibranch *Acanthodoris nanaimoensis*. J Am Chem Soc 118:4701–4702

Green BJ, Li WY, Manhart JR, Fox TC, Summer EJ, Kennedy RA, Pierce SK, Rumpho ME (2000) Mollusc–algal chloroplast endosymbiosis. Photosynthesis, thylakoid protein maintenance, and chloroplast gene expression continue for many months in the absence of the algal nucleus. Plant Physiol 124:331–342

Greene HW (2005) Organisms in nature as a central focus for biology. Trends Ecol Evol 20:23–27

Guzman del Proo SA, Reynoso-Granados T, Monsalvo-Spencer P, Serviere-Zaragoza E (2003) Larval and early juvenile development of the wavy turban snail *Megastraea undosa* (Wood, 1828) (Gastropoda: Turbinidae). Veliger 46:320–324

Hadfield MG, Paul VJ (2001) Natural chemical cues for settlement and metamorphosis of marine invertebrate larvae. In: McClintock JB, Baker BJ (eds.) Marine chemical ecology. CRC, Boca Raton, pp. 431–461

Hadfield MG, Scheuer PJ (1985) Evidence for a soluble metamorphic inducer in *Phestilla*: ecological, chemical and biological data. Bull Mar Sci 37:556–566

Hamann MT, Otto CS, Scheuer PJ (1996) Kahalalides: bioactive peptides from a marine mollusc. *Elysia rufescens* and its algal diet *Bryopsis* sp. J Org Chem 61:6594

Hanlon RT, Claes MF, Ashcraft SE, Dunlap PV (1997) Laboratory culture of the sepiolid squid *Euprymna scolopes*: a model system for bacteria–animal symbiosis. Biol Bull 192:364–374

Hansen B, Ockelmann KW (1991) Feeding behaviour in larvae of the opisthobranch *Philine aperta*. I. Growth and functional response at different developmental stages. Mar Biol 111:255–261

Harper MK, Bugni TS, Copp BR, James RD, Lindsay BS, Richardson AD, Schnabel PC, Tasdemir D, Van Wagoner RM, Verbitski SM, Ireland CM (2001) Introduction to the chemical ecology of marine natural products. In: McClintock JB, Baker BJ (eds.) Marine chemical ecology. CRC, Boca Raton, pp. 3–69

Herbert DG, Hamer ML, Mander M, Mkhize N, Prins F (2003) Invertebrate animals as a component of the traditional medicine trade in KwaZulu–Natal, South Africa. Afr Invertebr 44:327–344

Hickman CP, Roberts LS, Larson A (2002) Principios integrales de Zoología. McGraw-Hill/Interamericana, Madrid

Hubbard EJA (1988) Larval growth and the induction of metamorphosis of a tropical sponge-eating nudibranch. J Molluscan Stud 54:259–269

Ichida T, Higa T (1986) New cuparene derived sesquiterpenes with unprecedent oxygenation patterns from the sea hare *Aplysia dactylomela*. J Org Chem 51:3364–3366

Iijima R, Kisugi J, Yamazaki M (2003) L-Amino acid oxidase activity of an antineoplastic factor of a marine mollusk and its relationship to cytotoxicity. Dev Comp Immunol 27:505–512

Iken K, Avila C, Ciavatta ML, Fontana A, Cimino G (1998) Hodgsonal, a new drimane sesquiterpene from the mantle of the Antarctic nudibranch *Bathydoris hodgsoni*. Tetrahedron Lett 39:5635–5638

Iken K, Avila C, Fontana A, Cimino G (2002) Chemical ecology and origin of defensive compounds in the Antarctic nudibranch *Austrodoris kerguelenensis* (Opisthobranchia: Gastropoda). Mar Biol 141:101–109

Ishikura M, Hagiwara K, Takishita K, Haga M, Iwai K, Maruyama T (2004) Isolation of new *Symbiodinium* strains from tridacnid giant clam (*Tridachna crocea*) and sea slug

(*Pteraeolidia ianthina*) using culture medium containing giant clam tissue homogenate. Mar Biotechnol 6:378–385
Jansen BJM, de Groot A (2004) Occurrence, biological activity and synthesis of drimane sesquiterpenoids. Nat Prod Rep 21:449–477
Jimeno JM (2002) A clinical armamentarium of marine-derived anti-cancer compounds. Anti Cancer Drugs 13[Suppl 1]:15–19
Johnson PM, Willows AOD (1999) Defense in sea hares (Gastropoda, Opisthobranchia, Anaspidea): multiple layers of protection from egg to adult. Mar Freshwater Behav Physiol 32:147–180
Karuso P (1987) Chemical ecology of the nudibranchs. In: Scheuer PJ (ed.) Bioorganic marine chemistry, vol 1. Springer, Berlin Heidelberg New York, pp. 31–60
Kattner G, Hagen W, Graeve M, Albers C (1998) Exceptional lipids and fatty acids in the pteropod *Clione limacina* (gastropoda) from both polar oceans. Mar Chem 61:219–228
Klussmann-Kolb A, Brodie G (1999) Internal storage and production of symbiotic bacteria in the reproductive system of a tropical marine gastropod. Mar Biol 133:443–447
Kohn AJ (2005) The *Conus* biodiversity website. http://biology.burke.washington.edu/conus/
Koueta N, Boucaud-Camou E (2003) Combined effects of photoperiod and feeding frequency on survival and growth of juvenile cuttlefish *Sepia officinalis* L. in experimental rearing. J Exp Mar Biol Ecol 296:215–226
Kriegstein AR (1977) Stages in the post-hatching development of *Aplysia californica*. J Exp Zool 199:275–288
Krug PJ (2001) Bet-hedging dispersal strategy of a specialist marine herbivore: a settlement dimorphism among sibling larvae of *Alderia modesta*. Mar Ecol Progr Ser 213:177–192
Krug PJ, Manzi AE (1999) Waterborne and surface-associated carbohydrates as settlement cues for larvae of the specialist marine herbivore *Alderia modesta*. Biol Bull 197:94–103
Kuang SH, Regnier M, Goldberg RM (2002) Long-term culture of decapsulated gastropod embryos: a transplantation study. Biol Bull 203:278–288
Kubanek J, Faulkner DJ, Andersen RJ (2000) Geographic variation and tissue distribution of endogenous terpenoids in the Northeastern Pacific Ocean dorid nudibranch *Cadlina luteomarginata*: implications for the regulation of de novo biosynthesis. J Chem Ecol 26:377–389
Lafrance M, Cliché G, Haugum GA, Guderley H (2003) Comparison of cultured and wild sea scallops *Placopecten magellanicus*, using behavioral responses and morphometric and biochemical indices. Mar Ecol Progr Ser 250:183–195
Lambert WJ, Todd CD, Hardege JD (1997) Partial characterization and biological activity of a metamorphic inducer of the dorid nudibranch *Adalaria proxima* (Gastropoda: Nudibranchia). Invertebr Biol 116:71–81
Li KW, Smit AB, Geraerts WP (1992) Structural and functional characterization of neuropeptides involved in the control of male behaviour of *Lymnaea stagnalis*. Peptides 13:633–638
Lora-Vilchis MC, Ruiz-Velasco-Cruz E, Reynoso-Granados T, Voltolina D (2004) Evaluation of five microalgae diets for juvenile pen shells *Atrina maura*. J World Aquacult Soc 35:232–236
McClintock JB, Baker BJ (1997) A review of the chemical ecology of Antarctic marine invertebrates. Am Zool 37:329–342
McClintock JB, Baker BJ, Hamann MT, Yoshida W, Slattery M, Heine JN, Bryan PJ, Jayatilake GS, Moon BH (1994) Homarine as a feeding deterrent in common shallow-water antarctic lamellarian gastropod *Marseniopsis mollis*: a rare example of chemical defense in a marine prosobranch. J Chem Ecol 20:2539–2549
Melo VMM, Duarte ABG, Carvalho AF, Siebra EA, Vasconcelos IM (2000) Purification of a novel antibacterial and haemagglutinating protein from the purple gland of the sea hare, *Aplysia dactylomela* Rang, 1828. Toxicon 38:1415–1427

Moomjian L, Nystrom S, Rittschof D (2003) Behavioral responses of sexually active mud snails: kariomones and pheromones. J Chem Ecol 29:497–501

Mouries LP, Almeida MJ, Milet C, Berland S, Lopez E (2002) Bioactivity of nacre water-soluble organic matrix from the bivalve mollusk *Pinctada maxima* in three mammalian cell types: fibroblasts, bone marrow stromal cells and osteoblasts. Comp Biochem Physiol 132B:217–229

Mulcahy MF (2000) Culture of molluscan cells. In: Mothersill C, Austin B (eds) Aquatic invertebrate cell culture. Springer–Praxis, Chichester, pp. 165–181

Navarro JC, Villanueva R (2003) The fatty acid composition of *Octopus vulgaris* paralarvae reared with live and inert food: deviation from their natural fatty acid profile. Aquaculture 219:613–631

Navarte MA, Pascual MS (2003) Fertilization, larval rearing and post-larval growth of the Tehuelche scallop *Aequipecten tehuelchus* D'Orbigny, 1846. Aquaculture 217:259–274

Olivera BM, Cruz LJ (2001) Conotoxins in retrospect. Toxicon 39:7–14

Orlando P, Strazzullo G, Carretta F, De Falaco M, Grippo P (1996) Inhibition mechanisms of HIV-1, Mo-MULV and AMV reverse transcriptases by kelletinin-A from *Buccinulum corneum*. Experientia 52:812–813

Parnell PED (2002) Larval development, precompetent period and a natural spawning event of the Pectinacean Bivalve *Spondylus tenebrosus* (Reeve, 1856). Veliger 45:58–64

Paul VJ (1992) Ecological roles of marine natural products. Comstock, Ithaca

Paul VJ, Cruz-Rivera E, Thacker RW (2001) Chemical mediation of macroalgal–herbivore interactions: ecological and evolutionary perspectives. In: McClintock JB, Baker BJ (eds.) Marine chemical ecology. CRC, Boca Raton, pp. 227–265

Pawlik JR (1992a) Induction of marine invertebrate larval settlement: evidence for chemical cues. In: Paul VJ (ed.) Ecological roles of marine natural products. Comstock, Ithaca, pp 189–236

Pawlik JR (1992b) Chemical ecology of the settlement of benthic marine invertebrates. Oceanogr Mar Biol Annu Rev 30:273–335

Pawlik JR (1993) Marine invertebrate chemical defenses. Chem Rev 93:1911–1922

Pennec JP, Gallet M, Gioux M, Dorange G (2002) Cell culture of bivalves: tool for the study of the effects of environmental stressors. Cell Mol Biol 48:351–358

Pennings SC, Paul VJ (1993) Sequestration of dietary secondary metabolites by three species of sea hares: location, specificity and dynamics. Mar Biol 117:535–546

Pennings SC, Nastisch S, Paul VJ (2001) Vulnerability of sea hares to fish predators: importance of diet and fish species. Coral Reefs 20:320–324

Pernet F, Tremblay R, Langdon C, Bourget E (2004) Effect of additions of dietary tracylglycerol microspheres on growth, survival and settlement of mussel (*Mytilus* sp.) larvae. Mar Biol 144:693–703

Pettit GR, Kamano Y, Herald CL, Tuinman AA, Boettner FE, Kizu H, Schmidt JM, Baczynskyj L (1987) The isolation and structure of a remarkable marine animal antineoplastic constituent: dolastatin 10. J Am Chem Soc 109:6883–6885

Pettit GR, Kamano Y, Herald CL, Dufresne D, Bates RB, Schmidt JM, Cerny RL, Kizu H (1990) Antineoplastic agents. 190: isolation and structure of the cyclodepsipeptide dolastatin 14. J Org Chem 55:2989–2990

Pietra F (2001) Biodiversity and natural product diversity. Pergamon, Oxford

Plaut I, Borut A, Spira ME (1995) Growth and metamorphosis of *Aplysia oculifera* larvae in laboratory culture. Mar Biol 122:425–430

Plotnikov SV, Karpenko AA, Odintsova NA (2003) Comparative characteristic of *Mytilus* muscle cells developed in vitro and in vivo. J Exp Zool 298A:77–85

Poncet JM, Serpentini A, Boucaud-Camou E, Lebel JM (2002) Cryopreservation of mantle dissociated cells from *Haliotis tuberculata* (Gastropoda) and postthawed primary cell cultures. Cryobiology 44:38–45

Poncet JM, Serpentini A, Boucaud-Camou E, Lebel JM (2003) Use of mantle cell primary cultures from abalone (*Haliotis tuberculata*) to study in vitro extracellular matrix synthesis and biomineralization. Haliotis 32:21–30

Proksch P, Edrada RA, Ebel R (2002) Drugs from the seas – current status and microbiological implications. Appl Microbiol Biotechnol 59:125–134
Purcell SW, Amos MJ, Pakoa K (2004) Releases of cultured sub-adult *Trochus niloticus* generate broodstock for fishery replenishment in Vanuatu. Fish Res 67:329–333
Rajaganapathi J, Kathiresan K, Singh TP (2002) Purification of anti-HIV protein from purple fluid of the sea hare *Bursatella leachii* de Blainville. Mar Biotechnol 4:447–453
Rinehart KL, Shaw PD, Shield LS, Gloer JB, Harbour GC, Koker MES, Samain D, Schwart RE, Tymiak AA, Weller GT, Carter GT, Munro MHG (1981) Marine natural products as sources of antiviral, antimicrobial and antineoplastic agents. Pure Appl Chem 53:795–817
Rinkevich B (1999) Cell cultures from marine invertebrates: obstacles, new approaches and recent improvements. J Biotechnol 70:133–153
Rittschof D, Sawardecker P, Petry C (2002) Chemical mediation of egg capsule deposition by mud snails. J Chem Ecol 28:2257–2269
Rogers CN, Paul VJ (1991) Chemical defenses of three *Glossodoris* nudibranchs and their dietary *Hyrtios* sponges. Mar Ecol Progr Ser 77:221–232
Rogers CN, de Nys R, Steinberg PD (2002) Effects of algal diet on the performance and susceptibility to predation of the sea hare *Aplysia parvula*. Mar Ecol Progr Ser 236:241–254
Rupp GS, Thompson RJ, Parsons GJ (2004) Influence of food supply on postmetamorphic growth and survival of hatchery-produced lion's paw scallop, *Nodipecten nodosus* Linnaeus, 1758. J Shellfish Res 23:5–13
Sanina NM, Kostetsky EY (2002) Thermotropic behaviour of major phospholipids from marine invertebrates: changes with warm-acclimation and seasonal acclimatization. Comp Biochem Physiol 133B:143–153
Sarashina I, Endo K (2001) The complete primary structure of molluscan shell protein 1 (MSP-1), an acidic glycoprotein in the shell matrix of the scallop *Patinopecten yessoensis*. Mar Biotechnol 3:362–369
Sennett SH (2001) Marine chemical ecology: applications in marine biomedical prospecting. In: McClintock JB, Baker BJ (eds.) Marine chemical ecology. CRC, Boca Raton, pp. 523–542
Serpentini A, Poncet JM, Boucaud-Camou E, Lebel M (2002) Study of the global synthesis of collagen by hemocytes from a nacreous mollusc, the abalone *Haliotis tuberculata*. Evidence for an annual variation. Haliotis 31:45–52
Shie HY, Liu LL (1999) Positive effects of large concentration in culture on the development of the lecithotrophic larvae of *Babylonia formosae* (Sowerby) (Prosobranchs: Buccinidae). J Exp Mar Biol Ecol 241:97–105
Stachowicz JJ (2001) Chemical ecology of mobile benthic invertebrates: predators and prey, allies and competitors. In: McClintock JB, Baker BJ (eds.) Marine chemical ecology. CRC, Boca Raton, pp 157–194
Stanley-Samuelson DW (1994) The biological significance of prostaglandins and related eicosanoids in invertebrates. Am Zool 34:589–598
Steinberg PD (1992) Geographical variation in the interaction between marine herbivores and brown algal secondary metabolites. In: Paul VJ (ed.) Ecological roles of marine natural products. Comstock Publishing, Ithaca, pp 51–92
Sud D, Doumenc D, Lopez E, Millet C (2001) Role of water-soluble matrix fraction, extracted from the nacre of *Pinctada maxima*, in the regulation of cell activity in abalone mantle cell culture (*Haliotis tuberculata*). Tissue Cell 33:154–160
Takaya Y, Uchisawa H, Ichinohe H, Sasaki J, Ishida K, Matsue H (1998) Antotumor glycogen from scallops and the interrelationship of structure and antitumor activity. J Mar Biotechnol 6:208–213
Tamburri MN, Finelli CM, Wethey DS, Zimmer-Faust RK (1996) Chemical induction of larval settlement behaviour in flow. Biol Bull 191:367–373
Thoms C, Ebel R, Hentschel U, Proksch P (2003) Sequestration of dietary alkaloids by the spongivorous marine mollusc *Tylodina perversa*. Z Naturforsch 58:426–32

Thornton RS, Kerr RG (2002) Induction of pseudopterosin biosynthesis in the gorgonian *Pseudopterogorgia elisabethae*. J Chem Ecol 28:2083–2090
Turner T, Jackson WH, Pettit GR, Wells A, Kraft AS (1998) Treatment of human prostate cancer cells with dolastatin-10, a peptide isolated from a marine shell-less mollusc. Prostate 34:175–181
Vaz-Pires P, Seixas P, Barbosa A (2004) Aquaculture potential of the common octopus (*Octopus vulgaris* Cuvier, 1797): a review. Aquaculture 238:221–238
Vidal EAG, DiMarco FP, Wormuth JH (2002) Optimizing rearing conditions of hatchling loliginid squid. Mar Biol 140:117–127
Wägele H (2004) Potential key characters in Opisthobranchia (Gastropoda, Mollusca) enhancing adaptive radiation. Organ Divers Evol 4:175–188
Whiters NH, Kokke WCMC, Fenical W, Djerassi C (1982) Sterols patterns of cultured zooxanthellae isolated from marine invertebrates; synthesis of gorgosterol and 23-desmethylgorgosterol by aposymbiotic algae. Proc Natl Acad Sci USA 79:3764–3768
Williams SI, Walker DI (1999) Mesoherbivore–macroalgal interactions: feeding ecology of sacoglossan sea slugs (Mollusca, Opisthobranchia) and their effects on their food algae. Oceanogr Mar Biol Annu Rev 37:87–128
Xu PN, Distel DL (2004) Purification and characterization of an endo-1,4-α-D-glucanase from the cellulolytic system of the wood-boring marine mollusc *Lyrodus pedicellatus* (Bivalvia: Teredinidae). Mar Biol 144:947–953
Yamazaki M (1993) Antitumor and antimicrobial glycoproteins from sea hares. Comp Biochem Physiol 105C:141–146
Yan HY (2004) Harvesting drugs from the seas and how Taiwan could contribute to this effort. Changhua J Med 9:1–6
Yang DM, Hu KP, Li CY, Wu CH, Zhou PA (2000) Neural electrophysiological effect of crude venom of *Conus textile* from the South China Sea. Toxicon 38:1607–1612
Yoneda C, Nazmul Ahsan M, Nakaya M, Matsubara Y, Ebihara T, Irie S, Uno Y, Hatae K, Watabe S (2000) Abalone collagens: immunological properties and seasonal changes of their mRNA levels. Comp Biochem Physiol 126B:59–68
Yvin JC, Chevolot L, Chevolot-Magueur AM, Cochard JC (1985) First isolation of jacaranone from an alga, *Delesseria sanguinea*. A metamorphosis inducer of *Pecten* larvae. J Nat Prod 48:814–816
Zatylny C, Durantou F, Boucaud-Camou E, Henry J (2000) Evidence of 5-hydroxytryptamine synthesis in the follicles of *Sepia officinalis* and direct involvement in the control of egg-laying. Mol Rep Dev 55:182–188
Zentz F, Bédouet L, Almeida MJ, Milet C, Lopez E, Giraud M (2001) Characterization and quantification of chitosan extracted from nacre of the abalone *Haliotis tuberculata* and the oyster *Pinctada maxima*. Mar Biotechnol 3:36–44
Zhuang S, Zhang M, Zhang X, Wang Z (2004) The influence of body size, habitat and diet concentration on feeding of *Laternula marilina* Reeve. Aquacult Res 35:622–628
Zimmer RK, Butman CA (2000) Chemical signalling processes in the marine environment. Biol Bull 198:168–187

Shellfish Poisons

M. Kita, D. Uemura

Abstract. In our ongoing search for bioactive metabolites from marine organisms, novel shellfish poisons have been isolated. Pinnatoxins, which are amphoteric polyether compounds, were purified from the Okinawan bivalve *Pinna muricata*. Pinnatoxins show acute toxicity against mice and activate Ca^{2+} channels. Two novel alkaloids, pinnamine and pinnaic acid, were also obtained from *P. muricata*. Pinnaic acid inhibits cytosolic phospholipase ($cPLA_2$). Pteriatoxins, which are pinnatoxin analogs, were isolated from the Okinawan bivalve *Pteria penguin*. A nanomole-order structure determination of pteriatoxins was achieved by the detailed analysis of 2D-NMR and ESI-TOF MS/MS. This review covers the isolation, structure determination, bioactivity, synthesis, and biogenesis of these shellfish poisons and related compounds.

2.1 Introduction

Seafood products are important both nutritionally and economically, but human intoxication resulting from the ingestion of shellfish occurs worldwide. The most important diseases, which are caused by red tides, are paralytic shellfish poisoning (PSP), diarrhetic shellfish poisoning (DSP), neurotoxic shellfish poisoning (NSP), amnesic shellfish poisoning (ASP), and ciguatera fish poisoning (CFP; Yasumoto and Murata 1993; Todd 1994; Lehane 2001). Severe cases of many different types of seafood poisonings can result in fatalities. A number of toxins from poisonous shellfish have been isolated, e.g., okadaic acid (Tachibana et al. 1981), saxitoxin (Schantz et al. 1957), and pinnatoxins (Uemura et al. 1995; Chou et al. 1996a). These compounds have attracted interest not only from pharmacologists but also from biochemists and chemists due to their extraordinary biological activities and complex structures. These bioactive marine metabolites are quite useful candidates for drugs or biological probes for physiological studies. For the purpose of preventing disease and safeguarding consumer health, it is quite important to clarify the structure and biological functions of these poisonous metabolites.

Fascinating compounds with unique chemical structures and biological activities have been found in marine organisms. However, the true origins

Daisuke Uemura
Department of Chemistry, Graduate School of Science, and Institute for Advanced Research, Nagoya University, Furo-Cho, Chikusa
464-8602 Nagoya, Japan

Progress in Molecular and Subcellular Biology
Subseries Marine Molecular Biotechnology
G. Cimino, M. Gavagnin (Eds.): Molluscs

or progenitors of these metabolites are not entirely clear. The possible primary producers of the secondary metabolites have been suggested to be microalgae, bacteria, and fungi. These metabolites are carried through symbiosis, association, food chain, and other forms of nutrient dependency (Shimizu 1993, 1996, 2003; Daranas et al. 2001). For instance, palytoxin (PTX) is a potent toxic polyol compound isolated from the zoanthid Palythoa sp. (Cha et al. 1982). The presence of related toxins in animals from distinct geographical areas was questioned due to seasonal and regional variations (Moore and Scheuer 1971; Moore and Bartolini 1981; Uemura et al. 1981). In 1995, a PTX analog, osteocin, was isolated from the dinoflagellate Ostreopsis siamensis. Its true origin was suggested from microorganisms (Usami et al. 1995). Therefore, related shellfish poisons may also be biosynthesized by marine microorganisms.

Shellfish poisons and related compounds have been well represented, with many fine reviews (Kuramoto et al. 1999, 2003, 2004; Brett 2003; Ciminiello et al. 2003; Ciminiello and Fattorusso 2004; Jeffery et al. 2004; Suenaga 2004; Kita and Uemura 2005). In our ongoing search for bioactive metabolites from marine organisms, novel shellfish poisons such as pinnatoxins, pteriatoxins, pinnamine, and turbotoxins have been isolated. In this chapter, details of our latest work on shellfish poisons are described, along with up-to-date topics.

2.2 Pinnatoxins, Ca^{2+} Channel-Activating Polyether Toxins from the Okinawan Bivalve *Pinna muricata*

Shellfish of the genus *Pinna* live mainly in shallow waters of the temperate and tropical zones of the Indian and Pacific Oceans (Rosewater 1961). The adductor muscle of this bivalve is eaten in Japan and China, and food poisoning resulting from its ingestion occurs frequently. Although this poisoning has been shown to be caused by bacterial infection or neurotoxins, the true causative agent is ambiguous. Chinese investigators have reported that a toxic extract, referred to as pinnatoxin, from *P. attenuata* is a Ca^{2+} channel activator (Zheng et al. 1990). We have successfully isolated pinnatoxin A (**1**), a mixture of pinnatoxins B (**2**) and C (**3**), and pinnatoxin D (**4**) from *P. muricata* (Fig. 2.1) as a major cause of food poisoning (Uemura et al. 1995; Chou et al. 1996a,c; Takada et al. 2001a).

2.2.1 Isolation and Structure of Pinnatoxin A

The viscera (45 kg) of *P. muricata* collected in Okinawa, Japan, were extracted with 75% ethanol. The extract was filtered and the concentrated filtrate was washed with ethyl acetate and concentrated. The oily residue was successively chromatographed on TSK-G3000S polystyrene gel, Sephadex LH-20, DEAE Sephadex A-25, and an ODS-AQ column, using a bioassay-guided (intraperitoneal injection against mice) fractionation. Final purification was achieved by reverse-phase HPLC to give pinnatoxin A (3.5 mg), a mixture of B and C (1.2 mg), and D (2.0 mg).

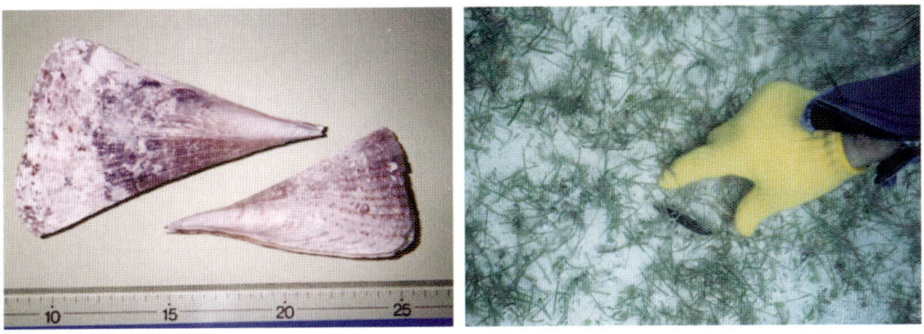

Fig. 2.1. The Okinawan bivalve *Pinna muricata*

The structures and stereochemistry of pinnatoxins have been clarified by extensive NMR experiments and positive ion ESI MS/MS spectra (Satake et al. 1991; Naoki et al. 1993). Pinnatoxins consist of a 20-membered ring, i.e., with 5,6-bicyclo, 6,7-azaspiro, and 6,5,6-triketal moieties in their structure. In particular, they contain a carboxylate anion and an iminium cation or an ammonium cation.

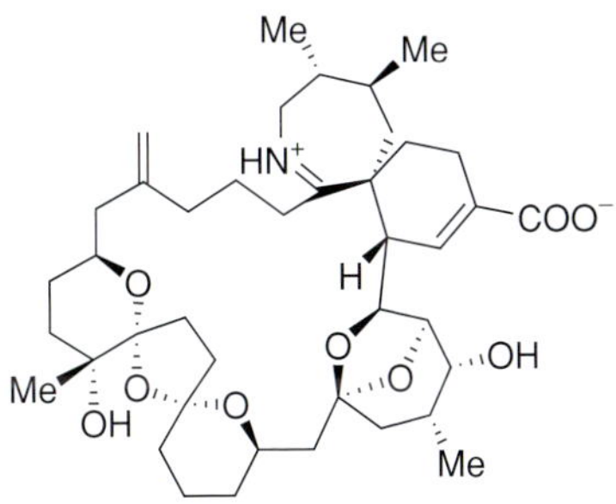

1

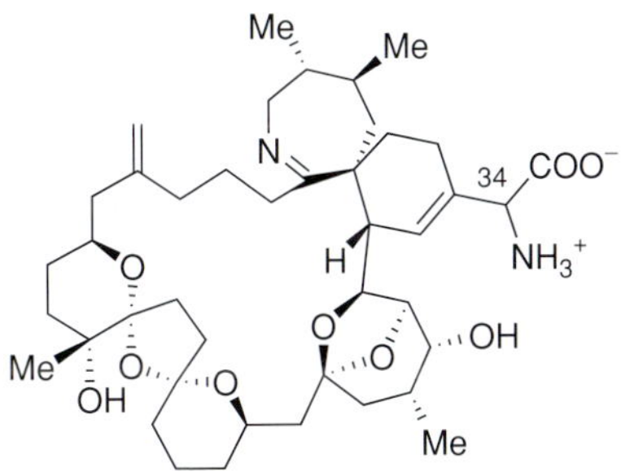

2 34*S* isomer
3 34*R* isomer

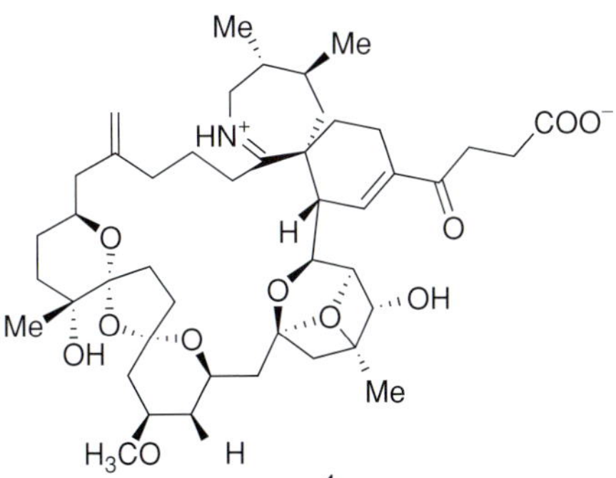

4

2.2.2 Structure of Pinnatoxins B and C

The molecular formula of both pinnatoxins B (**2**) and C (**3**) was determined to be $C_{42}H_{64}N_2O_9$ by ESI MS, which reflects a 29 MS unit (CH_3N) increase compared with that of pinnatoxin A (**1**). A positive ninhydrin test on a TLC plate for **2** and **3** suggested the presence of an amino group. Although they were obtained in small amounts, the structures and relative stereochemistries of the macrocycles in **2** and **3** were successively established as follows. A detailed analysis of the DQF-COSY and HOHAHA spectra of **2** and **3** supported almost all the carbon–carbon connectivities (Takada et al. 2001a). On the positive ion ESI MS/MS analysis, a series of prominent fragment ions was generated by a cyclohexane ring-opening reaction, the retro-Diels–Alder reaction, which was followed by bond cleavage of carbocycles (Fig. 2.2).

Reduction of the imino group in **2** and **3** with $NaBH_4$ followed by oxidative cleavage with $NaIO_4$ provided aldehyde **6** (Scheme 1). The same product was obtained from pinnatoxin A methyl ester (**5**) by reduction of both the iminium and carboxylic acid functionalities, followed by oxidation of the resulting alcohol. Thus, pinnatoxins **1**–**3** had the same relative stereochemistry in the macrocyclic core.

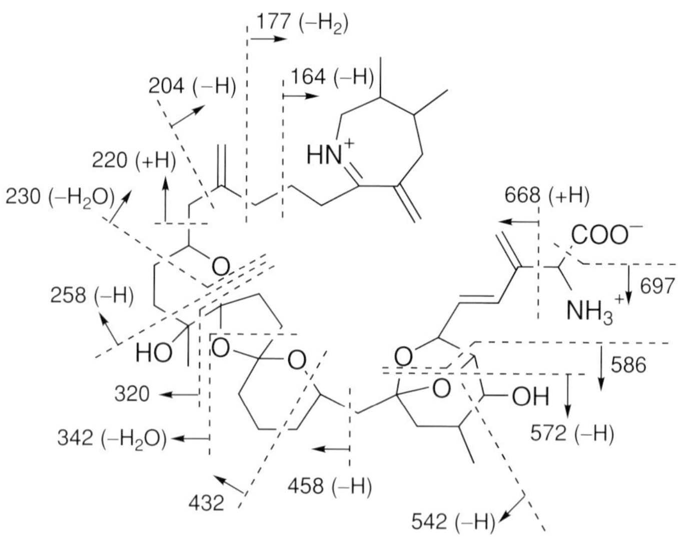

Fig. 2.2. Fragmentation patterns of pinnatoxins B (**2**) and C (**3**)

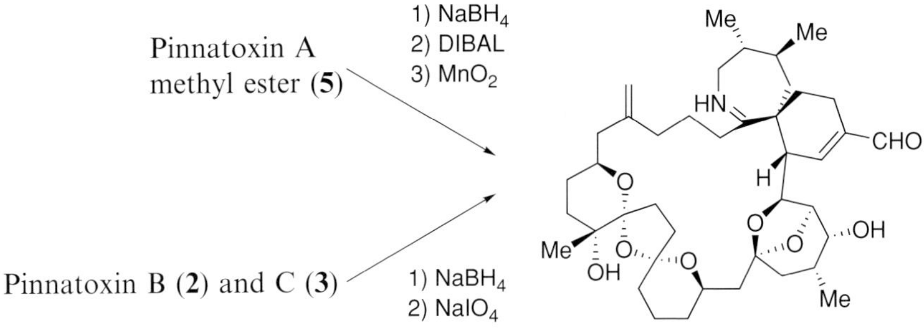

6

Scheme 1

2.2.3 Biological Activity of Pinnatoxins

Pinnatoxin A (1) showed potent acute toxicity against mice (LD_{99} 180 μg kg^{-1}, i.p.) with characteristic neurotoxic symptoms. Pinnatoxin A activated Ca^{2+} channels. Pinnatoxins B (2) and C (3), the most toxic constituents in the pinnatoxin series, have been isolated from *P. muricata* (as a 1:1 mixture; Takada et al. 2001a). The LD_{99} values of **2** and **3** were 22 μg kg^{-1}, which makes them as potent as tetrodotoxin. Although pinnatoxin D (4) showed weaker acute toxicity than the other pinnatoxins (LD_{50} > 10 μg $mouse^{-1}$ unit), **4** showed the strongest cytotoxicity against mouse leukemia cell line P388 (IC_{50} 2.5 μg ml^{-1}).

2.2.4 Biogenesis and Synthesis of Pinnatoxins

The backbone of pinnatoxins and their analogs could be configured from C1 to C34 in a single carbon chain, in a polyketide biogenetic pathway (Fig. 2.3; Uemura et al. 1995). This biosynthetic proposal entails an intramolecular Diels–Alder reaction to construct a G-ring as well as the macrocycle, followed by imine formation to establish a 6,7-spiro-ring system. Due to the structural similarity of the imine moiety adjacent to the spirocyclic core, other macrocyclic imines represented by pinnatoxin may also be biosynthesized via the same intramolecular Diels–Alder reaction.

Recently, Kishi's group achieved the total synthesis of **1** and *ent*-**1**, utilizing a biomimetic intramolecular Diels–Alder reaction (Scheme 2; McCauley et al. 1998). Interestingly, natural **1** showed significant acute toxicity, while its antipode *ent*-**1** was nontoxic (Nagasawa 2000). This investigation also confirmed the stereochemistry of **1**, including its absolute stereochemistry.

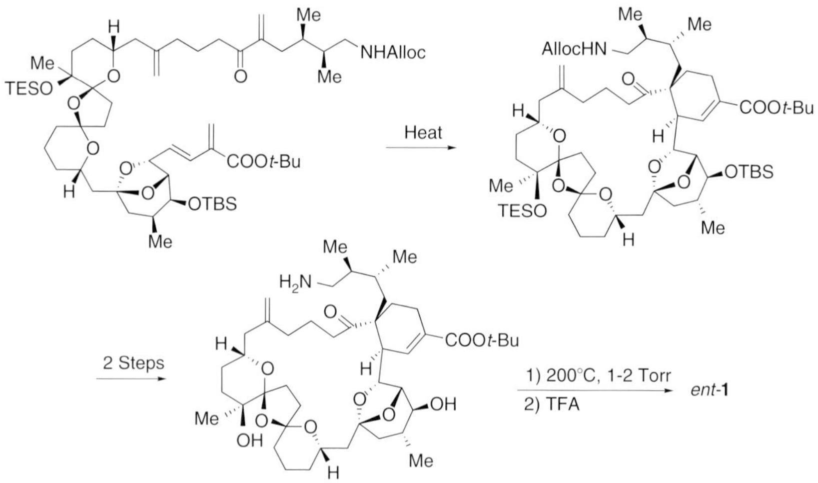

Scheme 2

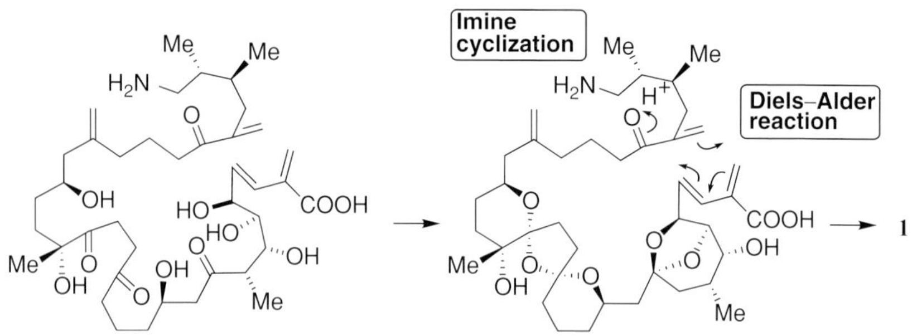

7 8

Fig. 2.3. Proposed biogenesis of pinnatoxin A (1)

2.2.5 Symbioimine, a Potential Antiresorptive Drug

Recently, we isolated a unique amphoteric iminium compound, named symbioimine (**9**), from the symbiotic marine dinoflagellate *Symbiodinium* sp. (Kita et al. 2004). This dinoflagellate is found in a wide range of marine invertebrates and produces several bioactive large polyol compounds, such as zooxanthellatoxins (ZTXs; Nakamura et al. 1993, 1995a,b) and zooxanthellamides (Onodera et al. 2003, 2004). A number of bioactive large polyol and polyether compounds, represented by PTX, halichondrin, ciguatoxin, and maitotoxin, have been reported (Uemura 1991;

Yasumoto 2001; Murata and Yasumoto 2000) and called "super-carbon -chain compounds" (Uemura 1991).

The structure of **9**, which consists of a characteristic 6,6,6-tricyclic iminium ring, was deduced by spectroscopic analysis and X-ray crystallographic analysis.

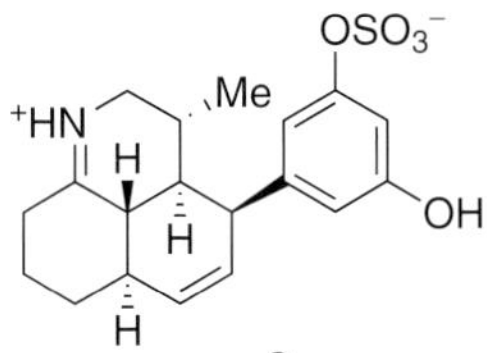

9

Symbioimine (**9**) inhibited osteoclastogenesis of the mouse monocytic cell line RAW264, which can differentiate into osteoclasts following treatment with a receptor activator of nuclear factor-κB ligand (RANKL; EC_{50} = 44 µg ml^{-1}; Kita et al. 2004). RANKL induces the formation of osteoclast-like multinucleated cells in cultures of bone marrow cells. Symbioimine (**9**) inhibited an increase in the sRANKL-induced TRAP activity of preosteoclast cells. Meanwhile, it did not affect cell viability, even at 100 µg ml^{-1}. Thus, symbioimine (**9**) is a potential antiresorptive drug for the prevention and treatment of osteoporosis in postmenopausal women. Symbioimine (**9**) also inhibited cyclooxygenase 2 (COX-2) activity (EC_{50} ~ 10 µM; Kita et al. 2005; Kita and Uemura 2005). Although the real ecological role of **9** in the symbiotic dinoflagellate is unknown yet, it can be hypothesized that it serves as a defense material which prevents their host animal's digestion.

A plausible biogenetic pathway for **9** could involve an intramolecular *exo*-transition state Diels–Alder reaction followed by imine cyclization, the same as that proposed for pinnatoxins (Fig. 2.4).

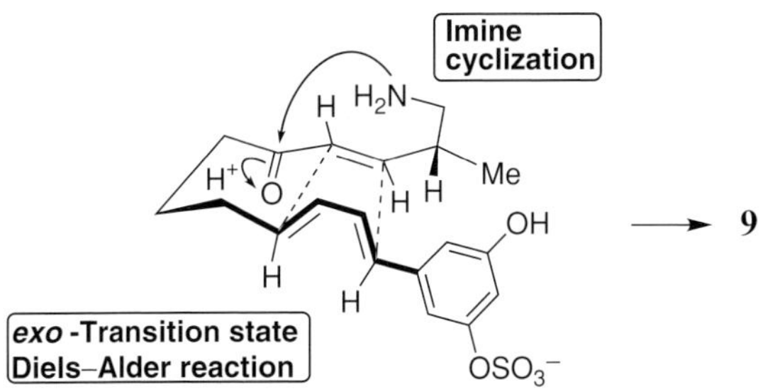

Fig. 2.4. Proposed biogenesis of symbioimine (**9**)

2.3 Pteriatoxins, Pinnatoxin Analogs from the Okinawan Bivalve *Pteria penguin* – Nanomole- Order Structure Determination

2.3.1 Isolation of Pteriatoxins

In our study of shellfish poisons, we observed that a moray eel vomits the viscera of the Okinawan bivalve *Pteria penguin* (Fig. 2.5). The aqueous 75% EtOH extract of viscera (82 kg) of *P. penguin*, which showed acute toxicity, was partitioned between EtOAc and H_2O. Guided by acute toxicity against mice, the aqueous fraction was chromatographed on TSK-G3000S polystyrene gel (50% EtOH), DEAE Sephadex A-25 (0.02 M phosphate buffer), CM Sephadex C-25 (0.2 M phosphate buffer), reversed-phase HPLC (Develosil 300 ODS, MeCN/H_2O/TFA), and reversed-phase HPLC (Develosil 300 C8, MeCN/H_2O/TFA) to isolate successfully pteriatoxins A (**10**) and B, C (**11**, **12** as a 1:1 mixture) as extremely toxic and minor components (Takada et al. 2001b). Since there was too little of these toxins to weigh, the weights of pteriatoxins A (20 μg) and B, C (8 μg) were estimated by comparison of the signal-to-noise (S/N) ratio in ^{1}H NMR spectra of 67 μM okadaic acid with those of pteriatoxins. Pteriatoxins (**10**–**12**) showed significant acute toxicity against mice, with LD_{99} values of 100 and 8 μg kg^{-1}, respectively. The toxic symptoms of pteriatoxins resemble those of pinnatoxins.

Fig. 2.5. The Okinawan bivalve *Pteria penguin*

2.3.2 Structure of Pteriatoxins

The molecular formula of pteriatoxin A (**10**) was determined by ESI MS to be $C_{45}H_{70}N_2O_{10}S$. As mentioned above, the positive ion ESI MS/MS of pinnatoxins showed a series of prominent fragment ions generated by G ring-opening reactions, followed by bond cleavage. The positive ion ESI MS/MS of **10** showed the same series of prominent fragment ions as the carbocyclic moiety in pinnatoxins (Fig. 2.6). The observation of fragment ion peaks (*m/z* 787, 744) suggested the presence of an α-amino acid moiety in the side-chain. Furthermore, the chemical shifts of H-35 (δ_H 2.80) and H-1′ (δ_H 3.04, 3.13) suggested the presence of a sulfide bond between C-35 and C-1′. The chemical shift of H-34 (δ_H 4.22) suggested the presence of an allylic hydroxyl group at C-34. Therefore, the gross structure of pteriatoxin A was determined, as shown in **10**.

The molecular formula of both **11** and **12** was determined by ESI MS to be $C_{45}H_{70}N_2O_{10}S$. Analysis of the ^{1}H NMR spectrum showed duplicate signals (1:1) for a set of protons (H-3, H-4, H-28 to H-37, H-40, H-41), suggesting the presence of epimeric isomers. Analysis of the ^{1}H NMR, COSY, and HOHAHA spectra suggested nine partial structures, as depicted in Fig. 2.7. The positive ion ESI MS/MS of **11** and **12** showed the same series of prominent fragment ions as the macrocyclic moiety in pinnatoxins and pteriatoxin A (**10**). Therefore, pteriatoxins B (**11**) and C (**12**) were also assumed to have the same polyether macrocycles as pinnatoxins. The observation of fragmentation ion peaks (*m/z* 787, 744, 710) suggested the presence of a cysteine moiety in the side-chain. Furthermore, the observation of another fragment ion peak (*m/z* 712), which was not observed in **10**, suggested the presence of a hydroxymethyl group. Therefore, the gross structure of pteriatoxins B and C was determined, as shown in **11** and **12**. The position of duplicate signals in the ^{1}H NMR spectrum suggested that pteriatoxins B (**11**) and C (**12**) are also C-34 epimers, like **2** and **3**.

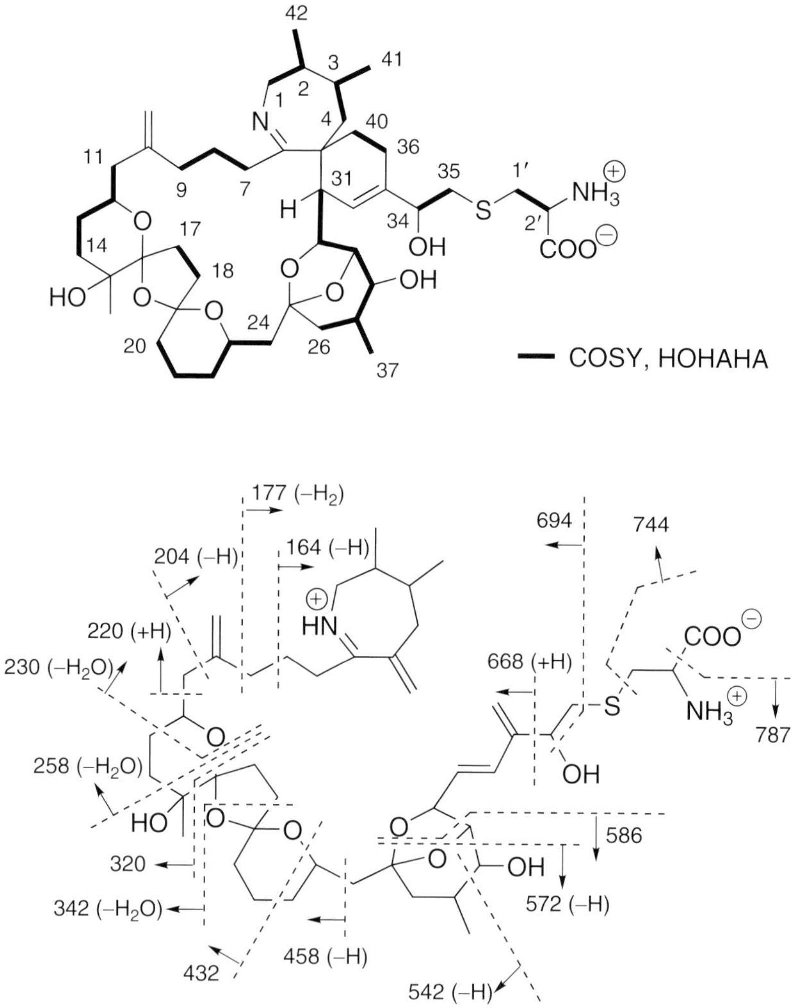

Fig. 2.6. Partial structures and fragmentation pattern of pteriatoxin A (10)

As described above, the absolute stereochemistries of a series of pinnatoxins have been confirmed by total synthesis (McCauley et al. 1998; Nagasawa 2000). By comparing the chemical shifts and coupling patterns in the ^{1}H NMR spectra of pteriatoxins with those of pinnatoxins, the same relative stereochemistry in the carbocyclic moieties was suggested for both series of compounds (Takada et al. 2001b). As a result, though the isolated yields of pteriatoxins were too small to deduce their structures by usual NMR analysis, a nanomole-order structure determination of pteriatoxins was achieved by the detailed analysis of ESI MS/MS.

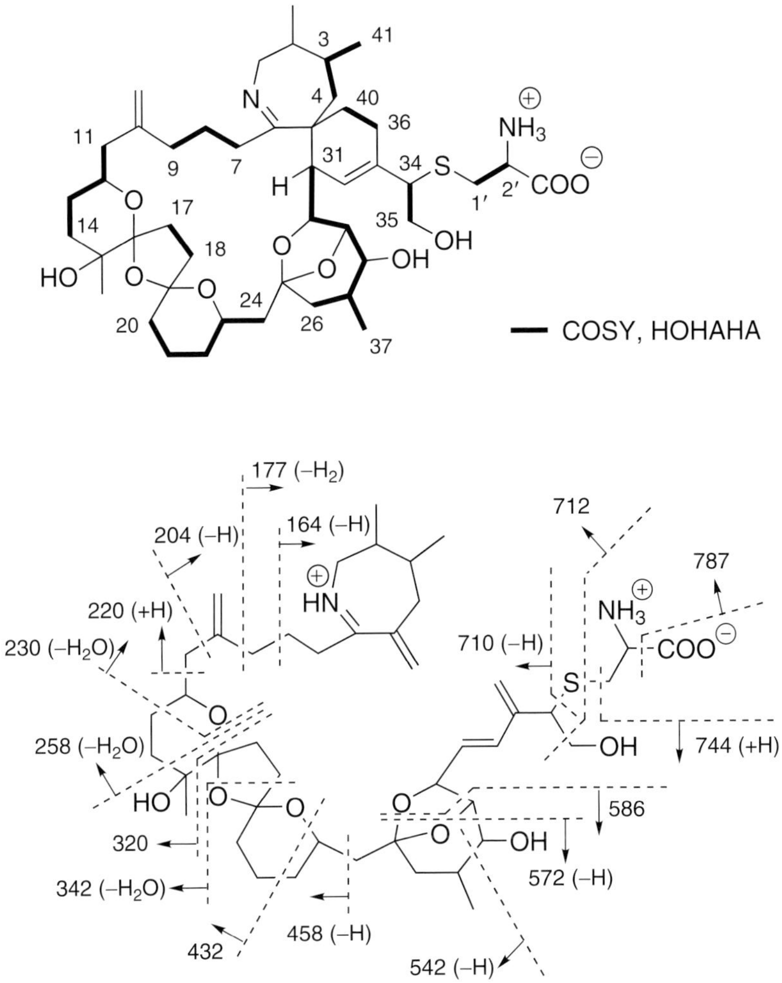

Fig. 2.7. Partial structures and fragmentation pattern of pteriatoxins B (**11**) and C (**12**)

All extracts from the digestive glands of several *Pinna* spp, including *P. muricata*, *P. attenuata*, *P. atropupurea*, and the commonly eaten shellfish *Atrina pectinata*, produced the same symptoms of poisoning in mice. These data suggested that *Pinna* shellfish might become toxic as a result of feeding on toxic organisms, such as dinoflagellates. Also, the presence of pinnatoxin analogs in both shellfish *Pinna* spp and *Pteria* spp suggested that the toxins are biosynthesized by microorganisms that are in the food chain of or in a symbiotic relationship with these shellfish.

2.3.3 Other Macrocyclic Iminium Toxins Related to Pinnatoxins

Several shellfish poisons containing nitrogen atoms, especially iminium moieties, have been isolated. Spirolides, a class of macrocyclic imines, were identified in extracts of the digestive glands of mussels and scallops from the Atlantic coast of Nova Scotia, Canada (Hu et al. 1995). The marine dinoflagellate *Alexandrium ostenfeldi* (Paulsen) Balech and Tangen was identified as the cause of spirolide toxicity in Nova Scotia in the early 1990s (Hu et al. 1996a, 2001; Cembella et al. 2000). Seven compounds, spirolides A–D (**13–16**) and 13-desmethyl C (**17**), which showed toxicity against mice, and the keto amine derivatives E and F (**18**, **19**), have been isolated and structurally characterized from the extracts of both shellfish and cultured dinoflagellate from Nova Scotia. The spirolide family contains a 5:5:6 trispiroketal ring system. Recently, the relative stereochemistry of spirolides B (**14**), D (**16**), and 13-desmethyl C (**17**), except for one chiral center, has been determined from 2D-NMR data analysis and a molecular modeling method, which showed that these compounds have the same relative stereochemistry as pinnatoxins in the region of their common structure (Falk et al. 2001).

Other marine toxins possessing a cyclic imine moiety elsewhere in the carbon skeleton are known as gymnodimine, centrimine, and prorocentrimine. Gymnodimine (**20**) was isolated from the New Zealand oyster *Tiostrea chilensis* and the dinoflagellate *Gymnodinium* sp. (Seki et al. 1995). The minimum lethal dose (MLD) of **20** was 450 $\mu g\ kg^{-1}$; and this compound also showed potent ichthyotoxicity against the fish *Tanichthys albonubes* at 0.1 ppm. The absolute stereostructure of gymnodimine (**20**) has been established by X-ray crystal structure analysis of the *p*-bromobenzamide derivative **22** derived from gymnodamine (**21**; Stewart et al. 1997). Gymnodimine B, which contains an exocyclic methylene at C-17 and an allylic hydroxyl group at C-18, was also isolated from the same dinoflagellate (Miles et al. 2000; Munday et al. 2004).

Prorocentrolide (**23**), a toxic marine macrolide that incorporates a hexahydroisoquinoline moiety, was isolated from the cultured dinoflagellate *Prorocentrum lima* (Torigoe et al. 1988; Hu et al. 1996b). This dinoflagellate produces DSP toxins, such as okadaic acid and dinophysistoxins (Tachibana et al. 1981; Yasumoto et al. 1985; Hu et al. 1992). Recently, a prorocentrolide derivative, spiro-prorocentrimine (**24**), was isolated from a cultured benthic *Prorocentrum* sp. in Taiwan and its relative stereochemistry was established by X-ray crystallographic analysis (Lu et al. 2001). Compound **24** is much less toxic than other cyclic iminium toxins.

It should be noted that both the keto amine derivatives spirolide E and F (**18**, **19**; Hu et al. 1996a), in which this ring is open, and the reduced form of gymnodamine (**21**; Stewart et al. 1997) were inactive. Although the pharmacological action of these iminium compounds has not been fully defined yet, the cyclic imine functionality may be essential and may act as a pharmacophore of macrocyclic iminium compounds, e.g., pinnatoxins and spirolides.

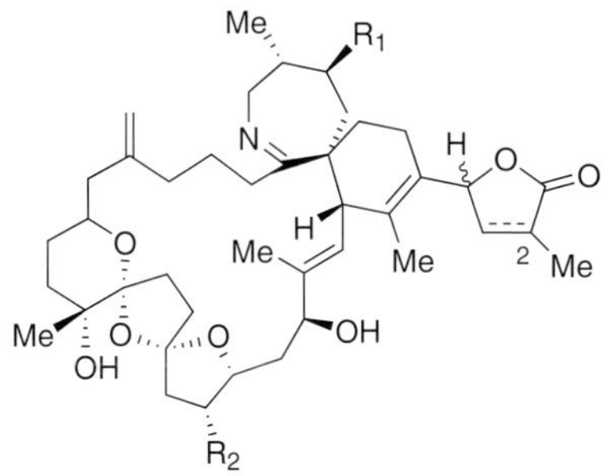

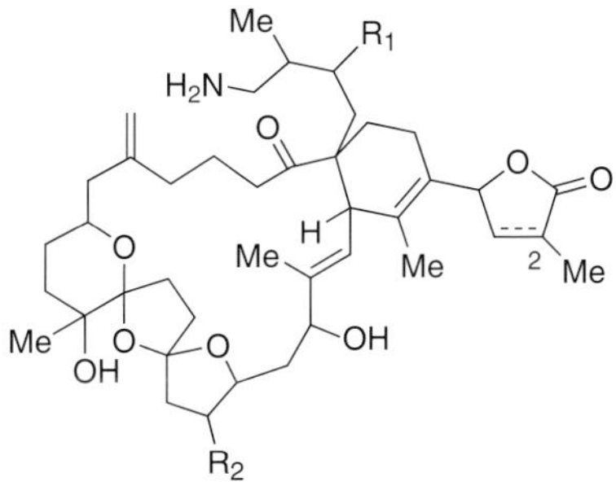

13 $R_1 = H$, $R_2 = CH_3$, $\Delta^{2,3}$
14 $R_1 = H$, $R_2 = CH_3$
15 $R_1 = CH_3$, $R_2 = CH_3$, $\Delta^{2,3}$

16 $R_1 = CH_3$, $R_2 = CH_3$
17 $R_1 = CH_3$, $R_2 = H$, $\Delta^{2,3}$

18 $R_1 = H$, $R_2 = CH_3$, $\Delta^{2,3}$
19 $R_1 = H$, $R_2 = CH_3$

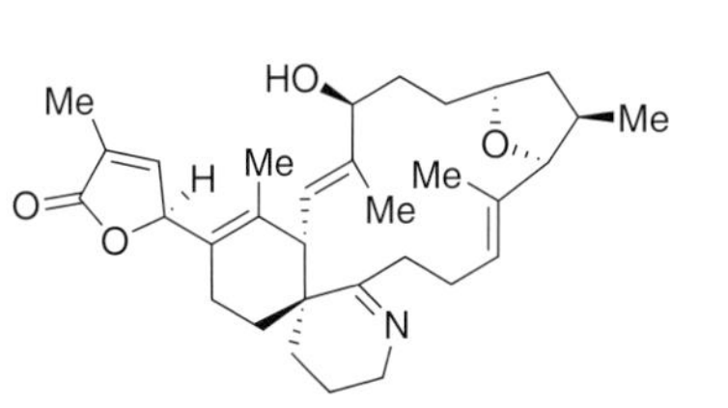

20

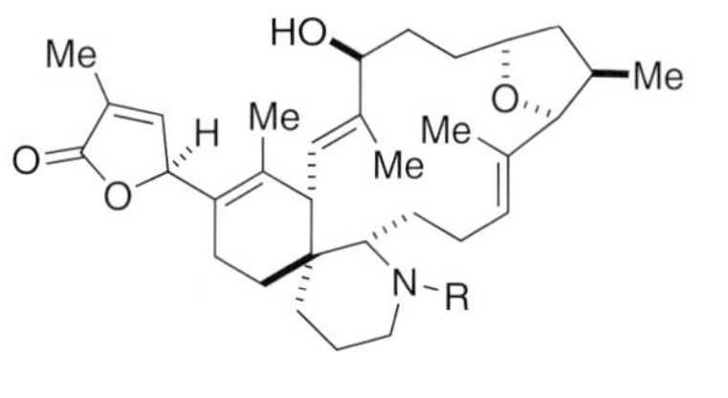

21 : R = H
22 : R = *p*-BrC_6H_4CO-

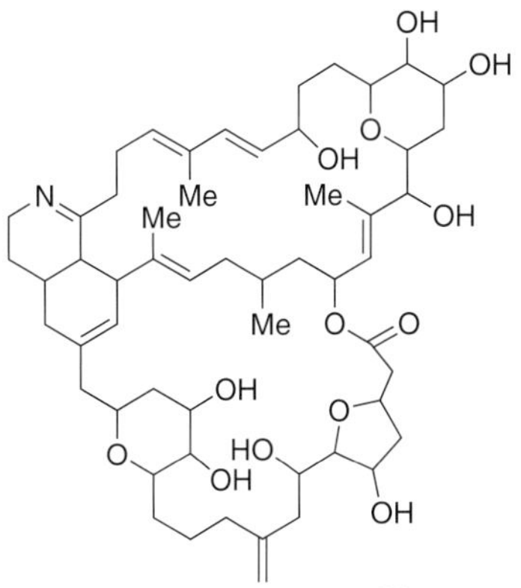

23

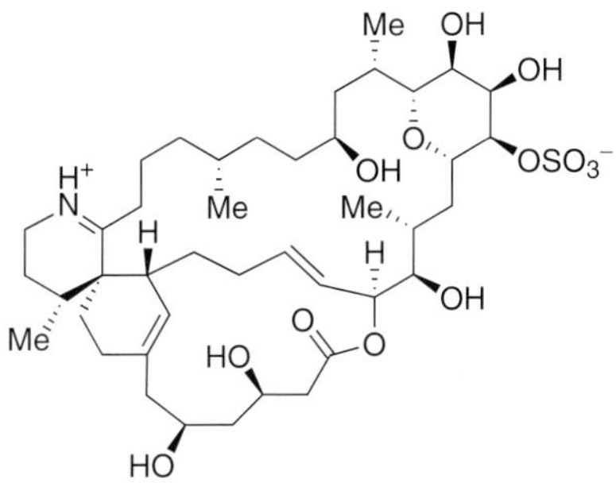

24

2.4 Turbotoxins, Diiodotyramine Derivatives from the Japanese Gastropod *Turbo marmorata*

In Japan, the gastropod *Turbo marmorata* (Fig. 2.8) is eaten after removal of the viscera, which cause intoxication. Yasumoto and coworkers studied the toxic components of *T. marmorata* and obtained several toxic fractions from this animal. They indicated the occurrence of saxitoxin and a minor toxin, iodomethyltrimethylammonium salt, in the water-soluble fraction of *T. marmorata* (Kanno et al. 1976; Yasumoto and Kotaki 1977; Kotaki et al. 1981). Although they mentioned the presence of other toxins in this animal (Kotaki and Yasumoto 1977), not all of them have been identified. We have isolated two diiodotyramine derivatives, turbotoxins A (**25**) and B (**26**), from the Okinawan gastropod *T. marmorata* (Kigoshi et al. 1999, 2000).

Fig. 2.8. The Okinawan gastropod *Turbo marmorata*

2.4.1 Isolation and Structure of Turbotoxins

The 75% aqueous ethanol extract of the viscera (4.5 kg, 36 individuals) of *T. marmorata* was partitioned between ethyl acetate and water. The aqueous layer was chromatographed using bioassay-guided (intraperitoneal mouse lethality) fractionation, to give two toxic fractions. The early toxic fraction was purified by HPLC to give turbotoxin A (**25**; 2.0 mg; LD_{99} 1.0 mg kg^{-1}) and the late toxic fraction was purified in the same way to give turbotoxin B (**26**; 0.9 mg; LD_{99} 4.0 mg kg^{-1}). Compounds **25** and **26** were isolated as trifluoroacetate salts because of the solvent system used for chromatographic purification.

Based on the HRFAB MS and 2D-NMR analyses including COSY, HSQC, and HMBC spectra, the structure of turbotoxin A (**25**) was confirmed as a diiodotyramine derivative. Turbotoxin B (**26**) was found to be a demethylated analogue of **25**, based on its molecular formula and NMR data, i.e., a trimethylammonium group at C9 in **25** was demethylated to a dimethylamino group in **26**. Turbotoxins A (**25**) and B (**26**) are structurally related to dakaramine (Xynas and Capon 1989) and dibromotyramine derivatives, such as aplysamine-1 (Hamann et al. 1993), moloka'iamine (Tsukamoto et al. 1996), and ceratinamine (Yasumoto and Endo 1974a,b), all of which were isolated from marine sponges.

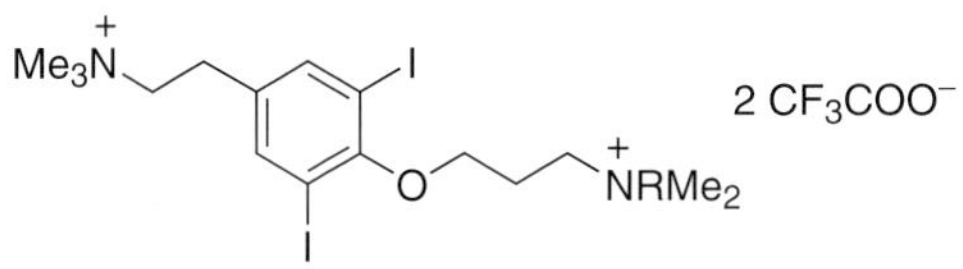

25 R = Me
26 R = H

2.4.2 Structure–Activity Relationship

To investigate the structure–toxicity relationships of turbotoxins, turbotoxins (**25**, **26**) and their 12 analogues **27–38** were synthesized (Fig. 2.9; Kigoshi et al. 2000). The relationships between the structure of turbotoxins and their acute toxicity (intraperitoneal mouse lethality) are shown in Table 2.1. Compound **28** shows the weakest toxicity among the compounds **25–28**, indicating that the quaternary ammonium group is responsible for the toxicity. The toxicities of turbotoxin B (**26**; 4.0 mg kg^{-1}) and isomer **27** (8.0 mg kg^{-1}) are weaker than that of **25** and stronger than that of **28**. This finding shows that the number of the quaternary ammonium groups is important to the toxicity of turbotoxins.

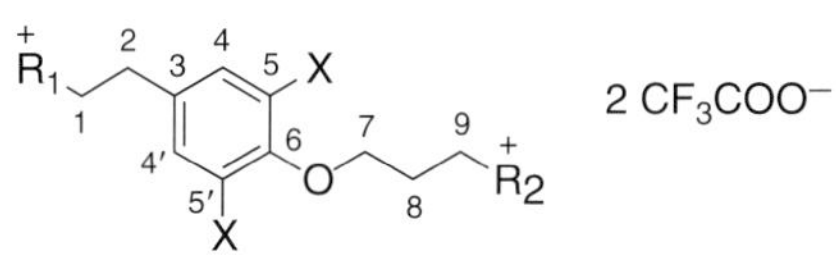

Turbotoxin A (**25**) X = I, $R_1 = R_2 = NMe_3$
Turbotoxin B (**26**) X = I, $R_1 = NMe_3$, $R_2 = NHMe_2$
27 X = I, $R_1 = NHMe_2$, $R_3 = NMe_3$
28 X = I, $R_1 = R_2 = NHMe_2$

29 X = Br, $R_1 = R_2 = NMe_3$
30 X = Cl, $R_1 = R_2 = NMe_3$
31 X = H, $R_1 = R_2 = NMe_3$
32 X = Me, $R_1 = R_2 = NMe_3$

33 X = I, $R_1 = R_2 = NMe_2Bn$
34 X = I, $R_1 = NMe_2Bn$, $R_2 = NMe_3$
35 X = I, $R_1 = NMe_3$, $R_2 = NMe_2Bn$

36 X = I, R_1 = R2 = NMe_2Ar
37 X = I, $R_1 = NMe_2Ar$, $R_2 = NMe_3$
38 X = I, $R_1 = NMe_3$, $R_2 = NMe_2Ar$

Ar = 4-phenylbenzyl

Fig. 2.9. Natural and artificial analogues of turbotoxins

Table 2.1. Acute toxicity of turbotoxins and their analogs

compound	acute toxicity (LD_{99}, mg/kg-1)
Turbotoxin A (**25**)	1.0
Turbotoxin B (**26**)	4.0
27	8.0
28	100
Bromo analog **29**	4.0
Chloro analog **30**	8.0
Hydro analog **31**	12
Methyl analog **32**	8.0
Dibenzyl analog **33**	4.0
Monobenzyl analog **34**	2.0
Monobenzyl analog **35**	0.5
Diphenylbenzyl analog **36**	>32
Monophenylbenzyl analog **37**	32
Monophenylbenzyl analog **38**	8.0

Upon intraperitoneal injection into ddY mice (n >4)

Comparison of the toxicities of turbotoxin A (**25**), bromo analogue **29**, chloro analogue **30**, and hydro analogue **31** (1.0, 4.0, 8.0, 12.0 mg kg^{-1}) indicates the importance of iodine atoms in the toxicity of **25**. The toxicity of methyl analogue **32** (8.0 mg kg^{-1}) is more potent than that of hydro analogue **31** and is the same as that of chloro analogue **30**, indicating the importance of the steric bulkiness of the substituents at C-5 and C-5′ positions.

The effect of the substituents of the quaternary ammonium moieties was also examined. While the toxicities of dibenzyl and monobenzyl analogues **33** and **34** were weaker than that of **25**, monobenzyl analogue **35** exhibited toxicity stronger than that of **25**. The benzyl substituent at the *N*-9 ammonium group increases its toxicity twofold. The phenylbenzyl analogues, **36**–**38**, exhibited weaker toxicities than turbotoxin A (**25**) and its benzyl analogues, **33**–**35**. These facts indicated that the bulky 4-phenylbenzyl group prevents the phenylbenzyl analogs from binding to target molecules.

The target biomolecule of turbotoxin A was investigated; and it was found that turbotoxin A (**25**) inhibits acetylcholinesterase, with an IC_{50} of 28 μM. Donepezil hydrochloride (Aricept), a drug for Alzheimer's disease, also inhibits degradation of acetylcholine and activates the central cholinergic system, with an IC_{50} of 5.7 nM (Sugimoto et al. 1995). X-ray crystallographic studies of complexes of acetylcholinesterase with small molecules, such as decamethonium bromide, tacrine, and edrophonium bromide, indicated that the aromatic gorge exists at the bottom of the active site (Harel et al. 1993). There as yet no data of relationships between the toxicity and affinity to acetylcholinesterase of turbotoxin analogues. The benzyl group in **35** might, however, be stacked against the aromatic gorge to increase its toxicity. Preliminary neuropharmacological experiments were affected for turbotoxin A (**25**); and **25** was proved not to interact with the peripheral nervous system.

2.5 Pinnamine and Pinnaic Acids, Alkaloidal Marine Toxins from *Pinna muricata*

2.5.1 Pinnamine

As described above, pinnatoxins (1–4) are Ca^{2+} channel activators that have been isolated from the Okinawan bivalve *Pinna muricata*. In a continuation of this work, we isolated a novel marine alkaloid, pinnamine (**39**), from aqueous ethanol extracts of *P. muricata*. Pinnamine exhibited significant acute toxicity against mice, with an LD_{99} of 0.5 mg kg^{-1}, and produced characteristic toxic symptoms, such as scurrying around (Takada et al. 2000).

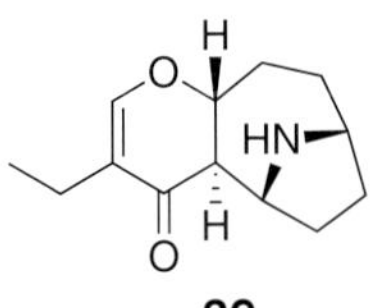

39

The structure of pinnamine (**39**) was determined by spectroscopic analysis to be an alkaloid containing a dihydropyrone ring. The relative stereostructure of pinnamine was determined by an analysis of coupling constants and NOE experiments (Fig. 2.10); and the absolute stereostructure was determined by an analysis of the circular dichroism spectrum (Djerassi et al. 1962; Snatzke 1965a–c). The structure and toxic symptoms of pinnamine resemble those of anatoxin (Devlin 1976), called Very Fast Death Factor, and atropine (Leete et al. 1954), a representative suppressor of the parasympathetic nervous system. This indicates that the toxic expression of pinnamine, similar to that of atropine, may result from excitability of the cerebrum.

Recently, we reported an enantioselective synthesis of pinnamine (**39**; Kigoshi et al. 2001). Synthetic pinnamine was found to correspond uniquely to natural **39** by comparison of their spectral data, including their CD spectra, and acute toxicity data. Further biological studies of pinnamine are currently underway.

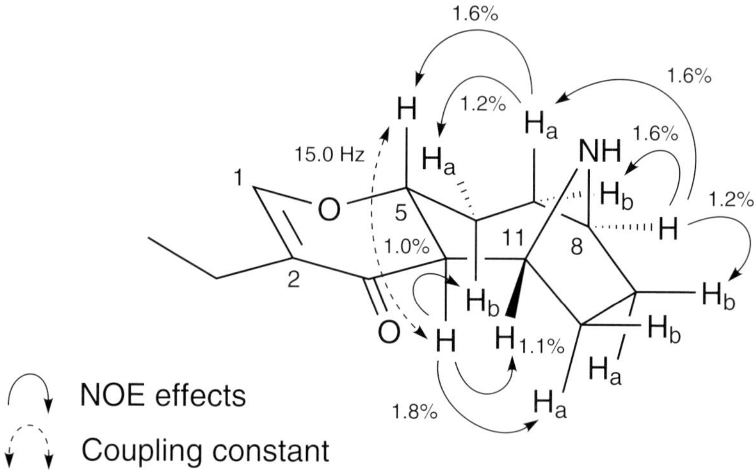

Fig. 2.10. A plausible conformation of pinnamine

2.5.2 Pinnaic Acids: $cPLA_2$ Inhibitors

Specific inhibitors of phospholipase A_2 (PLA_2) have been considered as potential drugs for the treatment of inflammation and other disease states, since PLA_2 is linked to the initial step in the cascade of enzymatic reactions that lead to the generation of inflammatory mediators (van den Bosch 1980; Arita et al. 1989). Marine natural products such as manoalide (Scheuer and de Silva 1980) and luffariellolide (Albizati et al. 1987) have been reported to be potent PLA_2 inhibitors (Potts et al. 1992a,b). A cytosolic 85-kDa phospholipase ($cPLA_2$; Kramer et al. 1990; Kramer and Sharp 1995) exhibits specificity for the release of arachidonic acid from membrane phospholipids (Kim et al. 1990). Therefore, compounds that inhibit $cPLA_2$ activity have been targeted as anti-inflammatory agents.

The 80% EtOH extract of the viscera (10 kg) of. *P. muricata* (3,000 individuals) was partitioned between EtOAc and water. The water layer was fractionated by column chromatography on TSK-G3000S polystyrene gel, Sephadex LH-20, DEAE Sephadex A-25, ODS-AQ, and silica gel to obtain two novel fatty acids, i.e., pinnaic acid (**40**; 1 mg) and tauropinnaic acid (**41**; 4 mg; Chou et al. 1996b).

The structure of **40** was determined by an analysis of NMR spectral data. Tauropinnaic acid (**41**) has a 6-azaspiro[4.5]decane unit and a taurine moiety. The relative stereochemistry of **41** was deduced from phase-sensitive NOE correlations. Furthermore, the gross structure of **40** was elucidated by a detailed comparison of the EI-MS fragment peaks with the corresponding peaks of **41**.

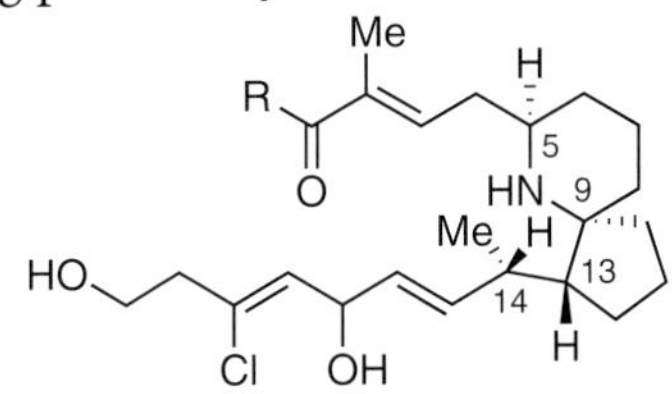

40 R = OH
41 R = $NHCH_2CH_2SO_3H$

Pinnaic acid (**40**) and tauropinnaic acid (**41**) inhibited $cPLA_2$ activity in vitro, with IC_{50} values of 0.2 and 0.09 mM, respectively. Inhibitors of $cPLA_2$ have rarely been reported. Therefore, though the activity of pinnaic acids was not as strong, we have great interest in the mode of action of these compounds.

2.5.3 Halichlorine: an Inhibitor of VCAM-1 Induction

Adhesion molecules are involved in the process of adhesion between cells and the extracellular matrix in the formation of multicellular bodies. In addition, it is known that the activity of adhesive molecules is very important for the maintenance of function and performance. A recent study suggested that the clinical application of adhesion molecules as anti-inflammatory agents and immunosuppressive agents may some day be possible, provided that the function of the adhesive molecules can be controlled (Kock et al. 1995).

A simple model of multistage adhesion between leukocyte and vascular cells is shown in Fig. 2.11. This process can be classified into four stages, i.e., (1) rolling, (2) triggering, (3) strong adhesion, and (4) transmigration. Vascular cell adhesion molecule-1 (VCAM-1; Osborn et al. 1988) is affected during (3), the phase of strong adhesion. Drugs that block the induced expression of VCAM-1 may be useful for treating atherosclerosis, coronary artery diseases, angina, and noncardiovascular inflammatory diseases.

The marine sponge *Halichondria okadai* Kadota (200 kg) was immersed in EtOH at room temperature. The ethanolic extract was filtered, concentrated under reduced pressure, and extracted with ethyl acetate. The ethyl acetate extract was partitioned between 70% aqueous methanol and hexane. The aqueous methanol layer was separated by column chromatography on TSK-G3000S, then on ODS, and then on SiO_2. Finally, the active fraction was purified by preparative TLC on SiO_2 to give halichlorine (**42**; 70.8 mg; Kuramoto et al. 1996).

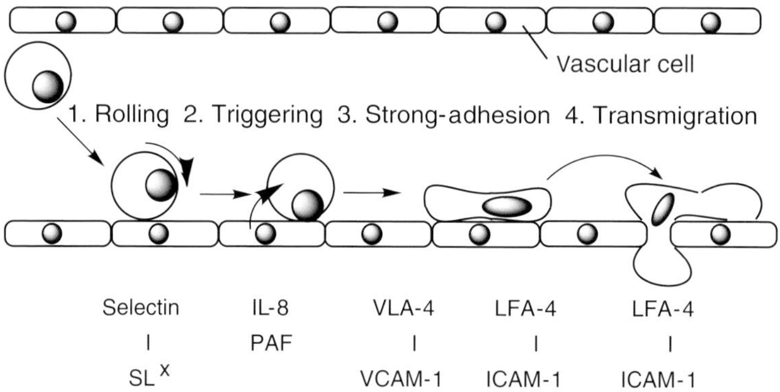

Fig. 2.11. A model of adhesion between leukocyte and vascular cells

The gross structure of **42** was elucidated by an analysis of MS, IR, and extensive 2D-NMR spectra. Halichlorine consists of a sterically hindered 15-membered lactone, an azabicyclo[4.4.0] ring, and a 5.6-spiro ring moiety. The relative stereochemistry of **42** was confirmed mainly by the coupling constants and NOESY spectral data. The Bohlmann band (Nikon 1957; Bohlmann, 1958) in the IR spectrum suggested the conformation around the tertiary amine functionality. Furthermore, oxidative degradation of **42**, as well as asymmetric synthesis of the degradation product **43** from D-(–)-tartaric acid, allowed us to determine the absolute stereochemistry of halichlorine (Scheme 3; Arimoto et al. 1998).

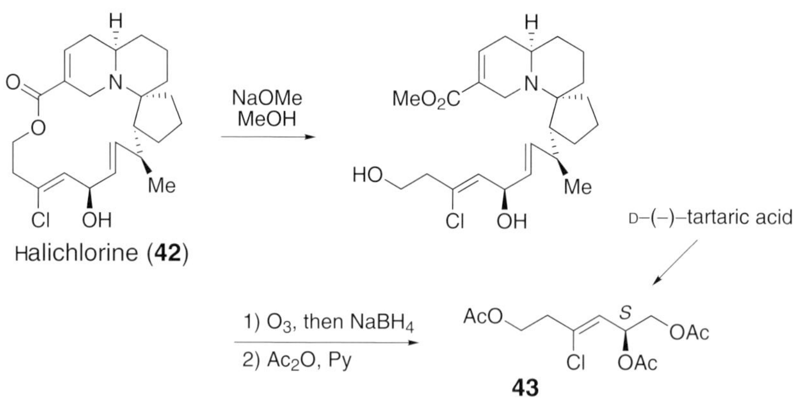

Scheme 3

Halichlorine inhibits the induction of VCAM-1 at IC_{50} 7 μg ml^{-1}. Although VCAM-1 and ICAM belong to the same immunoglobulin superfamily, halichlorine does not affect ICAM (IC_{50} > 100 μg ml^{-1}; Boschelli et al. 1995). It is largely unknown why halichlorine affects only VCAM-1. Additional research is needed to clarify the functions and mechanisms of action of VCAM-1.

These architecturally novel alkaloids have attracted the attention of synthetic chemists. (Clive et al. 2005) The Danishefsky group has achieved the total synthesis of pinnaic acid (Carson et al. 2001a,b) and halichlorine (Trauner and Danishefsky 1999; Trauner et al. 1999) in an asymmetric manner. We recently reported a racemic total synthesis of **40** (Hayakawa et al. 2003). Detailed comparison of the ^{1}H-NMR spectra of both synthetic and natural samples supported Danishefsky's revision of the configuration at C14. We also synthesized the tricyclic azadecaline core of halichlorine (**42**, Hayakawa et al. 2004). Further studies of the detailed chemistry of pinnaic acids and halichlorine, including biogenetic pathways and structure–activity relationships, are in progress.

2.5.4
Biogenesis of Pinnaic Acid

Interestingly, the structure of pinnaic acid (**40**) from the bivalve *P. muricata* has been shown to be closely similar to those of halichlorine (**42**) from the marine sponge *H. okadai*. Each carbon atom has been tentatively numbered according to the supposed biogenetic formation of the N–C23 bond (Fig. 2.12). These results suggest that both of these bioactive metabolites may be produced by symbiotic marine microorganisms.

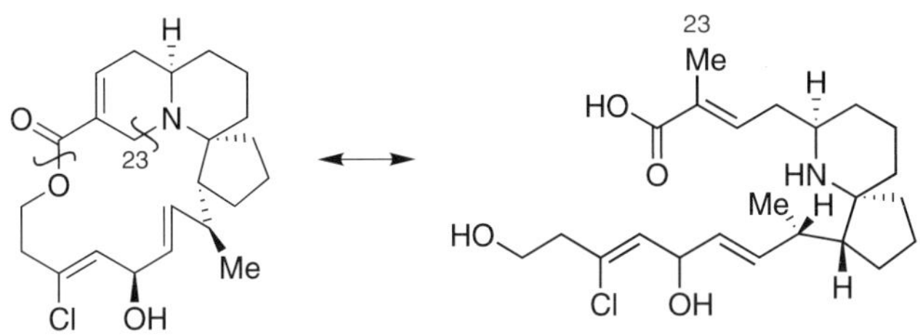

42 from *Halichondria okadai* **40** from *Pinna muricata*

Fig. 2.12. Biogenesis of pinnaic acids and halichlorine

2.6 Conclusions

The study of natural resources may lead to the further discovery of novel bioactive compounds. As described in this chapter, it is still important to clarify the structure and biological activity of shellfish poisons to prevent disease and safeguard consumer health. Further chemical and biological studies on these shellfish poisons should contribute a deeper understanding of their roles in nature. New natural products continue to fascinate us because of the extremely unexpected structure of molecules, their biologically relevant mode of action, and their sacred and inviolable roles in the ecosystem. However, a number of unexpected and unforeseen compounds may have been naturally secreted. To overcome difficulties in the isolation and characterization of such compounds, a new chemical approach for dynamic ecological systems will be required, together with new methods in organic chemistry and spectral analysis. Intensive studies involving the comprehensive evaluation of these molecules may lead to the creation of a new field in bioscience.

Acknowledgments. We would like to thank Prof. H. Kigoshi (University of Tsukuba), Prof. H. Arimoto (Tohoku University), Dr. K. Suenaga (University of Tsukuba), Dr. M. Kuramoto (Ehime University), Dr. N. Takada (Hokkaido University), and Dr. I. Hayakawa (University of Tsukuba) for their contributions to this work. We would also like to thank Dr. K. Yamada (Nagoya University) and Dr. K. Yamaguchi (Tokyo University of Fisheries) for testing biological activities. This study was supported in part by Grants-in-Aid for Creative Scientific Research from Japan Society for the Promotion of Science (JSPS). We are indepted to Wako Pure Chemical Industries, Ltd., Banyu Pharmaceutical Co., Ltd., Ono Pharmaceutical Co., Ltd., and the Naito Foundation for their financial support.

References

Albizati KF, Holman T, Faulkner DJ, Glaser KB, Jacobs RS (1987) Luffariellolide, an anti-inflammatory sesterterpene from the marine sponge *Luffariella* sp. Experientia 43:949–950

Arimoto H, Hayakawa I, Kuramoto M, Uemura D (1998) Absolute stereochemistry of halichlorine; a potent inhibitor of VCAM-1 induction. Tetrahedron Lett 39:861–862

Arita H, Nakano T, Hanasaki K (1989) Thromboxane A2: its generation and role in platelet activation. Prog Lipid Res 28:273–301

Bohlmann, F (1958) Lupinen-Alkaloide, VIII. Zur Konfigurationsbestimmung von Chinolizidin-Devivaten. Chem Ber 91:2157

Boschelli DH, Karmer JB, Khatana SS, Sorenson RJ, Connor DT, Ferin MA, Wright CD, Lesch ME, Imre K, Okonkwo GC, Schrie DJ, Conroy MC, Ferguson E, Woelle J, Saxena U (1995) Inhibition of E-selectin-, ICAM-1-, and VCAM-1-mediated cell adhesion by benzo[*b*]thiophene-, benzofuran-, indole-, and naphthalene-2-carboxamides: identification of PD 144795 as an anti-inflammatory agent. J Med Chem 38:4597–4614

Brett MM (2003) Food poisoning associated with biotoxins in fish and shellfish. Curr Opin Infect Dis 14:461–465

Carson MW, Kim G, Hentemann MF, Trauner D, Danishefsky SJ (2001a) Concise stereoselective routes to advanced intermediates related to natural and unnatural pinnaic acid. Angew Chem Int Ed Engl 40:4450–4452

Carson MW, Kim G, Danishefsky SJ (2001b) Stereochemistry of natural and unnatural pinnaic acids: a remarkable long-range stereochemical effect in the reduction of 17-oxo precursors of the pinnaic acids. Angew Chem Int Ed Engl 40:4453–4456

Cembella AD, Lewis NL, Quilliam MA (2000) The marine dinoflagellate *Alexandrium ostenfeldii* (Dinophyceae) as the causative organism of spirolide shellfish toxins. Phycologia 39:67–74

Cha JK, Christ WJ, Finan JM, Fujioka H, Kishi Y, Klein LL, Ko SS, Leder J, McWhorter WW Jr, Pfaff KP, Yonaga M, Uemura D, Hirata Y (1982) Stereochemistry of palytoxin 4. Complete structure. J Am Chem Soc 104:7369–7371

Chou T, Kamo O, Uemura D (1996a) Relative stereochemistry of pinnatoxin A, a potent shellfish poison from *Pinna muricata*. Tetrahedron Lett 37:4023–4026

Chou T, Kuramoto M, Otani Y, Shikano M, Yazawa, K, Uemura D (1996b) Pinnaic acid and tauropinnaic acid: two novel fatty acids composing a 6-azaspiro[4.5]decane unit from the Okinawan bivalve *Pinna muricata*. Tetrahedron Lett 37:3871–3874

Chou T, Haino T, Kuramoto M, Uemura, D (1996c) Isolation and structure of pinnatoxin D, a new shellfish poison from the Okinawan bivalve *Pinna muricata*. Tetrahedron Lett 37:4027–4030

Ciminiello P, Fattorusso E (2004) Shellfish toxins – chemical studies on northern Adriatic mussels. Eur J Org Chem 2004:2533–2551

Clive DLJ, Yu M, Wang J, Yeh VSC, Kang S (2005) Synthetic Chemistry of Halichlorine and the Pinnaic Acids. Chem Rev 105:4483–4514

Daranas AH, Norte M, Fernández JJ (2001) Toxic marine microalgae. Toxicon 39:1101–1132

Devlin JP, Edwards OE, Gorham PR, Hunter NR, Pike RK, Stavric B (1976) Anatoxin A, a toxic alkaloid from anabaena-flos-aquae NRC-44H. Can J Chem 55:1367–1371

Djerassi C, Records R, Bunnenberg E, Mislow K, Miscowitz A (1962) Inherently dissymetric chromophores. Optical rotatory dispersion of α, β-unsaturated ketones and conformational analysis of cyclohexenones. J Am Chem Soc 84:870–872

Falk M, Burton IW, Hu T, Walter JA, Wright JLC (2001) Assignment of the relative stereochemistry of the spirolides, macrocyclic toxins isolated from shellfish and from the cultured dinoflagellate *Alexandrium ostenfeldii*. Tetrahedron 57:8659–8665

Hamann MT, Scheuer PJ, Kelly-Borges M (1993) Biogenetically diverse, bioactive constituents of a sponge, order Verongida – bromotyramines and sesquiterpene-shikimate derived metabolites. J Org Chem 58:6565–6569

Harel M, Schalk I, Ehret-Sabatier L, Bouet F, Goeldner M, Hirth C, Axelsen PH, Silman I, Sussman JL (1993) Quaternary ligand-binding to aromatic residues in the active-site gorge of acetylcholinesterase. Proc Natl Acad Sci USA 90:9031–9035

Hayakawa I, Arimoto H, Uemura D (2003) Synthesis of (±)-pinnaic acid. Heterocycles 59:441–444

Hayakawa I, Arimoto H, Uemura D (2004) Synthesis of the tricyclic core of halichlorine. Chem Commun 1222–1223

Hu TM, Doyle J, Jackson D, Marr J, Nixon E, Pleasance S, Quilliam MA, Walter JA, Wright JLC (1992) Isolation of a new diarrehetic shellfish poison from Irish mussels. J Chem Soc Chem Commun 39:39–41

Hu T, Curtis JM, Walter JA, Wright JLC (1995) Spirolide-B and spirolide-D, 2 novel macrolides isolated from the digestive glands of shellfish. J Chem Soc Chem Commun 2159–2161

Hu T, Curtis JM, Walter JA, Wright JLC (1996a) Characterization of biologically inactive spirolides E and F: identification of the spirolide pharmacophore. Tetrahedron Lett 37:7671–7674

Hu T, DeFreitas ASW, Curtis JM, Oshima Y, Walter JA, Wright JLC (1996b) Isolation and structure of prorocentrolide B, a fast-acting toxin from *Prorocentrum maculosum*. J Nat Prod 59:1010–1014

Hu T, Burton IW, Cembella AD, Curtis JM, Quilliam MA, Walter JA, Wright JLC (2001) Characterization of spirolides A, C, and 13-desmethyl C, new marine toxins isolated from toxic plankton and contaminated shellfish. J Nat Prod 64:308–312

Jeffery B, Barlow T, Moizer K, Paul S, Boyle C (2004) Amnesic shellfish poison. Food Chem Toxicol 42:545–557

Kanno K, Kotaki Y, Yasumoto T (1976) Toxicity studies on marine snails. 6. Distribution of toxins in mollusks associated with coral reefs. Bull Jpn Soc Sci Fish 42:1395–1398

Kigoshi H, Kanematsu K, Uemura D (1999) Turbotoxins A and B, novel diiodotyramine derivatives from the Japanese gastropod *Turbo marmorata*. Tetrahedron Lett 40:5745–5748

Kigoshi H, Kanematsu K, Yokota K, Uemura D (2000) Turbotoxins A and B, novel diiodotyramine derivatives from the Japanese gastropod *Turbo marmorata*. Tetrahedron 56:9063–9070

Kigoshi H, Hayashi N, Uemura D (2001) Stereoselective synthesis of pinnamine, an alkaloidal marine toxin from *Pinna muricata*. Tetrahedron Lett 42:7469–7471

Kim DK, Kudo I, Fujimori Y, Mizushima H, Masuda M, Kikuchi R, Ikizawa K, Inoue K (1990) Detection and subcellular localization of rabbit platelet phospholipase A2 which preferentially hydrolyzes an arachidonoyl residue. J Biochem 108:903–906

Kita M, Kondo M, Koyama T, Yamada K, Matsumoto T, Lee KH, Woo JT, Uemura D (2004) Symbioimine exhibiting inhibitory effect of osteoclast differentiation, from the symbiotic marine dinoflagellate *Symbiodinium* sp. J Am Chem Soc 126:4794–4795

Kita M, Ohishi N, Washida K, Kondo M, Koyama T, Yamada K, Uemura D (2005) Symbioimine and neosymbiomine, amphpteric iminium metabolites form the symbiotic dinoflagellate *Symbiodinium* sp. Bioorg Med Chem 13:5253–5258

Kita M, Uemura D (2005) Iminium alkaloids from marine invertebrates: structure, biological activity, and biogenesis. Chem Lett 34:454–459

Kock AE, Halloran MM, Haskell CJ, Sah MR, Polverini PJ (1995) Angiogenesis mediated by soluble forms of E-selectin and vascular cell adhesion molecule-1. Nature 376:517–519

Kotaki Y, Yasumoto T (1977) Toxicity studies on marine snails. 8. Identification of iodomethyltrimethyl-ammonium as a minor toxin of green turban shell. Bull Jpn Soc Sci Fish 43:1467

Kotaki Y, Oshima Y, Yasumoto T (1981) Studies on paralytic shellfish poisoning in tropical waters. 1. Analysis of paralytic shellfish toxins of marine snails. Bull Jpn Soc Sci Fish 47:943–946

Kramer RM, Sharp JD (1995) Recent insights into the structure, function and biology of cPLA2. Agents Actions Suppl 46:65–67

Kramer RM, Johansen B, Hession C, Pepinsky RB (1990) Structure and properties of a secretable phospholipase A2 from human platelets. Adv Exp Med Biol 275:35–53

Kuramoto M, Chou T, Yamada K, Chiba T, Hayashi Y, Uemura D (1996) Halichlorine, an inhibitor of VCAM-1 induction from the marine sponge *Halichondria okadai* Kadota. Tetrahedron Lett 37:3867–3870

Kuramoto M, Arimoto H, Uemura D (1999) Structures and functions of marine bioactive alkaloids. J Synth Org Chem Jpn 57:105–115

Kuramoto M, Arimoto H, Uemura D (2003) Studies in bioactive marine alkaloids. J Synth Org Chem Jpn 61:1099–1105

Kuramoto M, Arimoto H, Uemura D (2004) Bioactive marine alkaloids. A review. Mar Drugs 1:39–54

Leete E, Marion L, Spenser ID (1954) The biogenesis of alkaloids. 12. The mode of formation of the tropine base of hyoscyamine. Can J Chem 32:1116–1123

Lehane L (2001) Paralytic shellfish poisoning: a potential public health problem. Med J Aust 175:29–31
Lu CK, Lee GH, Huang R, Chou HN (2001) Spiro-prorocentrimine, a novel macrocyclic lactone from a benthic *Prorocentrum* sp. of Taiwan. Tetrahedron Lett 42:1713–1716
McCauley JA, Nagasawa K, Lander PA, Mischke SG, Semones MA, Kishi Y (1998) Total synthesis of pinnatoxin A. J Am Chem Soc 120:7647–7468
Miles CO, Wilkins AL, Stirling DJ, MacKanzie AL (2000) New analogue of gymnodimine from a *Gymnodinium* species. J Agric Food Chem 48:1373–1376
Moore RE, Bartolini GJ (1981) Structure of palytoxin. J Am Chem Soc 103:2491–2494
Moore RE, Scheuer PJ (1971) Palytoxin – new marine toxin from a coelenterate. Science 172:495
Munday R, Towers NR, Mackenzie L, Beuzenberg V, Holland PT, Miles CO (2004) Acute toxicity of gymnodimine to mice. Toxicon 44:173–178
Murata M, Yasumoto T (2000) The structure elucidation and biological activities of high molecular weight algal toxins: maitotoxin, prymnesins and zooxanthellatoxins. Nat Prod Rep 17:293–314
Nagasawa K (2000) Total synthesis of pinnatoxin A. J Synth Org Chem Jpn 58:877–886
Nakamura H, Asari T, Murai A (1993) Structure of periodate oxidation products with characteristic partial structures of zooxanthellatoxin-A, a potent vasoconstrictive polyol from a symbiotic dinoflagellate. J Org Chem 58:313–314
Nakamura H, Asari T, Murai A, Kan Y, Kondo T, Yoshida K, Ohizumi Y (1995a) Zooxanthellatoxin-A, a potent vasoconstrictive 62-membered lactone from a symbiotic dinoflagellate. J Am Chem Soc 117:550–551
Nakamura H, Asari T, Fujimaki F, Maruyama K, Murai A, Ohizumi Y, Kan Y (1995b) Zooxanthellatoxin-B, vasoconstrictive congener of zooxanthellatoxin-A from a symbiotic dinoflagellate *Symbiodinium* sp. Tetrahedron Lett 36:7255–7258
Naoki H, Murata M, Yasumoto T (1993) Negative FAB tandem mass spectrometry for structural study on polyether compounds; structural verification of yessotoxin. Rapid Commun Mass Spectrom 7:179–182
Nikon A (1957) A relationship between conformation and infrared absorption in 1,2-halohydrins. J Am Chem Soc 79:243–247
Onodera K, Nakamura H, Oba Y, Ojika M (2003) Zooxanthellamide A, a novel polyhydroxy metabolite from a marine dinoflagellate of *Symbiodinium* sp. Tetrahedron 59:1067–1071
Onodera K, Nakamura H, Oba Y, Ojika M (2004) Zooxanthellamide B, a novel large polyhydroxy metabolite from a marine dinoflagellate of *Symbiodinium* sp. Biosci Biotechnol Biochem 68:955–958
Osborn L, Hession C, Tizard R, Vassallo C, Huhovoskyi S, Chi-Rosso G, Hobb R (1988) Direct expression cloning of vascular cell adhesion molecule 1, a cytokine-induced endothelial protein that binds to lymphocytes. Cell 59:1203–1211
Potts BCM, Faulkner DJ, Carvalho MS de, Jacobs RS (1992a) Chemical mechanism of inactivation of bee venom phospholipase A2 by the marine natural products manoalide, luffariellolide, and scalaradial. J Am Chem Soc 114:5093–5100
Potts BCM, Faulkner DJ, Jacobs RS (1992b) Phospholipase A2 inhibitors from marine organisms. J Nat Prod 66:1701–1717
Rosewater J (1961) The family Pinnidae in the Indo-Pacific. Indo-Pac Mollusca 1:53/501/632
Satake M, Murata M, Yasumoto T, Fujita T, Naoki H (1991) Amphidinol, a polyhydroxy-polyene antifungal agent with an unprecedented structure, from a marine dinoflagellate, *Amphidinium klebsii.* J Am Chem Soc 113:9859–9861
Schantz EJ, Mold JD, Stanger DW, Shavel J, Riel FJ, Bowden JP, Lynch JM, Wyler RS, Riegel B, Sommer H (1957) Paralytic shellfish poison. 4. A procedure for the isolation and purification of the poison from toxic clam and mussel tissues. J Am Chem Soc 79:5230–5235
Scheuer PJ, de Silva ED (1980) Manoalide, an antibiotic sesterterpenoid from the marine sponge *Luffariella variabilis* (polejaeff). Tetrahedron Lett 21:1611–1614

Seki T, Satake M, Mackenzie L, Kasper HF, Yasumoto T (1995) Gymnodimine, a new marine toxin of unprecendented structure isolated from New-Zealand oysters and the dinoflagellate, *Gymnodinium* sp. Tetrahedron Lett 36:7093–7096

Shimizu Y (1993) Microalgal metabolites. Chem Rev 93:1685–1698

Shimizu Y (1996) Microalgal metabolites: A new perspective. Annu Rev Microbiol 50:431–465

Shimizu Y (2003) Microalgal metabolites. Curr Opin Microbiol 6:236–243

Snatzke G (1965a) Circulardichroismus-VIII: modifizierung der octantenregel für, α, β-ungesättigte ketone: theorie. Tetrahedron 21:413–419

Snatzke G (1965b) Circulardichroismus-IX: modifizierung der octantenregel für, α, β-ungesättigte ketone: transoid enone. Tetrahedron 21:421–438

Snatzke G (1965c) Circulardichroismus-X: modifizierung der octantenregel für, α, β-ungesättigte ketone: cisoide enone, dienone und arylketone. Tetrahedron 21:439–448

Stewart M, Blunt JW, Munro MHG, Robinson WT, Hannah DJ (1997) The absolute stereochemistry of the New Zealand shellfish toxin gymnodimine. Tetrahedron Lett 38:4889–4890

Suenaga K (2004) Bioorganic studies on marine natural products with bioactivity, such as antitumor activity and feeding attractance. Bull Chem Soc Jpn 77:443–451

Sugimoto H, Iimura Y, Yamanishi Y, Yamatsu K (1995) Synthesis and structure–activity relationships of acethylcholinesterase inhibitors 1-benzyl-4-[(5,6-dimethoxy-1-oxoindan-2-yl)methyl]piperidine hydrochloride and related compounds. J Med Chem 38:4821–4829

Tachibana K, Scheuer PJ, Tsukitani Y, Kikuchi H, Engen DV, Clardy J, Gopichand Y, Schmitz FJ (1981) Okadaic acid, a cyto-toxic polyether from 2 marine sponges of the genus *Halichondria*. J Am Chem Soc 103:2469–2471

Takada N, Iwatsuki M, Suenaga K, Uemura D (2000) Pinnamine, an alkaloidal marine toxin, isolated from *Pinna muricata*. Tetrahedron Lett 41:6425–6428

Takada N, Uemura N, Suenaga K, Chou T, Nagatsu A, Haino T, Yamada K, Uemura D (2001a) Pinnatoxins B and C, the most toxic components in the pinnatoxin series from the Okinawan bivalve *Pinna muricata*. Tetrahedron Lett 42:3491–3494

Takada N, Uemura N, Suenaga K, Uemura D (2001b) Structural determination of pteriatoxins A, B and C, extremely potent toxins from the bivalve *Pteria penguin*. Tetrahedron Lett 42:3495–3497

Todd ECD (1994) Emerging diseases associated with seafood toxins and other water-borne agents. Ann NY Acad Sci 740:77–94

Torigoe K, Murata M, Yasumoto T, Iwashita T (1988) Prorocentrolide, a toxic nitrogenous macrocycle from a marine dinoflagellate, *Prorocentrum lima*. J Am Chem Soc 110:7876–7877

Trauner D, Danishefsky SJ (1999) Studies towards the total synthesis of halichlorine: asymmetric synthesis of the spiroquinolizidine subunit. Tetrahedron Lett 40:6513–6516

Trauner D, Schwarz JB, Danishefsky SJ (1999) Total synthesis of (+)-halichlorine: an inhibitor of VCAM-1 expression. Angew Chem Int Ed Engl 38:3542–3545

Tsukamoto S, Kato H, Hirota H, Fusetani N (1996) Ceratinamine: an unprecedented antifouling cyanoformamide from the marine sponge *Pseudoceratina purpurea*. J Org Chem 61:2936–2937

Uemura D (1991) Antitumor polyethers. In: Scheuer PJ (ed.) Bioorganic Marine Chemistry. Springer, Berlin Heidelberg New York, pp. 1–31

Uemura D, Ueda K, Hirata Y (1981) Further studies on palytoxin. 2. Structure of palytoxin. Tetrahedron Lett 22:2781–2784

Uemura D, Chou T, Haino T, Nagatsu A, Fukuzawa S, Zheng SZ, Chen H (1995) Pinnatoxin A: a toxic amphoteric macrocycle from the Okinawan bivalve *Pinna muricata*. J Am Chem Soc 117:1155–1156

Usami M, Satake M, Ishida S, Inoue A, Kan Y, Yasumoto T (1995) Palytoxin analogs from the dinoflagellate *Ostreopsis siamensis*. J Am Chem Soc 117:5389–5390

van den Bosch H (1980) Intracellular phospholipases A. Biochim Biophys Acta 604:191–246

Xynas R, Capon R (1989) Two new bromotyrosine-derived metabolites from an Australian marine sponge *Aplysina* sp. Aust J Chem 42:1427–1433

Yasumoto T (2001) The chemistry and biological function of natural marine toxins. Chem Rec 1:228–242

Yasumoto T, Endo M (1974a) Toxicity study on a marine snail, *Turbo afgyrostoma*. 3. Identification of (3-meyhylthiopropyl)trimethylammonuim chloride. Bull Jpn Soc Sci Fish 40:217–221

Yasumoto T, Endo M (1974b) Toxicity study on a marine snail *Turbo argyrostoma* – occurrence of candicine. Bull Jpn Soc Sci Fish 40:841–845

Yasumoto T, Kotaki Y (1977) Occurrence of saxitoxin in a green turban shell. Bull Jpn Soc Sci Fish 43:207–211

Yasumoto T, Murata M (1993) Marine toxins. Chem Rev 93:1897–1909

Yasumoto T, Murata M, Oshima Y, Sano M, Matsumoto GK, Clardy J (1985) Diarrhetic shellfish toxins. Tetrahedron 41:1019–1025

Zheng SZ, Huang FL, Chen SC, Tan XF, Zuo JB, Peng J, Xie RW (1990) The isolation and bioactivities of pinnatoxin. Zhongguo Haiyang Yaowu (Chin J Mar Drugs) 9:33–35

Bivalve Molluscs as Vectors of Marine Biotoxins Involved in Seafood Poisoning

P. Ciminiello, E. Fattorusso

Abstract. Molluscs of many sorts, which are high in protein and trace minerals, have always been a substantial portion of the human diet. A great variety of mollusc species are therefore of commercial importance throughout the world. Episodes of poisoning occasionally happen to the consumers of molluscs, the main hazard being represented by bivalve molluscs. These organisms are filter-feeders, feeding mainly on a wide range of phytoplankton species. Among the thousands of species of microscopic algae at the base of the marine food chain, there are a few dozen which produce potent toxins. One major category of impact occurs when toxic phytoplankton are filtered from the water as food by shellfish, which then accumulate the algal toxins to levels which can be lethal to humans. Incidences of poisoning related to marine algal toxins come under the main categories of paralytic shellfish poisoning (PSP), neurotoxic shellfish poisoning (NSP), diarrhetic shellfish poisoning (DSP), and amnesic shellfish poisoning (ASP), depending upon the toxins and the symptoms that they cause. Since the beginning of the 1990s, a research program has been initiated to examine the toxin profiles in mussels from the Adriatic Sea. Since then, a number of polyether toxins have been isolated and characterized, some of which represent new additions to the DSP class of biotoxins. During this investigation, new types of toxins have also been isolated. The recent application of LC–MS methods for the detection of Adriatic marine biotoxins made it possible to speed up the analysis of toxic samples.

3.1 Introduction

Molluscs, because of their ease of capture, edibility and beauty, have long been important to mankind. Molluscs of many sorts, which are high in protein and trace minerals, have always been a substantial portion of the human diet. Abalone, clams, cockles, muscles, octopus, oysters, periwinkles, scallops, snails, squid, whelks, winkles, and many more are all molluscs, and all make their contribution to the human diet.

Mankind has been deliberately culturing molluscs as food for a long time and the earliest known records of someone farming molluscs for food come from the Romans. It was in fact a Roman, a certain Sergius

Ernesto Fattorusso
Dipartimento di Chimica delle Sostanze Naturali, Università degli Studi di Napoli "Federico II", Via D. Montesano 49, 80131 Naples, Italy

Progress in Molecular and Subcellular Biology
Subseries Marine Molecular Biotechnology
G. Cimino, M. Gavagnin (Eds.): Molluscs

Orata, who established the first oyster farm in Lake Lucrinus near Naples, to the best of our knowledge the first mollusc farm in history, about 95 BC. Oysters were likely the first sea animal to be transported from one area to another and cultivated as food.

Although molluscs are generally considered desirable components of a healthy diet, numerous cases of poisoning, particularly frequent and recurring in bivalve molluscs, occur worldwide each year (Ahmed 1991). While most of these toxic events are associated with the consumption of bivalve molluscs contaminated with viral, bacterial, and parasitic microorganisms, a significant number of incidents are associated with natural toxins produced by microalgae.

All bivalves are filter-feeders, mainly feeding on a wide range of phytoplankton species. Mussels, like all filter-feeding bivalve molluscs, process large volumes of water. This is necessary because the amount of organic matter in seawater is low (average $1\,mg\,l^{-1}$). Mussels filter, on average, $7.5\,l\,h^{-1}$ of seawater. As a consequence of this, they accumulate and concentrate many pollutants in seawater, particularly those which are particulate or associated with particles. Like all bivalves, mussels are notorious for their ability to accumulate very high concentrations of metals. They also accumulate other pollutants, such as fecal bacteria and radionuclides, as well as all the metabolites produced by phytoplanktonic species used as food.

There are several thousand different phytoplanktonic species and of these some 60–80 algae, mainly belonging to the classes dinoflagellates and diatoms, are known to produce toxins (Tibbetts 1998). They are normally present in small quantities and do not represent a problem for public health. However, sometimes, algal proliferation occurs. The term harmful algal blooms (HABs) was initially coined to describe high concentrations of algae that produce extremely potent poisons. However, the scientific community recognizes now that, because a wide range of organisms is involved and some species have toxic effects at low cell densities, not all HABs are "algal" and not all occur as "blooms". How and why these blooms occur is a complex issue, depending on oceanographic currents, winds and other factors.

During toxic blooms, fish and shellfish consume these algae, then accumulate and concentrate the toxins without apparent harm. One major category of impact occurs when toxic phytoplankton is filtered from the water as food by edible shellfish such as clams, mussels, oysters, or scallops, which then accumulate the algal toxins to levels, which can be lethal to humans or other consumers (Shumway 1990). Typically, the shellfish contaminated by toxic phytoplankton are only marginally affected, even though a single clam can sometimes contain sufficient toxin to kill a human. In general, the natural marine toxins are tasteless, odorless, and are heat- and acid-stable. Therefore, normal screening and food preparation procedures do not prevent intoxication if the shellfish is contaminated. The myriad of toxic compounds that marine phytoplankton can produce are known as marine biotoxins (Botana 2000).

The number of documented toxic blooms has been found to be increasing globally over the past few decades (Hallegraeff 1993). This is undoubtedly the result of a number of factors, including increasing worldwide seafood consumption, increased public and scientific awareness of harmful events, improved detection and analytical capabilities, changing weather and global temperature, and coastal pollution. There is also a body of evidence to indicate human-induced transportation of the cysts or "seeds" of toxic marine and freshwater organisms such as dinoflagellates, or the dinoflagellates themselves located inside the "spat" (young bivalve shellfish sold commercially to global markets for aquaculture) and ship ballast water (Anderson 1989; Hallegraeff 1993). Dinoflagellate cysts are able to survive long journeys in the dark and cold ballast tanks, before being released into seas when the ships dump their ballast water prior to harbor entry. International regulations are now changing to require ship ballast water to be purged in the open ocean prior to docking.

It is also hypothesized that human-generated environmental changes, such as reef destruction and eutrophication, may be responsible for the apparent increase in reports of human cases of marine and freshwater toxin disease as well as the increased incidence of HABs reported worldwide. There is even evidence connecting the apparent global increase of algal blooms with global climate changes, as seen with the El Niño phenomenon (Maclean 1989).

3.2 Marine Biotoxins

Marine biotoxins, produced by phytoplankton usually during HAB events, are some of the most potent toxins in the world and extremely dangerous. For some toxins, doses at the microgram per kilogram level are more than sufficient to kill. When enough toxin is accumulated in fish or shellfish, small amounts of cooked or raw tissue, even the consumption of one or two small mussels, can kill a normal, healthy adult human. While some toxins are very potent, i.e., requiring only small amounts to produce illness or death, other less potent toxins may accumulate to such high levels that they can still cause harm.

The epidemiology of shellfish toxins around the world and their risk to human health have been well documented. In 1987 approximately 2,000 cases of human poisoning were thought to occur each year worldwide through the consumption of fish and shellfish contaminated with algal toxins, with a mortality rate of approximately 15% (Hallegraeff 1987), but more recently it was estimated to be approximately 60,000 people affected each year (Tibbetts 1998). Approximately 90% of all known poisoning incidents from seafood are associated with molluscs, mainly bivalves (Soames-Mraci 1995).

The risk of poisoning from the consumption of fish or shellfish is serious and of concern to public health authorities in all coastal environments. The risk and threat to public health is so great that many countries have instituted some form of risk management plan to deal with marine biotoxins. These "sanitation" plans are difficult to design and implement because the properties of the toxins are only poorly understood and, in addition to this, their origins may also not be known. Moreover, there is still a very poor understanding of the target organs for toxicity and the nature of any dose–response relationship associated with this toxicity. For these reasons, it is still difficult to identify a safe level of exposure to the respective toxins and, therefore, to provide an estimate of the margin of safety at various levels of exposure. In addition, not all countries have thorough monitoring systems: with the international transport and sale of seafood there is always a possibility of falling victim to a biotoxin.

Currently, the identified toxins are classified according to the poisoning syndromes they cause (Yasumoto and Murata 1993); and the incidences of poisoning related to marine algal toxins, depending upon the toxins and the symptoms that they cause, come under the main categories of:

- Paralytic shellfish poisoning (PSP)
- Neurotoxic shellfish poisoning (NSP)
- Diarrhetic shellfish poisoning (DSP)
- Amnesic shellfish poisoning (ASP)

Except for ASP, all are caused by biotoxins synthesized by dinoflagellates.

The toxins responsible for these syndromes are not single chemical entities, but are families of compounds having similar chemical entities and effects. Chemically, they can range from polar, low molecular weight compounds to high molecular weight, lipophilic substances. Most algal toxins cause human illness by disrupting electrical conduction, uncoupling communication between nerve and muscle, and impeding critical physiological processes. To do so, they bind to specific membrane receptors, leading to changes in the intracellular concentration of ions such as sodium or calcium.

3.2.1 Paralytic Shellfish Poisoning

PSP is the most studied and understood of all the shellfish poisoning syndromes. Incidents of PSP have been recorded throughout the world for many centuries. Historically, PSP incidents are associated with dinoflagellates of the *Alexandrium* species (Schantz 1986). However, with developments in technology and research on marine algae, more species and classes of microorganisms are now being found to produce these toxins. Marine bacteria such as *Moraxella* (Kodama 1988) and *Alteromonas tetraodonis* (Gallacher and Birkbeck 1995) and freshwater cyanobacteria such as *Aphanizomenon flos-aquae*, *Anabaena circinalis*, *Lyngbya wollei*, *Cylindrospermopsis raciborskii* (Humpage et al. 1994; Falconer 1996; Lagos et al. 1997; Onodera et al. 1997), and *Protogonyaulax* (Ogata et al. 1989) have all been found to produce or influence the production of these toxins in algae. Infection of *Ostreopsis lenticularis* by *Pseudomonas* species was also found to affect the production of toxins (Gonzalez et al. 1995).

Paralytic Shellfish Toxins

The toxins responsible for PSP are a suite of heterocyclic guanidines collectively called saxitoxins, of which there are currently over 29 known congeners (Shimizu 2000). Their structures vary, having different combinations of hydroxyl and sulfate substitutions at four sites on the molecule (Fig. 3.1). Based on substitutions at R4, the saxitoxins can be subdivided into four groups: (1) neurotoxic and highly potent carbamate toxins which include the non-sulfated saxitoxin (STX) and neosaxitoxin (NEO) and gonyautoxins (GTX1–GTX4), which are singly sulfated and more lethal than the non-sulfated carbamate toxins, (2) weakly toxic *N*-sulfocarbamoyl-11-hydroxysulfate toxins (B1, B2, C1–C4), which are the least toxic to mammals of all the PSP toxins, (3) decarbamoyl (dc-) analogs, which are thought to arise from the metabolism of dinoflagellate toxins within the shellfish and (4) deoxydecarbamoyl (do-) toxins, that have been detected until now only in Australian populations of *G. catenatum* (Oshima et al. 1993).

STX blocks neurotransmission at the neuromuscular junction. It causes a blockage of neuronal and muscular Na^+ channels, preventing the propagation of action potentials and causing a relaxant action on vascular smooth muscle cells (Falconer 1993). STX binds specifically to site 1 of voltage-sensitive sodium channels (VSSCs) and requires the presence of both the α and the $\beta 1$ subunit of the channel.

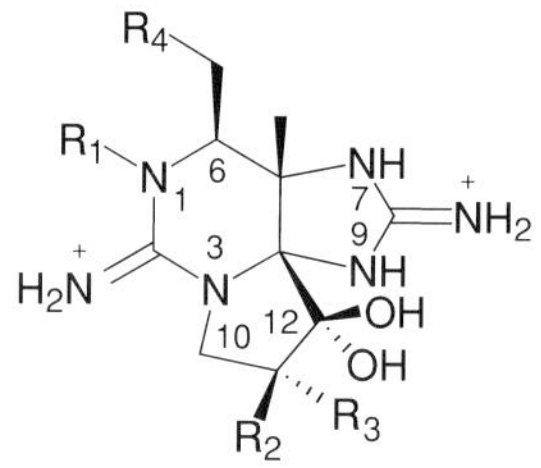

STX = saxitoxin
NEO = neosaxitoxin
GTX = gonyautoxin

			R4							
			carbamate toxins		N-sulfocarbamoyl toxins		decarbamoyl toxins		deoxydecarbamoyl toxins	
R_1	R_2	R_3	$-OC(=O)NH_2$		$-OC(=O)NHSO_3^-$		$-OH$		$-H$	
H	H	H	**1**	STX	**11**	GTX5, B1	**17**	dcSTX	**27**	doSTX
OH	H	H	**2**	NEO	**12**	GTX6, B2	**18**	dcNEO		
H	H	OSO_3^-	**3**	GTX2	**13**	C1	**19**	dcGTX2	**28**	doGTX2
H	OSO_3^-	H	**4**	GTX3	**14**	C2	**20**	dcGTX3	**29**	doGTX3
OH	H	OSO_3^-	**5**	GTX1	**15**	C3	**21**	dcGTX1		
OH	OSO_3^-	H	**6**	GTX4	**16**	C4	**22**	dcGTX4		
H	H	OH	**7**	11αOH-STX			**23**	11αOH-dcSTX		
H	OH	H	**8**	11βOH-STX			**24**	11βOH-dcSTX		
OH	H	OH	**9**	11αOH-NEO			**25**	11αOH-dcNEO		
OH	OH	H	**10**	11βOH-NEO			**26**	11βOH-dcNEO		

Fig. 3.1. Structure of paralytic shellfish toxins

Clinical Symptoms of PSP

PSP is a neurotoxic syndrome with a mortality rate of approximately 20% of those intoxicated. Symptoms of the disease develop fairly rapidly, within 0.5–2.0 h after ingestion of the shellfish, depending on the amount of toxin consumed. In humans, the peripheral nervous system is affected, with symptoms including tingling and numbness of extremities, muscular non-coordination, respiratory distress and muscular paralysis, leading to death by asphyxiation (Gessner et al. 1997). Transmission of nerve impulses to the muscles is inhibited and thus the diaphragm and other respiratory muscles in the lungs cease to assist breathing and the victim can die from respiratory arrest. When respiratory support is provided within 12 h of exposure, recovery is usually complete, with no lasting side-effects. In unusual cases, because of the weak hypotensive action of the toxin, death may occur from cardiovascular collapse despite respiratory support.

One of the most predominant toxins within this group is STX; and it is so potent that up to 50 humans can be poisoned from the level of poison contained in just one contaminated mussel (Soames-Mraci 1995 and references therein). There is currently no antidote for intoxication and prognosis for the patient is entirely based upon the amount ingested by the victim.

3.2.2 Diarrhetic Shellfish Poisoning

DSP is a human illness, associated with seafood consumption and characterized by acute gastrointestinal disturbance. It is caused by a class of acidic polyether toxins produced by dinoflagellates. DSP is widespread in its distribution, with essentially seasonal occurrence in Europe and Japan. The first incidence of human shellfish-related illness identified as DSP occurred in Japan in the late 1970s, when the dinoflagellate *Dinophysis fortii* was identified as the causative organism and the toxin responsible was termed dinophysistoxin 1 (DTX1; Yasumoto et al. 1980). DSP toxins are produced by several other *Dinophysis* species including *D. acuta*, *D. fortii*, *D. acuminata*, *D. norvegica*, *D. mitra* (Yasumoto and Murata 1990), and *D. caudata* (Eaglesham et al. 2000), in addition to being produced by benthic species such as *Prorocentrum lima* (Bravo et al. 2001).

Diarrhetic Shellfish Toxins

This toxin class consists of at least eight congeners, including the parent compound, okadaic acid (OA), which was first isolated from the black sponge, *Halichondria fortii* (Tachibana et al. 1981). OA, DTX1 (Murata et al. 1982) and dinophysistoxin 2 (DTX2; Hu et al. 1992) are the primary congeners involved in shellfish poisoning, with the other congeners believed to be either precursors or shellfish metabolites of the active toxins (Fig. 3.2).

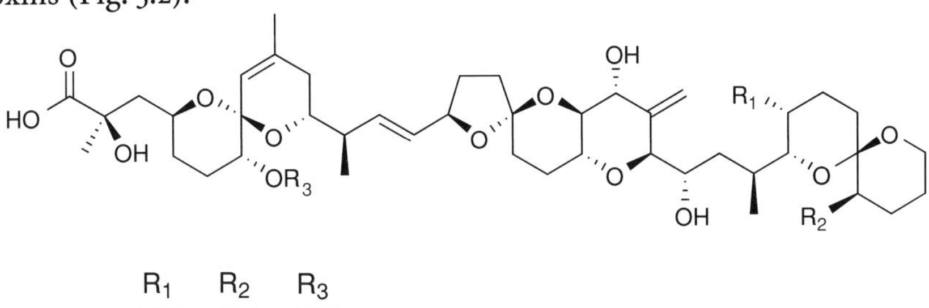

R_1	R_2	R_3	
CH_3	H	H	okadaic acid (OA)
CH_3	CH_3	H	dinophysistoxin 1 (DTX1)
H	CH_3	H	dinophysistoxin 2 (DTX2)
CH_3	CH_3	Acyl	dinophysistoxin 3 (DTX3)

Fig. 3.2. Chemical structure of main DSP toxins (okadaic acid group)

The OA class toxins are diarrhetic (Terao et al. 1986) and tumorigenic (Fujiki and Suganuma 1993). The mechanism of action underlying these activities is explained mainly by their potent inhibitory action against ser/thr protein phosphatases (Sasaki et al. 1994). Inhibitory activity is specific for classes PP2A and PP1, with PP2B being inhibited only at high concentrations and PP2C being insensitive.

Clinical Symptoms of DSP

Oral ingestion of the DSP toxins can lead to the gastrointestinal disturbances of acute diarrhea, nausea, vomiting, and abdominal pain, with symptoms often beginning within 30 min of consuming contaminated shellfish. No human mortalities to date have been reported from any cases of DSP poisoning, although there has been considerable morbidity resulting in hospitalization. The clinical symptoms of DSP may often have been mistaken for those of bacterial gastric infections and the problem may be much more widespread than currently thought.

OA is a potent tumor promoter (Fujiki and Suganuma 1993) and chronic exposure may promote tumor formation in the digestive system.

3.2.3 Toxins Found in Association with DSP Toxins

Other toxins have long been included in the DSP group for a number of reasons. First of all because they coexist in DSP-contaminated shellfish with OA and its congeners and, on account of their lipophilic nature, they are coextracted from the shellfish digestive gland together with okadaic acid group toxins. They are also included in the DSP group because they exert toxic effects following intraperitoneal (i.p.) injection in mice, thus producing positive results in the mouse bioassay, largely used as a screening method in monitoring programs. In addition, they are produced, as DSP toxins, by dinoflagellates; and indeed some of them are produced by the same toxinogenic algal species which also transmit OA and DTX1. However, they do not fit in the definition of DSP toxins, because they lack diarrhogenicity in mammals.

These toxins are represented by pectenotoxins (PTXs), yessotoxins (YTXs), and azaspiracids (AZAs).

Pectenotoxins

PTXs are a group of toxins isolated from algae commonly known to produce other DSPs such as OA and DTX1 (Yasumoto and Murata 1993). Examples of such algae include the dinoflagellates *D. acuta*, *D. fortii*,

D. acuminata, and *D. caudata*. The PTXs are often found in combination with other DSPs in shellfish and a debate exists over whether these toxins should be classified as DSP toxins. Some research groups have found mild diarrhetic effects caused by the administration of PTXs, while others have found no such evidence. Additionally, many DSPs have been found to be potent phosphatase inhibitors, but some PTX toxins were found to be inactive against PP1 and PP2A (Lun et al. 1993).

Structurally, PTXs resemble OA in molecular weight and in having cyclic ethers and a carboxylic group in the molecule (Fig. 3.3). Unlike in OA, however, the carboxyl moiety in PTX is in the form of a macrocyclic lactone (macrolide). It is believed that several of the pectenotoxins are derived from a parent pectenotoxin, where the parent molecule is metabolized within the scallops to form other pectenotoxin analogs.

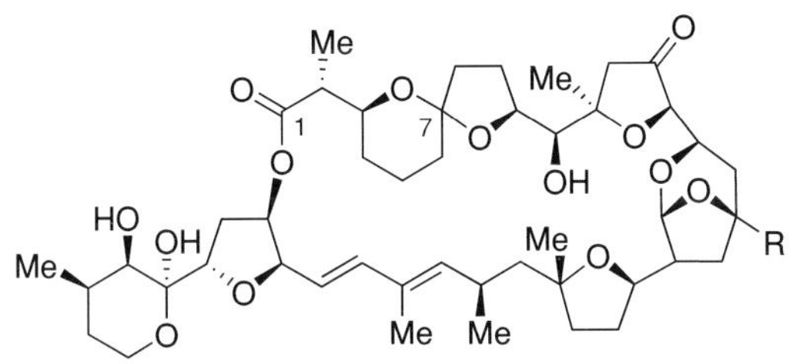

R	C-7		C-7	
CH_2OH	*R*	pectenotoxin 1 (PTX1)	*R*	pectenotoxin 2 seco acid (PTX2SA)
CH_3	*R*	pectenotoxin 2 (PTX2)	*S*	7-epi-pectenotoxin 2 seco acid (PTX2SA)
CHO	*R*	pectenotoxin 3 (PTX3)		
CH_2OH	*S*	pectenotoxin 4 (PTX4)		
COOH	*R*	pectenotoxin 6 (PTX6)		
COOH	*S*	pectenotoxin 7 (PTX7)		

Fig. 3.3. Structure of pectenotoxins (PTXs) and PTX2-seco acids (PTX2SAs)

Histopathological investigations of PTX2 to mice caused severe mucosal injuries and fluid accumulation in the small intestine and revealed that it is hepatotoxic and induces rapid necrosis of hepatocytes (Ishige et al. 1988). PTXs have a potent cytotoxicity (Jung et al. 1995) and probably inhibit actin polymerization (Spector et al. 1988).

Yessotoxins

YTXs are lipophilic polyether compounds. The group consists of yessotoxin, a ladder-shaped polycyclic ether toxin isolated for the first time from the scallop *Patinopecten yessoensis* (Murata et al. 1987) and a number of analogs including the homoyessotoxins, as shown in Fig. 3.4.

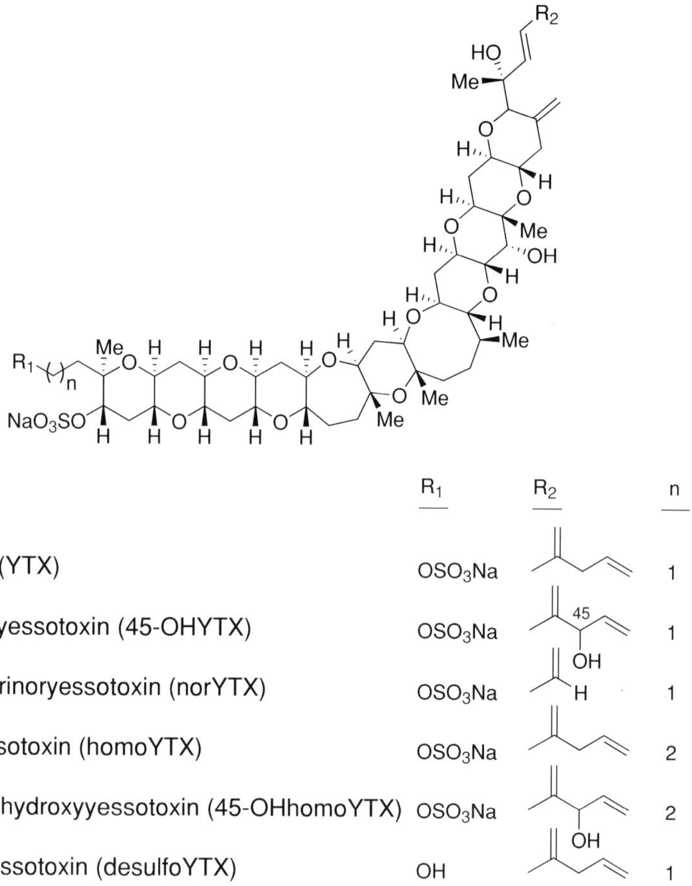

Fig. 3.4. Structure of some yessotoxins (YTXs)

Further yessotoxin analogs have been more recently isolated by our research group from Adriatic Sea mussels. Their structures are reported in Fig. 3.9.

Recent etiological study revealed that the origin of YTX is different from that of OA and DTX1. YTX is produced by the dinoflagellate *Protoceratium reticulatum* (Satake et al. 1997a), while OA and DTX1 are produced mainly by *Dinophysis* spp (Lee et al. 1989).

In several countries, YTXs are included within the class of DSP toxins for regulatory purposes because they coextract with other DSP toxins and are often found in association with DSPs in shellfish. However, studies on yessotoxin have shown it not to cause diarrhea or inhibit PP2A (Ogino et al. 1997). In the EU, the YTXs have been reclassified and are no longer included in recommended guidelines for DSPs, but are regulated within their own subgroup.

Yessotoxin was found to be more than ten times less toxic to mice via the oral route, compared with i.p. injections. Even at 10 mg kg^{-1} body weight, the highest dose ever tested orally, YTX did not kill the mice.

However, in spite of the wealth of data on OA, the molecular mechanism underlying the toxicity of YTXs is unknown. Indeed, very limited data are available regarding the effects of this group of components on cellular systems. Histopathological analysis revealed that a target organ of YTX is the heart: marked intracytoplasmatic edema in cardiac muscle cells was observed in mice after i.p. injection of the toxin (Terao 1990). An involvement of the nervous system in YTX toxicity can be also hypothesized on the basis of the chemical structure, since brevetoxins and ciguatoxins (Yasumoto and Murata 1993), both structurally strictly related to YTX, induce poisoning characterized by neurological and cardiovascular symptoms (Dechraoui et al. 1999).

As for the mechanism of action, by analogy with brevetoxins and ciguatoxins, YTX may act as a depolarizing agent, opening membrane channels of Na^+-permeable excitable cells and leading to a Na^+ influx (Gawley et al. 1992). It remains to be established, however, to what extent these toxins can be absorbed by the intestine and then gain access to the target organs.

Azaspiracids

Azaspiracid poisoning (AZP) is a newly identified syndrome. The causative toxin, azaspiracid, so named because of its unusual azaspiro ring assembly, was first identified from Irish mussel extracts in association with a shellfish poisoning incident that took place in the Netherlands during 1995 (Satake et al. 1998). In addition to AZA, four analogs, AZA2–AZA5, were isolated and their structures determined, as shown in Fig. 3.5 (Ofuji et al. 1999, 2001). The symptoms observed in the patients included nausea, vomiting, severe diarrhea and stomach cramps and thus resembled those of DSP. However, mouse symptoms induced by i.p. injection of acetone extracts of mussel hepatopancreas were distinctly different from those normally associated with DSP toxins, showing prominent neurological symptoms, such as respiratory difficulties, spasms, paralysis of the limbs and death within 20 min at higher doses (Ito et al. 2000).

3.2.4 Neurotoxic Shellfish Poisoning

A long history of toxic microalgal blooms exists in the Gulf of Mexico, blooms that have caused massive fish kills and respiratory irritation in humans. It was later realized that the toxin in these blooms could also be passed to humans via shellfish, to cause a syndrome named neurotoxic shellfish poisoning. Reports of NSP were limited for a long time to the west coast of Florida, where blooms of the dinoflagellate *Gymnodinium breve* initiate offshore and are subsequently carried inshore by wind and current conditions (Steidinger et al. 1998). In the early 1990s, outbreaks of shellfish toxicity were reported in New Zealand and Australia and resulted in the identification of additional *Gymnodinium* species which produce NSP-like toxins (Haywood et al. 1996). Recently, other fish-killing flagellate species, *Chattonella marina*, *C. antiqua*, *Fibrocapsa japonica*, and *Heterosigma akashiwo*, have also been reported as producers of this class of polyether toxins (Sagir Ahmed et al. 1995; Khan et al. 1997; Hallegraeff et al. 1998).

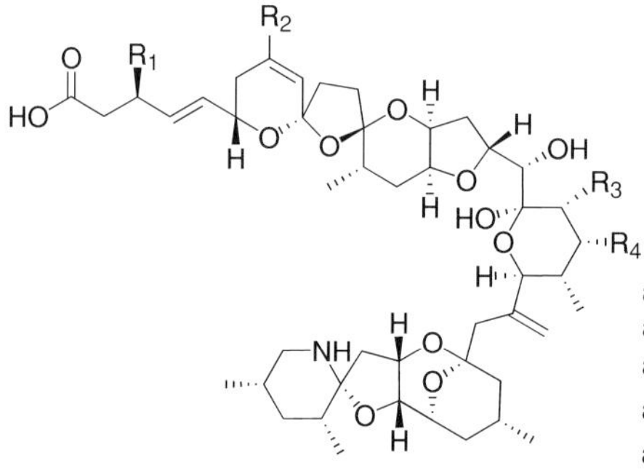

	R_1	R_2	R_3	R_4
azaspiracid (AZA)	H	H	Me	H
azaspiracid-2 (AZA2)	H	Me	Me	H
azaspiracid-3 (AZA3)	H	H	H	H
azaspiracid-4 (AZA4)	OH	H	H	H
azaspiracid-5 (AZA5)	H	H	H	OH

Fig. 3.5. Chemical structure of azaspiracids

Neurotoxic Shellfish Toxins

The toxins responsible for NSP are a suite of ladder-like polycyclic ether toxins collectively called brevetoxins (Fig. 3.6). Brevetoxin congeners fall into two types, based on backbone structure: the brevetoxin B backbone (type 1) and brevetoxin A backbone (type 2). Although the ring systems in the middle of the molecules differ somewhat, type 1 and type 2 toxins share a lactone in the A ring ("head" of the molecule) and a conserved structure on the "tail" ring, both of which are required for their toxicity (Baden 1989).

These toxins are depolarizing substances that open voltage-gated sodium (Na^+) ion channels in cell walls, leading to uncontrolled Na^+ influx into the cell (Baden 1983). This alters the membrane properties of excitable cell types in ways that enhance the inward flow of Na^+ ions into the cell; and this current can be blocked by external application of tetrodotoxin (Poli et al. 1986; Trainer et al. 1991).

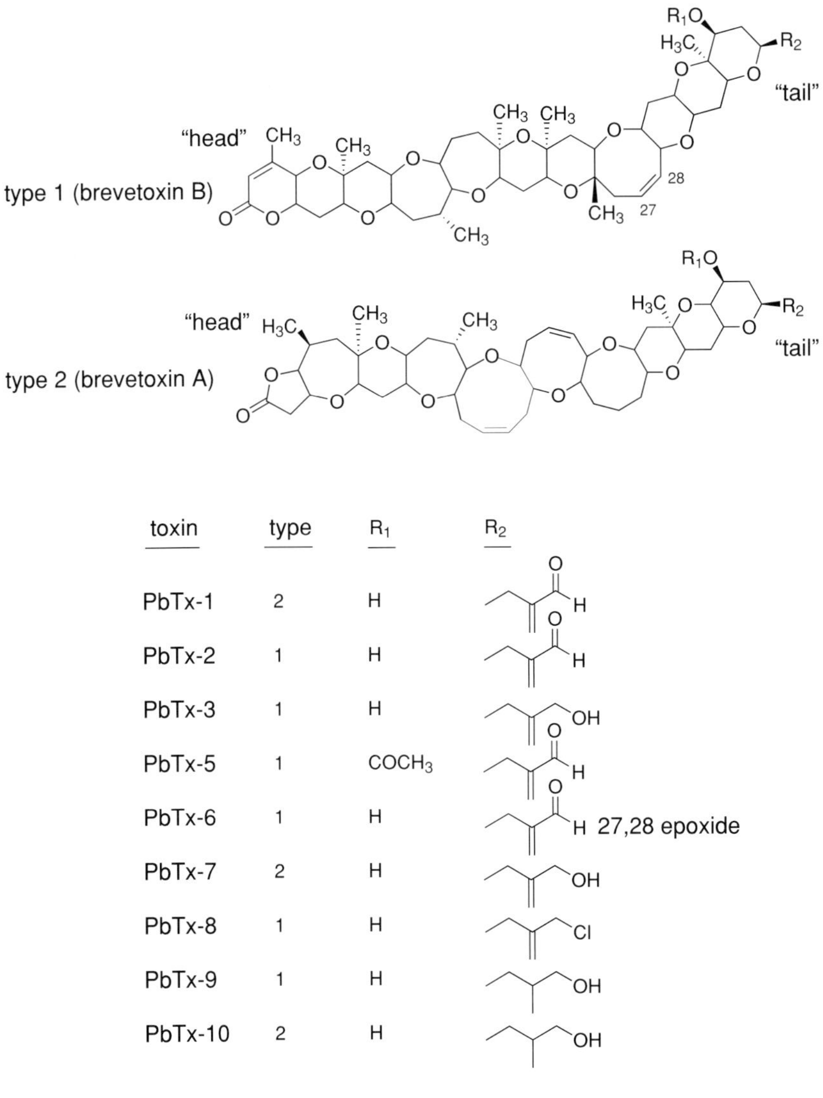

Fig. 3.6. Structure of brevetoxins

Clinical Symptoms of NSP

In humans, the symptoms of NSP intoxication include respiratory distress, as well as eye and nasal membrane irritation, caused principally by exposure to sea-spray aerosols and by direct contact with toxic blooms while swimming. There have been no reported fatalities from NSP, although the toxin kills test mammals when administered by various routes, including oral.

3.2.5 Amnesic Shellfish Poisoning

ASP is the only shellfish poisoning produced by a diatom. The syndrome of ASP was first recognized in 1987 on Prince Edward Island, Canada, where there were three deaths and 105 acute human poisonings from blue mussels (Perl et al. 1990; Teitelbaum et al. 1990). The chain-forming diatom *Pseudo-nitzschia multiseries* (formerly known as *Nitzschia pungens*) was recognized as the causative agent of that toxic event (Subba-Rao et al. 1988; Bates et al. 1989). It is now known that different diatom species induce ASP. These diatom species are distributed worldwide.

Amnesic Shellfish Toxins

The toxin responsible for ASP is domoic acid (DA; Fig. 3.7), a water-soluble excitatory tricarboxylic amino acid belonging to the kainoid class of compounds, which acts as a glutamate antagonist on the kainate receptors of the central nervous system. Several congeners of DA have been identified so far, of which three geometrical isomers, isodomoic acids D, E, and F and the C5′-diastereomer were found in small amounts in both the diatom and in shellfish tissue (Fig. 3.7; Wright et al. 1990; Walter et al. 1994).

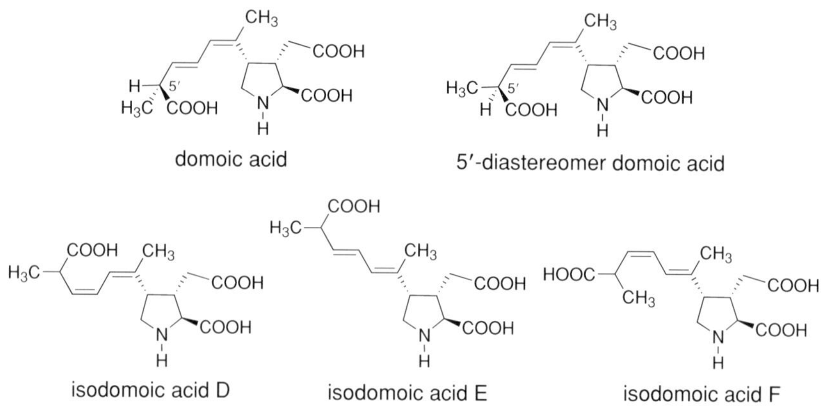

Fig. 3.7. Chemical structures of domoic acid and some of its congeners

Domoic acid binds with high affinity to both kainate and AMPA subtypes of glutamate receptor (Hampson et al. 1992). Persistent activation of the kainate glutamate receptor results in greatly elevated intracellular Ca^{2+} (Xi and Ramsdell 1996). This induces lesions in areas of the brain where glutaminergic pathways are heavily concentrated, particularly in the CA1 and CA3 regions of the hippocampus, areas responsible for learning and memory processing (Peng and Ramsdell 1996).

Clinical Symptoms of ASP

The clinical symptoms of ASP include abdominal cramps, vomiting, diarrhea, incapacitating headaches, disorientation and short-term memory loss. In the most severe case of poisoning, patients are victim to seizure, coma, profuse respiratory secretion, unstable blood pressure, and death. The loss of memory in patients intoxicated with mussel toxin appears to be similar to patients with Alzheimer's disease. However, the loss of memory in mussel-intoxicated patients is not affected by the age of patients, whereas symptoms of Alzheimer's disease intensify with advancing age and are generally noted in older people. Further, the findings that intellect and higher cortical functions are not influenced by DA intoxication distinguish the mussel-induced intoxication from Alzheimer's disease.

3.2.6 Spirolides and Shellfish Syndrome Related to Dinoflagellates

Spirolides are pharmacologically active macrocyclic imines that were first isolated and characterized from lipophilic extracts of scallop and mussel viscera harvested from aquaculture sites in Nova Scotia, Canada (Fig. 3.8; Hu et al. 1995, 1996). These "fast-acting toxins" cause rapid death upon i.p. injection into mice and also have a high oral potency with apparent neurotoxic symptomatology, but the mode of action is currently unknown. The symptoms include piloerection, abdominal muscle spasms, hyper-extensions of the back and arching of the tail to the point of touching the nose.

The biological origin of spirolides was unknown until recently, although the evidence (geographical extent, seasonality, occurrence in multiple shellfish species) strongly suggested a planktonic source (Cembella et al. 1998). This hypothesis was also supported by the high degree of structural homology between spirolides and other macrolides of marine dinoflagellate origin, including gymnodimine (from *Gymnodinium mikimoto*) and prorocentrolides (found in *Prorocentrum lima*; Wright and Cembella 1998). Using liquid chromatography–mass spectrometry (LC–MS) analyses, various spirolides were detected in fractions of planktonic material from Nova Scotian aquaculture sites. Particularly, the dinoflagellate *Alexandrium ostenfeldii* was shown to be the producing organism (Cembella et al. 2000).

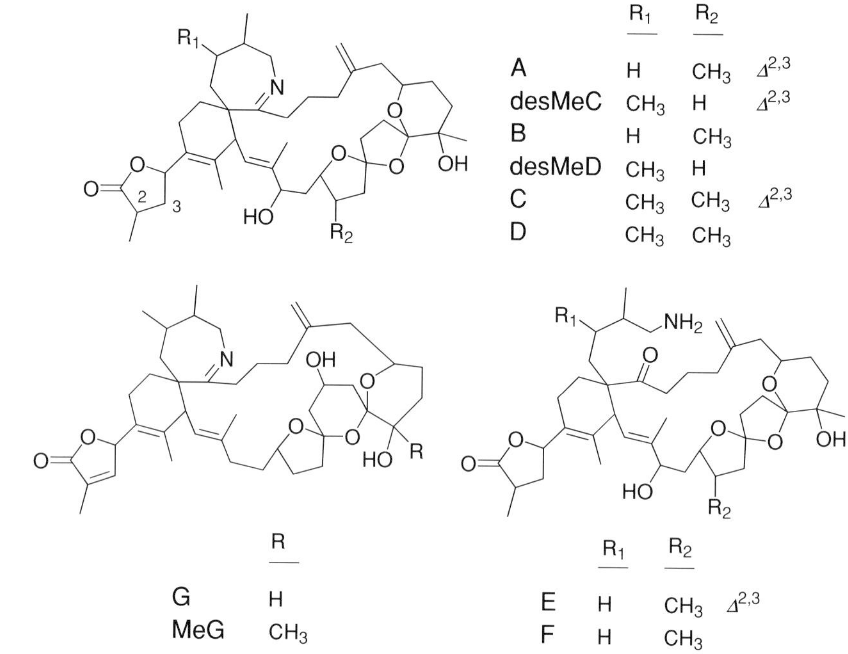

	R_1	R_2	
A	H	CH_3	$\Delta^{2,3}$
desMeC	CH_3	H	$\Delta^{2,3}$
B	H	CH_3	
desMeD	CH_3	H	
C	CH_3	CH_3	$\Delta^{2,3}$
D	CH_3	CH_3	

	R
G	H
MeG	CH_3

	R_1	R_2	
E	H	CH_3	$\Delta^{2,3}$
F	H	CH_3	

Fig. 3.8. Chemical structure of spirolides

3.3 DSP Toxins in Phytoplankton and Mussels from the Northwestern Adriatic Sea

The northwestern coast of the Adriatic Sea has been subject to recurring cases of red tides since 1975. DSP outbreaks associated with harmful algae blooms were, however, recognized as a problem only in 1989, when the first case of human gastroenteritis was related to the simultaneous presence of known producers of DSP toxins both in seawater and hepatopancreas of mussels (Boni et al. 1992). The evidence that certain cases of diarrhea in consumers of molluscs were not due to bacteria or virus but to biointoxication by DSP came from the isolation of lipid-soluble DSP-type toxins in mussel tissue collected in the coastal water of the Emilia Romagna region. This phenomenon has occurred in the Adriatic Sea with alarming frequency since then, subsequently extending over the coastal areas of Marche, Abruzzo, Veneto, and Friuli-Venezia Giulia.

The continuance of mussel toxicity causes a serious threat to human health and severe economic losses for the Adriatic shellfish industries, whose production areas, which cover 90% of the national total production of mussels, have been forced to remain closed for some months.

In order to prevent or minimize such damage, continuous monitoring of toxicity in shellfish and structural elucidation of the causative toxins are prerequisites. The most commonly used assay method is the mouse bioassay developed by the Japanese Ministry of Health and Welfare (Japanese Ministry of Health and Welfare 1981). One mouse unit (MU) is defined as the minimum quantity of toxin needed to kill a mouse within 24 h.

Major disadvantages of this assay are the lack of specificity (no differentiation between the various components of DSP toxins), the subjectivity of death time of the animals and the maintenance and killing of laboratory animals.

A research program based on instrumental analysis was, therefore, initiated in 1990 by our research group, to examine the toxic profiles in mussels from the northern Adriatic Sea. Until now, a number of toxic samples of shellfish collected along the Emilia Romagna coasts and corresponding to the highest level of toxicity have been analyzed.

In the first part of this study, OA was recovered as the causative toxin in the toxic episode of 1990, identified through ^{1}H NMR spectroscopy (Fattorusso et al. 1992). This result represented the first certain evidence of the presence of DSP toxins in mussels cultivated along the Italian coast.

Subsequently, DTX1 was also detected, using ionspray LC-MS (Draisci et al. 1995). However, recent research has demonstrated that other toxins are, at the moment, important contributors to DSP in Italy. In 1995, in fact, for the first time from Italian mussels, YTX was isolated in relatively large amounts, in addition to trace amounts of OA, by our research group (Ciminiello et al. 1997).

Later, besides a relatively large amount of 45-hydroxyyessotoxin (Ciminiello et al. 1999), two new analogs, homoYTX (which presents an extra methylene group to the structure of YTX in the western part of the molecule) and 45-hydroxyhomoYTX (Satake et al. 1997a) were also isolated from Italian mussels.

3.3.1 New YTX Analogs Isolated from Adriatic Mussels

Very recently, in our laboratory, we isolated from the hepatopancreas of Adriatic mussels and chemically characterized several new YTX-like structures (Fig. 3.9), such as adriatoxin (ATX; Ciminiello et al. 1998), carboxyyessotoxin (COOHYTX; Ciminiello et al. 2000a), carboxyhomoyessotoxin (COOHhomoYTX; Ciminiello et al. 2000b) and 42,43,44,45,46,47,55-heptanor-41-oxohomoyessotoxin (noroxohomoYTX; Ciminiello et al. 2001a).

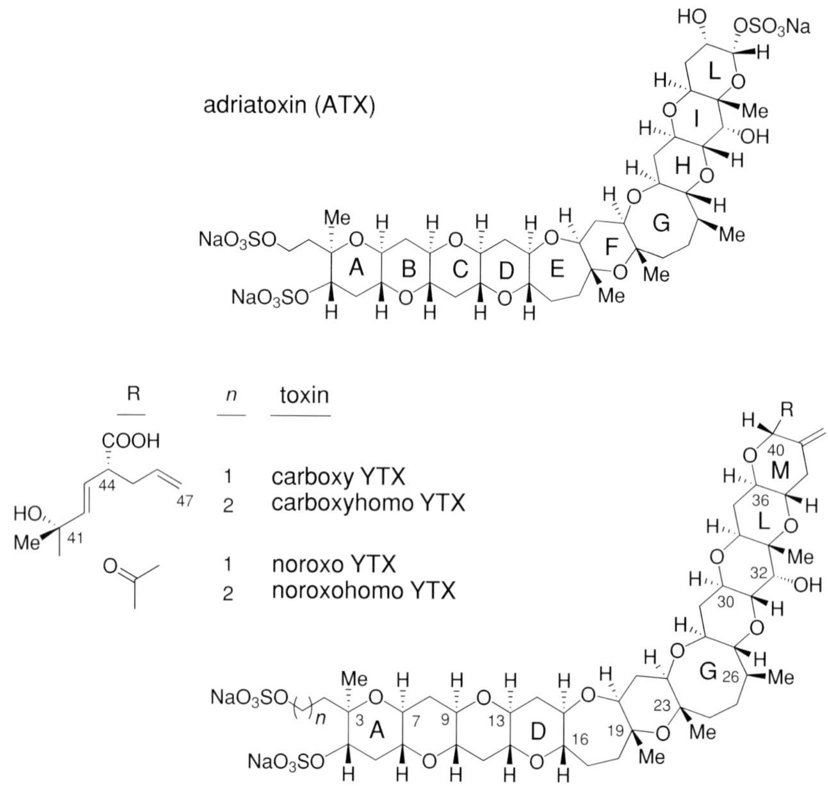

Fig. 3.9. New YTX analogs from Adriatic mussels

All the new analogs have been isolated in a pure form and show a close resemblance to YTX, the main differences being located in the eastern part of the molecule. Their chemical structures have been determined on the basis of spectral evidence, particularly mono-dimensional and two-dimensional ^{1}H NMR experiments, as well as MS/MS experiments.

The toxicology of these new biotoxins is unknown. The results of our studies indicate, however, that the composition and the relative abundance of YTXs in bivalves seem to vary regionally, seasonally and annually, as observed for other DSP toxins.

3.3.2 LC–MS Method for Analysis of YTXs

The chief obstacles in facing the acute and chronic risk associated with YTX-contaminated seafood are the limited availability of toxins for toxicological studies and the lack of a rapid and efficient analytical method for following the variation of the toxin profile in molluscs.

The combination of liquid chromatography and mass spectrometry (LC–MS) has proven to be the most powerful tool for the detection and quantitation of toxins in plankton and shellfish at trace levels, the identification of new toxins, the investigation of toxin production by plankton and the study of toxin metabolism in shellfish (Quilliam 1996).

This technique appears to be extremely useful for the detection and quantitation of DSP toxins. In fact, the most common analytical methods so far employed for the specific detection of DSPs provide for the derivatization of each toxin with an appropriate auxiliary reagent for fluorescence labeling followed by HPLC analysis. However, there is no reagent which fits all DSP toxins. 9-Anthryldiazomethane (ADAM) is used for OA, DTXs, and PTXs (Lee et al. 1987), while YTXs are derivatized with a dienophile reagent, 4-[2-(6,7-dimethoxy-4-methyl-3-oxo-3,4-dihydroquinoxalinyl)ethyl]-1,2,4-triazoline-3,5-dione (DMEQ-TAD; Yasumoto and Takizawa 1997). It has to be noted that, for the application of latter method, the presence of a conjugated diene functionality in the side-chain of YTX-like compounds is a prerequisite. Thus, the method is not reliable for the detection of those derivatives which lack a conjugated diene functionality in the molecule, such as the Adriatic analogs noroxohomoYTX, carboxyYTX, carboxyhomoYTX, and adriatoxin.

With the aim of setting up a suitable method for the rapid and unambiguous detection of all YTXs isolated so far and also the presence of OA which sometimes coexists in shellfish, we tested the suitability of the LC–MS method developed by Quilliam et al. (2001) for the detection of most lipophilic toxins (Ciminiello et al. 2002a). For this purpose, standard solutions at a known concentration of YTX and OA as well as solutions of a number of YTX analogs from North Adriatic mussels were employed.

Thus developed, this LC–MS technique allowed the determination of OA and all YTXs and homoYTXs derivatives so far isolated in a single chromatographic run of 25 min and showed itself to be both selective and sensitive with a detection limit of 68.4 pg for YTX (Fig. 3.10).

Together with the rapid detection of known compounds at parts per billion levels, the method makes it possible to highlight the potential presence of new analogs and can be usefully employed for the structural elucidation of new toxins whenever great structural analogies occur between the toxins under investigation and known compounds.

The potential of LC–MS analysis for the detection of new toxins can be illustrated by two new YTX analogs from Northern Adriatic mussels we recently identified in toxic mixtures obtained from *Mytilus galloprovincialis* collected in 1998 and 2001, respectively (Ciminiello et al. 2002a,b).

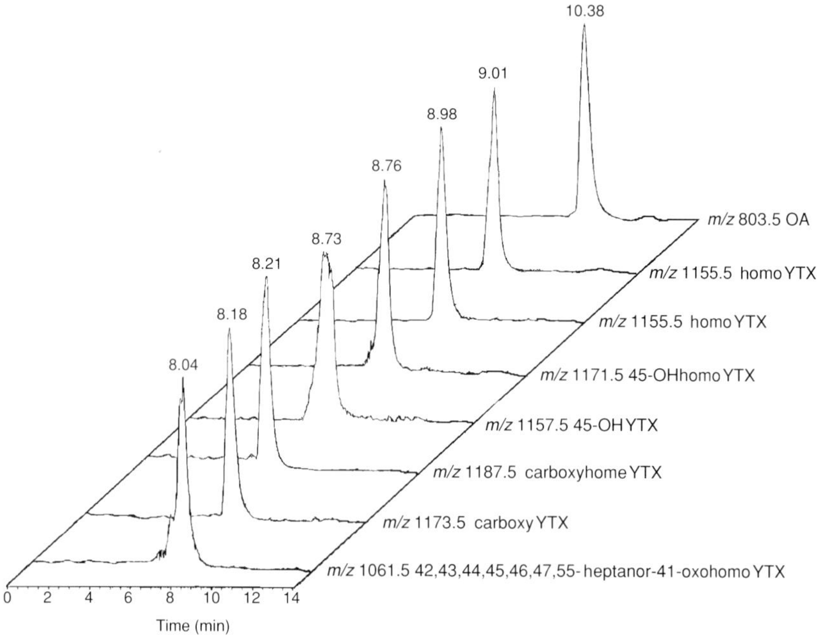

Fig. 3.10. LC-ESI (negative ion)-ion trap–MS analysis of a wide range of yessotoxins in a blend of mussel tissue extract added of OA standard solution. Selected monitoring of yessotoxins with different molecular masses was carried out by extracted ion chromatograms (XICs) of the $[M{-}2Na^{+}H]^{-}$ ions. For OA the XIC is relevant to the $[M{-}H]^{-}$ ion

Along with known YTXs identified by comparison of their retention times and mass spectra with those of appropriate standards, new marine toxins, a desulfoYTX and 42,43,44,45,46,47,55-heptanor-41-oxoyessotoxin, were detected. MS/MS experiments were used to gain structural information.

3.3.3 LC–MS Analysis of an Adriatic Strain of *P. reticulatum*

The LC–MS method developed for the determination of YTXs could be employed not only for the analysis of toxic mussels, but also for the screening of algal cultures in order to select the producer organisms of the compounds to be submitted to toxicological studies.

Identification of the organism(s) responsible for the production of YTX derivatives is of critical importance for the future regulation and management of toxic shellfish. In 1997, the marine dinoflagellate *Protoceratium reticulatum* collected in New Zealand was indicated as the biogenetic origin of YTX (Satake et al. 1997b). Subsequently, YTX was detected both in Adriatic *P. reticulatum* (Boni et al. 2001) and together with 45,46,47-trinoryessotoxin in strains of the same species collected in Japan (Satake et al. 1999). In 1999, the presence of YTX and homoYTX in *Gonyaulax polyedra* collected in the northwestern Adriatic Sea was reported (Draisci et al. 1999). However, the origin of all the other YTX analogs was still unknown, thus raising an issue whether they were metabolites of YTX formed in mussels or true products of different dinoflagellate species.

To ascertain their origin, a cultured strain of *P. reticulatum* (*G. grindley*) collected along the Cesenatico coasts (Emilia Romagna, Italy) in June 2001 was investigated (Ciminiello et al. 2003). Careful analysis of this strain obtained by high performance liquid chromatography coupled with electrospray ionization ion trap mass spectrometry (HPLC–ESI MS), suggests that *P. reticulatum* from the Northwestern Adriatic Sea is responsible for the production, together with YTX, of homoYTX, 45-OHYTX, carboxyYTX, and noroxoYTX.

This is the first identification of *P. reticulatum* as the producer of some of the YTX derivatives so far isolated from Italian mussels. Interestingly, Adriatic *P. reticulatum* is able to produce compounds belonging to both the YTX and homoYTX series, whereas previous studies were suggestive of two different organisms being responsible for production of the two homologous series.

Furthermore, these findings indicate that most of the Adriatic YTX derivatives are true products of the dinoflagellate and do not derive from the metabolic conversion of YTX in shellfish.

3.4
Detection of Domoic Acid in Adriatic Shellfish by Hydrophilic Interaction Liquid Chromatography–Mass Spectrometry

A hydrophilic interaction liquid chromatography-mass spectrometry (HILIC–MS) method has been very recently developed by us to allow for the rapid, unambiguous identification and quantitation of DA in shellfish sample (Ciminiello et al. 2005).

The obtained results showed that the HILIC–MS technique is suitable for combined analysis of DA and PSP toxins in a single 30-min chromatographic run, using gradient elution. Isocratic elution allows detection of DA in 10 min.

Application of the developed method to the analysis of a number of samples of *Mytilus galloprovincialis*, collected over the period 2000–2005 in the Adriatic Sea, indicated the presence of DA in some of the analyzed samples as a new toxin which has entered the Adriatic mussels' toxin profile.

This is the first time that DA has been detected in Adriatic shellfish, although in all analyzed samples the toxin appeared to be present at levels well below the regulatory limit (20 μg g^{-1} in edible tissue). The obtained results represent a warning for DA as one of the toxins which need to be carefully monitored in Adriatic shellfish.

3.5 Cytotoxins from Contaminated Adriatic Blue Mussels

In the course of our investigation into toxic Adriatic mussels, we have also isolated and structurally characterized, besides YTXs, new types of toxins, oxazinins and chlorosulfolipids, which are completely different in structure from the polyether DSP toxins isolated so far, but may represent a further alarm for public health, due to their cytotoxic activity.

3.5.1 Oxazinins

Three new compounds, oxazinin-1, oxazinin-2, and oxazinin-3, have been isolated from the toxic digestive glands of the Adriatic mussel *Mytilus galloprovincialis* (Ciminiello et al. 2001b; Fig. 3.11). They are characterized by unique structural features, which to the best of our knowledge have not been found in any other naturally occurring compound.

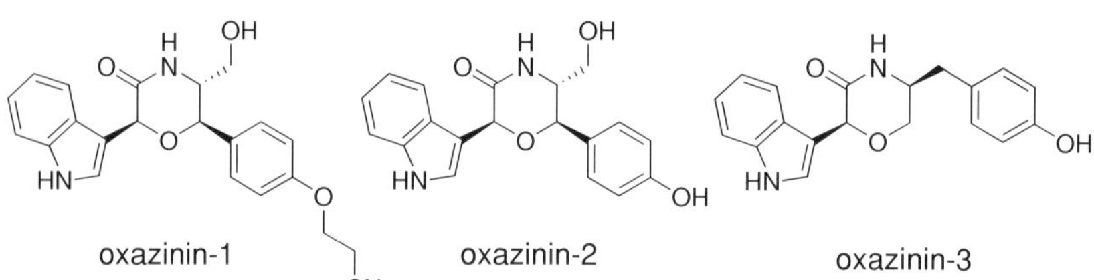

Fig. 3.11. Structure of oxazinins

The determination of the structure and the relative stereochemistry of the new molecules was based on spectroscopic evidence including extensive 2D-NMR experiments and molecular mechanical calculations. The absolute stereochemistry of oxazinin-1 has been assigned (Ciminiello et al. 2001c) by application of a method proposed by Latypov et al. (1998)

for the assignment of the absolute configuration of most β-chiral primary alcohols.

3.5.2 Chlorosulfolipids

Very recently, we reported the isolation from *M. galloprovincialis* of a new class of cytotoxins constituted by polychlorinated sulfolipids.

Some chlorosulfolipids were previously isolated from species of microalgae (Mercer and Davies 1979; Chen et al. 1994). These compounds, whose structure has so far been assigned devoid of stereochemical details, have been divided into two series: the polychlorodocosane 1,14-disulfates and the polychlorotetracosane 1,15-disulfates with a number of chlorine atoms, which range from zero to six in various combinations of positions on the aliphatic chain. In most cases, the exact location of the chlorine atoms could not be determined. They represent a unique class of products in that they are essentially polar at both ends of the molecule.

During our investigation into toxic Adriatic mussels, we succeeded in isolating three unique cytotoxic compounds (Fig. 3.12) which can be included in the class of chlorosulfolipids, even if they are structurally quite different from the previously reported ones.

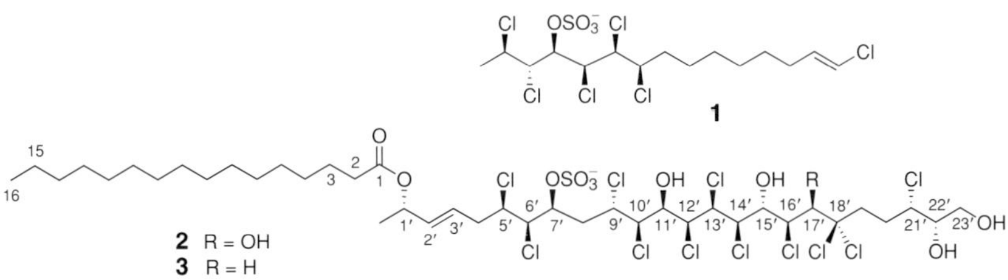

Fig. 3.12. Structure of chlorosulfolipids from Adriatic mussels

The first isolated compound is a hexachloromonosulfate, **1** (Ciminiello et al. 2001d), while **2** (Ciminiello et al. 2002c) and **3** (Ciminiello et al. 2004), very recently characterized, contain 11 chlorines, in addition to a fatty acid acyl moiety.

Structure determination of the new polychlorinated sulfolipids has been carried out by extensive use of 1D- and 2D-NMR techniques, supported by ESI MS and MS/MS experiments, as well as molecular mechanics and dynamics calculations.

Elucidation of the absolute stereochemistry of the three molecules appeared to change, particularly due to the presence of many stereogenic carbons, from 6 in compound **1** up to 15 in compound **2**.

The relative stereochemistry of the polychlorinated sulfolipids was determined by using the *J*-based configuration analysis method (Murata et al. 1999; Matsumori et al. 1999).

This method was first applied to assign the relative stereochemistry of 1; and in order to verify its confident applicability to our complex chlorosulfolipids, molecular mechanics calculations were carried out using the CHARMm force field. The obtained results fully substantiated the stereochemical assignments based on the Murata method.

The absolute stereochemistry of the chlorosulfolipids was defined by a modified Mosher's method (Ohtani et al. 1991).

3.6 Conclusions

The results of our studies have revealed a very interesting, complex and changeable scenario of shellfish toxicity in Italy. It is evident from our data that there is a variety of YTX analogs in some shellfish-producing areas. An important aspect to be considered is that the presence in Adriatic shellfish of several toxins of the YTX class creates complications due to the lack of toxicity data for this type of toxin and also makes quantification difficult in the absence of analytical reference compounds. It is indispensable to address toxicological investigations into all the YTX-like compounds. Much effort should therefore be directed at the accumulation of these toxins to be utilized in toxicological studies.

The great variety of closely related toxin structures and the varying toxicities present significant challenges to the analytical chemist interested in developing a method for their detection and quantitation.

Our results confirmed the LC–MS technique as being a very promising alternative tool to animal testing. This chemical analysis method has played an essential role in all phases of toxic investigations, including the identification of new toxins by bioassay-directed fractionation, the detection and quantitation of toxins in plankton and shellfish and the investigation of toxin production by plankton. However, improvement and inter-laboratory studies will be necessary before this technique can become a generally accepted tool in regulatory analysis. A serious problem hampering the further development and validation of analytical methodology for biotoxins is that pure analytical standards and reference materials are hardly or not readily available. This is particularly true for YTXs, which are really very rare materials: they are not commercially available and few laboratories possess even very small amounts of YTX and much less of its analogs.

Moreover, the presence of further toxic compounds in edible shellfish, such as oxazinin-1, chlorosulfolipids, and DA, in addition to contamination of DSP toxins, increases the potential risk to human health. To prevent the damage caused by pollution from harmful marine algae (both to public health and to the shellfish industries), it is necessary to implement careful monitoring, both at markets and at shellfish farms. Monitoring, in turn, cannot be run without good knowledge of the causative organisms and the nature of implicated toxins. Therefore, an accurate analysis of toxic mussels is indispensable in order to identify new toxins, even other than DSP polyether toxins, and to isolate a larger amount to clarify in depth their toxicological effects.

References

Ahmed FE (1991) Naturally occurring seafood toxins. J Toxicol Toxin Rev 10:263–287

Anderson DM (1989) Toxic algal blooms and red tides: a global perspective. In: Okaichi T, Anderson DM, Nemoto T (eds) Red tides: biology environmental science and toxicology. Elsevier, New York, pp. 11–16

Baden DG (1983) Marine food-borne dinoflagellate toxins. Int Rev Cytol 82:99–150

Baden DG (1989) Brevetoxins: unique polyether dinoflagellate toxins. FASEB J 3:1807–1819

Bates SS, Bird CJ, Defrietas ASW, Foxall R, Gilgan M, Hanic LA, Johnson GR, McCulloch AW, Odense P, Pocklington R, Quilliam MA, Sim PG, Smith JC, Subba-Rao DV, Todd ECD, Walter JA, Wright JLC (1989) Pennate diatom *Nitzschia pungens* as the primary source of domoic acid, a toxin in shellfish from eastern Prince Edward Island, Canada. Can J Fish Aquat Sci 46:1203–1215

Boni L, Mancini L, Milandri A, Poletti R, Pompei M, Viviani R (1992) First cases of DSP in the Northern Adriatic Sea. In: Vollenweider RA, Marchetti R, Viviani R (eds) Marine coastal eutrophication (Proc Int Conf Bologna, 21–24 March 1990). Sci Total Environ 1992[Suppl]:419–426

Boni L, Ceredi A, Guerrini F, Milandri A, Pistocchi R, Poletti R, Pompei M (2001) Toxic *Protoceratium reticulatum* (Peridiniales, Dinophyta) in the north-western Adriatic Sea (Italy). In: Hallegraeff GM, Blackburn SI, Bolch CJ, Lewis RJ (eds) Harmful algae. Intergovernmental Oceanographic Commission of UNESCO, Paris, pp. 37–40

Botana LM (2000) Seafood and freshwater toxins. Pharmacology, physiology and detection. Dekker, New York

Bravo I, Fernandez ML, Ramilo I, Martinez A (2001) Toxin composition of the toxic dinoflagellate *Prorocentrum lima* isolated from different locations along the Galician coast (NW Spain). Toxicon 39:1537–1545

Cembella AD, Quilliam MA, Lewis NI, Bauder AG, Wright JLC (1998) Identifying the planktonic origin and contribution of spirolides in coastal Nova Scotia waters. In: Reguera B, Blanco J, Fernandez ML, Wyatt T (eds) Harmful Algae. Xunta de Galicia and Intergovernmental Oceanography Commission of UNESCO, Santiago de Compostela, pp. 481–484

Cembella AD, Lewis NI, Quilliam MA (2000) The marine dinoflagellate *Alexandrium ostenfeldii* (Dinophyceae) as the causative organism of spirolide shellfish toxins. Phycologia 39:67–74

Chen JL, Proteau PJ, Roberts MA, Gerwick WH (1994) Structure of malhamensilipin A, an inhibitor of protein tyrosine kinase, from the cultured chrysophyte *Poterioochromonas malhamensis*. J Nat Prod 57:524–527

Ciminiello P, Fattorusso E, Forino M, Magno S, Poletti R, Satake M, Viviani R, Yasumoto T (1997) Yessotoxin in mussels of the northern Adriatic Sea. Toxicon 35:177–183

Ciminiello P, Fattorusso E, Forino M, Magno S, Poletti R, Viviani R (1998) Isolation of adriatoxin, a new analog of yessotoxin from mussels of the Adriatic Sea. Tetrahedron Lett 39:8897–8900
Ciminiello P, Fattorusso E, Forino M, Magno S, Poletti R, Viviani R (1999) Isolation of 45-hydroxyyessotoxin from mussels of the Adriatic Sea. Toxicon 37:689–693
Ciminiello P, Fattorusso E, Forino M, Poletti R, Viviani R (2000a) A new analogue of yessotoxin, carboxyyessotoxin, isolated from Adriatic Sea mussels. Eur J Org Chem 291–295
Ciminiello P, Fattorusso E, Forino M, Poletti R, Viviani R (2000b) Structure determination of carboxyhomoyessotoxin, a new yessotoxin analog isolated from Adriatic mussels. Chem Res Toxicol 13:770–774
Ciminiello P, Fattorusso E, Forino M, Poletti R (2001a) 42,43,44,45,46,47,55-Heptanor-41-oxohomoyessotoxin, a new biotoxin from mussels of the northern Adriatic Sea. Chem Res Toxicol 14:596–599
Ciminiello P, Dell'Aversano C, Fattorusso E, Forino M, Magno S, Ianaro A, Di Rosa M (2001b) Oxazinin-1, -2 and -3 – a novel toxic compound and its analogues from the digestive glands of *Mytilus galloprovincialis*. Eur J Org Chem 2001:49–53
Ciminiello P, Dell'Aversano C, Fattorusso C, Fattorusso E, Forino M, Magno (2001c) Assignment of the absolute stereochemistry of oxazinin-1: application of the 9-AMA shift-correlation method for β-chiral primary alcohols. Tetrahedron 57:8189–8197
Ciminiello P, Di Rosa M, Fattorusso E, Forino M, Ianaro A, Poletti R (2001d) Structural elucidation of a new cytotoxin isolated from mussels of the Adriatic Sea. J Org Chem 66:578–582
Ciminiello P, Dell'Aversano C, Fattorusso E, Forino M, Magno S, Poletti R (2002a) Direct detection of yessotoxin and its analogues by liquid chromatography coupled with electrospray ion trap mass spectrometry. J Chromatogr A 968:61–69
Ciminiello P, Dell'Aversano C, Fattorusso E, Forino M, Magno S, Poletti R (2002b) The detection and identification of 42,43,44,45,46,47,55-heptanor-41-oxoyessotoxin, a new marine toxin from Adriatic shellfish, by liquid chromatography–mass spectrometry. Chem Res Toxicol 15:979–984
Ciminiello P, Dell'Aversano C, Fattorusso E, Forino M, Magno S, Di Rosa M, Ianaro A, Poletti R (2002c) Structure and stereochemistry of a new cytotoxic polychlorinated sulfolipid from Adriatic shellfish. J Am Chem Soc 124:13114–13120
Ciminiello P, Dell'Aversano C, Fattorusso E, Forino M, Magno S, Guerrini F, Pistocchi R, Boni L (2003) Complex yessotoxins profile in *Protoceratium reticulatum* from north-western Adriatic sea revealed by LC–MS analysis. Toxicon 42:7–14
Ciminiello P, Dell'Aversano C, Fattorusso E, Forino M, Magno S, Di Meglio P, Ianaro A, Poletti R (2004) A new cytotoxic polychlorinated sulfolipid from contaminated Adriatic mussels. Tetrahedron 60:7093–7098
Ciminiello P, Dell'Aversano C, Fattorusso E, Forino M, Magno S, Tartaglione L, Quilliam MA, Tubaro A, Poletti R (2005) Hydrophilic interaction liquid chromatography–mass spectrometry for determination of domoic acid in adriatic shellfish. Rapid Commun Mass Spectrom 19:2030–2038
Dechraoui M-Y, Naar J, Pauillac S, Legrand A-M (1999) Ciguatoxins and brevetoxins, neurotoxic polyether compounds active on sodium channels. Toxicon 37:125–143
Draisci R, Lucentini L, Riannetti L, Boria P, Stacchini A (1995) Detection of diarrhetic shellfish toxins in mussels from Italy by ionspray liquid chromatography–mass spectrometry. Toxicon 33:1591–1603
Draisci R, Ferretti E, Palleschi L, Marchiafava C, Poletti R, Milandri A, Ceredi A, Pompei M (1999) High levels of yessotoxin in mussels and presence of yessotoxin and homoyessotoxin in dinoflagellates of the Adriatic sea. Toxicon 37:1187–1193
Eaglesham G, Brett S, Davis B, Holling N (2000) Detection of pectenotoxin-2 and pectenotoxin-2 seco acid in phytoplankton and shellfish from the Ballina region of New South Wales, Australia. Int IUPAC Symp Mycotoxins Phycotoxins 10
Falconer IR (1993) Algal toxins in seafood and drinking water. Academic, London

Falconer IR (1996) Potential impact on human health of toxic cyanobacteria. Phycologia 36:6–11

Fattorusso E, Ciminiello P, Costantino V, Magno S, Mangoni A, Milandri A, Poletti R, Pompei M, Viviani R (1992) Okadaic acid in mussels of Adriatic Sea. Mar Pollut Bull 24:234–237

Fujiki H, Suganuma M (1993) Tumor promotion by inhibitors of protein phosphatases 1 and 2A: the okadaic acid class of compounds. Adv Cancer Res 61:143–194

Gallacher S, Birkbeck TH (1995) Isolation of marine bacteria producing sodium channel blocking toxins and the seasonal variation in their frequency in seawater. In: Gallacher S, Birkbeck TH (eds) Harmful marine algal blooms. Intercept, Nantes

Gawley RE, Rein KS, Kinoshita M, Baden DG (1992) Binding of brevetoxins and ciguatoxin to the voltage-sensitive sodium channel and conformational analysis of brevetoxin B. Toxicon 30:780–785

Gessner B, Bell P, Doucette GJ, Moczydilowski E, Poli M, Van Dolah F, Hall S (1997) Hypertension and identification of toxin in human urine and serum following a cluster of mussels-associated paralytic shellfish poisoning outbreaks. Toxicon 35:711–722

Gonzalez L, Tosteson CG, Hensley V, Tosteson TR (1995) Associated bacteria and toxicity development in cultured *Ostreopsis lenticularis*. In: Gallacher S, Birkbeck TH (eds) Harmful marine algal blooms. Intercept, Nantes

Hallegraeff GM (1987) Red tides in the Australasian region. CSIRO Marine Laboratories, Hobart

Hallegraeff GM (1993) A review of harmful algal blooms and their apparent global increase. Phycologia 32:79–99

Hallegraeff GM, Munday BL, Baden DG, Whitney PL (1998) *Chattonella marina* raphidophyte bloom associated with mortality of cultured bluefin tuna (*Thunnus maccoyii*) in South Australia. In: Reguera B, Blanco J, Fernandez ML, Wyatt T (eds) Harmful algae. Xunta de Galicia and Intergovernmental Oceanography Commission of UNESCO, Santiago de Compostela, pp. 93–96

Hampson DR, Huang X, Wells JW, Walter JA, Wright JLC (1992) Interaction of domoic acid and several derivatives with kainic acid and AMPA binding sites in rat brain. Eur J Pharmacol 218:1–8

Haywood A, MacKenzie L, Garthwaite I, Towers N (1996) *Gymnodinium breve* 'look-alikes': three *Gymnodinium* isolates from New Zealand. In: Yasumoto T, Oshima Y, Fukuyo Y (eds) Harmful and toxic algal blooms. International Oceanographic Committee of UNESCO, Paris, pp. 227–230

Hu T, Doyle J, Jackson D, Mart J, Nixon E, Pleasance S, Quilliam MA, Walter JA, Wright JLC (1992) Isolation of a new diarrhetic shellfish poison from Irish mussels. J Chem Soc Chem Commun 30:39–41

Hu T, Curtis JM, Oshima Y, Quilliam MA, Walter JA, Watson-Wright WM, Wright JLC (1995) Spirolides B and D, two novel macrocycles isolated from the digestive glands of shellfish. J Chem Soc Chem Commun 2159–2161

Hu T, Curtis JM, Walter JA, Wright JLC (1996) Characterization of biologically inactive spirolides E and F: identification of the spirolide pharmacophore. Tetrahedron Lett 37:7671–7674

Humpage AR, Rositano J, Bretag AH, Brown R, Baker PD, Nicholson BC, Steffensen DA (1994) Paralytic shellfish poisons from Australian cyanobacterial blooms. Aust J Mar Freshwater Res 45:761–771

Ishige M, Satoh N, Yasumoto T (1988) Pathological studies on the mice administered with the causative agent of diarrhetic shellfish poisoning (okadaic acid and pectenotoxin-2). Bull Hokkaido Inst Public Health 38:15–19

Ito E, Satake M, Ofuji K, Kurita N, McMahon T, James KJ, Yasumoto T (2000) Multiple organ damage caused by a new toxin azaspiracid, isolated from mussels produced in Ireland. Toxicon 38:917–930

Japanese Ministry of Health and Welfare (1981) Method of testing for diarrhetic shellfish toxin. Food Sanit Res 7:60–65

Jung JH, Sim CJ, Lee CO (1995) Cytotoxic compounds from a two-sponge association. J Nat Prod 58:1722–1726
Khan S, Arakawa O, Onoue Y (1997) Neurotoxins in a toxic red tide of *Heterosigma akashiwo* (Raphidophyceae) in Kagoshima Bay, Japan. Aquacult Res 28:9–14
Kodama M (1988) Possible association of paralytic shellfish toxins-producing bacteria with bivalve toxicity. In: Kodama M (ed) Mycotoxins and phycotoxins. Elsevier Science, Tokyo
Lagos N, Liberona JL, Andrinolo D, Zagatto PA, Moraes SR, Azevedo MFQS (1997) First evidence of paralytic shellfish toxins in freshwater cyanobacterium *Cylindrospermopsis raciborskii* isolated from Brazil. Int Conf Harmful Algae 8
Latypov SK, Fereiro MJ, Quiñoá E, Riguera R (1998) Assignment of the absolute configuration of β-chiral primary alcohols by NMR: scope and limitations. J Am Chem Soc 120:4741–4751
Lee J-S, Yanagi T, Kenma R, Yasumoto T (1987) Fluorometric determination of diarrhetic shellfish toxins by high-performance liquid chromatography. Agric Biol Chem 51:877–881
Lee J-S, Igarashi T, Fraga S, Dahl E, Hovgaard P, Yasumoto T (1989) Determination of diarrhetic shellfish toxins in various dinoflagellate species. J Appl Phycol 1:147–152
Lun HA, Chen DZ, Magoon J, Worms J, Smith J, Holmes CF (1993) Quantification of diarrhetic shellfish toxins by identification of novel protein phosphatase inhibitors in marine phytoplankton and mussels. Toxicon 31:75–83
Maclean JL (1989) Indo-pacific red tides. Mar Pollut Bull 20:304–310
Matsumori N, Kaneno D, Murata M, Nakamura H, Tachibana K (1999) Stereochemical determination of acyclic structures based on carbon–proton spin-coupling constants. A method of configuration analysis for natural products. J Org Chem 64:866–876
Mercer EI, Davies CL (1979) Distribution of chlorosulfolipids in algae. Phytochemistry 18:457–462 and literature cited herein
Murata M, Shimatani M, Sugitani H, Oshima Y, Yasumoto T (1982) Isolation and structural elucidation of the causative toxin of the diarrhetic shellfish poisoning. Bull Jpn Soc Sci Fish 48:549–552
Murata M, Kumagai M, Lee JS, Yasumoto T (1987) Isolation and structure of yessotoxin, a novel polyether compound implicated in diarrhetic shellfish poisoning.Tetrahedron Lett 28:5869–5872
Murata M, Matsuoka S, Matsumori N, Kaneno D, Paul GK, Tachibana K (1999) Structure elucidation of marine natural products. J Am Chem Soc 121:870–871
Ofuji K, Satake M, McMahon T, Silke J, James KJ, Naoki H, Oshima Y, Yasumoto T (1999) Two analogs of azaspiracid isolated from mussels, *Mytilus edulis*, involved in human intoxication in Ireland. Nat Toxins 7:99–102
Ofuji K, Satake M, McMahon T, James KJ, Naoki H, Oshima Y, Yasumoto T (2001) Structures of azaspiracid analogs, azaspiracid-4 and azaspiracid-5, causative toxins of azaspiracid poisoning in Europe. Biosci Biotechnol Biochem 65:740–742
Ogata T, Sata S, Kodama M (1989) Paralytic shellfish toxins in bivalves which are not associated with dinoflagellates. Toxicon 27:1241–1244
Ogino H, Kumagai M, Yasumoto T (1997) Toxicologic evaluation of yessotoxin. Nat Toxins 5:255–259
Ohtani I, Kusumi T, Kashman Y, Kakisawa H (1991) High-field FT NMR application of Mosher method – the absolute configurations of marine terpenoids. J Am Chem Soc 113:4092–4096
Onodera H, Satake M, Oshima Y, Yasumoto T, Carmichael WW (1997) Detection of PSP toxins and six new saxitoxin analogs in the freshwater filamentous cyanobacterium *Lyngbya wollei*. Int Conf Harmful Algae 8
Oshima Y, Itakura H, Lee KC, Yasumoto T, Blackburn S, Hallegraeff G (1993) Toxin production by the dinoflagellate *Gymnodinium catenatum*. Dev Mar Biol 3:907–912
Peng YG, Ramsdell JS (1996) Brain fos induction is a sensitive biomarker for the lowest observed neuroexcitatory effects of domoic acid. Fund Appl Toxicol 31:162–168

Perl TM, Bedard L, Kosansky T, Hockin JC, Todd ECD, Remis RS (1990) An outbreak of toxic encephalopathy caused by eating mussels contaminated with domoic acid. N Engl J Med 322:1775–1780
Poli MA, Mende TJ, Baden DG (1986) Brevetoxins, unique activators of voltage-sensitive sodium channels, bind to specific sites in rat brain synaptosomes. Mol Pharmacol 30:129–135
Quilliam MA (1996) Liquid chromatography–mass spectrometry of seafood toxins. In: Barcelo D (ed) Application of LC–MS in environmental chemistry. Elsevier Science, Amsterdam, pp. 415–444
Quilliam MA, Hess P, Dell'Aversano C (2001) LC–MS method for the detection of lipophilic toxins. In: deKoe WJ, Samson RA, van Egmond HP, Gilbert J, Sabino M (eds) Mycotoxins and phycotoxins in perspective at the turn of the millennium. Elsevier, Wageningen, pp. 383–391
Sagir Ahmed MD, Arakawa O, Onoue Y (1995) Toxicity of cultured *Chattonella marina.* In: Lassus P, Arzul G, Erhard E, Gentien P, Marcaillou C (eds) Harmful marine algal blooms. Lavoisier, Paris, pp. 499–504
Sasaki K, Murata M, Yasumoto T, Mieskes G, Takai A (1994) Affinity of okadaic acid to type-1 and type-2A protein phosphatases is markedly reduced by oxidation of its 27-hydroxyl group. Biochem J 298:259–262
Satake M, Tubaro A, Lee J-S, Yasumoto T (1997a) Two new analogs of yessotoxin, homoyessotoxin and 45-hydroxyhomoyessotoxin, isolated from mussels of the Adriatic Sea. Nat Toxins 5:107–110
Satake M, MacKenzie L, Yasumoto T (1997b) Identification of *Protoceratium reticulatum* as the biogenetic origin of yessotoxin. Nat Toxins 5:164–167
Satake M, Ofuji K, Naoki H, James KJ, Furey A, McMahon T, Silk J, Yasumoto T (1998) Azaspiracid, a new marine toxin having unique spiro ring assemblies, isolated from Irish mussels, *Mytilus edulis.* J Am Chem Soc 120:9967–9968
Satake M, Ichimura T, Sekiguchi K, Yoshimatsu S, Oshima Y (1999) Confirmation of yessotoxin and 45,46,47-trinoryessotoxin production by *Protoceratium reticulatum* collected in Japan. Nat Toxins 7:147–150
Schantz EJ (1986) Chemistry and biology of saxitoxin and related toxins. Ann NY Acad Sci 479:15–23
Shimizu Y (2000) Chemistry and mechanism of action. Food Sci Technol 103:151–172
Shumway SE (1990) A review of the effects of algal blooms on shellfish and aquaculture. J World Aquacult Soc 21:65–104
Soames-Mraci CP (1995) Shellfish poisoning: public health risks, quality assurance and analytical detection. Chem Aust 1995:22–25
Spector I, Shcher NR, Bubb MR (1988) Actin binding marine natural products as investigative tools and anticancer agents. Proc Int Symp Mar Nat Prod 9
Steidinger KA, Vargo GA, Tester PA, Tomas CR (1998) Bloom dynamics and physiology of *Gymnodinium breve* with emphasis on the Gulf of Mexico. In: Anderson DM, Cembella AD, Hallegraeff GM (eds) Physiological ecology of harmful algal blooms (NATO Advanced Study Institute Series).Springer, Berlin Heidelberg New York, pp. 133–154
Subba-Rao DV, Quilliam MA, Pocklington R (1988) Domoic acid – a neurotoxic amino acid produced by the marine diatom *Nitzschia pungens* in culture. Can J Fish Aquat Sci 45:2076–2079
Tachibana K, Scheuer PJ, Tsukitani Y, Kikuchi H, Van Engen D, Clardy J, Gopichand Y, Schmitz EJ (1981) Okadaic acid, a cytotoxic polyether from two marine sponges of the genus *Halichondria.* J Am Chem Soc 103:2469–2471
Teitelbaum JS, Zatorre RJ, Carpenter S, Gendron D, Evans AC, Gjedde A, Cashman NR (1990) Neurologic sequelae of domoic acid intoxication due to the ingestion of contaminated mussels. N Engl J Med 322:1781–1787

Terao K (1990) Histopathological studies on experimental marine toxin poisoning 5. The effects in mice of yessotoxin isolated from *Patinopecten yessoensis* and of a desulfated derivative. Toxicon 28:1095–1104

Terao K, Ito E, Yanagi T, Yasumoto T (1986) Histopathological studies on experimental marine toxin poisoning. 1. Ultrastructural changes in the small-intestine and liver of suckling mice induced by dinophysistoxin-1 and pectenotoxin-1. Toxicon 24:1141–1151

Tibbetts J (1998) Toxic tides. Environ Health Perspect 106:A326–A331

Trainer VL, Thomsen WJ, Catterall WA, Baden DG (1991) Photoaffinity labeling of the brevetoxin receptor on sodium channels in rat brain synaptosomes. Mol Pharmacol 40:988–994

Walter JA, Falk M, Wright JLC (1994) Chemistry of the shellfish toxin domoic acid: characterization of related compounds. Can J Chem 72:430–436

Wright JLC, Cembella AD (1998) Ecophysiology and biosynthesis of polyether toxins. In: Anderson DM, Cembella AD, Hallegraeff GM (eds) Physiological ecology of harmful algal blooms (NATO Advanced Study Institute Series). Springer, Berlin Heidelberg New York, pp. 427–452

Wright JLC, Falk M, McInnes AG, Walter JA (1990) Identification of isodomoic acid D and two new geometrical isomers of domoic acid in toxic mussels. Can J Chem 68:22–25

Xi D, Ramsdell JS (1996) Glutamate receptors and calcium entry mechanisms for domoic acid in hippocampal neurons. Neuroreport 7:1115–1120

Yasumoto T, Murata M (1990) Polyether toxins involved in seafood poisoning. ACS Symp Ser Am Chem Soc 418:120–132

Yasumoto T, Murata M (1993) Marine toxins. Chem Rev 93:1897–1909

Yasumoto T, Takizawa A (1997) Fluorometric measurement of yessotoxins in shellfish by high-pressure liquid chromatography. Biosci Biotechnol Biochem 61:1775–1777

Yasumoto T, Oshima Y, Sugawara W, Fukuyo Y, Oguri H, Igarishi T, Fujita N (1980) Identification of *Dinophysis fortii* as the causative organism of diarrhetic shellfish poisoning. Bull Jpn Soc Sci Fish 46:1405–1411

Hyperhydroxylation: A New Strategy for Neuronal Targeting by Venomous Marine Molluscs

A. Franco, K. Pisarewicz, C. Moller, D. Mora, G.B. Fields, F. Marí

Abstract. Venomous marine molluscs belonging to the genus Conus (cone snails) utilize a unique neurochemical strategy to capture their prey. Their venom is composed of a complex mixture of highly modified peptides (conopeptides) that interact with a wide range of neuronal targets. In this chapter, we describe a set of modifications based upon the hydroxylation of polypeptidic chains that are defining within the neurochemical strategy used by cone snails to capture their prey. In particular, we present a differential hydroxylation strategy that affects the neuronal targeting of a new set of α-conotoxins, mini-M conotoxins, conophans, and γ-hydroxyconophans. Differential hydroxylation, preferential hydroxylation and hyperhydroxylation have been observed in these conopeptide families as a means of augmenting the venom arsenal used by cone snails for neuronal targeting and prey capture.

4.1 Introduction

Cone snails are predatory marine molluscs that utilize venom to capture their prey. These animals are among the most prolific and versatile peptide engineers known in nature. Their venom is an extremely complex concoction of 20–200 compounds (Fig. 4.1) mostly composed of modified peptides (conopeptides) which are part of a biochemical strategy used for predation. These unique marine organisms deliver their complex venom through a specialized radular tooth that serves as both harpoon and disposable hypodermic needle. Conopeptides elicit a wide range of strong neurophysiological responses (Myers et al. 1993; Adams et al. 1999; McIntosh and Jones 2001; Newcomb and Miljanich 2002; Olivera 2002); and in a few instances, human fatalities have resulted from mishandling these animals (Fegan and Andresen 1997). The development of such extremely potent and biochemically diverse venom is likely to be an evolutionary adaptation designed to compensate for the lack of mobility of cone snails when compared to other marine predators. The precise composition of cone snail venom is species-specific (Myers et al. 1993; Newcomb and Miljanich 2002; Olivera 2002). *Conus* venom is the product

Frank Marí
Department of Chemistry & Biochemistry and Centre of Excellence in Biomedical and Marine Biotechnology
Florida Atlantic University, 777 Glades Rd., Boca Raton, 33431 Florida, USA

Progress in Molecular and Subcellular Biology
Subseries Marine Molecular Biotechnology
G. Cimino, M. Gavagnin (Eds.): Molluscs

of 55 million years of evolutionary refinement that has yielded a complex library of over 100,000 neuroactive conopeptides, as this genus comprises over 1,000 species distributed in the tropical and subtropical areas of the Atlantic, Indian, and Pacific Oceans (Filmer 2001).

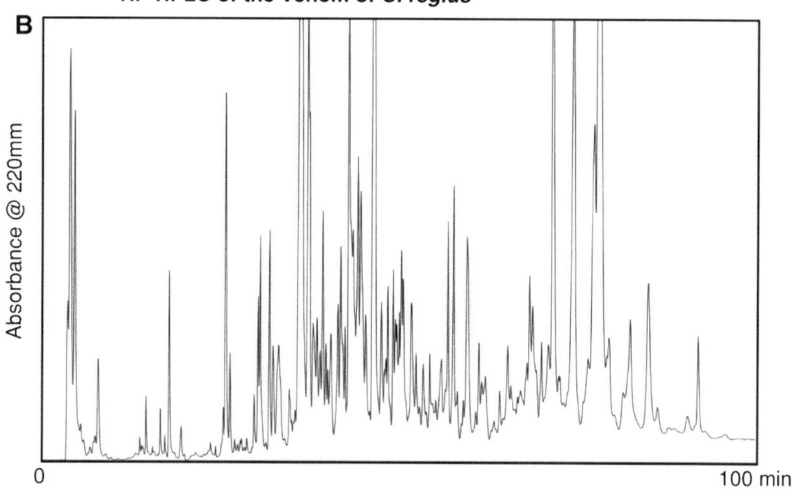

Fig. 4.1. Live specimen of *C. regius* (**a**) and the corresponding RP-HPLC chromatogram of its crude venom (**b**) C-18 Vydac, linear gradient 0–60% H_2O/ACN with 0.1% TFA in 100 min. *C. regius* is a widespread Western Atlantic worm-hunting species. This specimen was collected off the Florida Keys at a depth of 2 m. *C. regius* shows another shell variant known as *C. regius citrinus* whose shell is bright orange. This shell variation has little effect on the *C. regius* venom composition

Injecting venom that simultaneously acts on multiple components of the nervous system and that can be effective across a wide array of prey species has given *Conus* the edge needed to survive in the diverse and competitive marine environment they inhabit. The biochemical strategy developed by cone snails to target the multiplicity of the neuronal receptors has generated this immense conopeptide library with great potential therapeutic uses (Heading 1999, 2001, 2002). The first example is Prialt, formerly known as Ziconotide and SNX-111. Prialt, an ω-conotoxin isolated from *C. magus* (MVIIA), has been approved by the FDA in the USA and is one of the most powerful pain therapeutics known to date (Heading 2002; Miljanich 2004). Other conopeptides with similar or superior efficacy to Prialt have yet to be developed. For example, AM336, also known as ω-conotoxin CVIA isolated from *C. catus*, is being tested for severe morphine-resistant pain. AM336 has a higher selectivity than Prialt for N-type over P/Q-type calcium channels and a similar potency for inhibiting current through central splice variants of the rat N-type calcium channels (Lewis et al. 2000; Smith et al. 2002; Adams et al. 2003). An α-conotoxin, ACV1 (α-conotoxin Vc1.1 from *C. victoriae*), is being developed for neuropathic pain and for speeding the rate of functional recovery after a nerve injury (Sandall et al. 2003). χ-Conotoxins Mr-IA/B from *C. marmoreus*, which target the neuronal noradrenaline transporter (Sharpe et al. 2001; Harvey 2002), are being developed for strategies for the management of certain types of pain (for which there is currently a lack of effective treatment). Other conopeptides in clinical trials include two isolated from *C. geographus*. Contulakin-G, a 17-residue *O*-glycosylated linear peptide, targets the neurotensin receptor (Craig et al. 1999b). Conantokin-G, a 17-residue linear conopeptide that contains five γ-carboxyglutamate (Gla) residues, is an antagonist of the NMDA receptor. In addition to their therapeutic applications, conopeptides are currently being used in hundreds of research laboratories for a wide variety of physiological and pharmacological investigations. Some peptides have become well-established neurobiological tools.

Conopeptide precursors are ribosomally expressed proteins that subsequently undergo proteolytic cleavage and post-translational modifications to form the mature conopeptide (Santos et al. 2004). Conopeptides inherently contain high degrees of modified amino acids (usually combinations of them), such as cystines, hydroxyproline, γ-carboxyglutamate, Br-Trp, D-Trp, D-Leu, D-Phe (Buczek et al. 2005), pyro-Glu, glycosylated Ser/Thr, and sulfated Tyr (Craig et al. 1999a,b, 2001). These modifications confer conopeptides with unique stability and exquisite specificity toward neuronal targets (Myers et al. 1993; Craig et al. 1999a; McIntosh et al. 1999), enabling cone snails to capture prey.

Modification of polypeptide chains by epimerization of standard L-amino acids to produce their D-counterparts is a rare process. It is known to occur in position 2 of opionoid peptides such as dermorphine (Fujii 2002), in neuropeptides such as achatin (Fujii 2002) and in position 3 of the 72-residue crustacean hyperglycemic hormone (Soyez et al. 2000). D-Trp (and sometimes D-Leu) is found as the second residue after the first Cys in contryphans (Jimenez et al. 1996, 2001; Jacobsen et al. 1998). In these cases, the modifications of L-to D-amino acids are determinants of stability and potency. For example, D-Ser46 in the 48-residue funnel-web spider venom ω-agatoxin IVB provides more resistance to the major venom protease and is agatoxin IVC counterpart (Heck et al. 1994).

In contrast to epimerization, γ-hydroxylation of Pro residues is a more common process. γ-Hydroxyproline (γ-Hyp) is vital for collagen structural stability (Haading and Crabbe 1992; Perret et al. 2001) and is a commonly found modification in several conopeptide families. γ-Hydroxylation of non-Pro amino acids is an extremely rare process, since a hydroxyl group in the γ position of any amino acid (except Pro) could undergo nucleophilic attack at the contiguous peptide bond to form a stable five-membered ring lactone. Nonetheless, γ-hydroxyarginine has been found as part of the sequence of polyphenolic proteins that form the adhesive plaques of marine mussel species (Papov et al. 1995). The presence of γ-hydroxyarginine provides trypsin resistance to mussel glue proteins. γ-Hydroxylysine (γ-Hyk) has been reported within the sequence of crytonomad algae biliproteins (Sidler et al. 1983). However, the role of γ-Hyk in protein function has not been defined. The oxidation of Leu to produce hydroxyleucine (presumably in either the δ or γ position) has been described as an unusual post-translational modification present in unstable forms of hemoglobin associated with patients afflicted with hemolytic anemia (Brennan et al. 1992, 1993). The significance of this modification has not been established. Nonproteinogenous γ-hydroxylated amino acids have been found within enzymatically produced cyclic peptides (Shoji and Hinoo 1975; McGahren et al. 1977; Terui et al. 1990; Morita et al. 1997).

As part of our efforts on the analysis and characterization of conopeptides isolated from cone snail species from the Americas, we decided to undertake the isolation and structural analysis of conopeptides from *C. regius* (species code: reg), a widespread worm-hunting cone snail of the western Atlantic Ocean (Fig. 4.1). Here we present the results from the isolation and structural analysis of two hydroxylated conopeptide families from *C. regius*: α-conotoxins and the newly described mini-M conotoxins. These peptides exhibit a differential or preferential proline hydroxylation strategy that is likely to affect their neuronal targeting. Beyond the hydroxylation of Pro, we have discovered in the venom of *C. gladiator* and *C. mus* a new family of conopeptides that feature the modified amino acid D-γ-hydroxyvaline (Hyv = V*). This doubly modified amino acid is a novel structural motif, Ser-D-γ-OH-Xaa-Trp, which defines a new class of conopeptides that we have termed γ-hydroxyco nophans. Hydroxyconophans constitute the first examples of a polypeptide chain containing Hyv. We have also isolated analogous conopeptides containing D-Val instead of D-Hyv; and these are termed conophans. Additionally, we have isolated three related conopeptides from *C. villepinii* (vil-M̲, vil-I̲′ and vil-I̲′(O2P)) that are similar in sequence and properties to the gld/mus conophans. However, vil-M̲ and vil-I̲′ incorporate D-Met and D-*allo*Ile instead of D-Val.

4.2 Hydroxylation of α-Conotoxins

The venom of *C. regius* is an extremely complex mixture of peptides and proteins whose direct separation is shown in Fig. 4.1. More than 100 fractions can be obtained from this separation. However, most of these fractions show multiple components, for which we were required to adopt an improved separation scheme that included a prefractionation step using size exclusion chromatography on a Superdex-30 column, followed by a refined peptide-optimized size exclusion step on a Superdex peptide column. The resulting fractions are then separated by reversed phase on a peptide-optimized C18 Vydac Everest column. Most of the resulting fractions were single-component (Fig. 4.2) and were subsequently analyzed by mass spectrometry (MALDI-TOF, ESI-Q-TOF), NMR spectroscopy, and peptide sequencing by Edman degradation chemistry. Using this methodology, the most significant components of the venom of *C. regius* can be sequenced (Conopeptidome) and the components of the venom can be grouped in the different families of conopeptides. Three predominant families of hydroxylated conopeptides were found in the venom of *C. regius*: α-conotoxins (4/3 subtype), mini-M conotoxins and members of the P-superfamily of conotoxins.

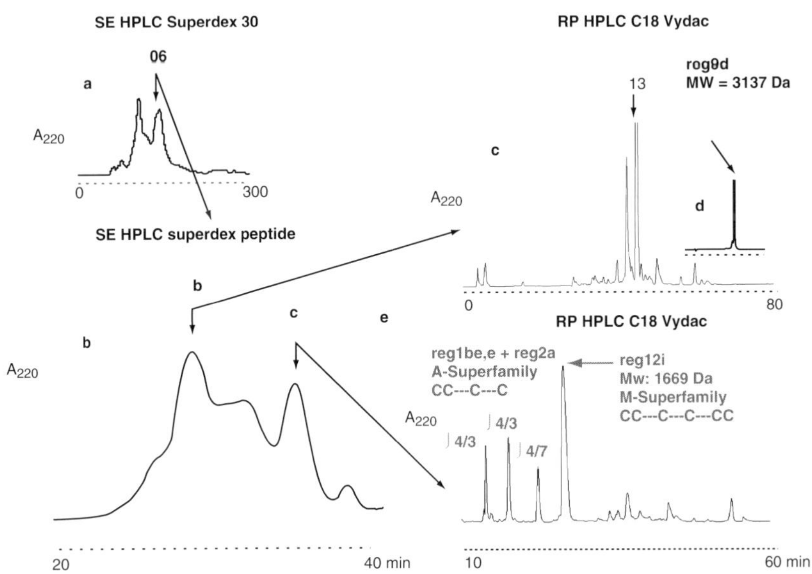

Fig. 4.2. Size exclusion (SE) and RP-HPLC separation of the crude venom of *C. regius*. (**a**) SE separation of *C. regius* crude venom. (**b**) Further separation by SE of fraction 06. (**c**) RP analysis of fraction b. (**d**) RP clean-up of fraction 13. (**e**) RP analysis of fraction C

Figure 4.3 shows the sequences of the α-conotoxins found in the venom of *C. regius*. Six of these α-conotoxins belong to the 4/3 subtype and one to the 4/7 subtype. The α4/3 subtypes are rare within the α-conotoxin family, as only ImI and ImII (isolated from *C. imperialis*, an Indo-Pacific worm-hunting species) have been characterized to date (McIntosh et al. 1994; Ellison et al. 2003). The regla–f are the only α4/3-conotoxins isolated from an Atlantic cone snail species with known sequences to date. The regla,b,e, and f conotoxins are also unusual as they are the only α4/3-conotoxins whose sequences are post-translationally hydroxylated to produce the modified amino acid hydroxyproline: reglf has two hydroxyproline residues in its first loop, whereas regla,b, and e have one Hyp. reglb and c have the same sequence, except that regld is not hydroxylated at Pro-6. The reglc and d only differ in residue 11 (Gln in reglc vs Glu in regld). The regla–e have significant sequence homology: they share the same sequence in the first loop (SDPR or SDOR) and a conserved positively charged residue (Arg or Lys) in the first position of the second loop. regla–e are homologous with the only other known α4/3 conotoxins, ImI and ImII; and these two toxins only differ in one residue (Hyp6 in ImI, Arg6 in ImII) and they both bind to the α7 subtype neuronal nAChR (Ellison et al. 2003). However, these two conotoxins bind at

α-4/3	**Sequence**	**Target:nAChR**
reg1a	--GCCSDORCRYX----C*	
reg1b	--GCCSDORCKHQ----C*	
reg1c	--GCCSDPRCKHQ----C*	
reg1d	--GCCSDPRCKHE----C*	
reg1e	--GCCSDORCRYR----C*	
reg1f	-DYCCRROOCTLI----C*	
ImI	--ACCSDPRCAWR----C*	α7
ImII	--GCCSDRRCAWR----C*	α7
α-4/7		
reg2a	--GCCSHPACNVNNPHIC*	
GID	RDγCCSNPACRVNNOHVC	α3β2, α7 >α4β2
GIC	--GCCSHPACAGNNQHIC*	α3β2
Vc1a	GCCSDORCNYDHPγIC*	α3α7β4/α3α5β4
AnIC	GGCCSHPACAANNQDŶC*	α3β2, α7
EI	RDOCCYHPTCNMSNPQIC*	α1γ = α1δβ2 musc.

Fig. 4.3. α-Conotoxins from *Conus regius*

different sites and their mechanisms of inhibition appear to be different in spite of their sequence homology. Likewise, the reg1a–e conotoxins are likely to show differential binding properties toward nAChR (Jacobsen et al. 1997). By way of contrast, reg1f has no sequence homology to the other known α4/3 conotoxins. Its sequence (1) has an extended N-terminal tail, (2) lacks the conserved Ser in the first residue of loop1 seen in virtually all α4/3 and 4/7 conotoxins (an Arg residue is in its place), (3) the first loop (RROO) has no homology with the other α4/3 conotoxins (only the first Hyp is common with other α4/3 and 4/7 conotoxins), and (4) the second loop (TLI) has no homology with the other α4/3 conotoxins and notably lacks positively charged residues. Presumably, this departure from the sequences of other α-conotoxins has profound consequences to its binding to the neuronal nAChR. All of these conotoxins are either hydroxylated or susceptible to hydroxylation, which is a feature relatively rare among α-conotoxins (Dutton and Craik 2001; Livett et al. 2004; Millard et al. 2004; Nicke et al. 2004). These modifications enhance the polarity and hydrogen-bonding capabilities to these conotoxins and are likely to define their mode of binding to nAChRs.

In addition to the α4/3 conotoxins, an α4/7 was found in the venom of *C. regius*: reg2a. This conotoxin, while not hydroxylated, shows extraordinary homology with GIC (McIntosh et al. 2002) and GID (Nicke et al. 2003), a pair of α4/7-conotoxins isolated from *C. geographus*, the quintessential Indo-Pacific fish-hunting cone snail species. This similarity in sequence is quite remarkable, considering that *C. regius* is a worm-hunting cone snail from the Western Atlantic. However, *C. geographus* does not contain α4/3 conotoxins within its venom. Perhaps, the proper combination of α-conotoxins subtypes is required for the effective targeting of neuronal receptors in order to produce a synergistic disabling effect on the different prey to be captured.

4.3 Hydroxylation of Mini-M Conotoxins

Within the venom of *C. regius*, another prevalent family of conopeptides is the mini-M (m1, m2, m3) subclass of conotoxins that belong to the M-superfamily (McDougal and Poulter 2004; Corpuz et al. 2005). Mini-M conotoxins share the same arrangement of Cys in their sequence as other members of the M-superfamily (CC–C–C–CC), such as the Maxi-M or M4 subclass. The loops in the mini-M are much shorter and the Cys pairing can be different from the classic Cys knot observed in the maxi-M subclass. Several targets have been identified in the maxi-M subclass: voltage-gated sodium channels for the μ-conotoxins (Li and Tomaselli

2004), voltage-gated potassium channels for the κM-conotoxins (Al-Sabi et al. 2004) and nAChR for the ψ-conotoxins (Shon et al. 1997). While the mini-M conotoxins are prevalent in many *Conus* species, details of their isolation and characterization are only starting to be fully disclosed (Corpuz et al. 2005); and few reports describe the presence of these toxins in cone snails. Mini-M conotoxins have been isolated from *C. quernicus* (Olivera et al. 1990), *C. pennaceus* (Fainzilber et al. 1995), *C. betulinus* (Chen et al. 1999; Zhao et al. 2005), *C. marmoreus*, and *C. textile* (McDougal and Poulter 2004; Corpuz et al. 2005). These conotoxins are the major venom components in many mollusc-hunting and worm-hunting *Conus* species. Recently, the NMR structure of a mini-M conotoxin was published (McDougal and Poulter 2004). Their molecular target has not been identified. However, it appears to be different from the maxi-M families.

To date, all reported maxi-M conotoxins are hydroxylated at the Pro residues of their sequences. In fact, it has been assumed to be a conserved feature of this superfamily (Ferber et al. 2003). We isolated eight mini-M conotoxins from the venom of *C. regius*: reg12a,e–i,k,l (Fig. 4.4). Unlike

Mini-M	**Sequence**	**Loops**	**Subtype**
reg12a	GCCOOQWCGOD--CTSOCC	4/3/3	m-3
reg12e	KCCMRPICT----C--OCCIGP*	4/1/1	m-1
reg12f	GCCPFPACTHTIICR--CC	4/5/1	m-1
reg12g	CCMAL-CS-RYHC-LPCC	3/4/2	m-2
reg12h	GCCSOWNCIQLRAC--OCCON	4/5/1	m-1
reg12i	-CCTAL-CSRYH-CL-PCC	3/4/2	m-2
reg12k	KCCMRPICM----C--OCCIGAG	4/1/1	m-1
reg12l	RCCPMPGCFAGPFC--PCCPP	4/5/1	m-1
Mr3a	GCC-GSFACRFG-CVO--CCV	4/3/2	m-2
RIIIK (κM Cntx)	LOSCCSLNNLRLCOVOACKRNOCCT	7/4/4	

from cDNA of *C. radiatus*
Shaker K+ channel blocker

maxi-M (M-4)		
μGIIIA *C. geographus*	RDCCTOOKKCKDRQCKOQRCCA	5/4/4
μPIIIA *C. purpurascens*	RLCCGFOKSCRSRQCKOHRCC	5/4/4
ψPIIIE *C. purpurascens*	HOCCLYGK-CRRYOGCSSASASCCQR	4/5/6

non comp. nAChR

Fig. 4.4. Mini-M-conotoxins from *Conus regius*

the reg1 α-conotoxins, there is no sequence homology among the reg12 mini-M conotoxins. When considering the size of their loops (a defining feature within the M-superfamily), there are four variants within the reg12 conotoxins: 3/4/2, 4/1/1, 4/3/3, and 4/5/1 subtypes. Just like the other members of the M-superfamily, reg12a and h are hydroxylated at all Pro residues of their sequences, just as the known maxi-M conotoxins. In contrast, the Pro residues of reg12f,g,i, and l are unmodified; and notably the latter did not exhibit hydroxylation in spite of having five Pro residues with its sequence. reg12e and k are partially hydroxylated as at least one of their prolines was found to be a Hyp residue. The sequence diversity found within these reg12 mini-M conotoxins suggests that their targeting of neuronal receptors and/or ion channels might be equally diverse. Perhaps in this case hydroxylation is part of a refinement strategy, where certain sequences within this family are required to have the additional polarity and hydrogen-bonding capability, whereas others do not. Differential hydroxylation has been observed within the same conopeptide sequence (Hopkins et al. 1995; see later); but this does not appear to be the case for the reg12 mini-M conotoxins.

4.4 Hyperhydroxylation of Conophans: D-γ-Hydroxyvaline and γ-Hydroxyconophans

We recently described a new family of linear short conopeptides that we termed γ-hydroxyconophans (Pisarewicz et al. 2005). These conopeptides have a high content of hydroxyl residues: Ser, Hyp, and the unprecedented presence of D-γ-hydroxyvaline in their sequence. These conopeptides are the first examples of polypeptide chains containing such an intriguing modification. We characterized four novel conopeptides from the venom of *C. gladiator* (gld-V̲*, gld-V̲*′) and *C. mus* (mus-V̲*, mus-V̲*′). These conopeptides contain the doubly modified amino acid D-γ-Hyv. We also isolated their respective peptidyl precursors that contain D-Val (gld-V̲, mus-V̲). γ-Hyv has been described as a free amino acid isolated from plants (Pollard et al. 1958) and bacteria (Hernandez et al. 2000). However, it would be an unpredicted post-translational modification in proteins and peptides, as its hydroxyl group would readily cleave the peptide bond by intraresidue cyclization to form a lactone (Hernandez et al. 2000).

We initially isolated three unusual conopeptides, gld-V̲*, gld-V̲*′, and gld-V̲ (Fig. 4.5a), from the venom of *C. gladiator* (species code gld), a cone snail species that inhabits the tropical Eastern Pacific region and preys on worms. Upon the identification of these conopeptides, the related conopeptides mus-V*, mus-V*', and mus-V were isolated from *C. mus* (species code mus), a cone snail species related to *C. gladiator* that inhabits the western Atlantic region (Fig. 4.5b). *C. gladiator* and *C. mus* share

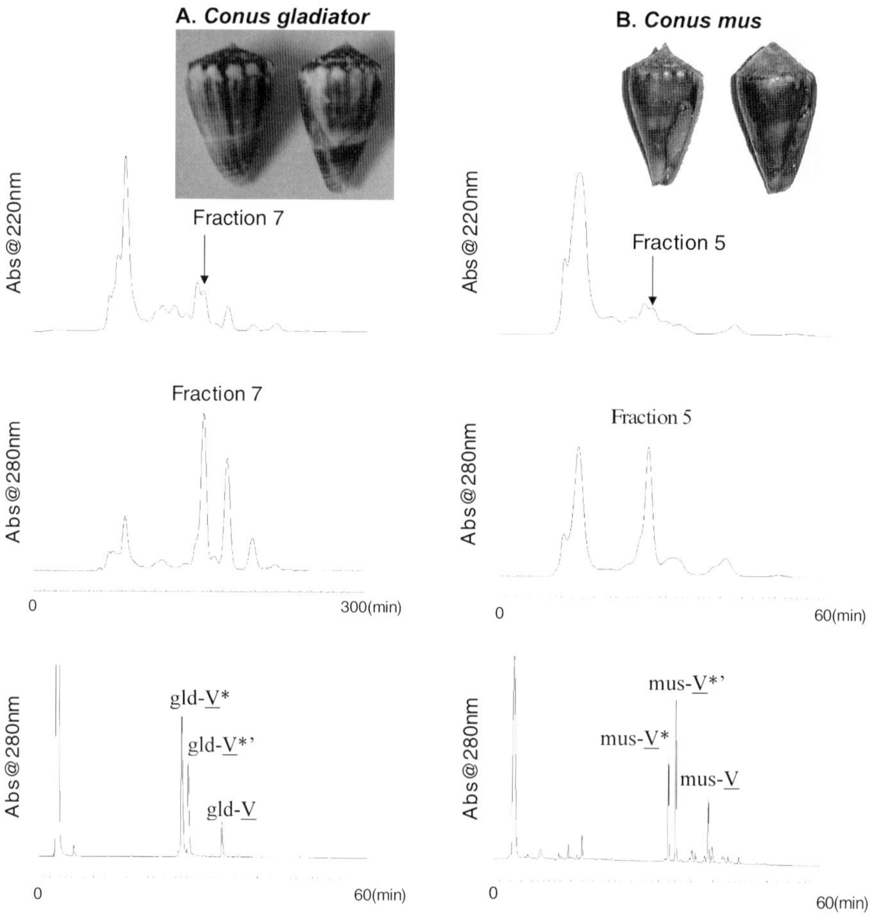

Fig. 4.5. Isolation of gld and mus conophans and γ-hydroxyconophans. Specimens of *C. gladiator* (**a**) were collected from several locations off the Pacific coast of Costa Rica, whereas specimens of *C. mus* (**b**) were collected off Plantation Key, Florida. The venom ducts were dissected and the venom was pooled and fractionated using size exclusion (SE) HPLC (Superdex-30, buffer = 0.1 M NH_4HCO_3). The elution SE-HPLC profiles are shown at $\lambda = 220$ nm (top) and $\lambda = 280$ nm (middle), respectively. The *arrows* indicate the selected fractions of Trp-containing gld and mus peptides. The Trp-containing fractions were further separated (bottom) using reversed-phase HPLC (Vydac C18, H_2O/60% CH_3CN linear gradient over 100 min with 0.1% TFA). The gld-V*/gld-V*' and mus-V*/mus-V*' pairs have the same covalent structures. NMR determined that they are diastereomeric conopeptides at the βH of their D-γ-Hyv residue

an ancestral origin that split three Mya upon the raise of the Isthmus of Panama (Duda and Palumbi 1999). The gld conopeptides were isolated in nanomolar quantities, whereas the mus conopeptides were isolated in picomolar quantities. Nano/Pico-NMR techniques (Barbara 1994; Barbara and Bronnimann 1999) allowed the acquisition of their spectra (Fig. 4.6) and revealed almost identical compositions for these octapeptides, including an unusual amino acid for gld-V* and gld-V*′, whereas gld-V showed Val in its place (Fig. 4.6). The mass spectra of gld-V*/gld-V*′ and gld-V gave molecular ions of 863.3 and 847.3 Da, respectively. gld-V* and gld-V*′ have the same covalent structures. The mus octapeptides gave information identical to their gld counterparts, except that their molecular masses were shifted by 16 Da. Combined Edman degradation sequencing, MS/MS and NMR analyses revealed the sequences shown in Fig. 4.7.

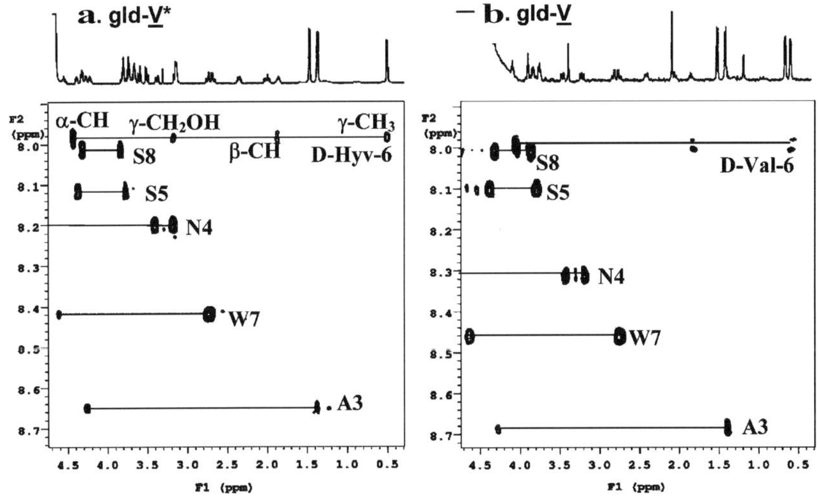

Fig. 4.6. NMR spectra of the gld peptides. (**a**) The 1D proton spectrum along with its corresponding 2D-TOCSY spectrum of 35 nmol gld-V* recorded at 25°C, using a gHX HR-MAS probe (Barbara and Bronnimann 1999) on a Varian Inova 500 MHz spectrometer. The resonance assignments were carried out using standard biomolecular NMR procedures. (**b**) The 1D and the corresponding 2D-TOCSY spectra of gld-V, putative precursor of gld-V*. The stereochemistry of the Val-6 residue was determined by comparing the NMR and MS/MS spectra of gld-V* with synthetic peptides made using several combinations of L and D amino acids at the Val-6 and Trp-7 positions. The stereochemistry of the Hyv was determined by extended and comparative NMR analysis of the gld-V analogs

```
mus-V*         SOANSV*WS       V* = γ-Hydroxyvaline
mus-V*'        SOANSV*WS       O  = Hyp
mus-V          SOANSV WS       V  = D-Valine

gld-V*         AOANSV*WS
gld-V*'        AOANSV*WS
gld-V          AOANSV WS

vil-M          EO-NSM WS       M = D-Methionine
vil-I'         EO-NSI'WS       I'= D-alloIsoleucine
vil-I' (O2P)   EP-NSI'WS
```

Fig. 4.7. Hydroxyconophan and conophan library

The D configuration of the Val-6α carbon in gld-V* and mus-V was determined by comparing the chromatographic profiles, NMR and MS/MS spectra of gld-V with synthetic analogs. D-amino acids are known in contryphans (Jimenez et al. 1996; Jacobsen et al. 1998; Pallaghy et al. 1999). However, here the situation is significantly different, as D-Val has never been found in ribosomally expressed polypeptide chains and its location (near the C-terminal, in close contact with L-Trp) has important structural connotations not associated with contryphans. The configuration of the α carbon in Hyv was determined by NMR analysis, revealing that its stereochemistry had been preserved upon hydroxylation.

The Ser-D-γ-Hyv-Trp triad is an unusually stable structural motif that incorporates the unprecedented modification of a D-amino acid and has produced the first examples of hydroxyvaline within polypeptide chains. This motif defines the new class of conopeptides: γ-hydroxyconophans. The corresponding precursors, such as gld-V, are termed conophans. The presence of γ-hydroxyvaline in gld-V* as opposed to valine in gld-V suggests the existence of a corresponding enzyme capable of D-Val oxidation. This putative enzyme could be using gld-V, or its precursor protein, as a substrate to modify the specified D-Val to generate the final form of the toxin. This process challenges previous understanding of homochirality in living organisms, as all known enzymatic reactions acting on peptides and proteins are stereoselective for the L-amino acids within them (Petsko 1992) (Fig. 4.8).

This unusual transformation appears to be part of a hyperhydroxylation strategy used by cone snails to optimize their venom efficacy by increasing the hydrogen-bonding capabilities of the toxin (see later). In addition to the three Ser residues (for mus-V*'), the hydroxylation of Pro and then D-Val yields a 63% content of hydroxylated residues. The relative stability of the hydroxyconophan may be attributed to a specific interaction with the Trp residue that follows the D-γ-hydroxyvaline by CH–π interactions and another with the Ser residue that precedes the D-Hyv by hydrogen bonding (Pisarewicz et al. 2005). NMR evidence and molecular modeling of gld-V* suggest that the D configuration of the Hyv residue allows the stabilizing interaction with Trp (Mitchell and Smith 2003) and Ser, therefore disfavoring intraresidue cleavage of the peptide bond by the γ-hydroxyl group.

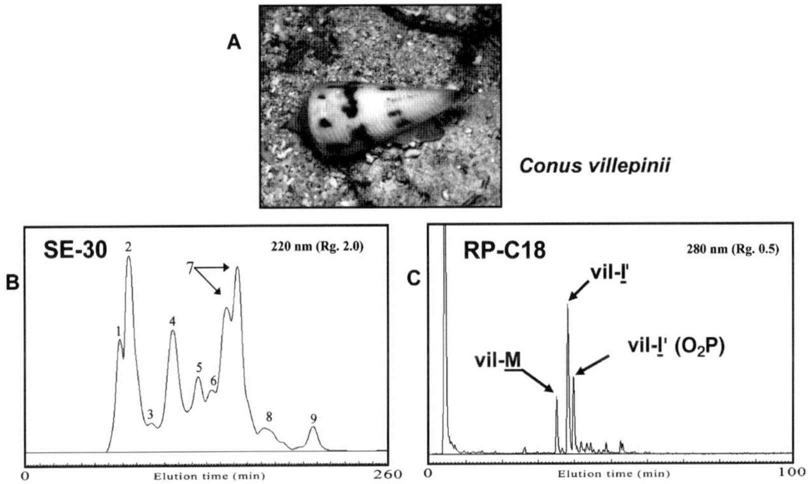

Fig. 4.8. Isolation of vil conophans. Specimens of *C. villepinii* (**a**) were collected off Marathon Key, Florida. The venom ducts were dissected and the venom was pooled and fractionated using size exclusion (SE) HPLC (Superdex-30, buffer = 0.1 M NH_4HCO_3). The elution SE-HPLC profiles are shown at $\lambda = 220$ (**b**). The *arrows* indicate the selected fractions of Trp-containing vil peptides. The Trp-containing fractions were further separated (**c**) using reversed-phase HPLC (Vydac C18, H_2O/60% CH_3CN linear gradient over 100 min with 0.1% TFA)

The epimerization and subsequent hydroxylation of Val provides further diversity to the venom by adding a new protein scaffold that, so far, is unique to *Conus*. However, just as other post-translational modifications found in *Conus* venom were previously described in other organisms (Olivera 2002; Mari and Fields 2003), it will not be surprising to find the γ-hydroxyconophan scaffold in other organisms. In addition to the unprecedented presence of D-Hyv in their sequence, γ-hydroxyconophans are unusual because (1) they are linear conopeptides and not constrained like the conotoxin and contryphan families, (2) they are extremely short in length, (3) they have a high content of hydroxylated residues, and (4) their primary structure has no close match in the sequence databases.

In the conophan family, the motif that is defining is the Ser-D-Xaa-Trp triad; where for the conophans of *C. gladiator* and *C. mus* the D-amino acid is Val. We have recently isolated three conophans from the venom of *C. villepinii* (species code vil), a western Atlantic deep-water cone snail species (Fig. 4.5). The sequences of these three conophans, vil-I̲′, vil-I̲′(O2P), and vil-M̲ are shown in Fig. 4.7. Their characterization is analogous to the gld and mus conophans (Fig. 4.9).

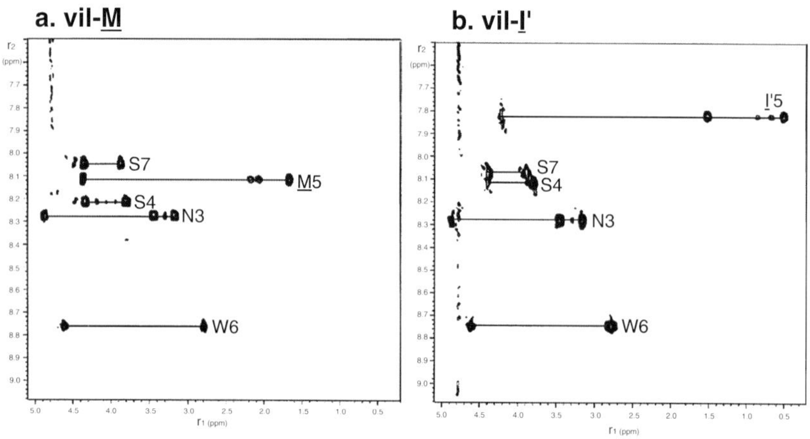

Fig. 4.9. 2D-NMR Spectra of the vil conophans. (**a**) The 2D-TOCSY spectrum of 35 nmol vil-M̲ recorded at 25°C using a 1.7-mm insert in a gHCN 3-mm probe on a Varian Inova 500 MHz spectrometer. The resonance assignments were carried out using standard biomolecular NMR procedures. (**b**) The 2D-TOCSY spectra of vil-I̲′, the putative precursor of vil-I̲′(O2P). The stereochemistries of the Met-5 and *allo*Ile-5 residues were determined by comparing the NMR and MS/MS spectra of the native peptides with synthetic ones

The vil conophans are among the major components of the venom of *C. villepinii*. However, unlike in the case of *C. gladiator* and *C. mus*, γ-hydroxyconophans are not found in *C. villepinii*. The vil conophans indicate that epimerization is not restricted to Val. Ile and Met can undergo epimerization as well, presumably by the same epimerizase (or an isotype) operating in the case of Val (see later). The presence of just conophans in the venom of *C. villepinii* indicates that this family of conopeptides has a neurochemical function within the arsenal of toxins of cone snails. However, this particular species is using "less hydroxylated" toxins compared to their γ-hydroxyconophan counterparts found in *C. gladiator* and *C. mus*, as they have one less Ser residue (in position 3; Fig. 4.7) and neither the D-*allo*Ile nor the D-Met are hydroxylated. This case is reminiscent of the one observed in the α-conotoxins discussed earlier, where in some instances, cone snails utilize the hydroxylated toxins, in others the nonhydroxylated ones and in other times a mixture of both is utilized (see later). Within this scheme of differential hydroxylation, the sequence of vil-I̲'(O2P) suggests a hierarchical sequence of modifications that lead to the hyperhydroxylation observed in the γ-hydroxyconophan family. vil-I̲'(O2P) has a Pro residue in position 2, instead of the Hyp observed in the other conophans and hydroxyconophans found so far. This indicates that epimerization of Ile in this case precedes hydroxylation and suggests that the sequence of events that leads to hyperhydroxylation proceeds as follows: epimerization of the Xaa within the Ser-Xaa-Trp triad, followed by hydroxylation of Pro2 and finally hydroxylation of D-Xaa.

The epimerization of Val by cone snails has produced the first examples of D-Val within ribosomally expressed polypeptide chains. D-Met has been found in dermenkephalin, a heptapeptide isolated from frog skin that targets the δ-opionid receptor (Amiche et al. 1989; Mor et al. 1989). D-*allo*Ile has been found in three 20-residue peptides isolated from the skin of *Bombina variegata* (yellowbelly toad) that have shown antibacterial and hemolytic properties (Mignogna et al. 1993). Most epimerizations found in small linear peptides occur near the N-terminal and preferentially at the second position. The D-Val, D-*allo*Ile, and D-Met in conophans are at the third amino acid from the C-terminal, at the same relative position as in the larger disulfide-constrained ω-agatoxin (Heck et al. 1994) and the r11a I-superfamily conotoxin (Buczek et al. 2005), which have 48 and 46 residues, respectively. Apparently, the epimerization has a strong preference at this position near the C-terminal regardless of the nature of the amino acid (D-Ser in ω-agatoxin, D-Phe in the r11a conotoxin or D-Val, D-*allo*Ile, and D-Met in conophans) or size and nature of the expressed protein. In fact, it is likely that the two-base enzymatic mechanism proposed for the epimerization of D-Ser in ω-agatoxin is in effect in all these cases, as the epimerase in the funnel web spider is also known to epimerize other amino acids, such as Ala, Cys, and *O*-methylserine (Heck et al. 1996). However, the substrate for this epimerase has a recognition site, Leu-Xaa-Phe-Ala, not observed in the r11 conotoxin nor in the conophans. Furthermore, the spider epimerase is capable of converting Xaa in small peptides at several positions within the polypeptide chain (Heck et al. 1996). Therefore, it is likely that different epimerases with distinct specificities are operating in each of these cases.

As discussed earlier, epimerization is a necessary step to hyperhydroxylation. Once epimerization occurs, differential hydroxylation, already described in other conopeptide families, is observed. In *C. villepinii*, modification of Pro2 leads to mixtures of vil-I̲'(O2P) and vil-I̲' within the venom, which are involved in differential neuronal targeting through hydroxylation. In *C. gladiator* and *C. mus*, modification of D-Val6 leads to mixtures of gld/mus-V̲* and gld/mus-V̲*/V̲*' in the venom, which are responsible for targeting.

The presence of D-Hyv in gld-V̲*/gld-V̲*', as opposed to D-Val in gld-V̲, suggests the existence of an enzyme capable of D-Val oxidation. This putative enzyme could be using gld-V, or its precursor protein, as a substrate to modify D-Val and generate the D-Hyv form of the toxin. This process would be analogous to Glu γ-carboxylation of certain conopeptides, which requires the action of a specific carboxylase on the precursor form of the peptide (Bandyopadhyay et al. 1998, 2002; Walker et al. 2001; Czerwiec et al. 2002) to produce conantokins, and other related Gla-containing conopeptides. The isolation and identification of a hydroxylase with D-amino acid specificity is under investigation. The function of this enzyme appears to be the augmentation of molecular functionality to enhance neuronal targeting and it is part of a novel neurochemical strategy used by cone snails to capture their prey.

4.5 Conclusions

Partial hydroxylation in conopeptide families can be observed. In principle, it could be considered that finding partially hydroxylated peptides is the consequence of using *Conus* venom dissected from the ducts. Hence, in these cases, nonhydroxylated conopeptides can be considered immature, as they have not been fully processed for final delivery to the prey. However, partially hydroxylated conopeptides have been found in milked *Conus* venom (Shon et al. 1997). PIVA, an αA conotoxin that targets nAChR, is found in the milked venom of *C. purpurascens* as a mixture of three peptides whose difference lies in the differential hydroxylation of residues. Therefore, just as in the case of the α4/3 conotoxins, αA conotoxins, the conophans/γ-hydroxyconophans, and other conopeptide families could be capable of using this differential hydroxylation and hyperhydroxylation strategy to enhance the venom complexity and range of action.

Why are cone snails hydroxylating and hyperhydroxylating several families of conopeptides in such an extensive way? Perhaps, just as in the case of collagen, hydroxylation is the preferred strategy used by the cones snails to increase hydrogen-bonding capabilities. However, in the *Conus* case, hydrogen bonding is directed toward increasing binding strength and selectivity toward their neuronal targets. Polyhydroxylation is a recurrent feature in natural products of marine and land origins (Danieli and Riva 1994; Zeng et al. 1999; Cui et al. 2000; Watson et al. 2001). Polyhydroxylated compounds, ranging from taxol to dermostatin, rely on the hydrogen-bonding capabilities of their hydroxyl groups to interact with their molecular targets. In the case of most polyhydroxylated natural products, their complex molecular scaffolds are the product of an intricate multienzymatic biosynthetic pathway which incorporates diverse metabolites in manners unique to the organisms that produce the compounds. In contrast, cone snails have relied on a universal mechanism for protein synthesis and modification for the production of small molecule-like structural scaffolds that fulfill the specialized task of targeting specific neuronal receptors. These versatile peptide engineers have developed a wide range of protein processing schemes that allow them to efficiently carry out multiple post-translational modifications on polypeptide chains across the many conopeptide families. While this process is reminiscent of the biosynthetic production of complex natural products, cone snails are using a reduced set of enzymes capable of modifying a wide range of ribosomally expressed polypeptide chains. This strategy is aimed at enhancing the molecular complexity and diversity of their venom. The need for the development of such biochemically diverse venom is likely to be an evolutionary adaptation designed to compensate for the lack of mobility of cone snails when compared to other marine predators.

Acknowledgments. We thank Prof. Anton Oleinik and his group for their help in collecting specimens. We thank Janelle Lauer-Fields, Mare Cudic, and Jose Rivera-Ortiz for synthesis of peptides. We thank the government of the Republic of Costa Rica for providing the permits to collect the necessary specimens. This work was supported by the Florida Sea Grant College Program (R/LR-MB-18, R/LR-MB-74), the NIH (GM 066004 to F.M., CA 77402 to G.B.F.), and private funding from William Cargile. This work is contribution P200514 from the Center of Excellence in Biomedical and Marine Biotechnology.

References

Adams DJ, Alewood PF, Craik DJ, Drinkwater RD, Lewis RJ (1999) Conotoxins and their potential pharmaceutical applications. Drug Dev Res 46:219–234

Adams DJ, Smith AB, Schroeder CI, Yasuda T, Lewis RJ (2003) ω-Conotoxin CVID inhibits a pharmacologically distinct voltage-sensitive calcium channel associated with transmitter release from preganglionic nerve terminals. J Biol Chem 278:4057–4062

Al-Sabi A, et al. (2004) κM-conotoxin RIIIK, structural and functional novelty in a K^+ channel antagonist. Biochemistry 43:8625–8635

Amiche M, Sagan S, Mor A, Delfour A, Nicolas P (1989) Dermenkephalin (Tyr-D-Met-Phe-His-Leu-Met-Asp-NH_2) – a potent and fully specific agonist for the δ-opioid receptor. Mol Pharmacol 35:774–779

Bandyopadhyay PK, Colledge CJ, Walker CS, Zhou L-M, Hillyard DR, Olivera BM (1998) Conantokin-G precursor and its role in γ-carboxylation by a vitamin K-dependent carboxylase from a *Conus* snail. J Biol Chem 273:5447–5450

Bandyopadhyay PK, Garrett JE, Shetty RP, Keate T, Walker CS, Olivera BM (2002) γ-Glutamyl carboxylation: an extracellular post-translational modification that antedates the divergence of molluscs, arthropods, and chordates. Proc Natl Acad Sci USA 99:1264–1269

Barbara TM (1994) Cylindrical demagnetization fields and microprobe design in high-resolution NMR. J Magn Reson Ser A 109:265–269
Barbara TM, Bronnimann CE (1999) Target field design for magic angle gradient coils. J Magn Reson 140:285–288
Brennan SO, Shaw J, Allen J, George PM (1992) β141-Leu is not deleted in the unstable hemoglobin Atlanta–Coventry but is replaced by a novel amino-acid of mass 129 daltons. Br J Haematol 81:99–103
Brennan SO, Shaw JG, George PM, Huisman THJ (1993) Post-translational modification of β141 Leu associated with the β75(E19)Leu-Pro mutation in Hb Atlanta. Hemoglobin 17:1–7
Buczek O, Yoshikami D, Bulaj G, Jimenez EC, Olivera BM (2005) Post-translational amino acid isomerization: a functionally important D-amino acid in an excitatory peptide. J Biol Chem 280:4247–4253
Chen JS, Fan CX, Hu KP, Wei KH, Zhong MN (1999) Studies on conotoxins of *Conus betulinus*. J Nat Toxins 8:341–349
Corpuz GP, et al. (2005) Definition of the M-conotoxin superfamily: characterization of novel peptides from molluscivorous *Conus* venoms. Biochemistry 44:8176–8186
Craig AG, Bandyopadhyay P, Olivera BM (1999a) Post-translationally modified neuropeptides from *Conus* venoms. Eur J Biochem 264:271–275
Craig AG, et al. (1999b) Contulakin-G, an O-glycosylated invertebrate neurotensin. J Biol Chem 274:13752–13759
Craig AG, Park M, Fischer WH, Kang J, Compain P, Piller F (2001) Enzymatic glycosylation of contulakin-G, a glycopeptide isolated from *Conus* venom, with a mammalian ppGalNAc-transferase. Toxicon 39:809–815
Cui J, Wang Y, Lin C, Zeng L (2000) Polyhydroxylated sterols with biological activities in marine organism. Tianran Chanwu Yanjiu Yu Kaifa 12:94–100
Czerwiec E, et al. (2002) Expression and characterization of recombinant vitamin K-dependent γ-glutamyl carboxylase from an invertebrate, *Conus textile*. Eur J Biochem 269:6162–6172
Danieli B, Riva S (1994) Enzyme-mediated regioselective acylation of polyhydroxylated natural products. Pure Appl Chem 66:2215–2218
Duda TF, Palumbi SR (1999) Developmental shifts and species selection in gastropods. Proc Natl Acad Sci USA 96:10272–10277
Dutton JL, Craik DJ (2001) α-Conotoxins: nicotinic acetylcholine receptor antagonists as pharmacological tools and potential drug leads. Curr Med Chem 8:327–344
Ellison M, McIntosh JM, Olivera BM (2003) α-Conotoxins ImI and ImII – similar alpha 7 nicotinic receptor antagonists act at different sites. J Biol Chem 278:757–764
Fainzilber M, et al. (1995) A new cysteine framework in sodium-channel blocking conotoxins. Biochemistry 34:8649–8656
Fegan D, Andresen D (1997) *Conus geographus* envenomation. Lancet 349:1672
Ferber M, et al. (2003) A novel *Conus* peptide ligand for K^+ channels. J Biol Chem 278:2177–2183
Filmer RM (2001) A catalogue of nomenclature and taxonomy in the living Conidae 1758–1998. Backhuys, Leiden
Fujii N (2002) D-Amino acids in living higher organisms. Origins Life Evol B 32:103–127
Haading JJ, Crabbe MJC (1992) Post-translational modifications of proteins. CRC Press, Boca Raton, FL
Harvey AL (2002) Toxins 'R' Us: more pharmacological tools from nature's superstore. Trends Pharmacol Sci 23:201–203
Heading C (1999) Ziconotide Neurex Corp. Curr Opin Cent Peripheral Nerv Syst Invest Drugs 1:153–166
Heading CE (2001) Ziconotide (Elan Pharmaceuticals). IDrugs 4:339–350
Heading CE (2002) *Conus* peptides and neuroprotection. Curr Opin Invest Drugs 3:915–920

Heck SD, et al. (1994) Functional consequences of posttranslational isomerization of Ser(46) in a calcium-channel toxin. Science 266:1464
Heck SD, Faraci WS, Kelbaugh PR, Saccomano NA, Thadeio PF, Volkmann RA (1996) Post-translational amino acid epimerization: enzyme-catalyzed isomerization of amino acid residues in peptide chains. Proc Natl Acad Sci USA 93:4036–4039
Hernandez ILC, Godinho MJL, Magalhaes A, Schefer AB, Ferreira AG, Berlinck RGS (2000) *N*-Acetyl-gamma-hydroxyvaline lactone, an unusual amino acid derivative from marine streptomycete. J Nat Prod 63:664–665
Hopkins C, et al. (1995) A new family of *Conus* peptides targeted to the nicotinic acetylcholine-receptor. J Biol Chem 270:22361–22367
Jacobsen R, et al. (1997) Differential targeting of nicotinic acetylcholine receptors by novel αA-conotoxins. J Biol Chem 272:22531–22537
Jacobsen R, et al. (1998) The contryphans, a D-tryptophan-containing family of *Conus* peptides: interconversion between conformers. J Pept Res 51:173–179
Jimenez EC, Olivera BM, Gray WR, Cruz LJ (1996) Contryphan is a D-tryptophan-containing *Conus* peptide. J Biol Chem 271:28002–28005
Jimenez EC, Watkins M, Juszczak LJ, Cruz LJ, Olivera BM (2001) Contryphans from *Conus textile* venom ducts. Toxicon 39:803–808
Lewis RJ, et al. (2000) Novel ω-conotoxins from *Conus catus* discriminate among neuronal calcium channel subtypes. J Biol Chem 275:35335–35344
Li RA, Tomaselli GF (2004) Using the deadly μ-conotoxins as probes of voltage-gated sodium channels. Toxicon 44:117–122
Livett BG, Gayler KR, Khalil Z (2004) Drugs from the sea: conopeptides as potential therapeutics. Curr Med Chem 11:1715–1723
Mari F, Fields GB (2003) Conopeptides: unique pharmacological agents that challenge current peptide methodologies. Chim Oggi 21:43–48
McDougal OM, Poulter CD (2004) Three-dimensional structure of the mini-M conotoxin mr3a. Biochemistry 43:425–429
McGahren WJ, Morton GO, Kunstmann MP, Ellestad GA (1977) C-13 nuclear magnetic-resonance studies on a new antitubercular peptide antibiotic Ll-Bm547beta. J Org Chem 42:1282–1286
McIntosh JM, Jones RM (2001) Cone venom – from accidental stings to deliberate injection. Toxicon 39:1447–1451
McIntosh JM, et al. (1994) A nicotinic acetylcholine-receptor ligand of unique specificity, α-conotoxin-ImI. J Biol Chem 269:16733–16739
McIntosh JM, Olivera BM, Cruz LJ (1999) *Conus* peptides as probes for ion channels. Methods Enzymol 294:605–624
McIntosh JM, Dowell C, Watkins M, Garrett JE, Yoshikami D, Olivera BM (2002) α-Conotoxin GIC from *Conus geographus*, a novel peptide antagonist of nicotinic acetylcholine receptors. J Biol Chem 277:33610–33615
Mignogna G, Simmaco M, Kreil G, Barra D (1993) Antibacterial and hemolytic peptides containing D-alloisoleucine from the skin of *Bombina variegata*. EMBO J 12:4829–4832
Miljanich GP (2004) Ziconotide: neuronal calcium channel blocker for treating severe chronic pain. Curr Med Chem 11:3029–3040
Millard EL, Daly NL, Craik DJ (2004) Structure–activity relationships of α-conotoxins targeting neuronal nicotinic acetylcholine receptors. Eur J Biochem 271:2320–2326
Mitchell JBO, Smith J (2003) D-Amino acid residues in peptides and proteins. Proteins Struct Funct Genet 50:563–571
Mor A, et al. (1989) Isolation of dermenkephalin from amphibian skin, a high-affinity δ-selective opioid heptapeptide containing a D-amino-acid residue. FEBS Lett 255: 269–274
Morita H, Kayashita T, Uchida A, Takeya K, Itokawa H (1997) Cyclic peptides from higher plants 33. Delavayins A–C, three new cyclic peptides from *Stellaria delavayi*. J Nat Prod 60:212–215

Myers RA, Cruz LJ, Rivier JE, Olivera BM (1993) *Conus* peptides as chemical probes for receptors and ion channels. Chem Rev 93:1923–1936

Newcomb R, Miljanich G (2002) Neurotoxins of cone snail venoms. Handb Neurotoxicol 1:617–651

Nicke A, et al. (2003) Isolation, structure, and activity of GID, a novel α4/7-conotoxin with an extended N-terminal sequence. J Biol Chem 278:3137–3144

Nicke A, Wonnacott S, Lewis RJ (2004) α-Conotoxins as tools for the elucidation of structure and function of neuronal nicotinic acetylcholine receptor subtypes. Eur J Biochem 271:2305–2319

Olivera BM (2002) *Conus* venom peptides: reflections from the biology of clades and species. Annu Rev Ecol Sys 33:25–47

Olivera BM, et al. (1990) Diversity of *Conus* neuropeptides. Science 249:257–263

Pallaghy PK, Melnikova AP, Jimenez EC, Olivera BM, Norton RS (1999) Solution structure of contryphan-R, a naturally occurring disulfide-bridged octapeptide containing D-tryptophan: comparison with protein loops. Biochemistry 38:11553–11559

Papov VV, Diamond TV, Biemann K, Waite JH (1995) Hydroxyarginine-containing polyphenolic proteins in the adhesive plaques of the marine mussel *Mytilus edulis*. J Biol Chem 270:20183–20192

Perret S, et al. (2001) Unhydroxylated triple helical collagen I produced in transgenic plants provides new clues on the role of hydroxyproline in collagen folding and fibril formation. J Biol Chem 276:43693–43698

Petsko GA (1992) On the other hand. Science 256:1403–1404

Pisarewicz K, Mora D, Pflueger FC, Fields GB, Mari F (2005) Polypeptide chains containing D-γ-hydroxyvaline. J Am Chem Soc 127:6207–6215

Pollard JK, Sondheimer E, Steward FC (1958) Hydroxyvaline in *Kalanchoe daigremontiana*. Nature 182:1356–1358

Sandall DW, et al. (2003) A novel α-conotoxin identified by gene sequencing is active in suppressing the vascular response to selective stimulation of sensory nerves in vivo. Biochemistry 42:6904–6911

Santos AD, McIntosh JM, Hillyard DR, Cruz LJ, Olivera BM (2004) The A-superfamily of conotoxins – structural and functional divergence. J Biol Chem 279:17596–17606

Sharpe IA, et al. (2001) Two new classes of conopeptides inhibit the α1-adrenoceptor and noradrenaline transporter. Nat Neurosci 4:902–907

Shoji J, Hinoo H (1975) Antibiotics from the genus *Bacillus*. II. Chemical characterization of new antibiotics, cerexins A and B. J Antibiot 28:60–63

Shon KJ, et al. (1997) A noncompetitive peptide inhibitor of the nicotinic acetylcholine receptor from *Conus purpurascens* venom. Biochemistry 36:9581–9587

Sidler W, Kumpf B, Suter F, Morisset W, Wehrmeyer W, Zuber H (1983) Structural studies on cryptomonad biliprotein subunits. Two different α-subunits in *Chroomonas* phycocyanin-645 and *Cryptomonas* phycoerythrin-545. Biol Chem Hoppe-Seyler 366:233–244

Smith MT, Cabot PJ, Ross FB, Robertson AD, Lewis RJ (2002) The novel N-type calcium channel blocker AM336 produces potent dose-dependent antinociception after intrathecal dosing in rats and inhibits substance P release in rat spinal cord slices. Pain 96:119–127

Soyez D, Toullec JY, Ollivaux C, Geraud G (2000) L to D amino acid isomerization in a peptide hormone is a late post-translational event occurring in specialized neurosecretory cells. J Biol Chem 275:37870–37875

Terui Y, Nishikawa J, Hinoo H, Kato T, Shoji J (1990) Structures of cepafungins I, II and III. J Antibiot 43:788–795

Walker CS, et al. (2001) On a potential global role for vitamin K-dependent γ-carboxylation in animal systems – evidence for a γ-glutamyl carboxylase in *Drosophila*. J Biol Chem 276:7769–7774

Watson AA, Fleet GWJ, Asano N, Molyneux RJ, Nash RJ (2001) Polyhydroxylated alkaloids: natural occurrence and therapeutic applications. Phytochemistry 56: 265–295

Zeng L, et al. (1999) Search for new compounds and biologically active substances from Chinese marine organisms. Pure Appl Chem 71:1147–1151

Zhao TY, Cao Y, Dai XD, Fan CX, Chen JS (2005) Purification, sequence and disulfide bonding pattern of a novel conotoxin BtIIIB. Acta Chim Sin 63:163–168

The Chemistry of Marine Pulmonate Gastropods

J. Darias, M. Cueto, A.R. Díaz-Marrero

Abstract. Secondary metabolites from pulmonate molluscs of the genera *Siphonaria*, *Onchidium*, and *Trimusculus* are described. *Siphonaria* and *Onchidium* biosynthesize mostly propionate-based metabolites whereas *Trimusculus* yields diterpene derivatives with a single type of labdane skeleton. The 42 regular polypropionates reported to date from *Siphonaria* are divided into two classes (class I, class II), based on their observed structural and stereochemical analogy. The strong resemblance between class I and cephalaspidean metabolites and between class II and onchidiid metabolites as well as the structural features of *Trimusculus*, in relation to the other pulmonates, encourage speculation about their biosynthetic and phylogenetic relationship. Class I metabolites could be suitable material to evidence that type I PKS modules are perhaps used iteratively in their biosynthesis.

5.1 Introduction

The gastropods form by far the largest and most diverse class of molluscs, comprising more than half of all mollusc species. The traditional division of gastropods into three subclasses, Prosobranchia, Opisthobranchia, and Pulmonata, was followed by several conflicting phylogenetic hypotheses and taxonomic classification systems (for a review of earlier work, see Bieler 1992). In an analysis of gastropod phylogeny based on a set of morphological and ultra-structural characters, Ponder and Lindberg (1997) demonstrated the probable monophyly of the gastropods and divided them into two major groups: subclass Eogastropoda, comprising the Patellogastropoda, and the remaining gastropods included in subclass Orthogastropoda that comprises the superorders Neritopsina, Vetigastropoda, Caenogastropoda, Heterobranchia, and the orders Opisthobranchia and Pulmonata. Patellogastropods and vetigastropods are all marine, while the neritopsines and caenogastropods are mostly marine, including a few freshwater and terrestrial groups. This newer classification, although currently recognized, will take some time to be widely applied.

Marine gastropods are the most diverse group of marine invertebrates. Within the gastropods, the subclass Pulmonata comprises marine groups, but the majority are freshwater or terrestrial. The marine members comprise six families in three orders, including air-breathing limpets (Siphonariidae, Trimusculidae), the mangrove onchidiid slugs

José Darias Jerez
Instituto de Productos Naturales y Agrobiología del CSIC, Avda Astrofísico F. Sánchez 3
Apdo 195, 38206 La Laguna, Tenerife, Spain

Progress in Molecular and Subcellular Biology
Subseries Marine Molecular Biotechnology
G. Cimino, M. Gavagnin (Eds.): Molluscs

(Onchidiidae), the estuarine ear shells (Ellobiidae), and mud snails (Amphibolidae). The pulmonates have no gills, but feature a vascularized mantle cavity which serves as a functional lung sac for air-breathing both above and under water, hence the name "pulmonates".

This chapter deals with the chemistry of marine pulmonates belonging to the families Siphonariidae, Onchidiidae (Order Systellomatophora), and Trimusculidae (Order Basommatophora).

5.2 Secondary Metabolites from *Siphonaria*

The Siphonariidae are regarded as the most primitive of pulmonates which probably have a marine ancestry and may represent an evolutionary link between marine and terrestrial gastropods (Hyman 1967; Purchon 1979). Nearly all siphonariids are unpalatable to predators since, when disturbed by a potential predator, they secrete a white mucus from lateral epidermal glands that contain polypropionate metabolites (Davies-Coleman and Garson 1998). The Siphonariidae are a diverse family, with over 60 recorded species; and their phylogeny is still in debate, even the number of genera being contentious. While only two genera, *Siphonaria* and *Williamia*, were previously recognized with two subgenera: *Liriola* and *Siphonaria*, other authors have since recognized four genera, *Siphonaria*, *Williamia*, *Kerguelenella*, and *Benhamina* (Hodgson 1999).

Of the marine pulmonates, the genus *Siphonaria* has been the most studied. The siphonariids are characterized by their ability to synthesize polypropionate metabolites with different types of skeletons, approximately one-third of which are acyclic while the remainder contain a 2-pyrone, 4-pyrone, or furanone ring. A defensive value of the polypropionates is not incompatible with the fact that *Siphonaria* spp. are physically protected by a shell (Faulkner 1992; Cimino and Ghiselin 1999). The polyketides isolated from *S. diemenensis* represented the first polypropionates obtained from the pulmonates since, until then in marine invertebrates, polypropionate metabolites had been isolated from sacoglossans (Ireland and Faulkner 1981).

Within the metabolites from *Siphonaria*, structural and stereochemical analogies are observed at specific positions. These analogies are used as criteria to assign the compounds from *Siphonaria* to two classes: class I comprises those metabolites that possess an identical configuration at all the comparable stereochemical centers, class II comprises those metabolites that possess an identical configuration at only a part of the comparable stereochemical centers of a complex pattern of functional groups. The compounds included in this class exhibit analogies with the model of

Celmer (1965), suggesting that the producer organism adopts a similar biosynthetic pattern.

5.2.1 Class I Siphonariid Polypropionates

Class I comprises approximately half of the polypropionate metabolites isolated from *Siphonaria* and includes essentially acyclic compounds, compounds with a 2-pyrone ring and those having a furanone ring (Fig. 5.1). Class I compounds have the following structural and stereochemical characteristics:

– The saturated linear alkyl chain is made up of at least three propionate units

– All the methyl groups of this portion of the linear chain have an *S* configuration

– The 2-pyrone ring and the furanone ring are characteristic of class I

The secondary metabolites comprised in class I have been isolated, as a whole, from seven species of the genus *Siphonaria*: *S. diemenensis*, *S. pectinata*, *S. grisea*, *S. virgulata*, *S. lessoni*, *S. capensis*, and *S. concina*, noting that no class II metabolite has been characterized from these species. Since many of these metabolites have been isolated from different species along the years, Table 5.1 summarizes class I siponariid polypropionates with attention to their species origin, locality, and bioactivity data when available.

The first polypropionates from *Siphonaria*, diemenensins A (**1**) and B (**2**), were isolated from *S. diemenensis* collected off the southeast coast of Australia. The identification of methyl (2*S*,4*S*)-2,4-dimethylheptanoate from the chemical degradation of **1** allowed its absolute configuration to be established (Hochlowski and Faulkner 1983).

Pectinatone (**3**) was isolated from the skin extract of *S. pectinata* (Florida). The absolute stereochemistry of **3** was initially determined by comparison of the value of the optical rotation of the methyl ester **4**, obtained by ozonolysis of **3** and subsequent methylation, with that described in the literature for methyl (2*S*,4*R*,6*S*)-2,4,6-trimethylnonanoate (Biskupiak and Ireland 1983). However, the true configuration of C-11 was resolved by X-ray diffraction analysis of **3** obtained from both *S. virgulata* (Garson et al. 1990) and *S. grisea* (Norte et al. 1990). The total synthesis of (+)-pectinatone confirmed the structure indicated in **3** (Birkbeck and Enders 1998).

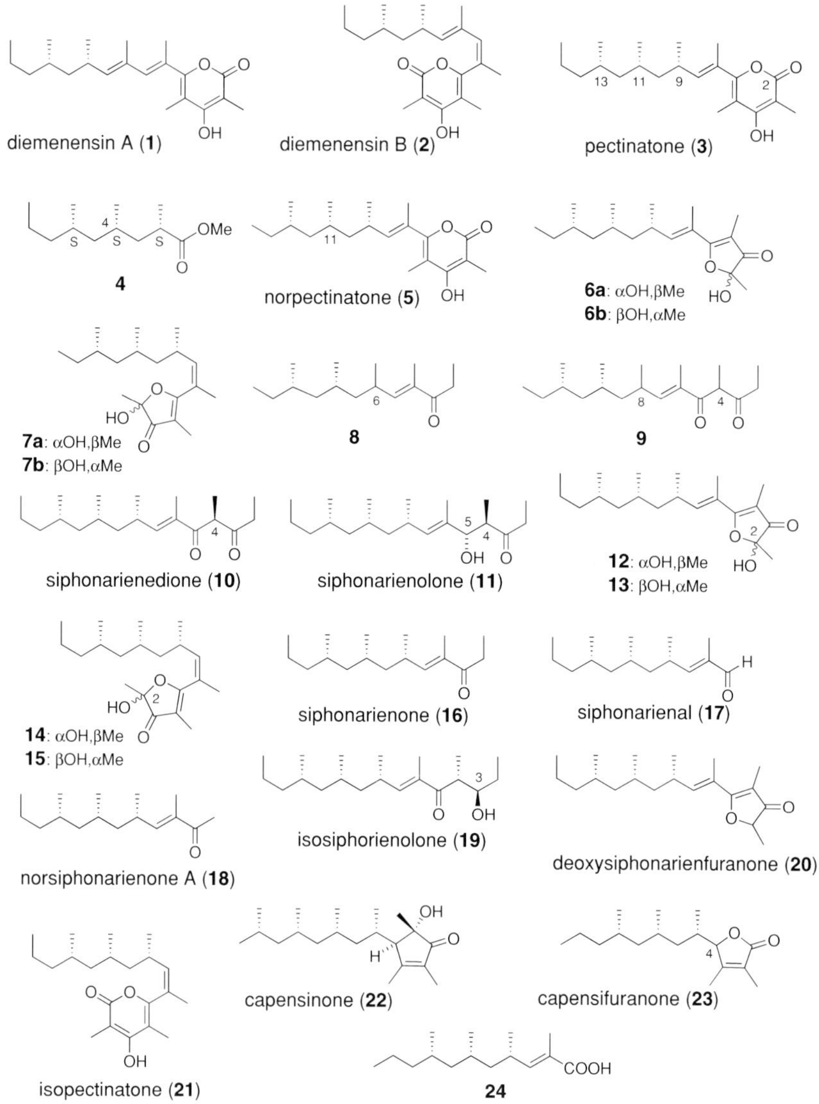

Fig. 5.1. Class I siphonariid polypropionates

Table 5.1. Class I siphonariid polypropionates

species	location	compounds	bioactivity	references
S. diemenensis	SE Australia	diemenensin A (**1**)	antimicrobial	Hochlowski and Faulkner (1983)
		diemenensin B (**2**)	cell division inhibition	
		6a, **6b**		Capon and Faulkner (1984)
S. pectinata	Florida	pectinatone (**3**)	antimicrobial	Biskupiak and Ireland (1983)
	Cádiz, Spain	**3**	cytotoxic	Paul et al. (1997)
		siphonarienedione (**10**)		
		siphonarienolone (**11**)	cytotoxic	
		12, **13**, **14**, **15**	cytotoxic	
		siphonarienone (**16**)		
		norsiphonarienone A (**18**)		
		isosiphorienolone (**19**)		
		deoxysiphonarienfuranone (**20**)		
		isopectinatone (**21**)		
S. lessoni	Chile	norpectinatone (**5**)		Capon and Faulkner (1984)
		6a, **6b**, **8**, **9**		Rovirosa et al. (1991)
S. grisea	Canary Islands	**3**, **10**, **11**, **12**, **13**, **14**, **15**, **16**	antimicrobial	Norte et al. (1988, 1990)
	Dakar, Senegal	**3**, **10**, **11**, **12**, **13**, **16**		Norte et al. (1994)
		siphonarienal (**17**)		
S. capensis	South Africa	**12**, **13**, **14**, **15**		Beukes and Davies-Coleman (1999)
		capensinone (**22**)		
		capensifuranone (**23**)		
		24		
S. virgulata		**3**, **5**		Garson et al. (1990)
S. concina		**3**		Davies-Coleman and Garson (1998)

Norpectinatone (5) was isolated from *S. lessoni* (Chile). This same extract afforded the 1:1 mixture of furanones **6a** and **6b** (Capon and Faulkner 1984). The 11*R* stereochemistry, originally proposed by comparison of the optical rotation of the methyl ester, obtained by degradation of compound 5, with that of four isomers of 2,4,6-trimethylnonanoate, was questioned by Oppolzer et al. (1986) since the spectroscopical data of norpectinatone, obtained by means of enantioselective synthesis, did not coincide with those of the natural product. Subsequently, X-ray diffraction analysis of pectinatone indicated that the configuration at C-11 of norpectinatone should be revised to 11*S* as indicated in 5 (Garson et al. 1990). The absolute stereochemistry initially proposed for the linear chain of the furanones **6a** and **6b** was revised to all *S* stereoisomers (Rovirosa et al. 1991). Moreover, the authors proposed that the mixture of furanones is formed by the epimers at C-2 and not by the geometric isomers *E* and *Z*, although it is possible that, on standing, **6a** and **6b** isomerize to **7a** and **7b** as occurred with the mixture of compounds **12** and **13** (Norte et al. 1990).

Compounds **8** and **9** were isolated from *S. lessoni* (Chile). The configuration of C-6 in **8** and those of C-4 and C-8 in **9** remain undetermined (Rovirosa et al. 1991). However, the stereochemical regularity observed in the methyl groups of the polypropionic chain of the group of metabolites comprising class I allows the prediction of an *S* configuration for Me-6 and Me-8 in **8** and **9**, respectively.

The acyclic polypropionates siphonarienedione (**10**) and siphonnarienolone (**11**) were isolated from *S. grisea* (Canary Islands). Oxidative degradation of **10** and **11** gave rise in both cases to the same 2,4,6-trimethylnonanoic acid, whose optical rotations indicated that they were in the same enantiomeric series as (2*S*,4*S*,6*S*)-trimethylnonanoic acid. However, the stereochemistry at C-4 and C-5 of **11**, assigned on the basis of the comparison of their spectral data with those of compound **29**, was incorrect (Norte et al. 1988), as was later verified by total synthesis of both natural compounds, establishing the correct configurations as those represented in **10** and **11** (Calter and Liao 2002; Magnin-Lachaux et al. 2004).

From another study of *S. grisea*, the following metabolites were obtained: siphonarienfuranones **12** and **13** as an inseparable mixture of epimers at C-2, the epimeric mixture of **14** and **15**, as well as siphonarienone (**16**). The absolute stereochemistry of the side-chain of **12** and **13** was established by chemical degradation (Norte et al. 1990). The enantioselective synthesis of (+)siphonarinenone (**16**) has been reported (Abiko and Masamune 1996).

S. grisea (Dakar) yielded siphonarienal (**17**). The absolute configuration was established by means of enantioselective synthesis (Norte et al. 1994).

The novel polypropionates **18–21** were isolated from *S. pectinata* (Cádiz, Spain; Paul et al. 1997). The absolute stereochemistry of **18** was proposed on the basis of biogenetic considerations. The configuration at

C-3 of **19** was established by comparison with the spectroscopical data of siphonarienolone (**11**).

S. capensis, endemic to South Africa, afforded the novel metabolites capensinone (**22**), capensifuranone (**23**), and (2*E*,4*S*,6*S*,8*S*)-2,4,6,8-tetramethyl-2-undecanoic acid **24**; (Beukes and Davies-Coleman 1999). The configuration of the side-chain chiral centers of **22** and **23** was assumed to be *S* on the basis of biogenetic considerations, the configuration of C-4 in **23** remaining undetermined. The oxidative degradation of **24** to (2*S*,4*S*,6*S*)-2,4,6 trimethylnonanoic acid allowed its absolute configuration to be established.

5.2.2
Structural Analogy Between Class I and Cephalaspidean Polypropionates

The siphonariid metabolites from this class possess an alkenyl chain that often contains an α-pyrone or a furanone ring. Remarkably, within the group, they share the same absolute configuration at all comparable stereocenters. It is interesting to observe that they are similar to propionates from *Bulla striata*, *B. gouldiana*, *B. speciosa*, *Aglaja depicta*, and *Navanax inermis* cephalaspidean molluscs (Cimino and Sodano 1993). Examples are shown in Fig. 5.2 in which representative linear alkenyl chain, alkenyl-furanone, and alkenyl-2-pyrone metabolites are compared. The comparison of class I siphonariids and cephalaspidean metabolites suggests that an olefin reductive enzymatic process occurred at certain positions along the chain to give all *S* stereocenters. However, this reductive process does not occur at the olefinic terminus of the linear chain (Fig. 5.2, dashed boxes). Aglajne-1 and aglajne-3 incorporate [1-^{14}C]-propionate when the animals are supplemented with the sodium salt of the precursor, evidencing that these compounds derive de novo from propionate (Fontana et al. 2004).

Polypropionates are biosynthesized by polyfunctional type-I polyketide synthases (PKSs), also called modular PKSs. These enzyme complexes use a wide range of organic acids as starter units; and the extenders are generally malonyl and methylmalonyl units. Polyketides are assembled by sequential decarboxylative condensation of short carboxylic acids and the PKSs contain a module for every cycle of chain extension; and this correlation is termed colinearity. In modular PKSs, typified by 6-deoxyerythronolide synthesis (Cortés et al. 1990; Donadio et al. 1991), each module contains the requisite enzymatic domains: ketosynthase (KS) catalyzing the chain elongation, acyl transferase (AT) selecting the extender and acyl carrier protein (ACP) mediating the correct transfer of the growing polyketide chain. These domains are covalently linked in the order: KS – AT – reduction domain loop – ACP. The three main reductive domain loops found in modules are: ketoreductase (KR), dehydratase (DH)-KR or DH-enoyl reductase (ER)-KR domains which determine, by a processive mechanism, the extent of β-ketone group processing in each cycle. Domain exchange experiments have shown that KR domains are responsible for determining the final stereochemistry at chiral centers derived from reduction of β-ketone to alcohols, but it is still unclear exactly how modular PKSs control alcohol stereochemistry in the nascent oligoketide chain. Predictive methods suggest that the DH domains of modular PKSs normally act on (3*R*)-hydroxyacyl chains to give *trans*-double bonds. DH and ER domains must determine alkyl stereochemistry when incorporation of a branched extender is followed by complete reduction of the β-ketone (Staunton and Weissman 2001).

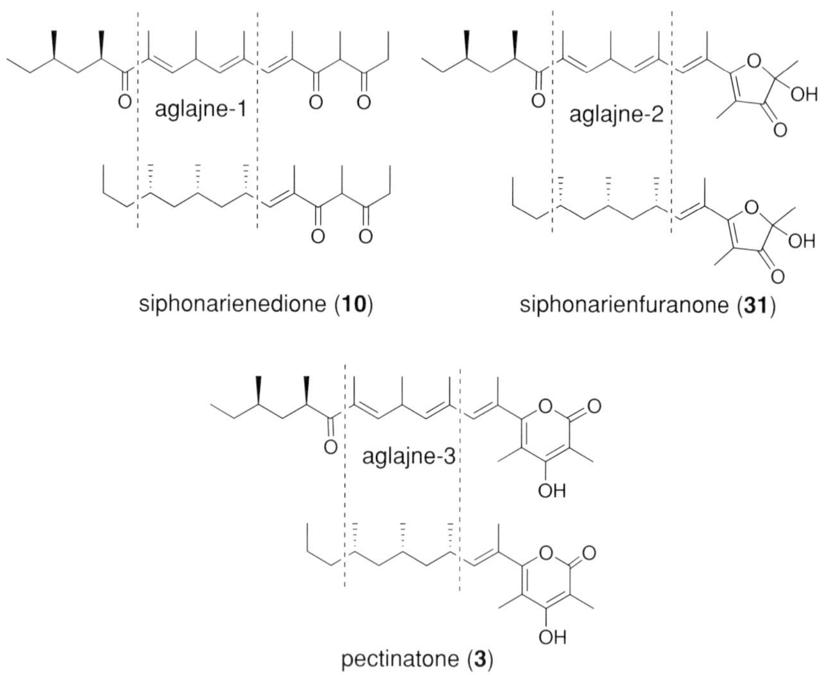

Fig. 5.2. Structural similarities between class I and cephalaspidean polypropionates

The striking properties of class I metabolites encourage speculation about whether the putative reductive loop leading to complete reduction of the β-ketone of the extender methylmalonate to methylene is repeated over and over again to elongate the chain. This would account for the secondary methyl groups at alternating positions and raises the question of whether, as occurs with bacterial type I, the PKSs include three modules, each containing the three core domains that are essential for formation of each of the C–C bonds, or whether there is a type I system in which iterative rounds of chain extension occur as a programmed event by using one module three times. Indeed, it has been proposed that the biosynthesis of borrelidin involves an iterative use of module 5 (BorA5) of its biosynthetic gene cluster (Olano et al. 2004). Because there is a 1,3, 5-trimethylhexyl moiety identity between borrelidin and class I metabolites (Fig. 5.3), this latter organization seems likely, therefore, for class I siphonariid metabolite biosynthesis.

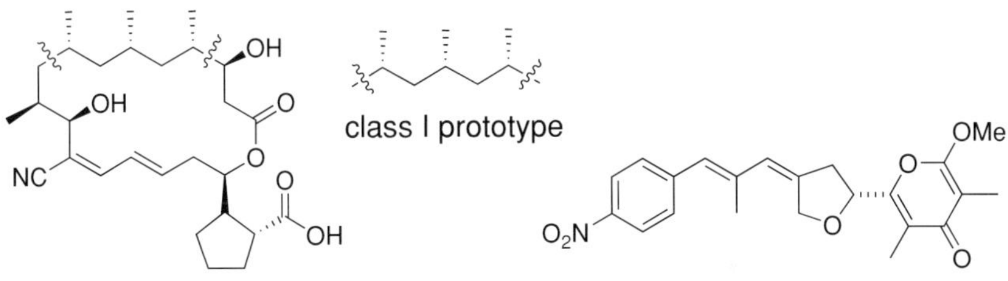

borrelidin aureothin

Fig. 5.3. Conserved configuration in the respective 1,3,5-trimethylhexyl moiety of Class I and borrelidin. Structure of aureothin

In the genesis of the cephalaspidean aglajne-1 to aglajne-3, the extent of the reductive carbonyl unit process in a comparable chain-extension cycle has been truncated twice by loss of the enoyl reductase domain, thus generating a double bond in the chain-extension product (Fig. 5.2). Although the generated couple of double bonds are located at nonalternating positions (one cycle is fully reduced), the possibility still exists of an iterative type I PKS by using a module twice, similar to module 1 (AurA1) of the biosynthetic gene cluster responsible for aureothin biosynthesis (He and Hertweck 2003), by considering the two alternate linear double bonds. Class I metabolites could thus be a source of suitable material to evidence that type I PKS modules are perhaps used iteratively in their biosynthesis.

It is unclear exactly how modular PKSs control methyl stereochemistry in the growing chain and the enoyl reductase domains occur relatively infrequently in PKS modules, being the least studied of all of the constituent domains. Because of this, the extended network analogy of the respective pairs: aglajne-1/siphonarienedione, aglajne-2/siponarienfuranone and aglajne-3/pectinatone (Fig. 5.2) and the fact that class I and cephalaspidean metabolites provide examples of modification of the level of reduction of the growing chain, it seemed that the tandem compounds could be good models for the cloning of the biosynthetic gene cluster of the producers. This might yield insights into our understanding of the molecular mechanistic basis of the enzymatic stereocontrol for determining the final stereochemistry at chiral centers derived from the reduction of β-ketone in the growing chain of the polypropionate biosynthesis. Also, the possibility to virtually convert syphonariids into producers of metabolites characteristic of cephalaspidean (and vice versa) by entire module exchange experiments in their respective PKSs, appears to constitute an exciting challenge. The core analogy of those pairs of compounds may preserve intact the acyl carrier protein–ketosynthase (ACP–KS) bi-domain that spans the junction between successive modules, conferring an advantage on any eventual goal undertaken in this sense (Gokhale et al. 1999; Ranganathan et al. 1999).

Although the aforesaid discussion on the class I/cephalaspidean polypropionates is speculative and there is no evidence to prove this biosynthesis, the rationalization of the observed structural and stereochemical analogies between these compounds may be useful in future related research.

5.2.3
Class II Siphonariid Polypropionates

Whilst class I secondary metabolites are structurally mundane in that almost the only real variation is in the length of the alkyl chain, a feature of class II is that it yields a profuse polyoxygenated network that frequently cyclizes to spiroacetal and/or γ-pyrone rings; and these functionalities are not observed in class I compounds. The striking structural and stereochemical correlation at comparable stereocenters between members of macrolide classes of actinomycete antibiotics and siphonariid metabolites of class II suggests that these compounds may share a common genetic origin (Garson et al. 1994b). The class II polypropionates (Fig. 5.4) have been isolated from the following ten species: *S. denticulata*, *S. australis*, *S. zelandica*, *S. normalis*, *S. lacinosa*, *S. baconi*, *S. atra*, *S. maura*, *S. funiculata* and *S. serrata*. None of the following described metabolites of this class has been found in class I. Table 5.2 summarizes class II siphonariid polypropionates with attention to their species origin, locality and bioactivity data when available.

The epimers denticulatins A (**25**) and B (**26**) were isolated from *S. denticulata* (Australia). The structure and relative stereochemistry of **26** was established by X-ray diffraction analysis and its absolute configuration was deduced on the basis that the levorotatory enantiomer **27**, obtained by degradation of both compounds, has an *R* configuration at C-4 (Hochlowski et al. 1983). A number of stereocontrolled syntheses of denticulatin A and denticulatin B have been reported; and pure samples of synthetic denticulatins A and B were found to interconvert on silica gel, indicating that the natural products from *S. denticulata* may actually be only a single compound which isomerizes at C-10 on chromatographic isolation (Ziegler and Becker 1990; Andersen et al. 1991a,b; Paterson and Perkins 1992, 1996; Oppolzer et al. 1995; De Brabander and Oppolzer 1997). Biosynthetic studies show that the denticulatins originate in the condensation of propionate units and not in the methylation of a standard polyacetate chain (Manker et al. 1988).

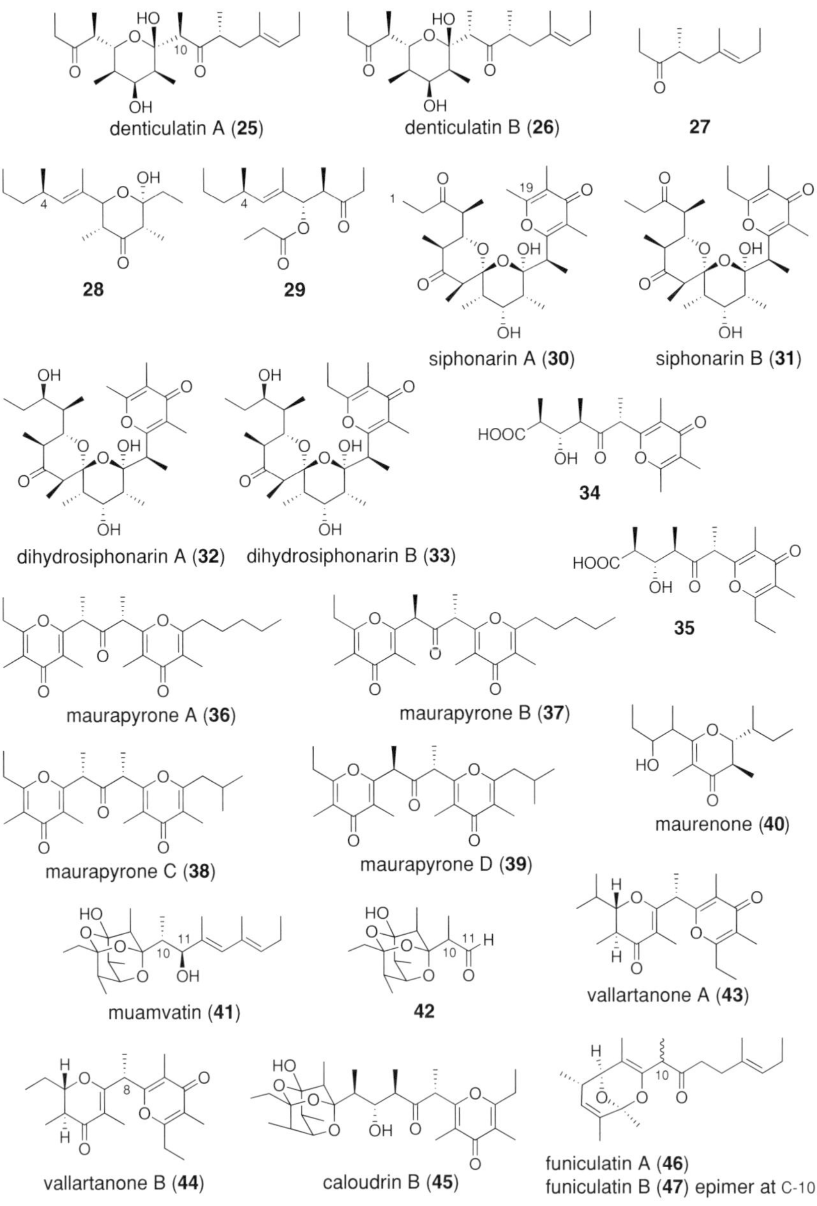

Fig. 5.4. Class II siphonariid polypropionates

Table 5.2. Class II siphonariid polypropionates

species	location	compounds	bioactivity	references
S. denticulata	New South Wales (Australia)	denticulatin A (**25**) denticulatin B (**26**)	ichthyotoxicity	Hochlowski et al. (1983)
S. australis	Auckland (New Zealand)	**28**, **29**		Hochlowski and Faulkner (1984)
S. zelandica	Sydney (Australia)	siphonarin A (**30**) siphonarin B (**31**)		Hochlowski et al. (1984)
	Queensland (Australia)	caloudrin B (**45**)		Blanchfield et al. (1994)
S. atra	Townsville (Australia)	**30**, **31**		Hochlowski et al. (1984)
S. baconi	Melbourne (Australia)	**30** baconipyrone A (**48**) baconipyrone B (**49**) baconipyrone C (**50**) baconipyrone D (**51**)		Manker et al. (1989)
S. normalis	Oahu (Hawaii)	dihydrosiphonarin A (**32**) dihydrosiphonarin B (**33**)		Hochlowski et al. (1984)
	Maumatu (Fiji)	maumvatin (**41**)	non-antimicrobial	Roll et al. (1986)
S. lacinosa	Townsville (Australia)	**32**, **33**		Hochlowski et al. (1984)
S. maura	Jaco Beach (Costa Rica)	maurapyrone A (**36**) maurapyrone B (**37**) maurapyrone C (**38**) maurapyrone D (**39**) maurenone (**40**)	antibacterial	Manker et al. (1986)
	Puerto Vallarta (Mexico)	vallartone A (**43**) vallartone B (**44**)	larval settlement inducer antifeeding	Manker and Faulkner (1989)
S. funiculata	Queensland (Australia)	funiculatin A (**46**) funiculatin B (**47**)		Blanchfield et al. (1994)
S. serrata	South Africa	siserrone A (**52**)		Brecknell et al. (2000)

Compounds **28** and **29** were isolated from *S. australis* (New Zealand; Hochlowski and Faulkner 1984). The synthesis of the linear compound allowed the establishment of its relative stereochemistry; and the absolute configuration, represented in **29**, was subsequently determined by circular dichroism. Since **29** can be derived from **28** by a retro-Claisen condensation process, both compounds possess the same stereochemistry and absolute configuration 4*R*, 7*S*, 8*R* (Sundram and Albizati 1992).

Siphonarins A (**30**) and B (**31**) were isolated from a mixed extract (~1:4) of *S. zelandica* and *S. denticulata* collected in Australia; and it was later shown that both compounds were present in the extract of *S. zelandica* (Hochlowski et al. 1984). The relative stereochemistry of **30** was established by X-ray diffraction analysis.

From *S. normalis* (Hawaii) and *S. lacinosa* (Australia) the dihydrosiphonarins A (**32**) and B (**33**) were obtained, together with the degradation products **34** and **35** (Hochlowski et al. 1984).

The absolute configuration of siphonarin A (**30**) was established based on an X-ray study of its *p*-bromophenyl boronate derivative. The methyls of the tetrahydropyrone ring of **30** have the same configuration as those in the tetrahydropyrone ring of **25** and **26** (Garson et al. 1994a,b). In contrast, the synthesis of the enantiomer of **35** shows that it is opposite to that obtained from the degradation of **33** and confirms the absolute stereochemistry of the siphonarins (Paterson and Franklin 1994). Biosynthetic experiments indicate a preference for an acetate chain starter unit in the biosynthesis of **30** and define the direction of chain assembly as from C-19 to C-1, demonstrating the presence of a functioning methylmalonyl-CoA mutase in *S. zelandica* (Garson et al. 1994a). The total synthesis of **31** and **33** has been achieved (Paterson et al. 2002).

Two pairs of racemic diastereoisomers maurapyrones A–D (**36–39**), together with maurenone (**40**), were isolated from *S. maura* (Costa Rica). The relative stereochemistry of maurapyrone A was established by X-ray diffraction analysis (Manker et al. 1986).

The triacetal muamvatin (**41**) containing an unusual 2,4,6-trioxa-adamantane ring system was isolated from *S. normalis* (Fiji). Its side-chain stereochemistry could not be fully determined (Roll et al. 1986). Later, a stereocontrolled synthesis of both epimers at C-10 of **42** and the comparison of their optical rotation with that of the corresponding aldehyde obtained by degradation of **41** allowed an *R* configuration to be established at C-10 (Hoffmann and Dahmann 1993; Dahmann and Hoffmann 1994). On the basis of Mosher analysis of the full synthetic C-11 muamvatin epimer, the absolute stereochemistry represented in **41** was established. During the synthesis of **41**, it was observed that the trioxaadamantane system is produced in silica gel by rearrangement, suggesting that muamvatin may be an artifact produced during purification (Paterson and Perkins 1993).

Vallartanones A (**43**) and B (**44**) were isolated from *S. maura* (Mexico). The absolute configuration of **43** was established by circular dichroism (Manker and Faulkner 1989). However, the total synthesis of **44** and circular dichroism studies indicate that the configuration at C-8 should be revised to *S*, suggesting the same configuration for **43** (Arimoto et al. 1996a,b).

A new collection of *S. zelandica* (Australia) yielded caloudrin B (**45**), while *S. funiculata* afforded funiculatins A (**46**) and B (**47**), epimeric at C-10 (Blanchfield et al. 1994). The relative stereochemistry of the side-chain and the absolute stereochemistry were inferred from biosynthetic comparison with the above known polypropionates and by correlation of funiculatin A with denticulatin A (**25**). The stereochemistry at C-10 of funiculatin A could not be unambiguously determined.

5.2.4 Class II Polypropionates with a Noncontiguous Propionate Skeleton

The polypropionates listed below possess a skeleton whose propionate units are noncontiguous, as could be expected from regular polyketide biosynthesis (Fig. 5.5). Interestingly, naturally occurring metabolites belonging to this class, membrenones, have also been found in Notaspidea (Opisthobranchia), a related but taxonomically distant taxa (Ciavatta et al. 1993).

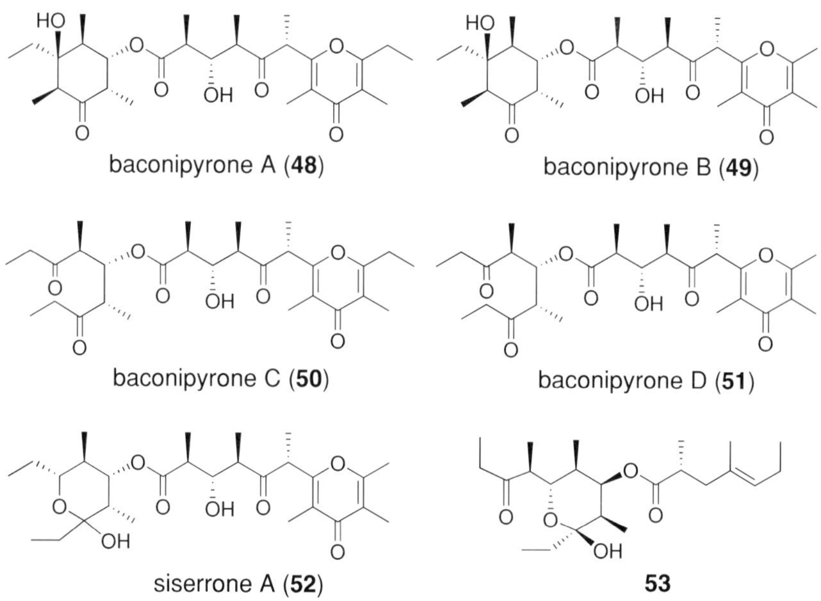

Fig. 5.5. Class II polypropionates with a noncontiguous propionate skeleton

Baconipyrones A–D (**48–51**) have been isolated from *S. baconi* (Australia). The structure of **49** was determined by X-ray diffraction studies. These compounds are presumed to be artifacts that derive from naturally occurring siphonarins (Manker et al. 1989). The enantioselective synthesis of **50** allowed the absolute configuration of the baconipyrones to be established and showed it to be in accord with that determined for the siphonarins (Paterson et al. 2000).

Specimens of *S. serrata* (South Africa) afforded siserrone A (**52**). The fact that the presence of baconipyrones was not detected in a sample of *S. baconi* from Sorrento (Australia) and that a second collection of *S. serrata* only gave **52** and no siphonarins leads to the conclusion that the siphonarins may be the precursors of these metabolites (Brecknell et al. 2000). However, it has been shown that base-catalyzed rearrangement of denticulatin A (**25**) yields the polypropionate ester **53** and funiculatin A (**46**). Those findings, together with the fact that denticulatins A and B undergo facile interconversion under mild conditions on silica gel (Paterson and Perkins 1992), appear to point to a non-natural origin for the polypropionate esters of siphonariids (Brecknell et al. 2000).

5.2.5 Structural Analogy Between Class II and Bacterial Metabolites

In 1965, Celmer noted that there are strong position-specific structural analogies between families of macrolides of various sizes. An intriguing feature, which emerges when compounds **25**, **30**, **36**, **41**, and **45** are compared with one another, is that they all share a common structural and stereochemical tetrapropionate unit, exemplified in the dashed box of the siphonarin A precursor, as shown in Fig. 5.6 (Garson et al. 1994b). This block is also present in the Cane–Celmer–Westley PAPA model (**56**) for polyethers, demonstrating that they all share a common fragment in their PKS products (Cane et al. 1983). The occurrence of common structural motifs in the bacterial macrolide, polyether antibiotics and siphonariid polypropionates first suggested they share a common biosynthetic origin, manifested in stereochemical control at the alkyl branching centers.

5.2.6 Siphonariid Nonpropionate-Derived Metabolites

A study of the composition of the fatty acids of *S. denticulata* (Queensland, Australia) allowed, in addition to the characterization of a high number of common fatty acids, the identification of two new acids (**54**, **55**; Fig. 5.7). Their structures were confirmed by total synthesis (Carballeira et al. 2001).

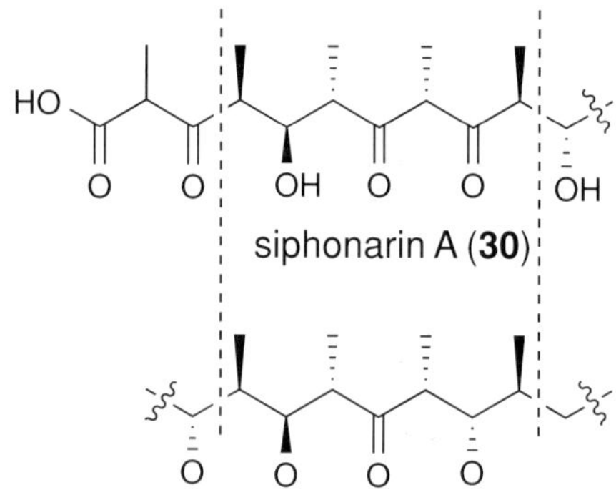

Fig. 5.6. Comparison of siphonarin open chain and PAPA model core

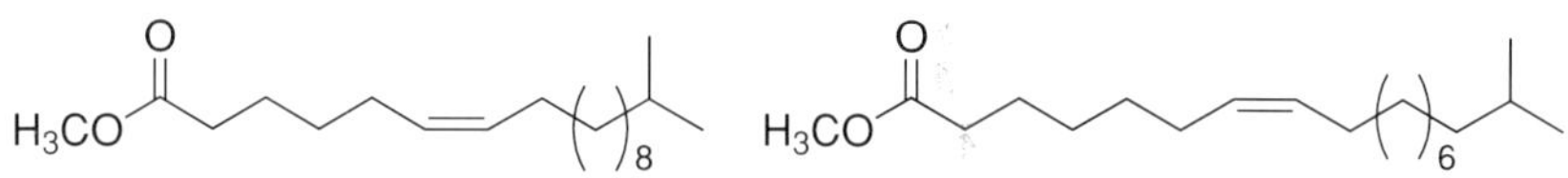

Fig. 5.7. Fatty acids from *Siphonaria*

5.3 Secondary Metabolites from *Onchidium*

The onchidioideans are shell-less marine molluscs. They are usually oval in shape with a dorsally arched notum bearing warts and papillae; and there are glands which apparently secrete noxious fluids at the sides of the body. They feed on the organic film of diatoms, other algae and bacteria which coat rocks and mud on sheltered intertidal shores.

Six species of the genus *Onchidium* have been studied to date. The greater part of the isolated compounds are isomeric polypropionates whose skeleton contains a γ-bispyrone ring and a great abundance of chiral centers (Fig. 5.8).

The first metabolite isolated from species of the genus *Onchidium* was the sesquiterpene onchidal (57; Ireland and Faulkner 1978). The isolation of an isoprenic derivative from this genus is unusual and noteworthy. This sesquiterpenic compound, presumed to play a defensive role, has also been isolated from *O. borealis* and *O. patelloides* (Manker and Faulkner 1987).

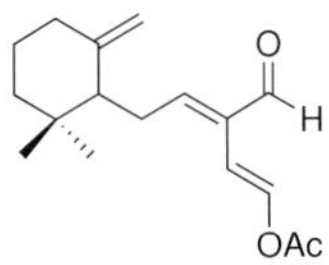

onchidal (**57**)

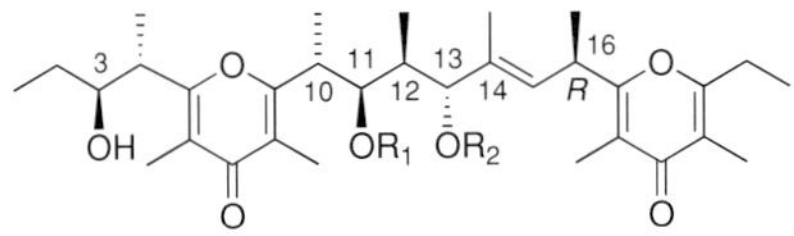

ilikonapyrone (**58**) : $R_1=R_2=H$
ilikonapyrone (**59**) : $R_1=R_2=COEt$
ilikonapyrone (**60**) : $R_1=R_2=MeCO_2$

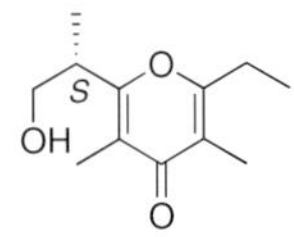

ilikonapyrone alcohol (**61**)

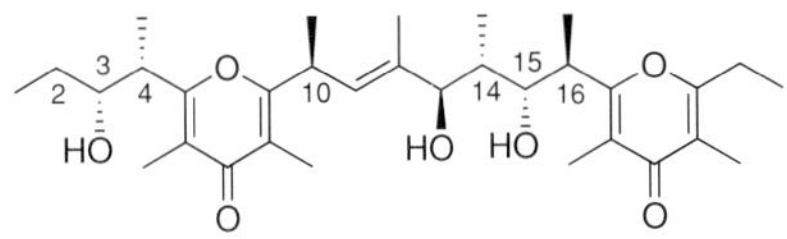

peroniatriol I (**62**)

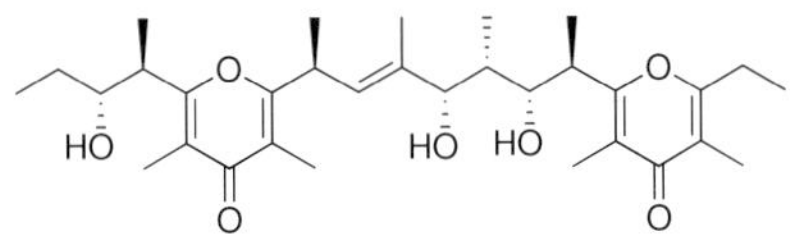

peroniatriol II (**63**)

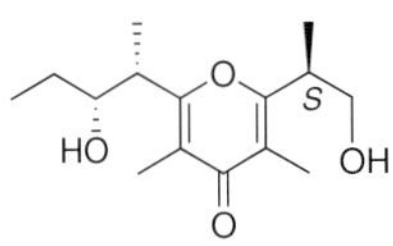

64

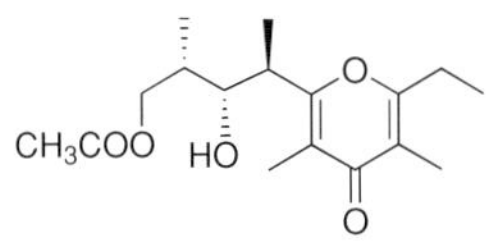

65

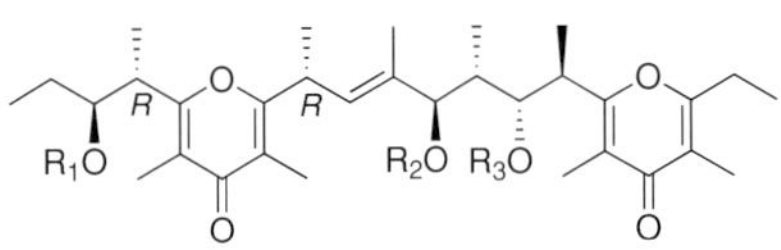

onchitriol I (**66**): $R_1=R_2=R_3=H$
onchitriol I A (**67**): $R_1=H$, $R_2=R_3=Ac$
onchitriol I B (**68**): $R_1=H$, $R_2=(CO)Et$, $R_3=Ac$
onchitriol I C (**69**): $R_1=H$, $R_2=Ac$, $R_3=(CO)Et$
onchitriol I D (**70**): $R_1=Ac$, $R_2=R_3=(CO)Et$

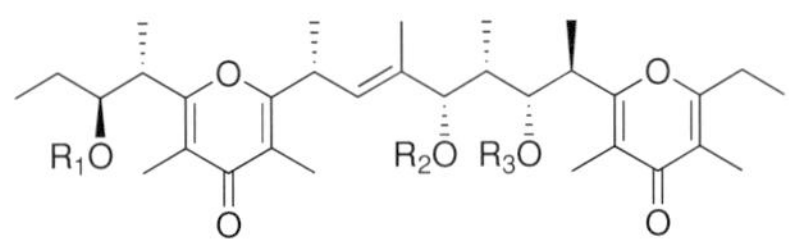

onchitriol II (**71**): $R_1=R_2=R_3=H$
onchitriol II A (**72**): $R_1=R_2=H$, $R_3=Ac$
onchitriol II B (**73**): $R_1=R_2=H$, $R_3=(CO)Et$
onchitriol II C (**74**): $R_1=R_2=R_3=Ac$
onchitriol II D (**75**): $R_1=R_3=Ac$, $R_2=H$

Fig. 5.8. Polypropionates from *Onchidium*

A mixture of esters based on the bispyrone alcohol ilikonapyrone (**58**) was isolated from *O. verruculatum* collected in Hawaii. Saponification of the mixture afforded the triol **58** whose structure, containing two γ-pyrones, was established by spectroscopical methods and chemical degradation. Its relative stereochemistry was established by X-ray analysis of the acetonide **60**. These compounds are considered defensive substances of Hawaiian *O. verruculatum*. The opposite signs of the optical rotations of fragment **61** obtained by: (a) oxidative degradation of **58** (Ireland et al. 1984) followed by reduction and (b) enantioselective synthesis (Arimoto et al. 1993) allowed the absolute configuration of **58** to be established.

Two cytotoxic metabolites, peroniatriols I (**62**) and II (**63**), have been isolated from the saponified extract of *Peronia peronii* (Guam; Biskupiak and Ireland 1985).

Comparison of the spectroscopical data and the optical activities of all the synthetic diastereoisomers of fragment **64** with those obtained from the natural products **62** and **63** allowed: (a) correction of the stereochemistry at

C-4 proposed for peroniatriol **62**, (b) confirmation of the 3*S* stereochemistry proposed for **62** and (c) proposal of the same configuration at C-10 for both compounds (Arimoto et al. 1990). The above-mentioned oxidative degradation of compounds **62** and **63** produced the same **65** fragment. Synthesis of the two enantiomeric forms of fragment **65** and comparison of the optical rotation of the synthetic fragments with those obtained from the natural products established the 14*R*, 15*R*, 16*R* configuration for both compounds.

In contrast, comparison of the spectral data and the optical rotation of the synthetic fragment **64** with an *S* configuration at the position equivalent to C-10 of peroniatriols I and II with the **64** fragment obtained from **62** confirmed the *S* configuration at C-10 for compounds **62** and **63** (Arimoto et al. 1993).

Eight new esters were isolated from an *Onchidium* sp. collected in New Caledonia. Four of them are esters of onchitriol I (**66**), onchitriols I A–D (**67**–**70**), and the remaining four are from onchitriol II (**71**), onchitriols II A–D (**72**–**75**). Onchitriols show in vitro cytotoxic activity (Rodríguez et al. 1992a,b). Although the absolute stereochemistries of these compounds were established by application of the Mosher–Trost method (Rodríguez et al. 1992a,b), a subsequent total synthesis of onchitriol II (**71**) and of some of its diastereoisomers confirmed the stereochemistry of **71** but suggested that the stereochemistry of onchitriol I (**66**) should be revised (Arimoto et al. 1994). A year later, the total synthesis of onchitriol I confirmed this supposition, allowing the *R* stereochemistry at C-4 and C-10 to be established as represented in **66** (Arimoto et al. 1995).

Finally, from the same New Caledonian collection of *Onchidium* spp, two depsipeptides, onchidin (**76**) and onchidin B (**77**; Fig. 5.9), have been isolated. Both compounds present a new β-amino acid unit. Their structure and absolute stereochemistry were determined by spectroscopic techniques, selective hydrolysis and chiral GC–MS (Rodríguez et al. 1994; Fernández et al. 1996).

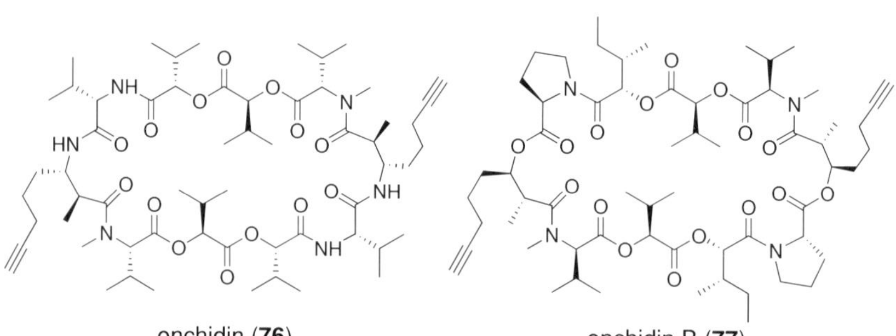

Fig. 5.9. Depsipeptides from *Onchidium*

5.3.1 Structural Analogy Between Onchidiid and Class II Siphonariid Polypropionates

When onchidiid polypropionates, drawn in ring-opened form, are compared with class II metabolites, a structural trend becomes apparent. PKS enzyme causes the polyketide chain to fold in specific fashions, leading to variations in ring formation patterns. Thus, it appears that a common biosynthetic origin for structurally related metabolites, for instance, siphonarin B (**31**) and the onchidiid ilikonapyrone (**58**), could be explained when the respective polyketide chain precursors folded in a different manner caused by slight differences at reductive and stereochemical levels in their network.

5.4 Secondary Metabolites from *Trimusculus*

Four species of the genus *Trimusculus* have been studied to date and all of them have a common feature: they produce diterpenes belonging to a unique class of labdane skeleton whose only structural variations are the degree and sites of oxidation. The most common functionalizations are acetoxy and isovaleroxy esters (Fig. 5.10).

The first species studied was *T. reticulatus* collected at San Nicolas Island, California. Diterpenes **78** and **79** were isolated from the extracts of both the whole animals and the mucus *T. reticulatus* produces to repel the starfish *Pisaster ocharaceus* (Rice 1985) and *Astromei* sp. (Manker and Faulkner 1996). These compounds have a labdane skeleton possessing four contiguous asymmetric carbons on ring B (Manker and Faulkner 1987). The structure of compound **78** was confirmed by total synthesis (Gao et al. 1996).

T. conica collected in New Zealand yielded the diterpene **80** and the steroid **81**. A careful investigation of the localization of the secondary metabolites of *T. reticulatus* from California and *T. conica* from New Zealand indicated that diterpenes **78**, **79**, and **80** were heavily concentrated in the mantle and foot of their respective organisms, while the viscera contained none of the compounds (Manker and Faulkner 1996).

The acetoxydiol **82** was isolated from *T. peruvianus*, collected in the intertidal area of Las Cruces (V Region, Chile; Rovirosa et al. 1992), while four new diterpenes **83–86** were isolated from another collection from El Tabo (V Region, Chile; San-Martín et al. 1996). Compounds **83–86** are the only labdane compounds isolated from *Trimusculus* that possess *Z* geometry on the double bond of the acyclic chain. Compounds **82–86** were evaluated for antimicrobial activity but only compound **84** exhibits modest activity.

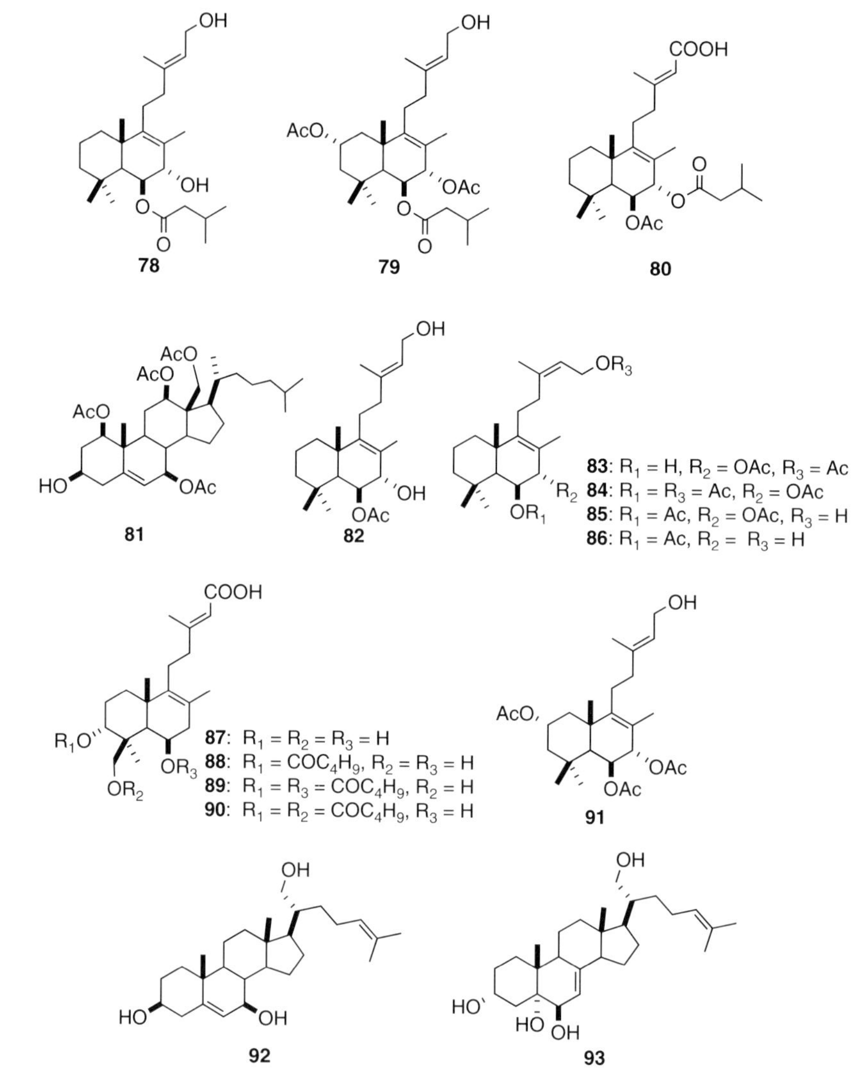

Fig. 5.10. Labdane and steroidal metabolites from *Trimusculus*

From another sample of *T. peruvianus* collected in Antofagasta (II Region, Chile) four new diterpenes were isolated, compounds **87–90**, presenting a new oxidation pattern. The absolute stereochemistry of **90** was established by application of the modified Mosher method. These compounds exhibit in vitro moderate cytotoxic activity (Díaz-Marrero et al. 2003a).

From *T. costatus*, endemic to South Africa, two diterpenes were obtained: **85** isolated previously from *T. peruvianus* and the new compound **91**, both exhibiting antifeeding activity against the predatory fish *Pomadasys commersonnii* and proving toxic to *Artemia salina* (Gray et al. 1998).

Finally, also from *T. peruvianus*, two new steroids, **92** and **93**, were isolated. The Δ^7-3α,5α,6β-triol nucleus of **93** is unprecedented in naturally occurring marine steroidal metabolites. Compounds **92** and **93** possess in vitro cytotoxic activity (Díaz-Marrero et al. 2003b).

Marine pulmonates of the genus *Trimusculus* are unusual not only in habitat and behavior but also in economizing the biosynthesis of diterpenes to a single type of labdane skeleton. Neither carbon rearrangement nor any functional group that induces significant structural modification has been observed in the diterpene metabolites they produce. The selectivity at the oxidation site along the skeleton resembles the mode of action of certain fungi on diterpene substrata, suggesting that some symbiotic microorganism interaction should not be excluded (Díaz-Marrero 2003a).

5.5 Conclusions

Marine pulmonates inhabit the intertidal zone and, with the adoption of fairly elaborate behavioral or chemical defenses against predators, have served as study organisms in chemical, physiological, ecological, and evolutionary research (Hisano et al. 1972; Maeda et al. 1998; Hodgson 1999; Katagiri et al. 2002; Grande et al. 2004a,b). Pulmonates share several synapomorphies with opisthobranchs (both conform to the clade Euthynera) but the monophyly of opisthobranchs with respect to pulmonates remains unclear, according to many phylogenetic hypotheses based on morphological characters (Ponder and Lindberg 1997; Dayrat and Tillier 2002). In most animal taxa, changes to the mtDNA gene order are rare, making these markers useful for higher-level phylogenetics, although one exception might be the gastropod molluscs, where the mtDNA gene order is extremely variable (Rokas and Holland 2000). However, Kurabayashi and Ueshima (2000) found that a unique gene arrangement and highly compact genome organization are shared between opisthobranch and pulmonate gastropods, strongly suggesting their close phylogenetic affinity. Recently, a new phylogenetic hypothesis for Euthyneura was proposed, based on the analysis of primary sequence and the phylogenetic utility of two rare genomic changes (Rokas and Holland 2000). Both sources of phylogenetic information clearly rejected the monophyly of pulmonates, supported so far by morphological evidence. In this analysis, the marine basommatophoran *Siphonaria* was placed within the opisthobranchs and shared with them the insertion of a glycine in the Cox 1 protein; and the marine systellommatophoran *Onchidella* was recovered at the base of the opisthobranch plus *Siphonaria* clade. Opisthobranchs, *Siphonaria* and *Onchidella* shared the relative position of the mitochondrial *trnP* gene between the mitochondrial *trnA* and *nad6* genes and warranted a more complete analysis of the phylogenetic relationships between opisthobranchs and pulmonates to test the monophyly of each group (Grande et al. 2004a,b).

The identification of structural and stereochemical similarity between siphonariid class I and cephalaspidean metabolites and also the shared polypropionates of siphonariid class II and onchidiid metabolites are in line with the above genetic studies. An interesting feature of the Trimusculidae, however, is that they produce no polypropionates but only isoprenoid metabolites. The labdane diterpenes from *Trimusculus* resemble the labdane-type diterpenoids isolated from opisthobranchs (Ciavatta et al. 1995), supporting their close pulmonate–opisthobranch relationship. Nevertheless, sperm ultrastructural studies of marine pulmonates (*T. costatus, T. reticulatus*) show characteristic heterobranch sperm features. Taxonomically useful differences in the shape and dimensions of the acrosome, nucleus and midpiece occur between the species. These results support the recent decision to transfer the Trimusculidae from the Siphonarioidea to a separate superfamily: Trimusculoidea (Hodgson and Healy 1998). The quite different chemistry of *Trimusculus* and *Siphonaria* appears to support this decision.

The similarity of the biosynthetic pathway within taxonomically distant but related taxa with those of bacterial metabolisms poses the question of whether polypropionates of class I and class II are microbial in origin; and it makes the pairs aglajne-1/siphonarienedione, aglajne-2/siponarienfuranone, and aglajne-3/pectinatone ideal models for the study of natural product symbiosis in molluscs, since those biosynthetic similarities could provide relatively rapid access to invertebrate symbiont genes putatively involved in the biosynthesis of polypropionates (Moffitt and Neilan 2003).

In conclusion, an understanding of how genetic information is correlated with chemical structures would prove useful in regard to secondary metabolites as taxonomic characters and evolutionary markers, as well as providing a valuable tool for designing modifications of the biosynthetic process through genetic engineering (McDaniel et al. 1999), with the goal of producing novel biomedically useful non-natural compounds.

References

Abiko A, Masamune S (1996) Synthesis of (+)-siphonarienone: asymmetric alkylation using a chiral benzopyrano-isoxazolidine auxiliary. Tetrahedron Lett 37:1081–1084

Andersen MW, Hildebrandt B, Hoffmann RW (1991a) Efficient stereoselective total synthesis of denticulatins A and B. Angew Chem Int Ed Engl 30:97–99

Andersen MW, Hildebrandt B, Dahmann G, Hoffmann RW (1991b) Stereoselective total synthesis of denticulatins. Chem Ber 124:2127–2139

Arimoto H, Nishiyama S, Yamamura S (1990) Synthetic studies on fully substituted γ-pyrone-containing natural products: synthesis of γ-pyrone derivatives obtained by decomposition of peroniatriols. Tetrahedron Lett 31:5491–5494

Arimoto H, Cheng J-F, Nishiyama S, Yamamura S (1993) Synthetic studies on fully substituted γ-pyrone-containing natural products: the absolute configurations of ilikonapyrone and peroniatriols I and II. Tetrahedron Lett 34:5781–5784

Arimoto H, Nishiyama S, Yamamura S (1994) Synthetic studies on fully substituted γ-pyrone-containing natural products: the first total synthesis of onchitriol II. Tetrahedron Lett 35:9581–9584

Arimoto H, Okumura Y, Nishiyama S, Yamamura S (1995) Synthetic studies on fully substituted γ-pyrone-containing natural products: total synthesis and structural revision of onchitriol I. Tetrahedron Lett 36:5357–5358

Arimoto H, Yokoyama R, Okumura Y (1996a) Synthesis and revised structure of vallartanone B. Tetrahedron Lett 37:4749–4750

Arimoto H, Yokoyama R, Nakamura K, Okumura Y, Uemura D (1996b) Vallartanone B: synthesis and related studies. Tetrahedron 52:13901–13908

Beukes DR, Davies-Coleman MT (1999) Novel polypropionates from the South African marine mollusc *Siphonaria capensis*. Tetrahedron 55:4051–4056

Bieler R (1992) Gastropod phylogeny and systematics. Annu Rev Ecol Syst 23:311–338

Birkbeck AA, Enders D (1998) The total synthesis of (+)-pectinatone: an iterative alkylation approach based on the SAMP-hydrazone method. Tetrahedron Lett 39:7823–7826

Biskupiak JE, Ireland CM (1983) Pectinatone, a new antibiotic from the mollusc *Siphonaria pectinata*. Tetrahedron Lett 24:3055–3058

Biskupiak JE, Ireland CM (1985) Cytotoxic metabolites from the mollusc *Peronia peronii*. Tetrahedron Lett 26:4307–4310

Blanchfield JT, Brecknell DJ, Brereton IM, Garson MJ, Jones DD (1994) Caloundrin B and funiculatin A: new polypropionates from siphonariid limpets. Aust J Chem 47:2255–2269

Brecknell DJ, Collett LA, Davies-Coleman MT, Garson MJ, Jones DD (2000) New non-contiguous polypropionates from marine molluscs: a comment on their product status. Tetrahedron 56:2497–2502

Calter MA, Liao W (2002) First total synthesis of a natural product containing a chiral, β-diketone: synthesis and stereochemical reassignment of siphonarienedione and siphonarienolone. J Am Chem Soc 124:13127–13129

Cane DE, Celmer WD, Westley JW (1983) Unified stereochemical model of polyether antibiotic structure and biogenesis. J Am Chem Soc 105:3594–3600

Capon RJ, Faulkner DJ (1984) Metabolites of the pulmonate *Siphonaria lessoni*. J Org Chem 49:2506–2508

Carballeira NM, Cruz H, Hill CA, De Voss JJ, Garson M (2001) Identification and total synthesis of novel fatty acids from siphonarid limpet *Siphonaria denticulata*. J Nat Prod 64:1426–1429

Celmer WD (1965) Macrolide stereochemistry. III. A configurational model for macrolide antibiotics. J Am Chem Soc 87:1801–1802

Ciavatta ML, Trivellone E, Villani G, Cimino G (1993) Membrenones: new polypropionates from the skin of the Mediterranean mollusc *Pleurobranchus membranaceus*. Tetrahedron Lett 34:6791–6794

Ciavatta ML, Villani G, Trivellone E, Cimino G (1995) Two new labdane aldehydes from the skin of the nostapidean *Pleurobranchaea meckelii*. Tetrahedron Lett 36:8673–8676

Cimino G, Ghiselin MT (1999) Chemical defense and evolutionary trends in biosynthetic capacity among dorid nudibranchs (Mollusca: Gastropoda: Opisthobranchia). Chemoecology 9:187–207

Cimino G, Sodano G (1993) Biosynthesis of secondary metabolites in marine molluscs. In: Scheuer PJ (ed) Topics in current chemistry, vol 167. Springer, Berlin Heidelberg New York, pp. 7–115

Cortés J, Haydock SF, Roberts GA, Bevitt DJ, Leadlay PF (1990) An unusually large multifunctional polypeptide in the erythromycin-producing polyketide synthase of *Saccacharopolyspora erythraea*. Nature 348:176–178

Dahmann G, Hoffmann RW (1994) The absolute configuration of muamvatin. Liebigs Ann Chem 837–845

Davies-Coleman MT, Garson MJ (1998) Marine polypropionates. Nat Prod Rep 15:477–493

Dayrat B, Tillier, S (2002) Evolutionary relationships of euthyneuran gastropods (Mollusca): a cladistic re-evaluation of morphological characters. Zool J Linn Soc 135:403–470

De Brabander J, Oppolzer W (1997) Enantioselective total synthesis of (−)-denticulatins A and B using a novel group-selective aldolization of a *meso* dialdehyde as a key step. Tetrahedron 53:9169–9202

Díaz-Marrero AR, Dorta E, Cueto M, Rovirosa J, San-Martín A, Loyola A, Darias J (2003a) Labdane diterpenes with a new oxidation pattern from the marine pulmonate *Trimusculus peruvianus*. Tetrahedron 59:4805–4809

Díaz-Marrero AR, Dorta E, Cueto M, Rovirosa J, San-Martín A, Loyola A, Darias J (2003b) New polyhydroxylated steroids from the marine pulmonate *Trimusculus peruvianus*. ARKIVOC 10:107–117

Donadio S, Staver MJ, McAlpine JB, Swanson SJ, Katz L (1991) Modular organization of genes required for complex polyketide biosynthesis. Science 252:675–679

Faulkner DJ (1992) Chemical defenses of marine molluscs. In: Paul VJ (ed) Ecological roles of marine natural products. Comstock, New York, pp. 11–163

Fernández R, Rodríguez J, Quiñoá E, Riguera R, Muñoz L, Fernández-Suárez M, Debitus C (1996) Onchidin B: a new cyclodepsipeptide from the mollusc *Onchidium* sp. J Am Chem Soc 118:11635–11643

Fontana A, Cutignano A, Giordano A, Coll AD, Cimino G (2004) Biosynthesis of aglajnes, polypropionate allomones of the opisthobranch mollusc *Bulla striata*. Tetrahedron Lett 45:6847–6850

Gao W-G, Sakaguchi K, Isoe S, Ohfune Y (1996) Stereoselective synthesis of a marine natural product, (+)-6β-isovaleroxylabda-8,13-diene-7α,15-diol. Tetrahedron Lett 37:7071–7074

Garson MJ, Small CJ, Skelton BW, Thinapong P, White AH (1990) Structural and stereochemical correlations of polypropionate metabolites from marine pulmonates: revision of the relative stereochemistry of pectinatone by X-ray structure analysis. J Chem Soc Perkin Trans I 805–807

Garson MJ, Jones DD, Small CJ, Liang J, Clardy J (1994a) Biosynthetic studies on polypropionates: a stereochemical model for siphonarins A and B from the pulmonate limpet *Siphonaria zelandica*. Tetrahedron Lett 35:6921–6924

Garson MJ, Goodman JM, Paterson I (1994b) A configurational model for siphonariid polypropionates derived from structural and biosynthetic considerations. Tetrahedron Lett 35:6929–6932

Gokhale RS, Tsuji SY, Cane DE, Khosla C (1999) Dissecting and exploiting intermodular communication in polyketide synthases. Science 284:482–485

Grande C, Templado J, Cervera JL, Zardoya R (2004a) Molecular phylogeny of euthyneura (mollusca: gastropoda). Mol Biol Evol 21:303–313

Grande C, Templado J, Cervera JL, Zardoya R (2004b) Phylogenetic relationships among Opisthobranchia (Mollusca: Gastropoda) based on mitochondrial *cox1*, *trnV*, and *rrnL* genes. Mol Phyl Evol 33:378–388

Gray CA, Davies-Coleman MT, McQuaid C (1998) Labdane diterpenes from the South African marine pulmonate *Trimusculus costatus*. Nat Prod Lett 12:47–53

He J, Hertweck C (2003) Iteration as programmed event during polyketide assembly; molecular analysis of the aureothin biosynthesis gene cluster. Chem Biol 10:1225–1232

Hisano N, Tateda H, Kuwabara M (1972) Photosensitive neurones in the marine pulmonate mollusc *Onchidium verruculatum*. J Exp Biol 57:651–660
Hochlowski JE, Faulkner DJ (1983) Antibiotics from the pulmonate *Siphonaria diemensis*. Tetrahedron Lett 24:1917–1920
Hochlowski JE, Faulkner DJ (1984) Metabolites of the marine pulmonate *Siphonaria australis*. J Org Chem 49:3838–3840
Hochlowski JE, Faulkner DJ, Matsumoto GK, Clardy J (1983) The denticulatins, two polypropionate metabolites from the pulmonate *Siphonaria denticulata*. J Am Chem Soc 105:7413–7415
Hochlowski JE, Coll JC, Faulkner DJ, Biskupiak JE, Ireland CM, Qi-tai Z, Cun-heng H, Clardy J (1984) Novel metabolites of four *Siphonaria* species. J Am Chem Soc 106:6748–6750
Hodgson AN (1999) The biology of siphonariid limpets (Gastropoda: Pulmonata). Oceanogr Mar Biol Annu Rev 37:245–314
Hodgson A, Healy J (1998) Comparative sperm morphology of the pulmonate limpets *Trimusculus costatus*, *T. reticulatus* (Trimusculidae) and *Burnupia stenochorias* and *Ancylus fluviatilis* (Ancylidae). J Molluscan Stud 64:447–460
Hoffmann RW, Dahmann G (1993) The absolute and relative configuration of muamvatin. Tetrahedron Lett 34:1115–1118
Hyman LH (1967) Mollusca. In: Hyman LH (ed) The invertebrates. MacGraw-Hill, New York, pp. 1–792
Ireland CM, Faulkner DJ (1978) The defensive secretion of the opisthobranch mollusc *Onchidella binneyi*. J Bioorg Chem 7:125–131
Ireland CM, Faulkner DJ (1981) The metabolites of the marine molluscs *Tridachiella diomedea* and *Tridachia crispata*. Tetrahedron 37:233–240
Ireland CM, Biskupiak JE, Hite GJ, Rapposch M, Scheuer PJ, Ruble JR (1984) Ilikonapyrone esters, likely defense allomones of the mollusc *Onchidium verruculatum*. J Org Chem 49:559–561
Katagiri N, Suzuki T, Yuichi S, Yasuo H (2002) Localization of retinal proteins in the stalk and dorsal eyes of the marine gastropod, *Onchidium*. Jpn Zool Sci 19:1231–1240
Kurabayashi A, Ueshima R (2000) Complete sequence of the mitochondrial DNA of the primitive opisthobranch gastropod *Pupa strigosa*: systematic implication of the genome organization. Mol Biol Evol 17:266–277
Maeda Y, Pakdaman K, Nomura T, Dois S, Sato S (1998) Reduction of a model for a *Onchidium* pacemarker neuron. Biol Cybern 78:265–276
Magnin-Lachaux M, Tan Z, Liang B, Negishi EI (2004) Efficient and selective synthesis of siphonarienolone and related reduced polypropionates via Zr-catalyzed asymmetric carboalumination. Organic Lett 6:1425–1427
Manker DC, Faulkner DJ (1987) Diterpenes from the marine pulmonate *Trimusculus reticulatus*. Tetrahedron 43:3677–3680
Manker DC, Faulkner DJ (1989) Vallartanones A and B, polypropionate metabolites of *Siphonaria maura* from Mexico. J Org Chem 54:5374–5377
Manker DC, Faulkner DJ (1996) Investigation of the role of diterpenes produced by marine pulmonates *Trimusculus reticulatus* and *T. conica*. J Chem Ecol 22:23–35
Manker DC, Faulkner DJ, Chang-fu X, Clardy J (1986) Metabolites of *Siphonaria maura* from Costa Rica. J Org Chem 51:814–816
Manker DC, Garson MJ, Faulkner DJ (1988) De novo biosynthesis of polypropionate metabolites in the marine pulmonate *Siphonaria denticulata*. J Chem Soc Chem Commun 1061–1062
Manker DC, Faulkner DJ, Stout TJ, Clardy J (1989) The baconipyrones: novel polypropionates from the pulmonate *Siphonaria baconi*. J Org Chem 54:5371–5374
McDaniel R, Thamchaipenet A, Gustafsson C, Fu H, Betlach M, Betlach M, Ashley G (1999) Multiple genetic modifications of the erythromycin polyketide synthase to produce a library of novel "unnatural" natural products. Proc Natl Acad Sci USA 96:1846–1851

Moffitt MC, Neilan BA (2003) Evolutionary affiliations within the superfamily of ketosynthases reflect complex pathway associations. J Mol Evol 56:446–457
Norte M, Cataldo F, González AG (1988) Siphonarienedione and siphonarienolone, two new metabolites from *Siphonaria grisea* having a polypropionate skeleton. Tetrahedron Lett 29:2879–2880
Norte M, Cataldo F, González AG, Rodríguez ML, Ruiz-Perez C (1990) New metabolites from the mollusc *Siphonaria grisea*. Tetrahedron 46:1669–1678
Norte M, Fernández JJ, Padilla A (1994) Isolation and synthesis of siphonarienal, a new polypropionate from *Siphonaria grisea*. Tetrahedron Lett 35:3413–3416
Olano C, Wilkinson B, Sánchez C, Moss SJ, Sheridan R, Math V, Weston AJ, Braña AF, Martin CJ, Oliynyk M, Méndez C, Leadlay PF, Salas JA (2004) Biosynthesis of the angiogenesis inhibitor borrelidin by *Streptomyces parvulus* Tü4055: cluster analysis and assignment of functions. Chem Biol 11:87–97
Oppolzer W, Moretti, R, Bernardinelli, G (1986) Enantioselective synthesis of the alleged structure of norpectinatone. Tetrahedron Lett 27:4713–4716
Oppolzer W, De Brabander J, Walther E, Bernardinelli G (1995) Asymmetric synthesis of (−)-denticulatins A and B via group-selective aldolization of a *meso* dialdehyde with a chiral *N*-propionylsultam. Tetrahedron Lett 36:4413–4416
Paterson I, Franklin AS (1994) Assignment of the absolute configuration of the siphonarins and bacopyrones. Enantiocontrolled synthesis of a γ-pyrone subunit. Tetrahedron Lett 35:6925–6928
Paterson I, Perkins MV (1992) Studies in polypropionate synthesis: stereoselective synthesis of (−)-denticulatins A and B. Tetrahedron Lett 33:801–804
Paterson I, Perkins MV (1993) Total synthesis of the marine polypropionate (+)-muamvatin. A configurational model for siphonariid metabolites. J Am Chem Soc 115:1608–1610
Paterson I, Perkins MV (1996) Total synthesis of (−)-deticulatins A and B using efficient methods of acyclic stereocontrol. Tetrahedron 52:1811–1834
Paterson I, Chen DY-K, Aceña JL, Franklin AS (2000) Studies in marine polypropionate synthesis: total synthesis of (−)-baconipyrone C. Org Lett 2:1513–1516
Paterson I, Chen DY-K, Franklin AS (2002) Total synthesis of siphonarin B and dihydrosiphonarin B. Org Lett 4:391–394
Paul MC, Zubía E, Ortega MJ, Salvá J (1997) New polypropionates from *Siphonaria pectinata*. Tetrahedron 53:2303–2308
Ponder WF, Lindberg DR (1997) Towards a phylogeny of gastropod molluscs: an analysis using morphological characters. Zool J Linn Soc 119:83–265
Purchon RD (1979) The biology of the Mollusca. Pergamon, Oxford
Ranganathan A, Timoney M, Bycroft M, Cortés J, Thomas IP, Wilkinson B, Kellenberger L, Hanefeld U, Galloway IS, Staunton J, Leadlay PF (1999) Knowledge-based design of bimodular and trimodular polyketide synthases based on domain and module swaps: a route to simple statin analogues. Chem Biol 6:731–741
Rice SH (1985) An anti-predator chemical defense of the marine pulmonate gastropod *Trimusculus reticulatus* (Sowerby). J Exp Mar Biol Ecol 93:83–89
Rodríguez J, Riguera R, Debitus C (1992a) New marine cytotoxic bispyrones. Absolute stereochemistry of onchitriols I and II. Tetrahedron Lett 33:1089–1092
Rodríguez J, Riguera R, Debitus C (1992b) The natural polypropionate-derived esters of the mollusc *Onchidium* sp. J Org Chem 57:4624–4632
Rodríguez J, Fernández R, Quiñoá E, Riguera R, Debitus C, Bouchet P (1994) Onchidin: a cytotoxic depsipeptide with C_2 symmetry from a marine mollusc. Tetrahedron Lett 35:9239–9242
Rokas A, Holland PWH (2000) Rare genomic changes as a tool for phylogenetics. Trends Ecol Evol 15:454–459
Roll DM, Biskupiak JE, Mayne CL, Ireland CM (1986) Muamvatin, a novel tricyclic spiro ketal from the Fijian mollusc *Siphonaria normalis*. J Am Chem Soc 108:6680–6682

Rovirosa, J, Quezada E, San-Martín A (1991) Nuevos polipropionatos de *Siphonaria lessoni* de la costa de Chile. Bol Soc Chil Quim 36:233–238
Rovirosa J, Quezada E, San-Martín A (1992) Un nuevo diterpeno del molusco *Trimusculus peruvianus*. Bol Soc Chil Quim 37:143–145
San-Martín A, Quezada E, Soto P, Palacios Y, Rovirosa J (1996) Labdane diterpenes from the marine pulmonate gastropod *Trimusculus peruvianus*. Can J Chem 74:2471–2475
Staunton J, Weissman KJ (2001) Polyketide biosynthesis: a millennium review. Nat Prod Rep 18:380–416
Sundram UN, Albizati KF (1992) Synthesis and absolute configuration of polypropionate metabolites of *Siphonaria australis*. Tetrahedron Lett 33:437–440
Ziegler FE, Becker MR (1990) Total synthesis of (–)-denticulatins A and B: marine polypropionates from *Siphonaria denticulata*. J Org Chem 55:2800–2805

Secondary Metabolites from the Marine Gastropod Molluscs of Antarctica, Southern Africa and South America

M.T. Davies-Coleman

Abstract. Despite their perceived inaccessibility, the marine intertidal and benthic environments of Antarctica, southern Africa and South America have continued to provide marine natural products for chemists with unique opportunities to study the secondary metabolite constituents and chemical ecology of a diverse array of marine gastropod molluscs. This review covers the literature up to 31 January 2005 and describes the structures and, where applicable, biological activities of 100 secondary metabolites isolated from 21 species of marine gastropod molluscs. Not unexpectedly, the chemistry of chemically defended shell-less opisthobranchs dominates the natural product studies of molluscs collected from these regions of the southern hemisphere.

6.1 Introduction

The majority of molluscan species are assigned to the class Gastropoda, the second most species-diverse class of animals after the class Insecta. For over 70 years, taxonomists have traditionally divided the class Gastropoda into three sub-classes: Prosobranchia, Opisthobranchia and Pulmonata. Unfortunately, not all gastropod molluscs are readily accommodated by this higher classification system and the trichotomy of gastropod sub-classes has recently been treated with circumspection by molluscan taxonomists (Kay et al. 1998). However, the general division of gastropods into prosobranchs, opisthobranchs and pulmonates is universally accepted within the marine natural products literature and this traditional taxonomic triad provides a useful framework for reviewing the secondary metabolite diversity reported from marine gastropod molluscs. The further classification of the gastropod molluscs described here to the level of order and family has been adopted from Beesly et al. (1998) and Cimino et al. (2001).

Marine natural products, including marine molluscan metabolites, are regularly reviewed (Faulkner 2002; Blunt et al. 2005). This review, which covers the molluscan chemistry literature up to 31 January 2005, focuses on the marine molluscs of the southern hemisphere and provides details

Mike Davies-Coleman
Department of Chemistry, P.O. Box 94, Rhodes University
6140 Grahamstown, South Africa

Progress in Molecular and Subcellular Biology
Subseries Marine Molecular Biotechnology
G. Cimino, M. Gavagnin (Eds.): Molluscs

of the chemical structures and, where reported, the ecological role or general bioactivity of the secondary metabolites isolated from molluscs collected from the oceans surrounding Antarctica, southern Africa and South America. Australian molluscan secondary metabolites are described elsewhere in this volume. Metabolites isolated from molluscs collected from islands near southern Africa, for e.g. the dolastatins, first obtained from Mauritian specimens of the sea hare *Dolabella auricularia* (Pettit 1997), will not be discussed here. Conversely, the chemistry of common molluscan species with a circumpolar distribution collected from Antarctic islands are included, e.g. *Austrodoris kerguelenensis* from the South Shetland Islands (Gavagnin et al. 2000). The rationale for the inclusion of the latter is based on the assumption by McClintock and Baker (1997) that the effective isolation of Antarctic benthic ecosystems from those of the surrounding oceans by the strong Antarctic circumpolar current results in the chemical defence strategies employed by Antarctic marine invertebrates remaining reasonably consistent over their geographic range.

6.2 Prosobranch Secondary Metabolites

The subclass Prosobranchia is the largest of the three traditionally recognised subclasses of gastropods; and prosobranch molluscs are renowned for their adaptive radiation and concomitant diversity of both morphological and physiological characteristics (Fretter et al. 1998). Marine prosobranchs are typically shelled snails and the physical defence provided by a robust external shell obviates an additional acquired chemical defence to protect their soft tissues. However, the presence of a chemical defence system, sequestered from other marine invertebrates, is occasionally observed in prosobranchs where the shell is reduced, e.g. the Antarctic prosobranch *Marseniopsis mollis* (Family Velutinidae, formerly Lamellariidae). *M. mollis* is similar to other lamellarian gastropods, in that, this species possesses a vestigial internal shell enveloped by a large fleshy mantle. There are no reports of the isolation of secondary metabolites from either southern African or South American prosobranch molluscs.

6.2.1 Antarctic Marine Prosobranchs

The conspicuous bright yellow colour of the mantle tissue of *M. mollis* probably provided the first indication of the possible utilization of a chemical defence system by this species. Seastars are the main marine invertebrate predators in Antarctic benthic communities (McClintock

1994) and feeding deterrent studies, mostly utilizing the large predatory Antarctic seastar *Odantaster validus*, have been regularly used to investigate the chemical defence strategies employed by Antarctic marine invertebrates. From the results of a series of *O. validus* feeding deterrent assays, McClintock et al. (1994a) proposed that the ubiquitous osmolyte, *N*-methyl picolinic acid or homarine (**1**), was the principal feeding deterrent utilised by *M. mollis* as a form of chemical defence. In common with most other chemically defended gastropod molluscs, *M. mollis* sequesters **1** from its diet. Paradoxically, McClintock et al. (1994a) discovered that homarine was not present, as expected, in the tunic of *M. mollis*' primary food source, the large solitary ascidian *Cnemidocarpa verrucosa*, but instead occurred in the small epizoites (predominantly hydroids and bryozoans) that reside on the exterior surface of *C. verrucosa*.

6.3 Opisthobranch Secondary Metabolites

Opisthobranch molluscs are almost exclusively marine and there are no terrestrial opisthobranch species. Only one or two opisthobranch species could be considered to be freshwater (Smith and Stansic 1998). Unlike shelled marine snails (prosobranchs), marine opisthobranch molluscs exhibit an evolutionary trend towards elimination of the shell (Faulkner and Ghiselin 1983; Gosliner 1987; Cimino and Ghiselin 1998, 1999). In shell-less opisthobranchs, the physical defence offered by a shell has generally been replaced with a chemical defence system incorporating bioactive metabolites either sequestered from the opisthobranch's diet or, less commonly, derived from de novo biosynthesis. The chemistry and biological activity of metabolites isolated from marine opisthobranch molluscs have been reviewed by Cimino and Ghiselin (1998), Cimino et al. (1999, 2001) and Gavagnin and Fontana (2000).

6.3.1 Antarctic Marine Opisthobranchs

McClintock and Baker (1997) summarised their contribution to chemical ecological studies of Antarctic opisthobranch molluscs in their comprehensive review of the chemical ecology of Antarctic marine invertebrates. Natural product and associated chemical ecology studies have been conducted on opisthobranch molluscs collected from several regions around the continent of Antarctica, including under the sea ice in McMurdo Sound (McClintock et al. 1994b, McClintock and Baker 1997), the adjacent Italian Antarctic base at Terra Nova Bay (Gavagnin et al.

2003a,b), Tethys Bay (Gavagnin et al. 1995), the Weddell Sea (Avila et al. 2000) and the South Shetland Islands (Gavagnin et al. 1999a,b, 2000).

Pteropods (Order Gymnosomata)

The evolution of the pteropod foot into a pair of "wings", adapted for both flotation and swimming through the water column, enables pteropod molluscs to adopt a pelagic lifestyle (Rudman and Willan 1998). The wing-like structure of the foot also contributes to the common name of sea butterfly given to this group of gastropods. Only a few species of pteropods belong to the order Gymnosomata and one of these species, the yellow, shell-less pteropod *Clione antarctica*, loses its shell, mantle and mantle cavity in the adult stage of its life cycle (Cimino et al. 2001). *C. antarctica* is common in McMurdo Sound, Antarctica, near the undersurface of the sea ice and is rarely preyed upon by pelagic predators. McClintock and Janssen (1990) were the first to report the abduction and transport of *C. antarctica* on the dorsal surface of the amphipod *Hyperiella dilatata* as a novel form of physically acquired chemical defence, on the part of the amphipod, against fish predators. The feeding deterrent, polypropionate-derived pteroenone (**2**), responsible for the chemical defence of *C. antarctica*, was later isolated and identified by Yoshida et al. (1995) following a bioassay guided fractionation of an extract of *C. antarctica*, in which various zooplanktivorous Antarctic fish species, e.g. *Pagothenia borchgrevinki*, were used as the bioassay test organisms. The 6*S* configuration of **2**, established in the usual manner via the modified Mosher's method, provided the key to determining the configuration at C-5. Large (10.1 Hz) diaxial coupling constants between H-4 and H-5 and between H-6 and H-5, observed in the ^{1}H NMR spectrum of the *syn*-acetonide (**3**) prepared from one of the epimeric sodium borohydride reduction products of **2**, required the C-5 methyl group to be equatorial and thus secured the *R* absolute configuration at the homoallylic chiral centre.

Nudibranchs (Order Nudibranchia)

Three dorid nudibranch species (Suborder Doridina) viz. *Tritoniella belli*, *Bathydoris hodgsoni* and *A. kerguelenensis* have exclusively dominated the natural product and chemical ecology studies of Antarctic marine opisthobranch molluscs over the past decade. Dorid nudibranchs are carnivorous and prey on other marine invertebrates, including sponges, octocorals, bryozans and ascidians (Cimino et al. 2001), which act as a basic food source and in many instances provide the dorid nudibranchs with a source of bioactive metabolites for their chemical defence systems. Ironically, of the three Antarctic species of dorid nudibranchs studied thus far, two species, *B. hodgsoni* and *A. kergulensis*, have been reported to obtain their chemical defence metabolites through de novo biosynthesis.

The feeding deterrent chimyl alcohol (**4**), ubiquitous in tropical and temperate molluscs, was found to be the major glycerol ether present in extracts of the dorid nudibranch, *T. belli* collected from Ross Island, McMurdo Sound (McClintock et al. 1994b). Chimyl alcohol was also isolated from the stoloniferan octocoral *Clavularia frankliniana*, which forms a major component of *T. belli*'s diet, suggesting that this octocoral was the source of **4** used in *T. belli*'s sequestered chemical defence system. The increasingly widely used and ecologically relevant seastar (*O. validus*) tube foot retraction assay confirmed the feeding deterrent properties of **4** (McClintock et al. 1994b).

B. hodgsoni (Eliot 1907) is a member of a primitive group of polar dorid nudibranchs belonging to the superfamily Bathydoridae. An extract of the mantle tissue of *B. hodgsoni* from animals dredged at depths >200 m in the eastern Weddell Sea yielded the first 2-substituted drimane sesquiterpene, hodgsonal (**5**) to be isolated from the marine environment (Iken et al. 1998, Gavagnin et al. 2000). The relative configurations of the four chiral centres in **5** were initially assigned from NOE data. Cyclization of deacetylated **5**, in the presence of manganese dioxide, afforded a tricyclic γ-lactone (**6**). The 2*S* configuration in **6**, established using Mosher's method, thus provided the absolute configuration of the remaining three chiral centres and unequivocally confirmed the drimane skeleton of the native hodgsonal, which proved to be repugnant to *O. valdivus* in a modified version of McClintock's predatory seastar feeding deterrent assay (Avila et al. 2000).

The absence of any detectable amounts of **5** in the gut of *B. hodgsoni* and the constancy of its concentration in the mantle tissues of a number of specimens collected from different depths and localities around Antarctica led Avila et al. (2000) to suggest that **5** was a product of de novo biosynthesis in *B. hodgsoni* and was not sequestered by this organism from its diet. The structures of compounds **1**–**6** are presented in Fig. 6.1.

A. kerguelenensis (Bergh 1884) is a large nudibranch varying in colour from white to yellow. *A. kerguelenensis* is not confined to Antarctica and is also reported from the southern tip of South America and some sub-Antarctic islands, e.g. the Kerguelen Islands, whence it takes its name. The apparent chemical defensive properties of the mantle tissue of *A. kerguelenensis* initially reported by McClintock et al. (1990) was followed by a study of the natural product chemistry by Davies-Coleman and Faulkner (1991) of specimens of this species collected in McMurdo Sound (note: the species name "*kerguelensis*" was incorrectly spelt in the paper by Davies-Coleman and Faulkner). Five *ent*-labdane diterpene glycerides (**7**–**11**) were isolated from the McMurdo Sound *A. kerguelenensis* extracts.

The assignment of diterpenes **7–11** to the *ent*-labdane series followed from comparison of the optical rotation of the methyl ester of saponified **7** ($[\alpha]_D$-49) with that reported for methyl (5*R*, 10*R*, 13*R*) labda-8-en-15-oate ($[\alpha]_D$-48). Biosynthetic arguments were used to extrapolate the absolute stereochemistry of the diterpene moiety in **7** to compounds **8** and **9**, while the diketone obtained from ozonolysis of **7** was found to be identical with **10** thus confirming the absolute stereochemistry of the latter compound and also, by further recourse to biosynthetic arguments, the absolute stereochemistry of **11** (Davies-Coleman and Faulkner 1991).

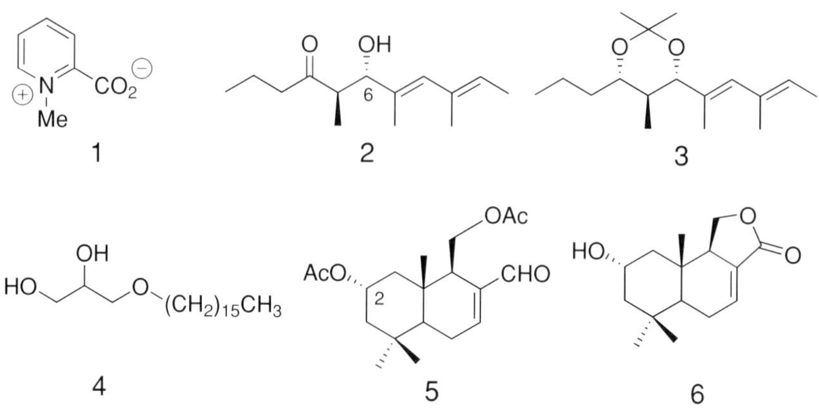

Fig. 6.1. Secondary metabolites isolated from the Antarctic prosobranch *Marseniopsis mollis*, the opisthobranchs *Tritoniella belli* and *Bathydoris hodgsoni* and two key derivatives (**3**, **6**) used to determine the absolute stereochemistry of **2** and **5**, respectively

From Dayton et al.'s (1974) field observations of the feeding habits of *A. kerguelenensis*, Davies-Coleman and Faulkner (1991) postulated that **7–11** were the products of de novo biosynthesis given the known paucity of organic biomass in the "glass sponges" (Order Hexactinellida) which predominate in *A. kergeulenensis*' diet. Davies-Coleman and Faulkner's initial de novo biosynthetic hypothesis was recently corroborated by Iken et al. (2002). In a meticulous chemical ecology study involving 117 specimens of *A. kerguelenensis* collected from 32 different locations in the Weddell Sea at depths of 65–1550 m, Iken et al. (2002) were able to clearly demonstrate that the diacylglycerides **7** and **8** were only present in the mantle tissue and not the viscera. The complete absence of these bioactive metabolites in the gut unequivocally confirmed that these compounds are not sequestered from the nudibranch's diet. Compounds **7** and **8** and a cohort of simple fatty acid monoacylglycerides co-occurring in the mantle tissue of *A. kerguelenensis* exhibited feeding deterrence activity towards the seastar *O. validus*. Iken et al. (2002) were also able to confirm that **7** and **8** were absent, as suspected by Davies-Coleman and Faulkner (1991), from the Antarctic hexactinellid sponges preyed upon by *A. kerguelenensis*.

The biogenesis of terpenoic acid glycerides in nudibranchs has been the subject of conjecture over the past two decades (e.g. Cimino et al. 1983; Gustafson and Andersen 1985; Graziani et al. 1996; Fontana et al. 1998). The early studies of Cimino et al. (1983) and Gustafson and Andersen (1985) showed low levels of incorporation of ^{14}C-labelled mevalonic acid into a sesquiterpene dialdehyde from the dorid nudibranch *Dendrodoris limbata* and diterpenoic acid glycerides from *Archidoris montereyensis* and *A. odheneri*, respectively. Frustratingly, evidence for the incorporation of ^{14}C-labelled mevalonic acid in the de novo biosynthesis of mantle metabolites from other dorid nudibranchs was elusive and a different approach to elucidating the biosynthesis of these metabolites was required (Graziani et al. 1996). Using NMR detection of stable isotopes, Graziani et al. (1996) were able to unequivocally elucidate the incorporation of [1,2-$^{13}C_2$]acetate into the diterpenoid skeleton of a diterpenoic acid glyceride biosynthesised by *A. montereyensis* and *A. odheneri*. More recently, Fontana et al. (1998) used [6-^{13}C]- and [5-^{13}C]glucose and [2-^{13}C]pyruvate to explore the de novo biosynthesis of a series of diterpenoic acid glycerides (verrucosins) by *Doris verrucosa*. Although incurporation levels into the diterpene skeleton were low, Fontana et al. (1998) were able to tentatively postulate an acetate/ mevalonate pathway for the biosynthesis of the diterpenoid moiety in the veruccosins. Conversely, they were able to unambiguously determine the biosynthetic origin of the *sn*-glyceride moiety in the verrucosins from D-glyceraldehyde 3-phosphate derived from the glycolysis of D-glucose.

The absolute stereochemistry at C-2 in the 1,3 diacylglyceride moiety of **8**, originally unassigned by Davies-Coleman and Faulkner, was later established as *R* by Gavagnin et al. (1999a) by applying Mosher's method to **8** isolated from the mantle tissue of two *A. kerguelenensis* specimens collected from the Weddell Sea. The 2*R* configuration was similarly assigned (Gavagnin et al. 1999a) to the 1,3 diacylglyceride moiety in the diterpene glyceride (**12**) isolated as a minor metabolite from two *A. kerguelenensis* specimens collected near the South Shetland Islands. The major metabolite in the South Shetland Island specimens of *A. kerguelenensis* was the 1,2-diacylglyceride (**13**; Gavagnin et al. 1999a). Both **12** and **13** are analogues of the halimane diterpene glyceride (**14**) first isolated as its diacetyl derivative (**15**) from *A. kerguelenensis* specimens from Tethys Bay (Gavagnin et al. 1995).

Surprisingly, the 2*R* configuration of **8** and **12** was at variance with the 2*S* stereochemistry reported for all other diacylglycerols previously isolated from dorid nudibranchs (Fontana et al. 1998; Gavagnin et al. 1999a). At a loss for an alternative explanation, Gavagnin et al. (1999a) tenuously suggested that the harsh Antarctic environmental conditions might be influencing de novo biosynthesis of diacylglycerols in *A. kerguelenensis* and resulting in this puzzling stereochemical anomaly. Gavagnin et al. (2003a) finally resolved the outstanding glyceride stereochemical question four years later when they returned to the primary NMR spectroscopic data acquired for **7** and **13**, and unearthed a fundamental flaw in the structures originally proposed for these two compounds. Definitive heteronuclear multi-bond correlation (HMBC) connectivities, including those between the glyceride oxymethine proton and the diterpene ester carbonyl carbon in **7** and **13**, placed the diterpene moiety at C-2 and the acetate at C-1 in the glycerol moiety in these two compounds, thus confirming that they were indeed 1,2-*sn* diacylglycerides as expected. After revising the structures of **7** and **13** to **16** and **17**, respectively, Gavagnin et al. (2003a) deduced that the 1,3 diacylglycerides **8** and **12** were not natural products but rather artefacts of the extraction and chromatographic workup where the terpenoid acyl group in **16** and **17** undergoes facile migration from C-2 to C-3 in the glycerol moiety, resulting in the 2*R* configuration observed in **8** and **12**.

Careful analysis of the NMR data of an inseparable mixture of minor products from *A. kerguelenensis* specimens collected from Terra Nova Bay suggested the presence of clerodane diterpene (**18**) as the main component in this mixture (Gavagnin et al. 2003a). NMR resonances attributed to the diterpene residue of **18** were consistent with those assigned to the clerodane moiety in archidorin (**19**) previously isolated from the Atlantic nudibranch *A. tuberculata*. Further evidence for the structural affinity between **18** and archidorin was provided by methanolysis of the inseparable diterpene mixture followed by high performance liquid chromatography (HPLC) and NMR comparison of the methylated products with an authentic sample of methyl ester of archidorin (**20**) (Gavagnin et al. 2003a).

The propensity of *A. kerguelenensis* extracts to provide investigators with a steady stream of new terpenoid variants was further exemplified by the isolation of tricyclic isocopalane-type diterpenes austrodorins A (**21**) and B (**22**; Gavagnin et al. 1999b) and the *nor*-sesquiterpenes austrodoral (**23**) and austrodoric acid (**24**) (Gavagnin et al. 2003b) from *A. kerguelenensis* extracts. The absolute stereochemistry of **21** and **22** followed from comparison of the circular dichroism (CD) spectra of these compounds with the CD spectra of synthetic isocopalane and *ent*-isocopalane diterpenoids of known absolute configuration (Gavagnin et al. 1999b). The ^{13}C NMR data of **23** and **24** were compatible with the literature values quoted for a drimane skelton with a *trans*-A,B ring junction, while NOE data supported the assignment of a β-equatorial methyl substituent at C-8 in both these compounds. Interestingly, the concentrations of **23** and **24** in the *A. kerguelenensis* mantle tissue were significantly higher when these animals were kept in an aquarium for an extended period, suggesting that their biosynthesis may be induced by exposure to stress (Gavagnin et al. 2003b). The structures of the *A. kerguelenensis* metabolites **7**–**24** are presented in Fig. 6.2.

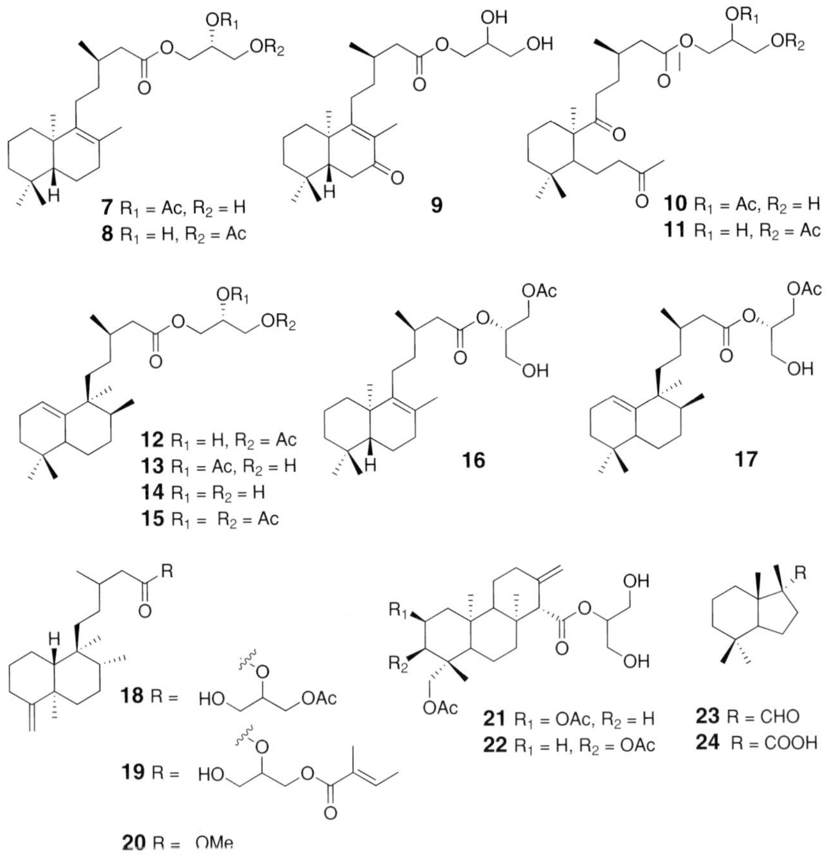

Fig. 6.2. Secondary metabolites isolated from the Antarctic opisthobranch *Austrodoris kerguelenensis* and the Atlantic opisthobranch *Archidoris tuberculata*

6.3.2 Southern African Marine Opisthobranchs

The southern African coastline, approximately 3,000 km long and stretching from Namibia in the west to southern Mozambique in the east, is broadly subdivided into three bio-geographical zones, the cool temperate west coast, the warm temperate south east coast and the subtropical east coast (Branch and Branch 1981). Each of the southern African bio-geographical zones sustains distinctive populations of marine flora and fauna and a large proportion of the over 10,000 species of marine organisms recorded off the southern African coast are reported to be endemic (Branch et al. 1994). The studies of the natural product chemistry of African marine molluscs have thus far been confined to southern African species collected predominantly on the warm temperate south east coast and the subtropical and tropical east coast of South

Africa and Mozambique. Gosliner (1987) estimates that more than 250 opisthobranch mollusc species occur off the southern African coast, of which many species remain undescribed.

Nudibranchs (Order Nudibranchia)

The endemic southern African nudibranch *Leminda millecra* (Griffiths 1985) is the only representative of the family Charcotiidae (formerly Lemindidae) occurring off the coast of South Africa. Pika and Faulkner (1994) isolated four sesquiterpene metabolites, millecrone A (**25**), millecrone B (**26**), millecrol A (**27**) and millecrol B (**28**) from an extract of four *L. millecra* specimens collected off Coffee Bay on the Wild Coast of South Africa. The data provided from a series of 1D NOE difference experiments combined with an analysis of selective coupling constants provided the relative stereochemistry of compounds **25**–**28**. From an examination of the gut contents from the *L. millecra* specimens that provided **25**–**28**, Pika and Faulkner (1994) identified spicules from three soft coral species, *Alcyonium foliatum*, *A. valdivae* and *Capnella thyrsoidea*. From this evidence, they tentatively proposed that *L. millecra* sequesters **25**–**28** from one or more of these species. Of the four *L. millecra* metabolites, only **25** inhibited the growth of *Candida albicans*, while **26** was active against *Staphylococcus aureus* and *Bacillus subtilis* and **28** only exhibited antibiotic activity against the latter bacterium (Pika and Faulkner 1994).

Although initially purported to be rare off the coast of southern Africa (Gosliner 1987), *L. millecra* was found to be abundant in Algoa Bay, South Africa, ca. 500 km southwest of the Wild Coast (McPhail et al. 2001). From a combined extract of 32 specimens of *L. millecra* collected in Algoa Bay, McPhail et al. (2001) isolated **25**, **26**, isofuranodiene (**29**), (+)-8-hydroxycalamenene (**30**), algoafuran (**31**), cubebenone (**32**) and a cohort of seven triprenylquinones and hydroquinones (**33**–**39**). Standard spectroscopic techniques were used to determine the chemical structures of the *L. millecra* metabolites with NOESY data, providing the relative stereochemistry of **32**. In accordance with similar trends in the ^{13}C chemical shift data reported for analogous compounds, the relatively deshielded allylic methyl carbon (C-14′ δ_C 26) in the triprenyl side-chain of **36**–**38** suggested a *Z* configuration for the α,β unsaturated ketone in these three compounds. This latter assignment was corroborated by a NOESY correlation between the 3H-14′ methyl protons and the adjacent vinylic proton. There was no evidence to support the occurrence of millecrols A and B in the combined Algoa Bay *L. millecra* extract, suggesting some geographical variation in the sequestered chemistry of this species. While the bulk extraction of large numbers of a single nudibranch species provides an opportunity to rapidly survey the diversity of metabolites sequestered by that species in a particular area, it does not provide details of the dietary selectivity, if any, of individual nudibranchs. Once isolated in sufficient quantities for structure elucidation studies, nudibranch metabolites can subsequently be used as analytical standards in gas chromatography (GC) or HPLC analyses of extracts of individual nudibranchs. Accordingly, GC analysis of extracts from eight individual specimens of *L. millecra* collected from a large reef in Algoa Bay revealed that each nudibranch contained **26** and **32** as minor and major metabolites, respectively. Extrapolation of the GC method to extracts of 21 octocorals collected in Algoa Bay confirmed that the sea fan *Leptogorgia palma,* on which *L. millecra* had often been observed feeding, was the source of **26** and **32** (McPhail et al. 2001). The structures of the structurally diverse metabolites isolated from *L. millecra* are presented in Fig. 6.3.

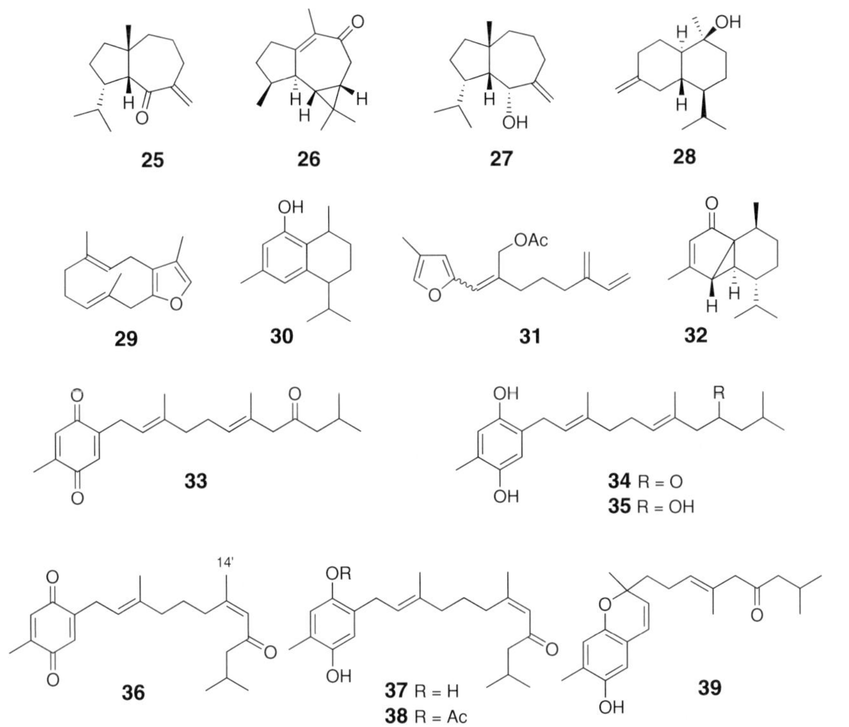

Fig. 6.3. Secondary metabolites isolated from the southern African opisthobranch *Leminda millecra*

The secondary metabolites of only two species of southern African dorid nudibranchs have been investigated. The first of these established that the brightly coloured *Chromodoris hamiltoni* (Rudman 1977) (Family Chromodorididae), collected from the Aliwal Shoal, a large sub-tropical reef system off the coast of KwaZulu Natal, South Africa, was the source of four unusual chlorinated homoditerpenes, hamiltonins A–D (**40–43**), the sesterterpene hamiltonin E (**44**) and the relatively common sponge toxins latrunculin A and B (**45**, **46**; Pika and Faulkner 1995). The relative stereochemistry of **40** and **41**, established from coupling constants and NOE data, confirmed that both these compounds possessed the same unprecedented 3-homo-4,5-*seco*spongian skeleton. A paucity of **42** and **43** required structure elucidation of these two compounds, predominantly through comparison of their spectral data with those of the more abundant hamiltonins A and B. Compound **44** was also isolated as a minor metabolite from the *C. hamiltoni* extracts. The relative stereochemistry of the tricyclic ring system in **44** was proposed from comparison of the spectroscopic data of this compound with those of known compounds, while the magnitude of the negative Cotton effect observed at 218 nm in the CD spectrum of **44** compared favourably with the Cotton effects observed in the CD spectra of luffarin-I and 4*R*-manolalide, and led to the assignment of an *R* absolute stereochemistry to the asymmetric oxymethine carbon in **44**. Of these seven metabolites, only **46** was present in extracts of *C. hamiltoni* collected from the reefs off southern Mozambique (400 km north of the Aliwal Shoal), possibly reflecting geographical variation in the organisms that make up *C. hamiltoni*'s diet in this region of the southern African coast (McPhail and Davies-Coleman 1997). Although devoid of hamiltonins, the extract of the Mozambique specimens of *C. hamiltoni* yielded two new spongian diterpene lactones (**47**, **48**; McPhail and Davies-Coleman 1997). The relative stereochemistry of **47** and **48** followed from the NOESY and 1D NOE difference data acquired for these two compounds.

The genus *Hypselodoris* differs from the closely related genus *Chromodoris* both in its colouration and the structure of the radula, a chitinous ribbon of teeth used for feeding by most molluscs (Gosliner 1987). The endemic southern African species, *Hypselodoris capensis* (Barnard 1927), is a colourful member of the family Chromodorididae; and an investigation by McPhail et al. (1998) of the metabolites present in an extract of 16 specimens of *H. capensis*, collected in the Tsitsikamma Marine Reserve, situated on the warm temperate southeastern coast of South Africa, afforded the linear β-substituted sesterterpenes (18*R*)-variabilin (**49**), 22-deoxyvariabilin (**50**) and 22-deoxy-23-hydroxymethylvariabilin (**51**), in addition to the known sesquiterpenes nakafurans 8 and 9 (**52**, **53**). The assignment of an 18*R* configuration to the variabilin isolated from *H. capensis* followed from comparison of the optical rotation obtained for **49** with published values for this ubiquitous bioactive metabolite. The absolute configuration at C-18 in **50** and **51** remains unassigned. McPhail et al. (1998, 2000) provided evidence for the sequestration of these compounds by *H. capensis* from a *Fasciospongia* sponge (the source of 49–51) and a *Dysidea* sponge (the source of 52, 53). Field observations indicated that both sponges form part of *H. capensis*'s diet in the Tsitsikamma Marine Reserve.

The dark blue nudibranch, *Tambja capensis* (Bergh 1907) (Family Polyceridae), is endemic to the cool temperate coastal waters off the south-eastern coast of South Africa (Gosliner 1987). A recent comparative study by Rapson (2004) of the distribution of 4-methoxypyrrolic metabolites in three populations of this species collected from three well dispersed locations (False Bay, Algoa Bay, East London), revealed that all three populations contained the tetrapyrrole pigment (54) and the tambjamines A (55) and E (56) as major metabolites. In Algoa Bay, *T. capensis* has been observed feeding extensively on the blue bryozoan *Bugula dentata*. HPLC analysis of Algoa Bay specimens of *B. dentata* revealed that this species contained all three 4-methoxypyrrolic metabolites and thus suggested that *B. dentata* is the primary source of the sequestered chemistry of *T. capensis* off the coast of southern Africa. The structures of the metabolites isolated from *C. hamiltoni*, *H. capensis* and *T. capensis* are presented in Fig. 6.4.

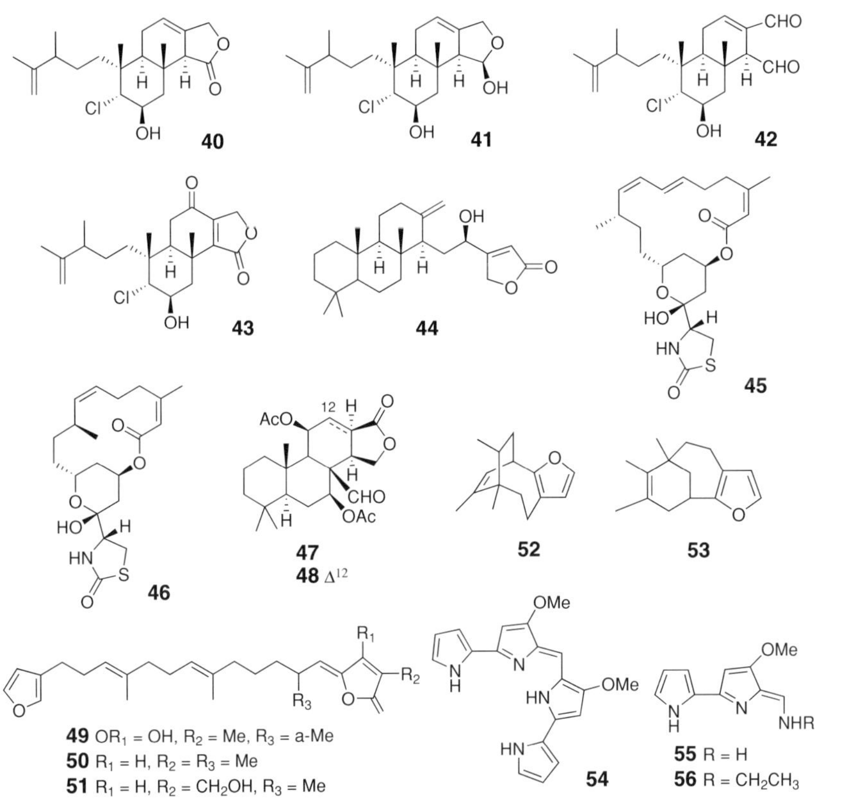

Fig. 6.4. Secondary metabolites isolated from the southern African opisthobranchs *Chromodoris hamiltoni*, *Hypselodoris capensis* and *Tambja capensis*

Sea Hares (Order Anaspidea)

Anaspideans are large herbivorous molluscs with either a fragile, transparent external shell (Family Akeridae) or a reduced or absent internal shell (Family Aplysiidae; Cimino et al. 2001). The large circumtropical sea hare *Aplysia dactylomela* (Rang 1882) (Family Aplysiidae) typically stores a plethora of halogenated terpenes from its red algal diet in a large internal digestive gland. Although Algoa Bay is at the extreme southern end of *A. dactylomela*'s range off the east coast of southern Africa (Gosliner 1987), an uncommon prolonged influx of warm water into Algoa Bay during the late summer of 1998 resulted in a proliferation of this species on the reefs on the western edge of the bay. Six halogenated sesquiterpenes, algoane (**57**), 1-deacetoxyalgoane (**58**), 1-deacetoxy-8-deoxyalgoane (**59**), ibhayinol (**60**), nidificene (**61**) and prepacifenol epoxide (**62**) were isolated from extracts of the excised digestive glands of four specimens of *A. dactylomela* collected from Algoa Bay during this warm-water event (McPhail et al. 1999). An X-ray structure of **57** provided the initial entry into the structures and relative stereochemistry of **58** and **59**. A further X-ray analysis of ibhayinol (Copley et al. 2002) corrected an erroneous stereochemistry originally assigned to five of the seven chiral centres in this compound (McPhail et al. 1999) and necessitated a revision of the structure of ibhayinol to **63**. The revised stereochemistry of ibhayinol also enabled Copley et al. (2002) to postulate a putative biosynthetic link between **63** and compounds **55–57** through a hypothetical 8-hydroxy-1-deacetoxyalgoane precursor (**64**). Loss of the equatorial bromine substituent at C-10 in **64** followed by a 1,2-alkyl shift of the axial methyl group on C-11 facilitated ring closure at the resultant carbocation generated at C-11 via nucleophilic attack of the hydroxyl moiety resident on C-1 with *si* facial selectivity. The original red algal source of the halogenated sesquiterpenes isolated from *A. dactylomela* is unknown.

The cosmopolitan sea hare *A. parvula* (Mörch 1863) is the smallest of the *Aplysia* sea hares found off the coast of South Africa. McPhail and Davies-Coleman (2005) isolated (3*Z*)-bromofucin (**65**) from a combined extract of 49 specimens of *A. parvula* collected from the Tsitsikamma Marine Reserve on the temperate south coast of South Africa. Although (3*Z*)-bromofucin is analogous to several *Laurencia* algal metabolites, the algal source of **62** was not established. A combination of NOESY data and molecular modelling studies provided the relative stereochemistry of **65**. The chemical structures of the southern African *Aplysia* secondary metabolites are presented in Fig. 6.5.

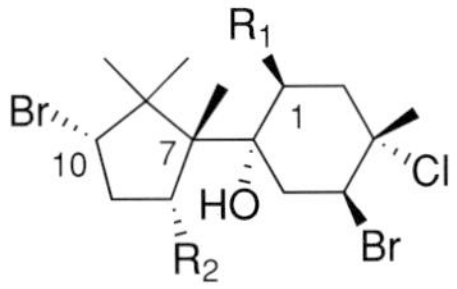

57 R_1 = OAc, R_2 = OH
58 R_1 = H, R_2 = OH
59 R_1 = R_2 = H
64 R_1 = OH, R_2 = H

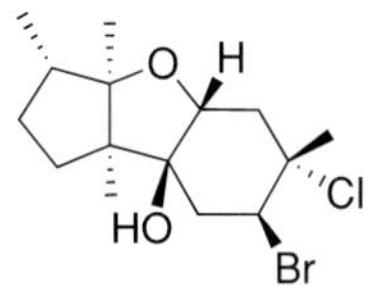

60

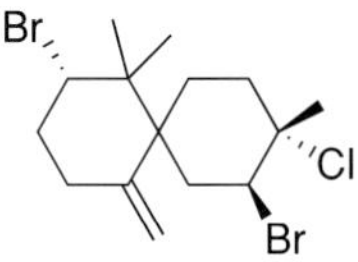

61

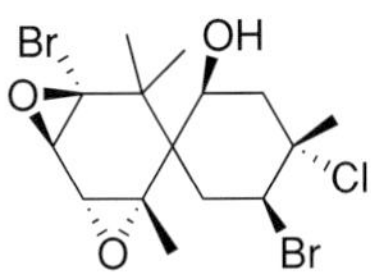

62

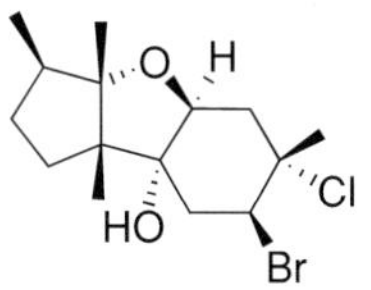

63

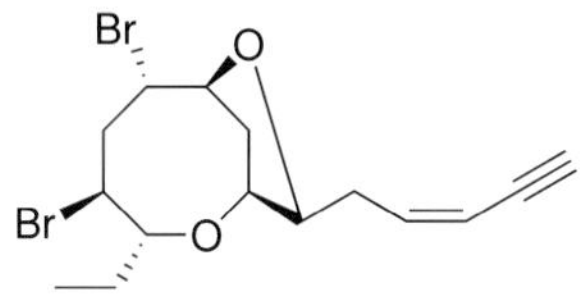

65

Fig. 6.5. Secondary metabolites isolated from the southern African opisthobranchs *Aplysia dactylomela* and *A. parvula* and a hypothetical biosynthetic precursor (**64**) of ibhayinol (**63**)

6.3.3 South American Marine Opisthobranchs

The natural product chemistry of Brazilian marine organisms, including molluscs, has been recently reviewed (Berlinck et al. 2004). The natural product studies of southern South American opisthobranch molluscs appear to have been confined to two Brazilian nudibranch species, *Doris* aff. *verrucosa* and *Tambja eliora*, and the sea hare *A. dactylomela* and two dorid nudibranchs, *Tyrinna nobilis* and *Anisodoris fontaini* collected off the Patagonian coast.

Nudibranchs (Order Nudibranchia)

Granato et al. (2000) isolated a plethora of common sterols and 9-[5′-(methylthio)-D-xylofuranosyl]adenine (xylosyl-MTA; **66**) from the mantle tissue of the dorid nudibranch *D.* aff. *verrucosa* (Family Chromodorididae) collected near São Paulo, Brazil. Xylosyl-MTA was previously isolated from a related Mediterranean nudibranch species (Cimino and Sodano 1993). Although *D.* aff. *verrucosa* preys extensively on the sponge *Hymeniacidon* aff. *heliophila*, the absence of **66** in this sponge's tissue led Granato et al. (2000) to conclude that *D.* aff. *verrucosa* does not sequester **66** from this sponge species. An extract of nine specimens of *Tambja eliora* (Family Polyceridae) also collected off the coast of Brazil afforded tambjamine A (**55**) and tambjamine D (**67**) as their imino salts, e.g. **67**

(Berlinck et al. 2004). Both **55** and **67** were originally isolated by Carte and Faulkner (1983) from *Tambja eliora* collected in the Gulf of Mexico.

The dorid nudibranch *Tyrinna nobilis*, collected of the cost of Patagonia, yielded a novel *seco*-11,12-spongiane diterpene, tyrinnal (**68**) and the known compounds dendrolasin (**69**), pallescensin A (**70**) and dehydropallescensin-2 (**71**) (Fontana et al. 1998). NOE data provided the relative stereochemistry of **68**. The probable sponge source of the sequestered chemistry of *T. nobilis* is unknown (Fontana et al. 1998). A second Patagonian nudibranch *Anisodoris fontaini* yielded a series of five new minor metabolites, anisodorins 1–5 (**72–76**), and two known metabolites (**77**, **78**) as the major mantle metabolites in this dorid nudibranch (Gavagnin et al. 1999c). The structure elucidation of **72–78** proved to be relatively straightforward, given the similarities in the isocopalane skeleton common to all seven metabolites isolated from *A. fontaini* and the already well established structures of **77** and **78**. Semi-syntheses of both *ent*-anisodorin 1 (Gavagnin et al. 1999c) and *ent*-anisodorin 5 (Ungur et al. 1999) from the commercially available terrestrial plant diterpene (–)-sclareol unequivocally confirmed the structures and absolute stereochemistry of **72** and **76**. The absolute stereochemistry of **76** is opposite to that of similar isocopalane diterpenoids typically isolated from sponges of the genus *Spongia*, adding credence to the hypothesis that all the compounds **72–78** are the products of de novo biosynthesis and not sequestered by *A. fontaini* from its diet. Circular dichroism, supported by NOE data, suggested that anisiodorin 3 (**74**) shares the same absolute stereochemistry as **72** and **73**.

Sea Hares (Order Anaspidea)

An examination of extracts of the mantle tissue and viscera of specimens of *A. dactylomela* collected from the intertidal zone on the coast of Rio de Janeiro State, Brazil, yielded four known halogenated chamigrane sesquiterpenes: prepacifinol epoxide (**62**), johnstonol (**79**), pacifidiene (**80**) and the diol (**81**; Pitombo et al. 1996). Although the former two compounds were originally isolated from the red marine algae *Laurencia nidifida* and *L. johnstoni* collected from Hawaii and the Gulf of California, respectively, the Brazilian algal source of **62** and **79–81** was not established (Pitombo et al. 1996). In the original reports describing the structures of **62** and **79–81**, the NMR data had been unassigned and the isolation of these compounds from the Brazilian specimens of *A. dactylomela* provided an opportunity for these spectral assignments to be established and the preferred conformation of chamigrane sesquiterpenes in solution to be explored (Kaiser et al. 1998, 2000, 2001). The chemical structures of secondary metabolites isolated from South American opisthobranch molluscs are presented in Fig. 6.6.

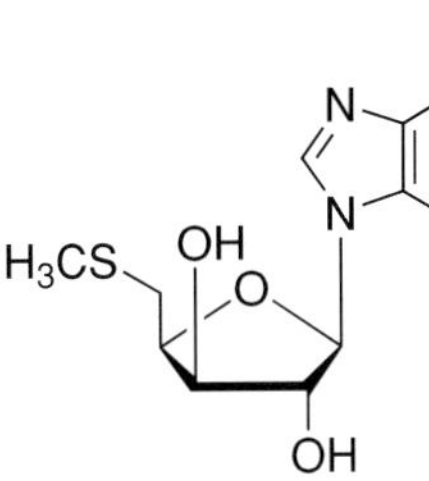

66

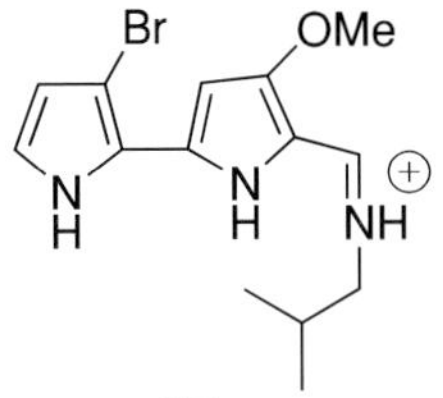

67

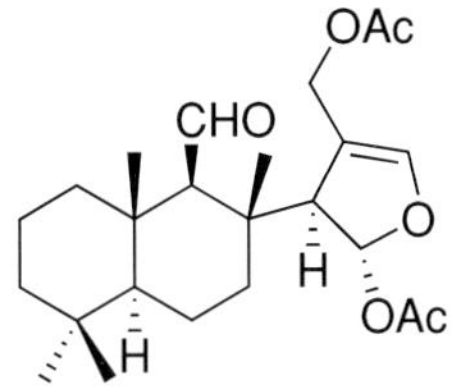

68

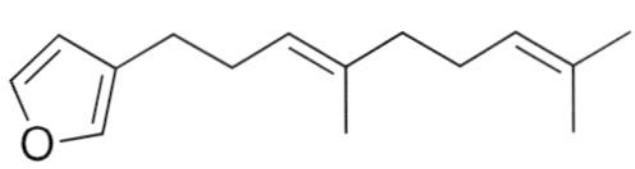

69

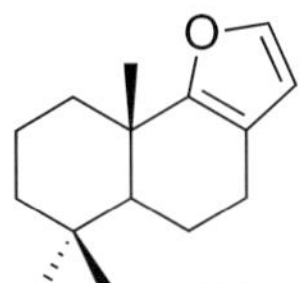

70

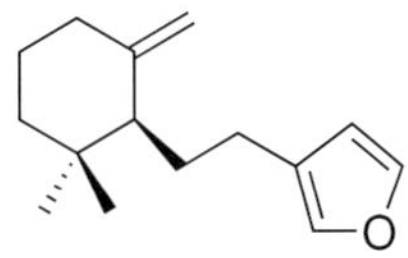

71

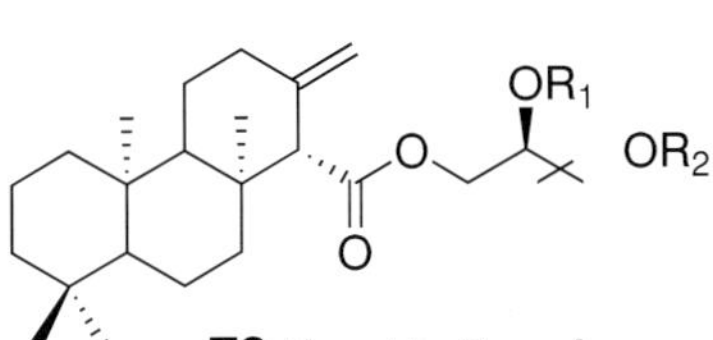

72 R_1 = H, R_2 = Ac
73 R_1 = Ac, R_2 = H

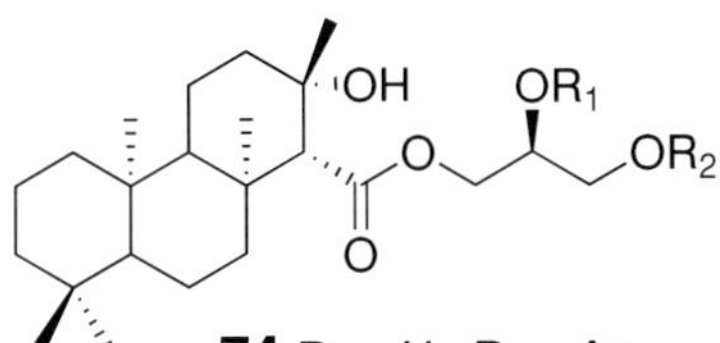

74 R_1 = H, R_2 = Ac
75 R_1 = Ac, R_2 = H

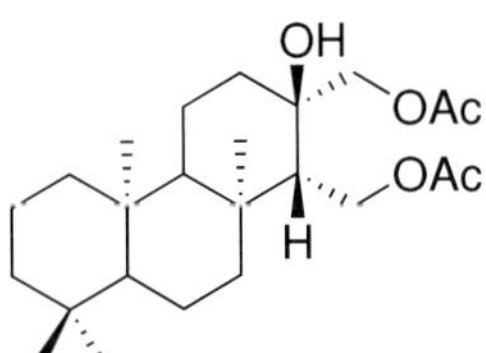

76

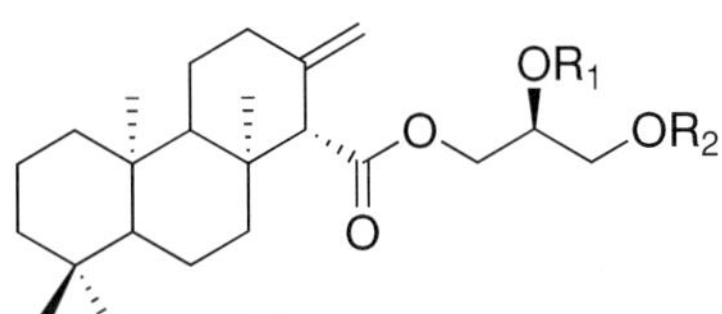

77 R_1 = H, R_2 = Ac
78 R_1 = Ac, R_2 = H

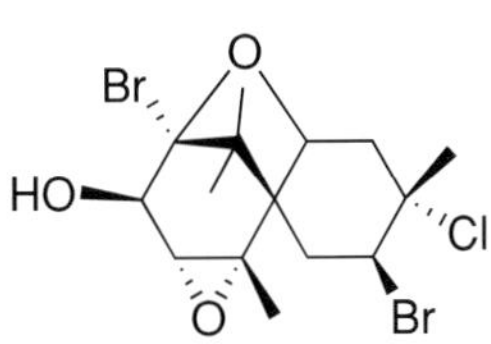

79

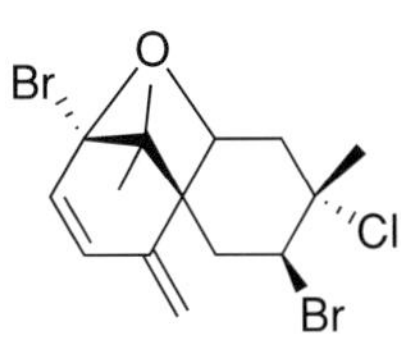

80

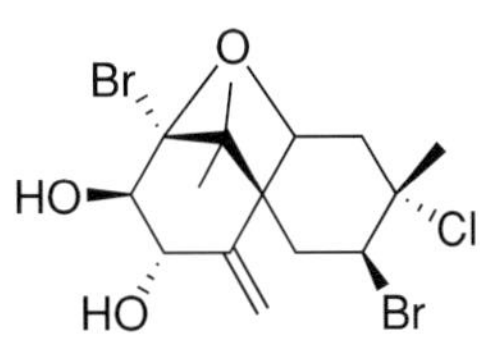

81

Fig. 6.6. Secondary metabolites isolated from the South American opisthobranchs *Doris* aff. *verrucosa*, *Tambja eliora*, *Tyrinna nobilis*, *Anisodoris fontaini* and *Aplysia dactylomela*

6.4 Pulmonate Secondary Metabolites

In contrast to other gastropods, the sub-class Pulmonata is dominated by terrestrial and to a lesser extent freshwater species. As a requirement for their terrestrial existence, pulmonates do not use gills to breathe but instead possess a pulmonary cavity within their mantle tissue which functions as a primitive lung. The roof of the pulmonary cavity is highly vascularised and provides the necessary respiratory surface for the uptake of oxygen from air. This clear evolutionary adaption to terrestrial life within the sub-class Pulmonata means that the few marine pulmonate genera are regarded as primitive members of this sub-class (Smith and Stansic 1998). Marine pulmonates are typically shelled inter-tidal molluscs and are very similar in appearance and habits to their distant relatives, the true limpets (Prosobranchia). In the southern hemisphere, studies of the secondary metabolites produced by pulmonate gastropods have been confined to the genera *Siphonaria* and *Trimusculus* occurring off the coasts of southern Africa, South America and Australia.

6.4.1 Southern African Marine Pulmonates

The natural products chemistry of four species of South African inter-tidal pulmonate molluscs has been reported.

Siphonarids (Order Basommatophora)

Molluscs of the genus *Siphonaria* (Family Siphonariidae) are air-breathing, shelled, inter-tidal herbivores often referred to as "false" limpets. Siphonariids have re-evolved a set of gill-like structures in the mantle cavity, enabling them to breathe effectively when they are submerged during high tides (Branch and Branch 1981). As the tide recedes, siphonariids revert to an air-breathing lifestyle by drawing air into the pulmonary cavity through a siphon situated on the right-hand side of the foot. During low tide, siphonariids graze on the algae and micro-organisms living on the rocks in the inter-tidal zone. In contrast to their distant "true" limpet relatives, siphonariids can be relatively easily displaced from the rock surface and are thus prone to predation by both terrestrial and aquatic predators at low and high tides, respectively (Beukes and Davies-Coleman 1999). In response to any perceived predatory pressure, siphonariids produce copious amounts of white mucus from lateral pedal glands. Although the mucus produced by siphonariids often contains a plethora of acyclic and cyclic propionate-derived metabolites, the role of these compounds or their acyclic precursors in the chemical ecology of *Siphonaria* species is not clear (Davies-Coleman and Garson 1998).

S. capensis is the most common of the nine *Siphonaria* species known to occur off the coast of southern Africa. A C-2 epimeric mixture of *E* and *Z*-siphonarienfuranone (**82**, **83**), capensinone (**84**), capensifuranone (**85**) and a known polypropionate biogenetic precursor (2*E*, 4*S*, 6*S*, 8*S*)-2,4,6, 8-tetramethyl-2-undecenoic acid (**86**) were isolated from specimens of *S. capensis* collected off the eastern Cape coast of South Africa (Beukes and Davies-Coleman 1999). The biogenesis of polypropionates has been extensively reviewed (Davies-Coleman and Garson 1998) and is also discussed in this volume by Darias et al. (in the chapter titled "The Chemistry of Marine Pulmonate Gastropods"). The 1,3-*syn* arrangement of methyl substituents in the aliphatic chains of siphonariid metabolites has proved to be an interesting synthetic challenge; and the recent synthesis of **85** (Williams et al. 2004) first confirmed the (*S*)-configuration at each of the three chiral centres in the side-chain of **85**, originally proposed from biosynthetic arguments, and second provided the (4*S*)-configuration in this compound which was unassigned by Beukes and Davies-Coleman (1999). Acetone extracts of two other *Siphonaria* species, *S. concinna* and *S. costatus* collected from the same region of the South African coast, afforded the ubiquitous siphonariid metabolite pectinatone (**87**) and siserrone A (**88**) as the major metabolites, respectively, (Beukes and Davies-Coleman 1999; Brecknell et al. 2000). The non-contiguous polypropionate skeleton of **88** led Brecknell et al. (2000) to question the natural product status of this compound and to propose that **88** was possibly a product of the facile rearrangement of another *S. serrata* metabolite dihydrosiphonarin A (**89**), during chromatographic work-up of the *S. serrata* extract. Both dihydrosiphonarin A and its ethyl homologue, dihydrosiphonarin B (**90**), were present as minor metabolites in the *S. serrata* extract. The ecological role of compounds **82–90** in southern African *Siphonaria* species was not established. The chemical structures of compounds **82–90** are presented in Fig. 6.7.

Trimusculids (Order Eupulmonata)

Marine pulmonate gastropods of the family Trimusculidae congregate in large colonies on the under-surface of inter-tidal rocky overhangs on exposed shores (Gray et al. 1998). In response to the constraints of their sedentary way of life, trimusculids secrete a mucous net, which they use to filter out food particles present in the water column at high tide. In common with other shelled pulmonates, trimusculids are subject to predation by a variety of sub-tidal and inter-tidal predators. Diterpenes predominate in the secondary metabolites produced by trimusculid molluscs as a chemical defence against predation and the two bioactive diterpene acetates (**91**, **92**) isolated by Gray et al. (1998) from the only trimusculid species known to occur off the southern African coast, *Trimusculus costatus*, were found to deter the feeding of *Pomadasys commersonni*, a common, omnivorous inter-tidal and sub-tidal southern African fish, at natural concentrations (ca. 2.5 mg per pellet).

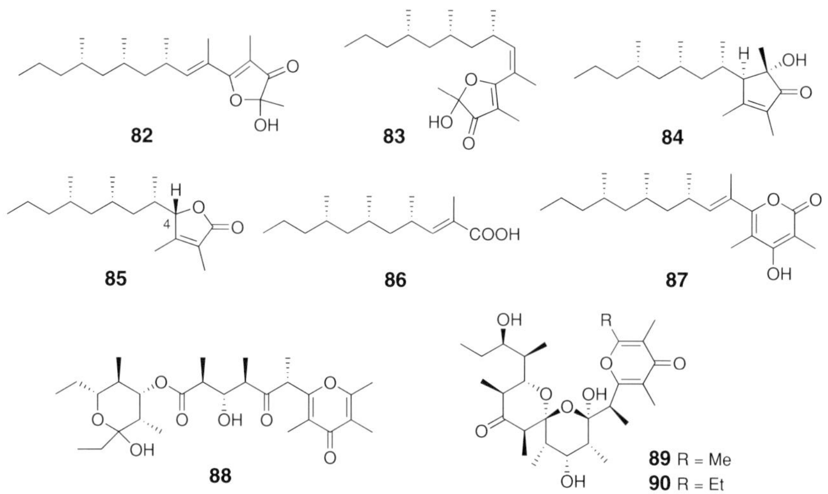

Fig. 6.7. Secondary metabolites isolated from the southern African pulmonates *S. capensis*, *S. costatus* and *S. concinna*

6.4.2 South American Marine Pulmonates

Studies of South American marine pulmonates have focussed exclusively on the trimusculid *T. peruvianus*.

Trimusculids (Order Eupulmonata)

A preliminary study of the diterpene secondary metabolites present in extracts of *T. peruvianus* specimens collected from the coast of Chile (Rovirosa et al. 1992) yielded the diterpene acetate (**93**). Additional investigations of the diterpene metabolites produced by this species afforded the four diterpene acetates (**94–97**) with a *Z* configuration assigned to the exocyclic olefin in these compounds (San Martin et al. 1996. Interestingly, the recent isolation of further four diterpene esters (**98–101**) containing a Δ^{13} *E*-olefin (Díaz-Marrero et al. 2003a) further confirmed that *T. peruvianus* biosynthesises both geometric isomers. All the diterpenes isolated thus far from *Trimusculus* species have been assumed to possess a labdane, as opposed to *ent*-labdane, stereochemistry and this assignment still requires corroboration. Finally, two polyhydroxylated steroids (**102**, **103**) with similar moderate in vitro cytotoxicity (IC_{50} 2.5 and 12.5 μg ml^{-1}, respectively) in two human colon tumour cell lines (H-116, HT-29) were also isolated from *T. peruvianus* (Díaz-Marrero et al. 2003b). The chemical structures of the diterpene and sterol metabolites isolated from both *T. costatus* and *T. peruvianus* are presented in Fig. 6.8.

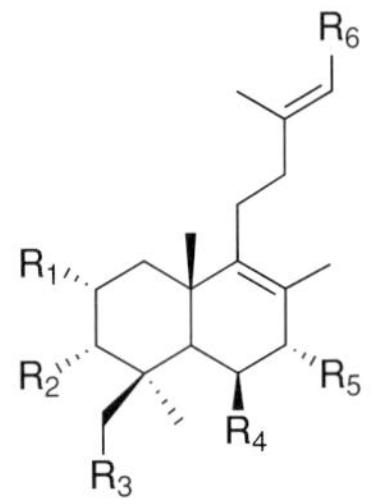

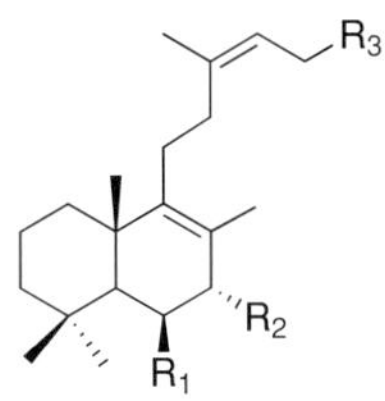

91 $R_1 = R_2 = R_3 = H$, $R_4 = R_5 = OAc$, $R_6 = CH_2OH$
92 $R_1 = R_4 = R_5 = OAc$, $R_2 = R_3 = H$, $R_6 = CH_2OH$
93 $R_1 = R_2 = R_3 = H$, $R_4 = OAc$, $R_5 = OH$, $R_6 = CH_2OH$
98 $R_1 = R_5 = H$, $R_2 = R_3 = R_4 = OH$, $R_6 = COOH$
99 $R_1 = R_5 = H$, $R_2 = OCOC_4H_9$, $R_3 = R_4 = OH$, $R_6 = COOH$
100 $R_1 = R_5 = H$, $R_2 = R_4 = OCOC_4H_9$, $R_3 = OH$, $R_6 = COOH$
101 $R_1 = R_5 = H$, $R_2 = R_3 = OCOC_4H_9$, $R_4 = OH$, $R_6 = COOH$

94 $R_1 = OH$, $R_2 = R_3 = OAc$
95 $R_1 = R_2 = R_3 = OAc$
96 $R_1 = R_2 = OAc$, $R_3 = OH$
97 $R_1 = OAc$, $R_2 = H$, $R_3 = OH$

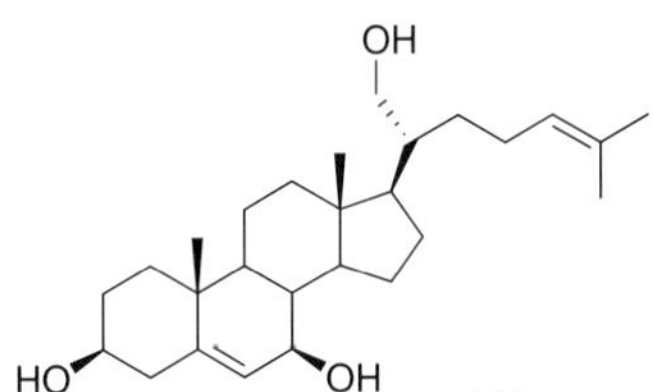

102

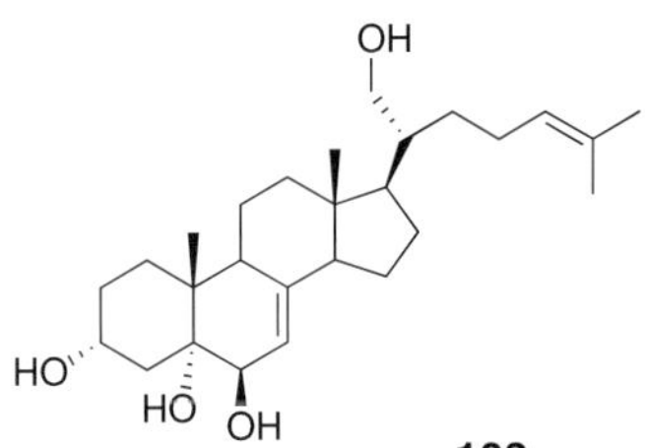

103

Fig. 6.8. Secondary metabolites isolated from the southern African and South American pulmonates *Trimusculus costatus* and *T. peruvianus*

6.5 Conclusions

Ironically, despite the relative inaccessibility of Antarctica and the obvious logistic difficulties in collecting gastropod molluscs from under the sea ice or from great depths in harsh and unpredictable climatic conditions that prevail in the southern Oceans, more appears to be known about the chemical ecology of Antarctic gastropod molluscs than the chemical ecology of the relatively more accessible marine molluscs of southern Africa and South America. Studies of the natural product chemistry of the marine gastropod fauna of South America and, to a lesser extent, southern Africa are still in a state of relative infancy. From the handful of natural products investigations of South American gastropod molluscs already carried out, it would appear that the extensive South American coastline (exceeding 140,000 km in length) holds enormous promise for future marine mollusc natural product studies. Supporting evidence for a putative evolutionary link between cool temperate, shallow-water South American marine invertebrate fauna and those of Antarctica, first alluded to by Dayton et al. (1994), could possibly be achieved from a study of the diterpene metabolites produced by specimens of *A. kerguelenensis* known to inhabit the cold waters off southern South America.

References

Avila C, Iken K, Fontana A, Cimino G (2000) Chemical ecology of the Antarctic nudibranch *Bathydoris hodgsoni* Eliot, 1907: defensive role and origin of its natural products. J Exp Mar Biol Ecol 252:27–44

Beesly PL, Ross GJB, Wells A (1998) Mollusca: the southern synthesis. Part B (Fauna of Australia, vol 5). CSIRO, Canberra, pp. 565–1234

Berlinck RGS, Hajdu E, Rocha RM da, Oliveira JHHL de, Hernandez ILC, Seleghim MHR, Granato AC, Almeida EVR de, Nunez CV, Muricy G, Peixinho S, Pessoa C, Moraes MO, Cavalcanti BC, Nascimento GGF, Thiemann O, Silva, M, Souza AO, Silva CL, Minarini PRR (2004) Challenges and rewards of research in marine natural products chemistry in Brazil. J Nat Prod 67:510–522

Beukes DR, Davies-Coleman MT (1999) Novel polypropionates from the South African marine mollusk *Siphonaria capensis*. Tetrahedron 55:4051–4056

Blunt JW, Copp BR, Munro MHG, Northcote PT, Prinsep MR (2005) Marine natural products. Nat Prod Rep 22:15–61

Branch G, Branch M (1981) The living shores of southern Africa. Struik, Cape Town

Branch GM, Griffiths CL, Branch ML, Beckley LE (1994) Two Oceans. David Philip, Cape Town

Brecknell DJ, Collett LA, Davies-Coleman MT, Garson MJ, Jones DD (2000) New non-contiguous polypropionates from marine molluscs: a comment on their natural product status. Tetrahedron 56:2497–2502

Carte B, Faulkner DJ (1983) Defensive metabolites from three nembrothid nudibranchs. J Org Chem 48:2314–2318

Cimino G, Ghiselin MT (1998) Chemical defense and evolution in the Sacoglossa (Mollusca: Gastropoda: Opisthobranchia). Chemoecology 8:51–60

Cimino G, Ghiselin MT (1999) Chemical defense and evolution in biosynthetic capacity among dorid nudibranchs (Mollusca: Gastropoda: Opisthobranchia). Chemoecology 9:187–207

Cimino G, Sodano G (1993) Biosynthesis of secondary metabolites in marine mollusks. Top Curr Chem 167:77–115

Cimino G, De Rosa S, De Stefanon S, Sodano G, Villani G (1983) Dorid nudibranch elaborates its own chemical defense. Science 219:1237–1238

Cimino G, Fontana A, Gavagnin M (1999) Marine opisthobranch molluscs: chemistry and ecology in sacoglossans and dorids. Curr Org Chem 3:327–372

Cimino G, Ciavatta ML, Fontana A, Gavagnin M (2001) Metabolites of marine opisthobranchs: chemistry and biological activity. In: Tingali C (ed) Bioactive compounds from natural sources. Taylor and Francis, London, pp. 577–637

Copley RCB, Davies-Coleman MT, Edmonds DR, Faulkner DJ, McPhail KL (2002) Absolute stereochemistry of ibhayinol from a South African sea hare. J Nat Prod 65:580–582

Davies-Coleman MT, Faulkner DJ (1991) New diterpenoic acid glycerides from the Antarctic nudibranch *Austrodoris kerguelensis*. Tetrahedron 47:9743–9750

Davies-Coleman MT, Garson MJ (1998) Marine polypropionates. Nat Prod Rep 15:477–493

Dayton PK, Robilliard GA, Paine RT, Dayton LB (1974) Biological accommodation in the benthic community at McMurdo Sound, Antarctica. Ecol Monogr 44:105–128

Dayton PK, Mordida BJ, Bacon F (1994) Polar marine communities. Am Zool 34:90–99

Díaz-Marrero AR, Dorta E, Cueto M, Rovirosa J, San-Martin A, Loyola A, Darias J (2003a) Labdane diterpenes with a new oxidation pattern from the marine pulmonate *Trimusculus peruvianus*. Tetrahedron 59:4805–4809

Díaz-Marrero AR, Dorta E, Cueto M, Rovirosa J, San-Martin A, Loyola A, Darias J (2003b) New polyhydroxylated steroids from the marine pulmonate *Trimusculus peruvianus*. ARKIVOC 10:107–117

Faulkner DJ (2002) Marine natural products. Nat Prod Rep 19:1–48

Faulkner DJ, Ghiselin MT (1983) Chemical defense and evolutionary ecology of dorid nudibranchs and some other opisthobranch gastropods. Mar Ecol Prog Ser 13:295–301

Fontana A, Muniain C, Cimino G (1998) First chemical study of patagonian nudibranchs: a new seco-11,12-spongiane, tyrinnal, from the defensive organs of *Tyrinna nobilis*. J Nat Prod 61:1027–1029

Fretter V, Graham A, Ponder WF, Lindberg DR (1998) Prosobranchs. In: Beesly PL, Ross GJB, Wells A (eds) Mollusca: the southern synthesis. Part B (Fauna of Australia, vol 5). CSIRO, Canberra, pp. 605–638

Gavagnin M, Fontana A (2000) Diterpenes from marine opisthobranch molluscs. Curr Org Chem 4:1201–1248

Gavagnin M, Trivellone E, Castelluccio F, Cimino G, Cattaneo-Vietti R (1995) Glyceryl ester of a new halimane diterpenoic acid from the skin of the Antarctic nudibranch *Austrodoris kerguelensis*. Tetrahedron Lett 36:7319–7322

Gavagnin M, De Napoli A, Cimino G, Iken K, Avila C, Garcia FJ (1999a) Absolute configuration of diterpenoid diacylglycerols from the Antarctic nudibranch *Austrodoris kerguelenensis*. Tetrahedron Asymm 10:2647–2650

Gavagnin M, De Napoli A, Castelluccio F, Cimino, G (1999b) Austrodorin-A and -B: first tricyclic diterpenoid 2′-monoglyceryl esters from an Antarctic nudibranch. Tetrahedron Lett 40:8471–8475

Gavagnin M, Ungur N, Castelluccio F, Muniain C, Cimino G (1999c) New minor diterpenoid diacylglycerols from the skin of the nudibranch *Anisodoris fontaini*. J Nat Prod 62:269–274

Gavagnin M, Fontana A, Ciavatta ML, Cimino G (2000) Chemical studies on Antarctic nudibranch mollusks. Ital J Zool 67[Suppl 1]:101–109

Gavagnin M, Carbone M, Mollo E, Cimino G (2003a) Further chemical studies on the Antarctic nudibranch *Austrodoris kerguelenensis*: new terpenoid acylglycerols and revision of the previous stereochemistry. Tetrahedron 59:5579–5583

Gavagnin M, Carbone M, Mollo E, Cimino G (2003b) Austrodoral and austrodoric acid: *nor*-sesquiterpenes with a new carbon skeleton from the Antarctic nudibranch *Austrodoris kerguelenensis*. Tetrahedron Lett 44:1495–1498

Gosliner T (1987) Nudibranchs of southern Africa: a guide to the opisthobranch molluscs of southern Africa. Sea Challengers, Monterey

Granato AC, Berlinck RGS, Magalhaes A, Schefer AB, Ferreira AG, De Sanctis B, De Freitas JC, Hajdu E, Migotto AE (2000) Natural products from the marine sponges *Aaptos* sp. and *Hymeniacidon* aff. *heliophila*, and from the nudibranch *Doris* aff. *verrucosa*. Quim Nov 23:594–599

Gray CA, Davies-Coleman MT, McQuaid C (1998) Labdane diterpenes from the South African marine pulmonate *Trimusculus costatus*. Nat Prod Lett 12:47–53

Graziani EI, Andersen RJ, Krug PJ, Faulkner DJ (1996) Stable isotope incorporation evidence for the de novo biosynthesis of terpenoic acid glycerides by dorid nudibranchs. Tetrahedron 52:6869–6878

Gustafson K, Andersen RJ (1985) Chemical studies of British Columbia nudibranchs. Tetrahedron 41:1101–1108

Iken K, Avila C, Ciavatta ML, Fontana A, Cimino G (1998) Hodgsonal, a new drimane sesquiterpene from the mantle of the Antarctic nudibranch *Bathydoris hodgsoni*. Tetrahedron Lett 39:5635–5638

Iken K, Avila C, Fontana A, Gavagnin M (2002) Chemical ecology and the origin of defensive compounds in the Antarctic nudibranch *Austrodoris kerguelenensis* (Opisthobranchia: Gastropoda). Mar Biol 141:101–109

Kaiser CR, Pitombo LF, Pinto AC (1998) C-13 and H-1 NMR assignments of the chamigrenes prepacifenol and dehydroxyprepacifenol epoxides. Spec Lett 31:573–585

Kaiser CR, Pitombo LF, Pinto AC (2000) NMR analysis of a complex spin system from a spiro-chamigrene. Spec Lett 33:457–467

Kaiser CR, Pitombo LF, Pinto AC (2001) Complete ^{1}H and ^{13}C NMR assignments of chamigrenes from *Aplysia dactilomela*. Magn Reson Chem 39:147–149

Kay EA, Wells FE, Ponder WF (1998) Class Gastropoda. In: Beesly PL, Ross GJB, Wells A (eds) Mollusca: the southern synthesis. Part B (Fauna of Australia, vol 5). CSIRO, Canberra, pp. 565–604

McClintock JB (1994) The trophic biology of Antarctic echinoderms. Mar Ecol Progr Ser 111:191–202

McClintock JB, Baker BJ (1997) A review of the chemical ecology of Antarctic marine invertebrates. Am Zool 37:329–342

McClintock JB, Janssen J (1990) Pteropod abduction as a chemical defence in a pelagic Antarctic amphipod. Nature 346:462–464

McClintock JB, Heine J, Slattery M, Weston J (1990) Chemical bioactivity in common shallow-water Antarctic marine invertebrates. Antarct J US 25:204–206

McClintock JB, Baker BJ, Hamann MT, Yoshida W, Slattery M, Heine JN, Bryan PJ, Jayatilake GS, Moon BH (1994a) Homarine as a feeding deterrent in common shallow-water Antarctic lamellarian gastropod *Marseniopsis mollis*: a rare example of chemical defense in a marine prosobranch. J Chem Ecol 20:2539–2549

McClintock JB, Baker BJ, Slattery M, Heine JN, Bryan PJ, Yoshida W, Davies-Coleman MT, Faulkner DJ (1994b) Chemical defense of common Antarctic shallow-water nudibranch *Tritoniella belli* Eliot (Mollusca: Tritonidae) and its prey *Clavularia frankliniana* Rouel (Cnidaria: Octocorallia). J Chem Ecol 20:3361–3372

McPhail K, Davies-Coleman MT (1997) New spongian diterpenes from the east African nudibranch *Chromodoris hamiltoni*. Tetrahedron 53:4655–4660

McPhail KL, Davies-Coleman MT (2005) (3*Z*)-Bromofucin from a South African sea hare. Nat Prod Res 19:449–452

McPhail K, Davies-Coleman MT, Coetzee P (1998) A new furanosesterterpene from the South African nudibranch *Hypselodoris capensis* and a Dictyoceratida sponge. J Nat Prod 61:961–964

McPhail KL, Davies-Coleman MT, Copley RCB, Eggleston DS (1999) New halogenated sesquiterpenes from South African specimens of the circumtropical sea hare *Aplysia dactylomela*. J Nat Prod 62:1618–1623

McPhail KL, Rivett DEA, Lack DE, Davies-Coleman MT (2000) The structure and synthesis of tsitsikammafuran: a new furanosesquiterpene from a South African *Dysidea* sponge. Tetrahedron 56:9391–9396

McPhail KL, Davies-Coleman MT, Starmer J (2001) Sequestered chemistry of the arminacean nudibranch *Leminda millecra* in Algoa Bay, South Africa. J Nat Prod 64:1183–1190

Pettit GR (1997) The dolastatins. Fortschr Chem Organ Naturst 70:1–79

Pika J, Faulkner DJ (1994) Four sesquiterpenes from the South African nudibranch *Leminda millecra*. Tetrahedron 50:3065–3070

Pika J, Faulkner DJ (1995) Unusual chlorinated homo-diterpenes from the South African nudibranch *Chromodoris hamiltoni*. Tetrahedron 51:8189–8198

Pitombo LF, Kaiser CR, Pinto AC (1996) Occurrence of chamigrenes in *Aplysia dactylomela* from Brazilian waters. Bol Soc Chil Quim 41:433–436

Rapson TD (2004) Bioactive 4-methoxypyrrolic natural products from two South African marine invertebrates. MSc thesis, Rhodes University, Rhodes

Rovirosa J, Quezada E, San-Martin A (1992) New diterpene from the mollusc *Trimusculus peruvianus*. Bol Soc Chil Quim 37:143–145

Rudman WB, Willan RC (1998) Opisthobranchia. In: Beesly PL, Ross GJB, Wells A (eds) Mollusca: the southern synthesis. Part B (Fauna of Australia, vol 5). CSIRO, Canberra, pp. 915–1035

San-Martin A, Quezada E, Soto P, Palacios Y, Rovirosa J (1996) Labdane diterpenes from the marine pulmonate gastropod *Trimusculus peruvianus*. Can J Chem 74:2471–2475

Smith BJ, Stansic J (1998) Pulmonata. In: Beesly PL, Ross GJB, Wells A (eds) Mollusca: the southern synthesis. Part B (Fauna of Australia, vol 5). CSIRO, Canberra, pp. 1037–1125

Ungur N, Gavagnin M, Mollo E, Cimino G (1999) Absolute stereochemistry of anisodorin 5. Tetrahedron Asymm 10:1635–1636

Williams DR, Nold AL, Mullins RJ (2004) Asymmetric conjugate addition for the preparation of *syn*-1,3-dimethyl arrays: synthesis and structure elucidation of capensifuranone. J Org Chem 69:5374–5382

Yoshida WY, Bryan PJ, Baker BJ, McClintock JB (1995) Pteroenone: a defensive metabolite of the abducted Antarctic pteropod *Clione antarctica*. J Org Chem 60:780–782

Marine Mollusks from Australia and New Zealand: Chemical and Ecological Studies

M.J. Garson

Abstract. Marine mollusks contain structurally diverse terpenes, polyketides, polypropionates and nitrogenous metabolites that may confer an ecological advantage on the host organism. From a chemical perspective, the most studied Australian taxa include representatives of the nudibranchs and sea hares, which are characterised by terpenes acquired from their specialised diets of sponges and algae, respectively. In contrast, siphonariid limpets that are prevalent on temperate seashores carry out *de novo* biosynthesis of polypropionate metabolites. Nitrogenous compounds isolated from Australian marine mollusks include precursors to the first commercially significant marine bioproduct, Tyrian Purple, and metabolites that are characteristic of ingested cyanobacteria.

7.1 Introduction

Over the past 30 years, there has been a rich tradition of marine natural products chemistry in both Australia and New Zealand. Within Australia, work in this interdisciplinary research field was initiated in the 1960s by Maurice Sutherland and his (then) Ph.D. student Joe Baker at The University of Queensland. Subsequently, Baker became Director of the Roche Research Institute for Marine Pharmacology (RRIMP; 1974–1981). During the 1980s, a number of university research groups were active in Australia, while the Australian Institute of Marine Science developed a substantial marine biodiscovery program under first Peter Murphy and later Chris Battershill. For over 10 years, AstraZeneca have funded a major natural products discovery initiative, led by Ron Quinn at Griffith University, while more recently a Centre for Molecular Biodiscovery has been established at The University of Queensland. Meanwhile in New Zealand, over 20 years of marine natural products research has been led by John Blunt and Murray Munro. Together with Brent Copp, Peter Northcote and Michelle Prinsep, they now write the annual updates on marine natural products that were first initiated by the late John Faulkner (Faulkner 2001; Blunt et al. 2004).

The research from Australasian groups initially targeted algae that could be collected in quantity and then bioactive sponge and ascidian

Mary Garson
School of Molecular and Microbial Sciences, The University of Queensland,
QLD 4072, Australia

Progress in Molecular and Subcellular Biology
Subseries Marine Molecular Biotechnology
G. Cimino, M. Gavagnin (Eds.): Molluscs

metabolites, because of the industry links that could be harnessed through identifying the pharmaceutical potential of these sessile invertebrates. Scientific research on mollusks requires a specialised collection strategy. Even though chemical ecology studies reveal that mollusks assimilate biologically active metabolites (for reviews, see Karuso 1987; Paul 1992; Cimino et al. 1999), there have been few chemical studies on Australian and New Zealand marine mollusks. For example, despite a prolific publication record on algal and sponge metabolites, there are no publications on mollusks arising out of RRIMP activities in the primary literature. In this chapter, the published chemistry from Australian and New Zealand mollusks is reviewed, and the chemical ecology and biosynthesis of selected metabolites that are of interest to the author are described.

7.2 Polyketide and Polypropionate Metabolites

7.2.1 Polypropionate Metabolites in Australian and New Zealand Mollusks

A recent review considered the structures, stereochemistry and synthesis of polypropionates from siphonariid mollusks or sacoglossans (Davies-Coleman and Garson 1998).

Pulmonates of the genus *Siphonaria* are air-breathing, intertidal mollusks commonly found in temperate zones. From an evolutionary perspective, these mollusks are significant since they may represent an evolutionary link between marine and terrestrial gastropods. Although siphonariid limpets feed by moving around rock platforms and pools, grazing on algae at low tides, there is no evidence that they assimilate algal metabolites from their food. Instead, the chemistry of this genus is characterised by the presence of polypropionate metabolites that are now known to be products of *de novo* synthesis.

The first polypropionates reported from an Australian siphonariid mollusk were isolated from *Siphonaria diemenensis* collected at Phillip Island and near Mallacoota, in Victoria. The diemenensins A (1) and B (2), both containing an α-pyrone ring and differing only in the configuration of the Δ6,7 double bond, show modest antibiotic activity and inhibit cell division in a sea urchin egg assay (Hochlowski and Faulkner 1983). Pectinatone (3) and norpectinatone (4) are α-pyrones that were isolated from *S. pectinata* collected at Wollongong, NSW (Garson et al. 1990), but which were first isolated from *S. pectinata* (Florida) and *S. lessoni* (Chile), respectively (Biskupiak and Ireland 1983; Capon and Faulkner 1984). The relative stereochemistry of these two metabolites was shown to be incorrect when the alleged structure of norpectinatone (5) and the C-9 epimer (6) were synthesised (Oppolzer et al. 1986). Subsequently, the structure of pectinatone was corrected by two independent X-ray studies (Garson et al. 1990; Norte et al. 1990). A recent total synthesis of pectinatone used SAMP-hydrazone methodology to introduce the methyl stereocentres (Birkbeck and Enders 1998).

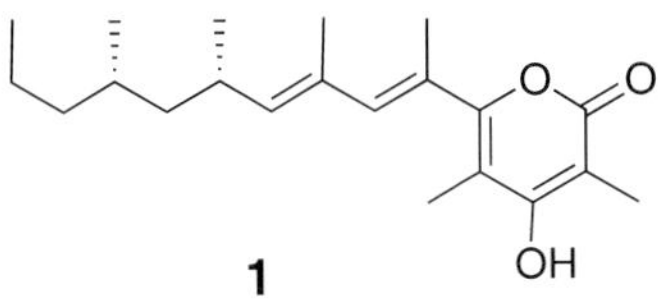

1

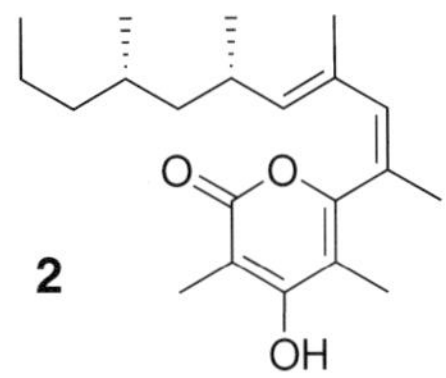

2

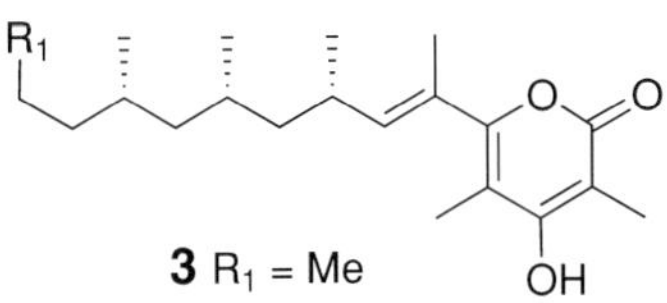

3 R_1 = Me
4 R_1 = H

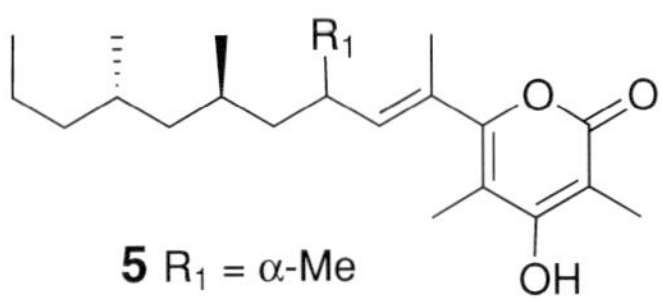

5 R_1 = α-Me
6 R_1 = β-Me

The siphonarins A (7) and B (8) are hemiacetal metabolites also containing a γ-pyrone unit that were first reported from *S. zelandica* collected in several parts of the Indo-Pacific. One collection was made from Bottle and Glass Rocks in Sydney Harbour, NSW, by the late John Faulkner while in Australia as a Queens Fellow in Marine Science. The relative stereochemistry of siphonarin A was deduced by single crystal X-ray analysis (Hochlowski et al. 1984), and the absolute stereochemistry deduced by X-ray analysis of a boronic acid derivative (Garson et al. 1994b) and by total synthesis of the γ-pyrone subunit (Paterson and Franklin 1994). The baconipyrones A–D (9–12) were isolated from *S. baconi* collected at Sorrento, Victoria, and are γ-pyrones that possess an unusual non-contiguous polypropionate structure (Manker et al. 1989). Although it was recognised that these metabolites originated from their siphonarin co-metabolites (Fig. 7.1), their isolation is now believed to represent post-collection chemistry rather than an enzymatic process (Brecknell et al. 2000). A total synthesis of (–)-baconipyrone C (12) established the absolute configuration of this series (Paterson et al. 2000), while asymmetric synthesis of the cyclohexanone subunit of baconipyrones A (10) and B (9) has been achieved (Turks et al. 2004).

Many siphonariid metabolites contain a hemiacetal ring. The denticulatins A (13) and B (14) (Fig. 7.2) are toxic hemiacetal metabolites first isolated from *S. denticulata* collected in Coledale or Eden, NSW, and differing from each other only in stereochemistry at C-10. The structure of denticulatin B was confirmed by single crystal X-ray analysis (Hochlowski et al. 1983). Several total syntheses of these metabolites have been reported (Ziegler and Becker 1990; Andersen et al. 1991a,b; Paterson and Perkins 1992, 1996; Oppolzer et al. 1995; De Brabander and Oppolzer 1997). Specimens of *S. australis* collected near Auckland were found to contain the hemiacetal **15** and a related ketoester **16** (Fig. 7.3) (Hochlowski and Faulkner 1984), whose absolute stereochemistry was later confirmed by total synthesis (Sundram and Albizati 1992). When the rearranged ester **17** was isolated from denticulatin A during an attempt to derivatise this compound using Sharpless dihydroxylation, it became further apparent that polypropionate esters (e.g. **9–12**, **16**) are artefacts that can be generated by mild basic treatment of hemiketal metabolites (Fig. 7.2) (Brecknell et al. 2000).

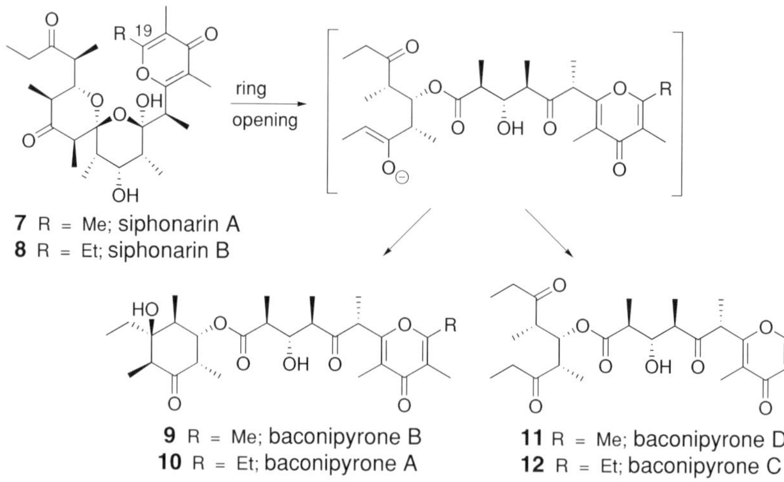

Fig. 7.1. Proposed conversion of siphonarins A and B into baconipyrones A–D in *Siphonaria zelandica*

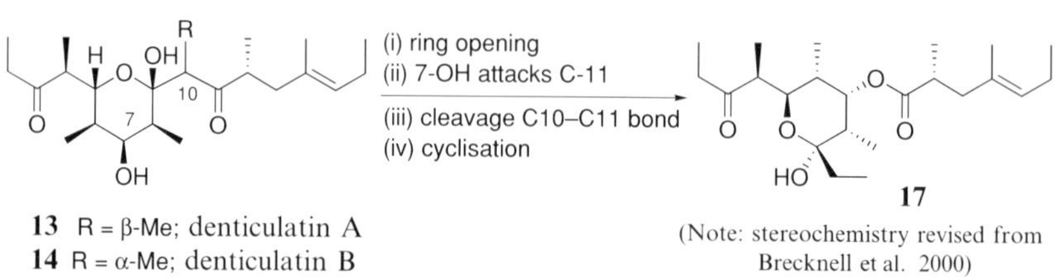

Fig. 7.2. Proposed conversion of denticulatin metabolites into polypropionate esters in *Siphonaria denticulata*

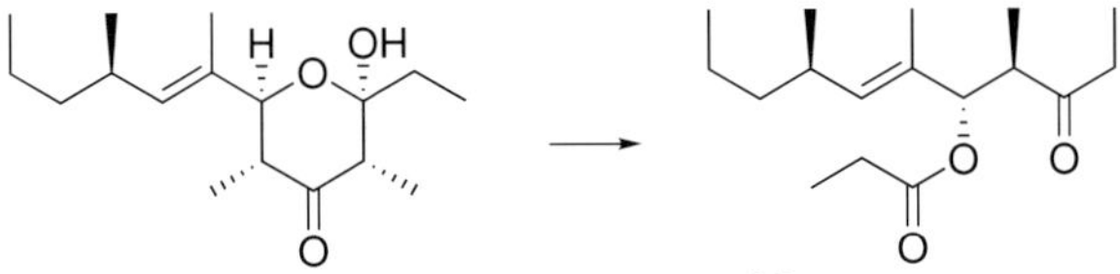

15 **16**

Fig. 7.3. Proposed conversion of a spiroacetal metabolite into a polypropionate ester in *Siphonaria australis*

Funiculatins A (**18**) and B (**19**) are polypropionates from *S. funiculata* collected on the Sunshine Coast of south-east Queensland that are structurally closely related to the denticulatins (Blanchfield et al. 1994). Two tricyclic polypropionate ketals have so far been identified, namely muamvatin (**20**) from a Fijian siphonariid (Roll 1986) and caloundrin B (**21**) from *S. zelandica* collected in south-east Queensland (Blanchfield et al. 1994), in which the unusual trioxaadamantyl moiety results from cyclisation of acyclic precursors (Blanchfield et al. 1994; Garson et al. 1994a).

The aposematically coloured sacoglossan *Cyerce nigricans* contains structurally related α– and γ-pyrones (**22**) and (**23**), related to metabolites of the cyercene family. The pyrones **22** and **23** were not responsible for the ichthyodeterrent properties of the whole-animal extract (Roussis et al. 1990). The original structure proposed for compound **23** has been revised to that shown here (Vardaro et al. 1992); and the structure suggested for pyrone **22** warrants further investigation.

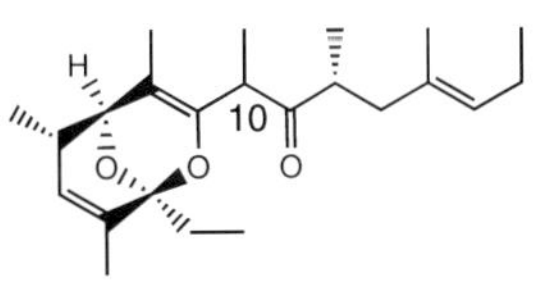

18, 19; epimers at C-10

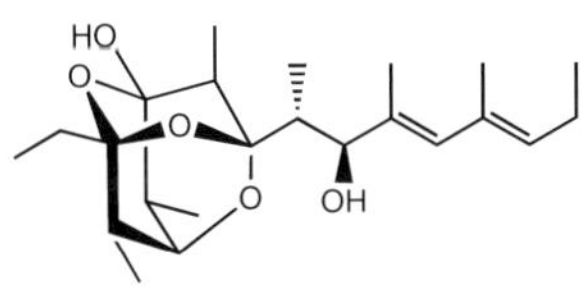

20

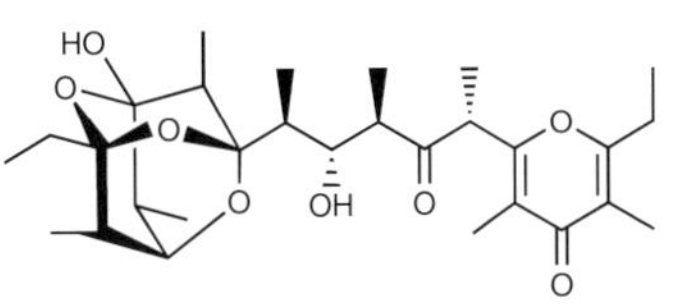

21

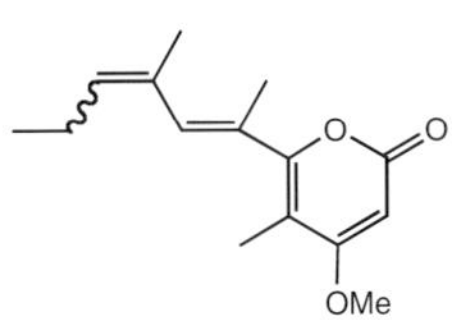

22

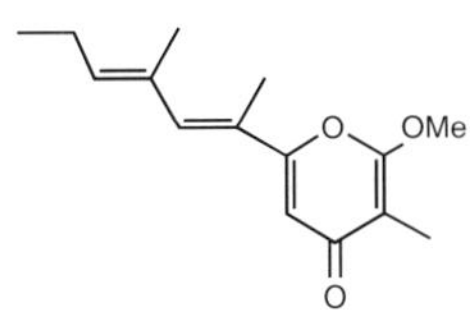

23

7.2.2 Polyketides

In Sydney, the sea hare *Aplysia parvula* selectively accumulates some fimbrolide metabolites (**24–27**) from the red alga *Delisea pulchra* at levels that are higher than in plant tissue (De Nys et al. 1996). The ecological role of dietary-derived metabolites sequestered by sea hares has been questioned, since these metabolites are frequently stored in the digestive gland rather than in skin tissue. Other functions that have been suggested include protection against infection, a role in chemical camouflage, or storage as part of a detoxification process (Pennings et al. 1999). For *A. parvula*, the major furanone **26** is concentrated in the digestive gland and is also present in skin tissue at concentrations that deter predators. However, the metabolite is not present in significant quantities in the opaline or ink secretions of this animal (De Nys et al. 1996; Rogers et al. 2000). Tasmanian specimens of *A. parvula* accumulate furanone **24** from a *Delisea* sp. (Jongaramruong et al. 2002). These brominated furanones have been intensively studied for their antifouling potential (Steinberg et al. 1998), but few studies on their synthesis have been reported in the primary literature (Beecham and Sims 1979; Manny et al. 1997).

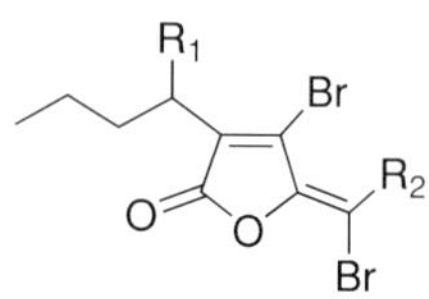

24 R_1 = H; R_2 = Br
25 R_1 = R_2 = Br
26 R_1 = OAc; R_2 = H
27 R_1 = OH; R_2 = Br

7.2.3 Biosynthetic Studies

The first studies on marine polypropionate biosynthesis investigated metabolites from siphonariid limpets and confirmed the operation of a polypropionate pathway, rather than an acetate–methionine route. The denticulatins A and B, **13** and **14**, were found to incorporate a radioactive label after a 6-day incubation period when [1-^{14}C] propionate was injected into specimens of *Siphonaria denticulata*. In contrast, incorporation of [1-^{14}C] acetate into these animals did not provide significantly labelled denticulatins. These data were consistent with a propionate-derived biosynthesis rather than the alternative addition of methyl groups from C_1-tetrahydrofolate metabolism onto a polyacetate backbone. Kuhn–Roth degradation revealed that only the carbons anticipated to originate from C-1 of propionate were labelled, but these data did not demonstrate a uniform distribution of ^{14}C label along the carbon chain. In an alternative incorporation protocol, the precursor was added directly to the aquarium water, from where it was absorbed directly through the skin tissue (Manker et al. 1988). Injection of [1-^{14}C] propionate into the foot tissue of *S. zelandica* followed by a 4-day incubation gave radiolabelled siphonarins A and B **7** and **8**, confirming their propionate origin (Garson et al. 1994b). The lower homologue, siphonarin A **7**, was shown to be assembled from an acetate chain starter unit with nine propionate-derived chain-building units, by determining that C-19 was not labelled by [1-^{14}C] propionate. The selective isolation of C-19 plus attached methyl carbon (as *p*-bromophenacyl acetate) was achieved by ozonolysis, hydrolysis of the ensuing anhydride and derivatisation (Garson et al. 1994b). Furthermore, acetate was utilised by *S. zelandica* for synthesis of **7**. The role of succinate in furnishing propionate units is consistent with the presence of a functioning methylmalonyl mutase in these mollusk.

7.3 Terpenes

7.3.1 Terpene Metabolites from Australian and New Zealand Mollusks

Several studies have explored the chemical relationships between Australian sea hares and the algal diets that contain sesquiterpenes. Studies on *A. parvula* from Sydney, NSW revealed that it accumulates *Laurencia obtusa* metabolites, such as the palisadins A **28** and B **29**, aplysistatin **30** and brasilenol **31**. Metabolite **28** is excreted in the mucous and opaline secretions of this animal consistent with a defensive role. The sea hare *A. dactylomela* also feeds on the same alga and stores low quantities of metabolites **28**, **29** and **30**. When fed on a diet of *L. obtusa* in laboratory experiments, the sea hare also accumulates brasilenol **31**, but this compound was never found in field-collected *A. dactylomela* (Rogers et al. 2000). Tasmanian specimens of *A. parvula* acquire the chamigrenes **32** and **33** and the pacifenols **34** and **35** from *L. filiformis* (Jongaramruong et al. 2002).

Nudibranchs frequently feed on terpene-containing sponges. A *Hypselodoris* sp. from south-east Queensland contains the well-known furanosesquiterpenes furodysinin **36** and dehydroherbadysidolide **37** (Garson 2004). Six sesquiterpene furans **38**–**43** have been reported from *Chromodoris epicuria* (previously named *Ceratosoma brevicaudatum*; Ksebati and Schmitz 1988). Furodysinin has also been isolated from *Asteronotus cespitosus* (Fahey and Garson 2002). All of these metabolites are characteristic of sponges of the genus *Lamellodysidea*, previously placed within the genus *Dysidea* (Cook and Bergquist 2002). A specimen of *Chromodoris coi* from Heron Island contained dendrillolide A **44** (Garson et al., unpublished data) while in south-east Queensland *C. elisabethina* consistently feeds on a grey-purple dictyoceratid sponge, as yet of unknown taxonomy, from which it sequesters the strongly cytotoxic terpene quinone metabolite puupehenone **45** (Garson 2004). The mollusk *Glossodoris atromarginata* collected at Mooloolaba, south-east Queensland, contains spongiane metabolites such as spongiadiol (**46**) and the acetate derivatives (**47**) and (**48**). On dissection, the mantle dermal formations contained spongiadiol-3α, 19-diacetate (**49**) (Garson et al., unpublished data). Nine spongiane diterpenes, including **50–52**, have been isolated from *C. epicuria* from Adelaide, South Australia (Ksebati and Schmitz 1987).

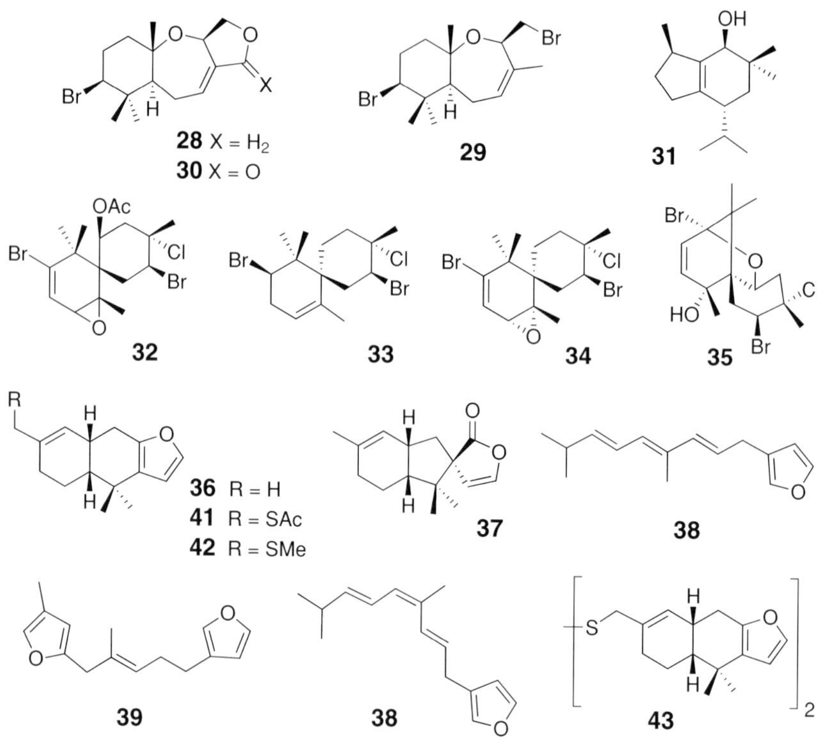

Soft corals can also be eaten by mollusks. On the Great Barrier Reef, the aeolid nudibranch *Phyllodesmium longicirra* feeds on the soft coral *Sarcophyton trocheliophorum* and accumulates diterpenes such as thunbergol **53**, trocheliophorol **54** and the epoxyalcohol **55** (Coll et al. 1985). The prosobranch mollusk *Ovula ovum* feeds on a *Sarcophyton* sp. and may detoxify ingested sarcophytoxide **56** by converting it into the deoxy derivative **57** prior to excretion in the faeces (Coll et al. 1983). The chemistry of the clam *Tridachna maxima* has been investigated and has yielded the sequiterpene germacrene C (**58**), previously isolated from a soft coral. This is suggestive of a zooxanthellar origin for the metabolite in both animals (Bowden et al. 1980).

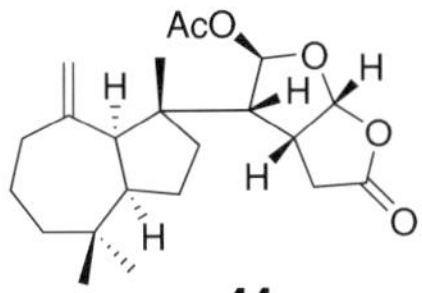

44

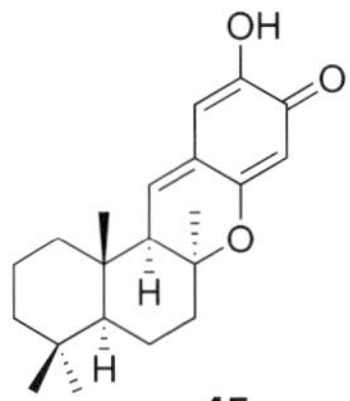

45

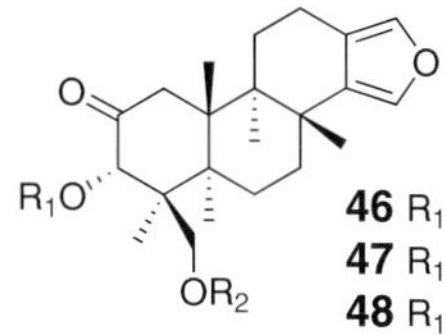

46 $R_1 = R_2 = H$
47 $R_1 = Ac; R_2 = H$
48 $R_1 = H; R_2 = Ac$
49 $R_1 = R_2 = Ac$

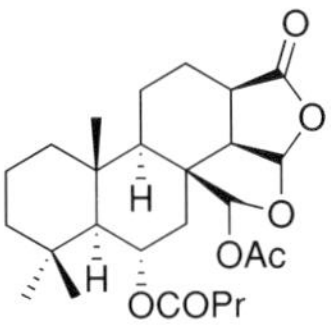

50

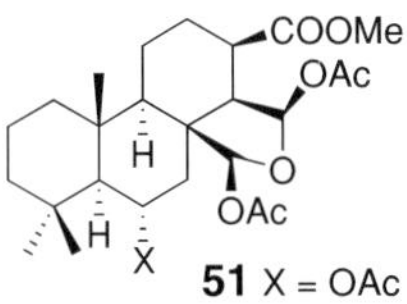

51 X = OAc
52 X = H

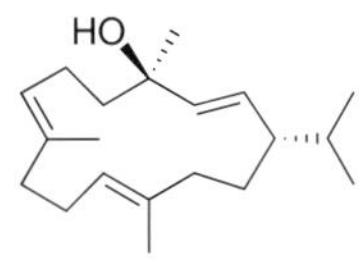

53

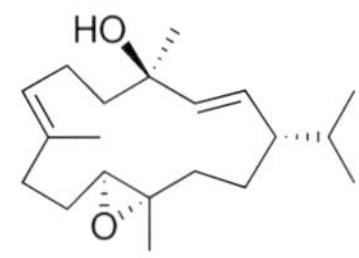

54

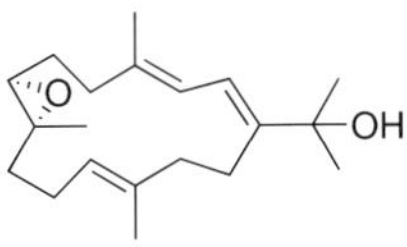

55

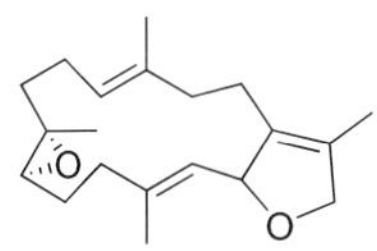

56

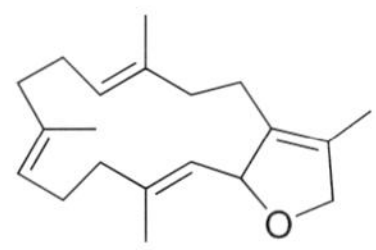

57

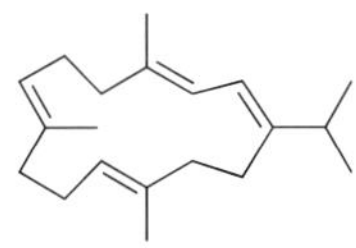

58

7.3.2 Chemical and Biosynthetic Studies on Phyllidid Nudibranchs

Nudibranchs from the genus *Phyllidia* consistently form associations with marine sponges that contain nitrogenous terpene metabolites. A N_1–C_1 functional group such as –NC, –NCS, or –NHCHO is common, but also found are the rarer –NCO, –SCN and –N=CCl_2 groups (Garson and Simpson 2004). Great Barrier Reef specimens of *Phyllidiella pustulosa* feed on the sponge *Acanthella cavernosa* from which they acquire terpene isocyanide and isothiocyanate metabolites (**59**–**62**) (Fig. 7.4). In aquaria, sponge samples converted ^{14}C-cyanide or -thiocyanate into the metabolites axisonitrile-3 (**59**) and axisothiocyanate-3 (**60**), while the mollusks, when injected with these precursors, were incapable of de novo synthesis. Mollusks were then allowed to feed on "labelled" sponges and subsequently contained radioactive metabolites that could only have come from the sponge (Dumdei et al. 1997). A useful chemical comparison is provided by the Fijian *P. pustulosa*, which accumulates axisonitrile-3 (**59**) and the isocyanide/isothiocyanate pair **63** and **64** from the sponge *Phakellia* (syn. *Acanthella*) *carduus* (Wright 2003).

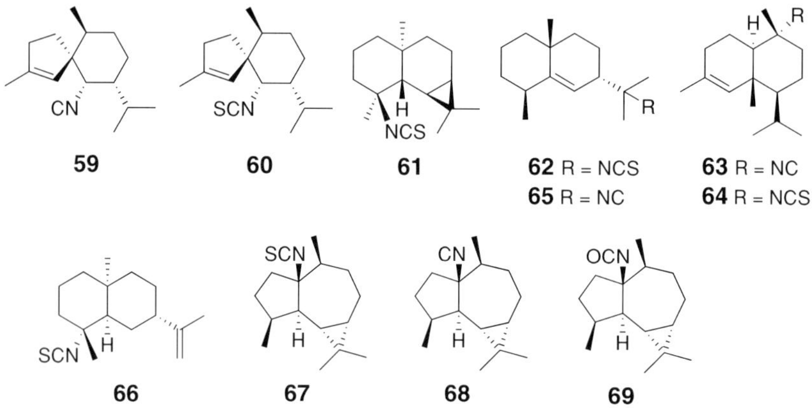

Fig. 7.4. Terpene metabolites isolated both from Phyllidid nudibranchs collected in Queensland, Australia, and in Fiji, and from their dietary sponge *Acanthella cavernosa*

Collections of "isocyanide" sponges and their associated phyllidid mollusks from south-east Queensland have provided great structural diversity. Specimens of *A. cavernosa* collected at Mooloolaba contained the known sesquiterpenes **59–60** and **62–69** (Fig. 7.4), three of which (**59**, **62**, **67**) were also found in the nudibranch *Phyllidia ocellata* that feeds on this sponge.

The rare isocyanate (**69**; Braekman et al. 1989) was found in trace quantities in the sponge (Stapleton 2002). Collections of *P. ocellata* from Rottnest Island in Western Australia have yielded isocyanides and isothiocyanate sesquiterpenes (e.g. **59**, **61**, **65**) that are again typical of an *A. cavernosa* diet (Stapleton 2002).

Study of the mollusk *P. varicosa* and its sponge prey *Axinyssa* n.sp. from south-east Queensland reveals that this mollusk also concentrates a range of sesquiterpene metabolites typical of its diet. Compounds isolated include tricyclic (**70–74**) and bicyclic (**75–76**) metabolites (Fig. 7.5) (Simpson et al. 1997; Stapleton 2002). Notably the metabolite, 9-thiocyanatopupukeanane (**71**), first reported together with its 9-epimer from Indonesian specimens of *P. varicosa* and its sponge source *A. aculeata*, (Yasman et al. 2003) was isolated from the Australian nudibranch, but has not yet been found in the sponge diet (Stapleton 2002). When tested in an ascidian larval assay (Degnan et al. 1996), the various sponge sesquiterpene fractions inhibited larval development, although individual *Acanthella* or *Axinyssa* metabolites were not effective inhibitors (Stapleton 2002). The reasons why the phyllidid mollusks selectively accumulate certain sponge chemicals, and the ecological implications of these specific sponge–nudibranch associations, require further study.

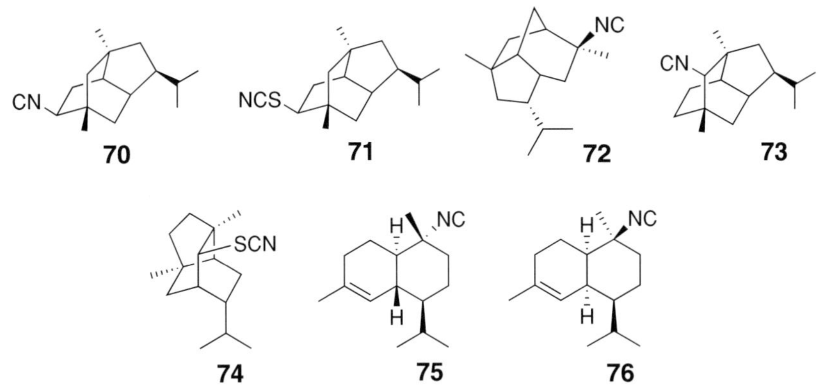

Fig. 7.5. Chemistry of *Phyllidiella varicosa* and its dietary sponge *Axinyssa* n.sp. from Mooloolaba, south-east Queensland. *Note*: compounds **75** and **76** have not been reported previously from this sponge

7.4 Miscellaneous Metabolites

In Sect. 7.3.1, the association of the tropical nudibranch *Asteronotus cespitosus* with the sponge *Lamellodysidea* (syn. *Dysidea*) *herbacea* was discussed. Further evidence of this dietary association is provided by the isolation of halogenated metabolites that are characteristic of this sponge. The bromophenols **77**, **78** and **79** have been isolated from Great Barrier Reef and West Australian samples of the mollusk, while a specimen collected under permit in the Philippines contained the chloropeptide **80** in addition to bromophenol metabolites (Fahey and Garson 2002).

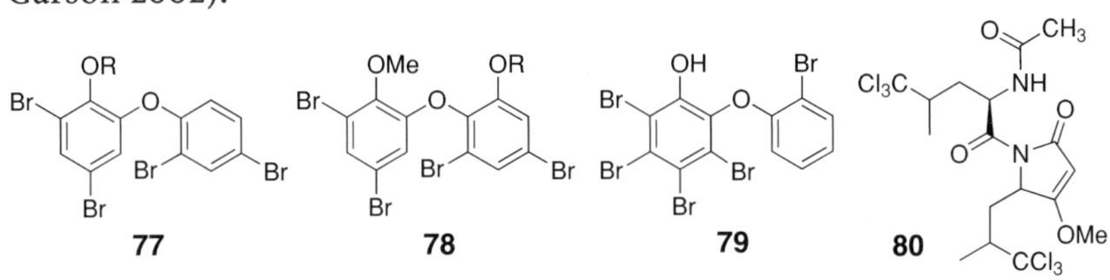

Mollusks of the families Muricidae and Thaisidae are sources of the purple dye Tyrian purple (**81**) that has been used since antiquity. The early chemical work of Baker with Sutherland (Baker and Sutherland 1968) and Duke (Baker and Duke 1973, 1976), which revealed the role of bromoindole sulphates (Fig. 7.6) in dye formation, has been reviewed (Baker et al, 1974; Benkendorff et al, 2000). A recent study revealed that fresh egg masses of the muricid *Dicathais orbita* are protected by the presence of tyrindoleninone (**82**). As the eggs mature and hatch into larvae, the monomer 6-bromoisatin (**83**) and the dimer tyriverdin (**84**) are formed. The antimicrobial or pathogenic properties of metabolites **82–84** appear to guard the eggs against infection during the ripening process (Benkendorff et al. 2000).

The nudibranch *Notodoris gardineri* is commonly found on sponges such as *Leucetta* sp. and *Pericharax* sp. in the northern Great Barrier Reef. Specimens of *N. gardineri* from Flynn Reef (Carroll et al. 1993) and from Lizard Island (Garson et al., unpublished data) contain the imidazole alkaloid clathridine **85**.

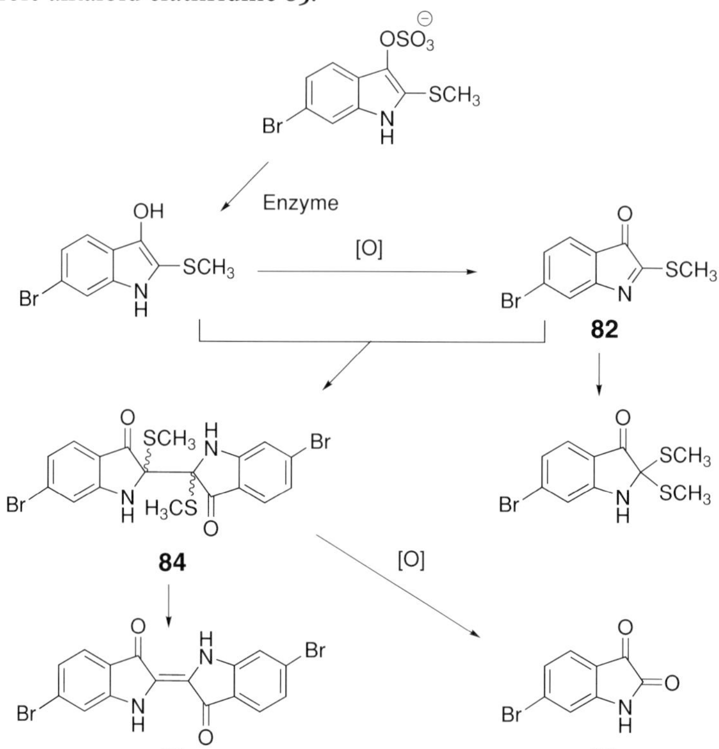

Fig. 7.6. The formation of Tyrian purple in *Dicathais orbita* (from Benkendorff et al. 2000)

Nitrogenous metabolites have also been isolated from New Zealand mollusks. The sea hare *Aplysia dactylomela* was shown to contain the tryptophan dipeptides **86** and **87** (Appleton et al. 2001), while *Bursatella leachii* contains a new biologically active malyngamide **88** (Appleton et al. 2002). These metabolites are likely to be of dietary origin, since the feeding preference of sea hares for chemically defended algae is well established. The isolation of the purple pigment aplysioviolin **89**, whose structure resembles the photosynthetic pigment phycoerythrin (Pennings et al. 1999), from a Tasmanian collection of the sea hare *A. parvula* in Tasmania must be related to its red algal diet (Jongaramruong et al. 2002).

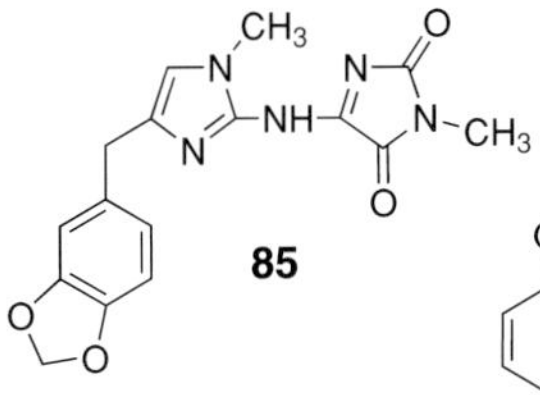

85

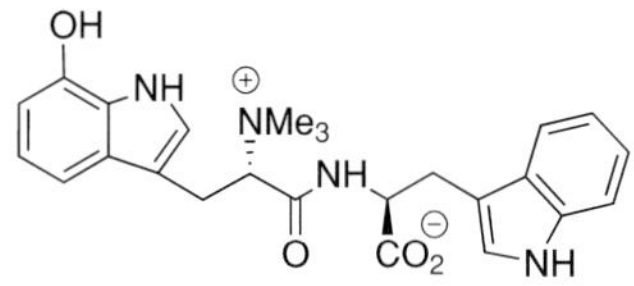

86

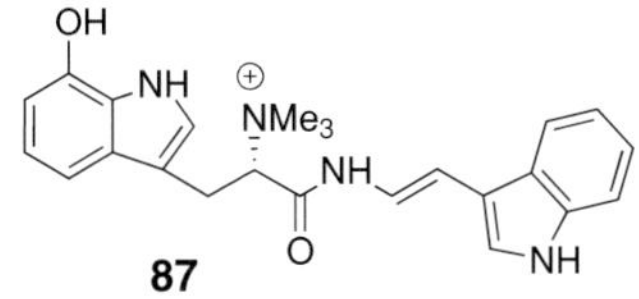

87

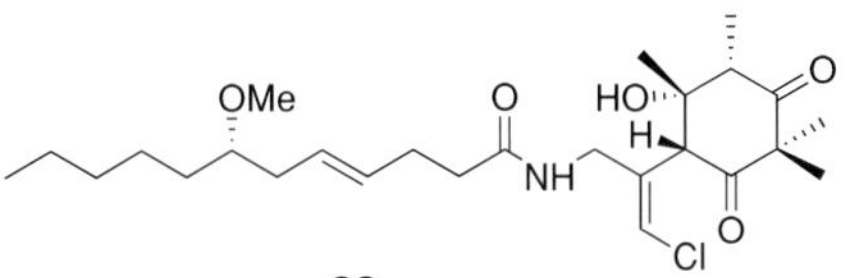

88

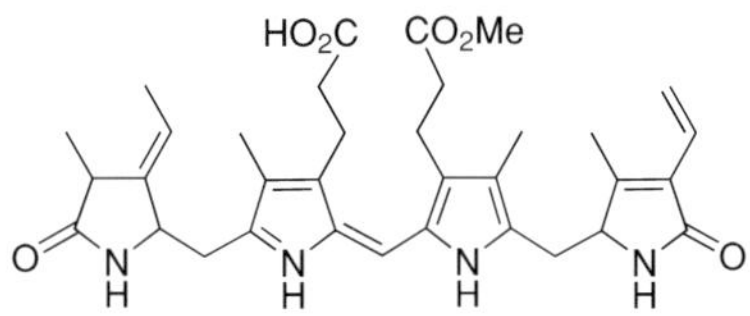

89

7.5
Conclusions

The survey above indicates that there continues to be much scope for chemical study of Australian and New Zealand mollusks. Although this group has received extensive taxonomic attention, current understanding of their natural products chemistry is limited. In contrast, there is a rich and diverse literature on the chemistry of Australian and New Zealand marine invertebrates (soft corals, sponges, ascidians) and algae (for a general overview, see Volkman 1999).

Experimental attention has focused largely on two classes of mollusk metabolites (i) the polypropionate metabolites of the *Siphonaria* and selected sacoglossans, (ii) the sesquiterpene metabolites from nudibranch mollusks and their dietary sources. The dietary associations between sea hares and algae have also been studied. The biosynthetic processes that lead to the metabolites, and an understanding of their natural biological roles, deserve detailed study.

Acknowledgments. Unpublished work from the author's laboratory was conducted under the appropriate permits by Bronwin Stapleton, Shireen Fahey, Respati Swasono and Michael Somerville, and is financed by grants from the Australia Research Council. Work in progress on *Glossodoris atromarginata* is conducted in collaboration with Drs. G. Cimino, E. Mollo and M. Gavagnin of the Instituto di Chimica Biomoleculare, CNR (Italy) and Dr. A. Kotze of CSIRO Livestock Industries (Australia).

References

Andersen MW, Hildebrandt B, Dahmann G, Hoffmann RW (1991a) Stereoselective total synthesis of the denticulatins. Chem Ber 124:2127–2139

Andersen MW, Hildebrandt B, Hoffmann RW (1991b) Efficient stereoselective total synthesis of denticulatins A and B. Angew Chem Int Ed Engl 30:97–99

Appleton DR, Babcock RC, Copp BR (2001) Novel tryptophan-derived dipeptides and bioactive metabolites from the sea hare *Aplysia dactylomela*. Tetrahedron 57:10181–10189

Appleton DR, Sewell MA, Berridge MV, Copp BR (2002) A new biologically-active malyngamide from a New Zealand collection of the sea hare *Bursatella leachii*. J Nat Prod 65:630–631

Baker JT (1974) Tyrian purple: ancient dye, modern problem. Endeavour 33:11–17

Baker JT, Duke CC (1973) Chemistry of the indolenines. II. Isolation from the hypobrancial gland of marine molluscs of 6-bromo-2,2-dimethylthioindolin-3-one and 6-bromo-2-methylthioindoleninone as alternative precursors to Tyrian purple. Aust J Chem 26:2153–2157

Baker JT, Duke CC (1976) Isolation of choline and choline ester salts of tyrindoxyl sulphate from the marine molluscs *Dicathais orbita* and *Mancinella keineri*. Tetrahedron Lett 1976:1233–1234

Baker JT, Sutherland MD (1968) Pigments of marine animals VIII. Precursors of 6,6′-dibromoindigotin (Tyrian purple) from the marine mollusc *Dicathais orbita* (Gmelin). Tetrahedron Lett 1968:43–46

Beecham CM, Sims JJ (1979) The first synthesis of fimbrolides, a novel class of halogenated lactone occuring in the red seaweed *Delisia fimbriata*. Tetrahedron Lett 1979:1649–1652

Benkendorff K, Davis AR, Bremner JB (2000) Tyrian purple precursors in the egg masses of the Australian muricid *Dicathais orbita*: a possible defense role. J Chem Ecol 26:1037–1050

Birkbeck AA, Enders D (1998) The total synthesis of (+)-pectinatone: an iterative alkylation approach based on the SAMP-hydrazone method. Tetrahedron Lett 39:7823–7826

Biskupiak JE, Ireland CM (1983) Pectinatone, a new antibiotic from the mollusc *Siphonaria pectinata*. Tetrahedron Lett 24:3055–3058

Blanchfield JT, Brecknell DJ, Brereton IM, Garson MJ, Jones DD (1994) Caloundrin A and funiculatin B: new polypropionates from siphonariid limpets. Aust J Chem 50:2255–2269

Blunt JW, Copp BR, Munro MHG, Northcote PT, Prinsep MR (2004) Marine natural products. Nat Prod Rep 21:1–49

Bowden BF, Coll JC, Mitchell SJ (1980) Studies of Australian soft corals. XXI. A new sesquiterpene from *Nepthea chabrolii* and an investigation of the common clam *Tridacna maxima*. Aust J Chem 33:1833–1839

Braekman JC, Daloze D, Moussiaux B, Stoller C, Deneubourg F (1989) Sponge secondary metabolites: new results. Pure Appl Chem 61:509–512

Brecknell DJ, Collett LA, Davies-Coleman MT, Garson MJ, Jones DD (2000) New non-contiguous polypropionates from marine molluscs: a comment on their natural product status. Tetrahedron 56:2497–2502

Capon RJ, Faulkner DJ (1984) Metabolites of the pulmonate *Siphonaria lessoni*. J Org Chem 49:2506–2508

Carroll AR, Bowden BF, Coll JC (1993) New imidazole alkaloids from the sponge *Leucetta* sp. and the associated predatory nudibranch *Notodoris gardineri*. Aust J Chem 46:1229–1234

Cimino G, Fontana A, Gavagnin M (1999) Marine opistobranch molluscs: chemistry and ecology in sacoglossans and dorids. Curr Org Chem 3:327–372

Coll JC, Tapiolas DM, Bowden BF, Webb L, Marsh H (1983) Transformation of soft coral (Coelenterata: Octocorallia) (*Sarcophyton* sp.) terpenes by *Ovula ovum* (Mollusca: Prosobranchia). Mar Biol 74:35–40

Coll JC, Bowden BF, Tapiolas DM, Willis RH, Djura P, Streamer M, Trott L (1985) Studies of Australian soft corals – XXXV. The terpenoid chemistry of soft corals and its implications. Tetrahedron 41:1085–1092

Cook SC, Bergquist PR (2002) Family Dysideidae Gray 1876. In: Hooper JNA, Van Soest RWM (eds) Systema Porifera: a guide to the classification of sponges. Plenum, New York, pp. 1061–1066

Davies-Coleman MT, Garson MJ (1998) Marine polypropionates. Nat Prod Rep 15:477–512

De Brabander J, Oppolzer W (1997) Enantioselective total synthesis of (–)-denticulatins A and B using a novel group-selective aldolisation of a *meso* dialdehyde as a key step. Tetrahedron 53:9169–9202

De Nys R, Steinberg PD, Rogers CN, Charlton TS, Duncan MW (1996) Quantitative variation of secondary metabolites in the sea hare *A. parvula* and its host plant *Delisea pulchra*. Mar Ecol Prog Ser 130:135–146

Degnan BM, Rohde PR, Lavin MF (1996) Normal development and embryonic gene activity of the ascidian *Herdmania momus*. Mar Freshwater Res 47:543–551

Dumdei EJ, Flowers AE, Garson MJ, Moore CJ (1997) The biosynthesis of sesquiterpene isocyanides and isothiocyanates in the marine sponge *Acanthella cavernosa*: evidence for dietary transfer to the dorid nudibranch *Phyllidiella pustulosa*. Comp Biochem Physiol 118B:1385–1392

Fahey SJ, Garson MJ (2002) Geographic variation of natural products of the tropical nudibranch *Asteronotus cespitosus*. J Chem Ecol 28:1773–1785

Faulkner DJ (2001) Marine natural products. Nat Prod Rep 18:1–49

Garson MJ (2004) Chemical associations between Australian nudibranchs and their dietary sponges. Chem Aust 71:16–19

Garson MJ, Simpson JS (2004) Marine isocyanides and related natural product structures – structure, biosynthesis and ecology. Nat Prod Rep 21:164–179

Garson MJ, Small CJ, Skelton BW, Thinapong P, White AH (1990) Structural and stereochemical correlations of polypropionate metabolites from marine pulmonates: revision of relative stereochemistry of pectinatone by X-ray structure analysis. J Chem Soc Perkin Trans 1:805–807

Garson MJ, Goodman JM, Paterson I (1994a) A configurational model for siphonariid polypropionates derived from structural and biosynthetic considerations. Tetrahedron Lett 35:6929–6932

Garson MJ, Jones DD, Small CJ, Liang J, Clardy J (1994b) Biosynthetic studies on polypropionates: a stereochemical model for siphonarins A and B from the pulmonate limpet *Siphonaria zelandica*. Tetrahedron Lett 35:6291–6294

Hochlowski JE, Faulkner DJ (1983) Antibiotics from the marine pulmonate *Siphonaria diemenensis*. Tetrahedron Lett 24:1917–1920

Hochlowski JE, Faulkner DJ (1984) Metabolites of the pulmonate *Siphonaria australis*. J Org Chem 49:3838–3840

Hochlowski JE, Faulkner DJ, Matsumoto GK, Clardy J (1983) The denticulatins, two polypropionate metabolites from the pulmonate *Siphonaria denticulata*. J Am Chem Soc 105:7413–7415

Hochlowski JE, Coll JC, Faulkner DJ, Biskupiak JE, Ireland CM, Qi-tai Z, Cun-heng H, Clardy J (1984) Novel metabolites of four *Siphonaria* species. J Am Chem Soc 106:6748–6750

Jongaramruong J, Blackman AJ, Skelton BW, White AH (2002) Chemical relationships between the sea hare *Aplysia parvula* and the red seaweed *Laurencia filiformis* from Tasmania. Aust J Chem 55:275–280

Karuso P (1987) Chemical ecology of the nudibranchs. In: Scheuer PJ (ed) Bioorganic marine chemistry. Springer, Berlin Heidelberg New York, pp. 31–60

Ksebati MB, Schmitz FJ (1987) New spongiane diterpenes from an Australian nudibranch. J Org Chem 52:3766–3773

Ksebati MB, Schmitz FJ (1988) Sesquiterpene furans and thiosesquiterpenes from the nudibranch *Ceratosoma brevicaudatum*. J Nat Prod 51:857–861

Manker DC, Garson MJ, Faulkner DJ (1988) De novo biosynthesis of polypropionate metabolites in the marine pulmonate *Siphonaria denticulata*. J Chem Soc Chem Commun 1061–1062

Manker DC, Faulkner DJ, Stout TJ, Clardy J (1989) The baconipyrones: novel polypropionates from the pulmonate *Siphonaria baconi*. J Org Chem 54:5371–5374
Manny AJ, Kjelleberg S, Kumar N, De Nys R, Read RW, Steinberg PD (1997) Reinvestigation of the sulfuric acid-catalysed cyclisation of brominated 2-alkyllevulinic acids to 3-alkyl-5-methylene-2(5*H*)-furanones. Tetrahedron 53:15813–15826
Norte M, Cataldo F, González AG, Rodríguez ML, Ruiz-Perez C (1990) New metabolites from the marine mollusc *Siphonaria grisea*. Tetrahedron 46:1669–1678
Oppolzer W, Moretti R, Bernadinelli G (1986) Enantioselective synthesis of the alleged structure of norpectinatone. Tetrahedron Lett 27:4713–4716
Oppolzer W, De Brabander J, Walther E, Bernadinelli G (1995) Asymmetric synthesis of (–)-denticulatins A and B via group-selective aldolization of a *meso* dialdehyde with a chiral *N*-propionylsultam. Tetrahedron Lett 36:4413–4416
Paterson I, Franklin AS (1994) Assignment of the absolute stereochemistry of the siphonarins and baconipyrones. Enantiocontrolled synthesis of a γ-pyrone unit. Tetrahedron Lett 35:6925–6928
Paterson I, Perkins MV (1992) Studies in polypropionate synthesis: stereoselective synthesis of (–)-denticulatins A and B. Tetrahedron Lett 33:801–804
Paterson I, Perkins MV (1996) Total synthesis of (–)-denticulatins A and B using efficient methods of acyclic stereocontrol. Tetrahedron 52:1811–1834
Paterson I, Chen DY-K, Aceña JL, Franklin AS (2000) Studies in marine polypropionate synthesis: total synthesis of (–)-baconipyrone C. Org Lett 2:1513–1516
Paul VJ (1992) Ecological roles of marine natural products. Comstock, Ithaca
Pennings SC, Paul VJ, Dunbar DC, Hamann MT, Lumbang WA, Novack B, Jacobs RS (1999) Unpalatable compounds in the marine gastropod *Dolabella auricularia*: distribution and effect of diet. J Chem Ecol 25:735–755
Rogers CN, De Nys R, Charlton TS, Steinberg PD (2000) Dynamics of algal secondary metabolites in two species of sea hare. J Chem Ecol 26:721–744
Roll DM, Biskupiak JE, Mayne CE, Ireland CM (1986) Muamvatin, a novel tricyclic spiro ketal from the Fijian mollusc *Siphonaria normalis*. J Am Chem Soc 108:6680–6682
Roussis V, Pawlik JE, Hay ME, Fenical W (1990) Secondary metabolites of the chemically rich ascoglossan *Cyerce nigricans*. Experientia 46:327–329
Simpson JS, Garson MJ, Hooper JNA, Cline EI, Angerhofer CK (1997) Terpene metabolites from the tropical marine sponge *Axinyssa* sp. Aust J Chem 50:1123–1127
Stapleton BL (2002) Structural studies of bioactive natural products from marine invertebrates. Ph.D. thesis, The University of Queensland, Brisbane
Steinberg PD, De Nys R, Kjelleberg S (1998) Chemical inhibition of epibiota by Australian seaweeds. Biofouling 12:227–244
Sundram UN, Albizati KF (1992) Synthesis and absolute configuration of polypropionate metabolites of *Siphonaria australis*. Tetrahedron Lett 33:437–440
Turks M, Murcia MC, Scopelliti R, Vogel P (2004) First asymmetric synthesis of the cyclohexanone subunit of baconipyrones A and B. Revision of its structure. Org Lett 6:3031–3034
Vardaro RR, Di Marzo V, Marin A, Cimino G (1992) α and γ-pyrone-polypropionates from the Mediterranean ascoglossan mollusc *Ercolania funerea*. Tetrahedron 48:9561–9566
Volkman JK (1999) Australasian research on marine natural products: chemistry, bioactivity and ecology. Mar Freshwater Res 50:761–779
Wright AD (2003) GC-MS and NMR analysis of *Phyllidiella pustulosa* and one of its dietary sources, the sponge *Phakellia carduus*. Comp Biochem Physiol 134A:307–313
Yasman, Edrada RA, Wray V, Proksch P (2003) New 9-thiocyanatopupukeanane sesquiterpenes from the nudibranch *Phyllidia varicosa* and its sponge prey *Axinyssa aculeata*. J Nat Prod 66:1512–1514
Ziegler FE, Becker MR (1990) Total synthesis of (–)-denticulatins A and B: marine polypropionates from *Siphonaria denticulata*. J Org Chem 55:2800–2805

Chemical Diversity in Opisthobranch Molluscs from Scarcely Investigated Indo-Pacific Areas

S. Wahidullah, Y.-W. Guo, I.M.I. Fakhr, E. Mollo

Abstract. Opisthobranch molluscs with both a rich variety of secondary metabolites and great biomedical potential represent the most intensively studied group of molluscs in natural product chemistry. We review here the chemical investigations into secondary metabolites of "sea slugs" from less-studied Indian, Chinese and Egyptian coasts, giving an overview of their most relevant biological activities. In addition to the biomedical interest of the metabolites, in which both structures and organisms often lose their own importance, this chapter emphasizes the phyletic and geographic distribution of the compounds in order to provide a further informational base for chemotaxonomical generalizations.

8.1 Introduction

The evolutionary reduction of the protective shell in the "sea slugs" (Mollusca: Gastropoda: Opisthobranchia) is correlated with different defensive adaptations, including autotomy, camouflage and the use of chemical weapons, often reinforced by aposematic colouration (Fig. 8.1).

Chemical defence is now understood to be the driving force behind the evolution of the group, preceding the abandonment of mechanical defence (Faulkner and Ghiselin 1983; Cimino and Ghiselin 1998, 1999, 2001). Opisthobranchs contain a majority of the known secondary metabolites within the Mollusca; and the chemicals that make them poisonous or at least extremely distasteful to potential predators have a wide range of structural characteristics, activities and origins. Due to this great chemical diversity, the potential of sea slugs to provide bioactive compounds is enormous and opisthobranchs have become a target for research leading to new drugs. In addition to their biomedical interest and from a less anthropocentric point of view, the chemical data on opisthobranchs provides the basis for studying chemotaxonomic and evolutionary aspects of the distribution of metabolites (Cimino and Ghiselin 1998, 1999, 2001) and a tool for the appreciation and the preservation of molecular diversity and biodiversity. Sea slug metabolites have been studied as "taxonomic markers" and also from the point of

Solimabi Wahidullah
National Institute of Oceangraphy (NIO-CSIR)
Dona Paula 403 004, Goa, India

Progress in Molecular and Subcellular Biology
Subseries Marine Molecular Biotechnology
G. Cimino, M. Gavagnin (Eds.): Molluscs

Fig. 8.1. From left: *Hypselodoris kanga*, *Glossodoris rufomarginata* and *Hexabranchus sanguineus*

view of geographic variation, but there has been little attempt to analyse the underlying causes of the geographical distribution patterns. Some of the defensive metabolites of opisthobranchs have dietary origin, while other compounds are biosynthesized *de novo* (Cimino et al. 1999, 2001a, 2004). Despite the evidence that many opisthobranchs, especially nudibranchs, have a highly specialized diet with a large number of monophagous species, the level of specialization is debatable (Todd et al. 2001), as well as the chemotaxonomic value of the isolated compounds. In this regard, chemical comparison of metabolites found in sea slugs and their prey as well as the study of intra-specific chemical variation assume great importance for the study of the phyletic distribution of compounds from opisthobranchs. Furthermore, comparative chemical studies between populations from different geographical locations have proved to be an informative tool in verifying the real specificity of the proposed chemical markers and/or clarifying dietary relationships. Faulkner et al. (1990) proposed that the variation in the chemical constituents of spongivorous species of nudibranchs could depend on the location, depth and habitat at which the collection was made and suggested that it is possible to predict the origin of defensive chemicals in molluscs by examining geographical variations in the secondary metabolites of any species. Unfortunately, most of the chemical data available on sea slugs comes from restricted areas, closer to specialized laboratories in the field of marine biology and natural product chemistry, while the data from other regions are scanty. Such ambiguities make it difficult to arrive at any definite conclusion regarding the relationship of chemical diversity to species diversity. Even though opisthobranch molluscs are widespread in the seas around India (Subba Rao and Dey 2000, Subba Rao 2003), along the South China Sea (Lin 1975, 1981, 1986; Rudman and Darvell 1990) and in the Red Sea (Yonow 1990, 2000), the literature on their chemistry is scanty. However, recently, an increasing number of papers have been published reporting chemical studies on opisthobranchs conducted in the frame of Indo-Italian (CSIR-CNR), Sino-Italian (CAS-CNR) and Egypto-Italian (ASRT-CNR) collaborative programmes. In this chapter, the chemical results on molluscs from Indian, Chinese and Egyptian coasts belonging to four opisthobranch orders (Table 8.1) have been reviewed, using a comparative approach with respect to both the geographical provenance of the animals and previous chemical findings. Biological activities have also been reviewed, to underline the biomedical interest of the isolated metabolites. To treat the topic exhaustively, we also include some chemical results, as yet unpublished. Even though the chemical data have been reviewed organizing the species studied into orders, for a more "natural" presentation we avoid further unnecessary subdivisions in other taxa, which are frequently subject to changes due to new data or interpretations. However, other categories have been mentioned in the final comparative discussion in assessing chemotaxonomic relationships among the species.

Table 1. Studied species and their provenance

Orders	From India	From China	From Egypt
Nudibranchia	*Jorunna funebris*	*Hexabranchus sanguineus*	*Hexabranchus sanguineus*
	Glossodoris atromarginata	*Glossodoris rufomarginata*	*Glossodoris cincta*
	Chromodoris mandapamensis	*Chromodoris reticulata*	*Chromodoris africana*
	Hypselodoris kanga	*Ceratosoma trilobatum*	
		Ceratosoma gracillimum	
		Phyllidiella pustulosa	
		Dermatobranchus ornatus	
Sacoglossa	*Volvatella* sp.		
	Plakobranchus ocellatus		
	Elysia grandifolia		
	Costasiella sp.		
Cephalaspidea	*Haminoea cymbalum*		
Anaspidea	*Bursatella leachii*	*Bursatella leachii*	
	Aplysia dactylomela	*Aplysia dactylomela*	

8.2 Studies on Nudibranchs

Nudibranchs are exclusively carnivorous with a broad range of alimentary habits. They appear to have the ability to sequester, accumulate and modify dietary metabolites as well as to synthesize defensive compounds *de novo* (Cimino and Ghiselin 1998; Cimino et al. 1999, 2001a, 2004) and this accounts for the vast array of secondary metabolites reported from them.

Hexabranchus sanguineus (Ruppell and Leuckart 1828) is a large nudibranch widespread throughout the tropical Indo-West Pacific Ocean. So-called "Spanish dancers" from Japan, Hawaii and the Marshall Islands are known to contain antitumour and antifungal macrolides (e.g. Kabiramide C, **1**; Matsunaga et al. 1986, 1989; Roesener and Scheuer 1986; Kernan et al. 1988) with a probable dietary origin from *Halichondria* sponges. Chemical studies on *H. sanguineus* collected along the Egyptian coast of Red Sea not only confirmed the presence of oxazole-containing macrolides but also resulted in the isolation of an unusual esterified carotenoid pigment, named hurghadin (**2**; Guo et al. 1998), which is structurally related to the sea anemone metabolite actinioerythrin (Hertzberg et al. 1969). More recently, chemical investigation on a specimen from Hainan (China) resulted surprisingly in the finding of a series of terpenoids possessing different carbon skeletons (e.g. **3**–**5**; Zhang et al., unpublished data). One of them, *ent*-isocopalane diterpene **3** is a novel molecule, while the sesquiterpene formamide **4** and the furanosesquiterpenoid (–)-furodysinin **5** have been previously isolated from sponges of the genera *Axinella* (Fattorusso et al. 1975) and *Dysidea* (Guella et al. 1985), respectively.

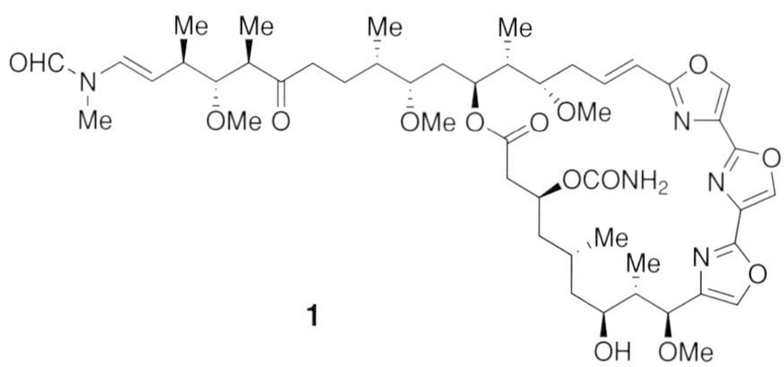

1

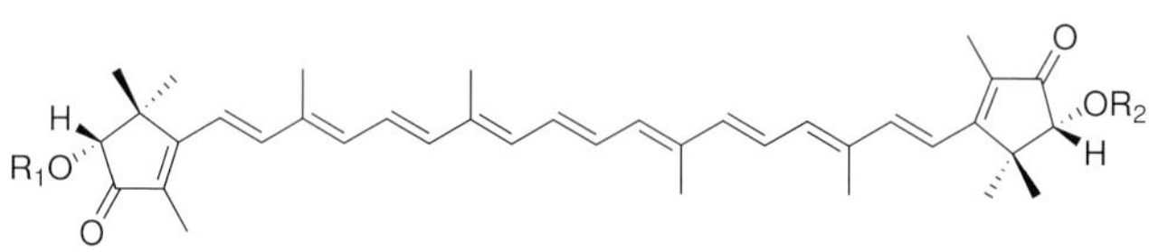

2 R1 + R2 = C27:2 + C20:1 or C20:2 + C18:0 or C20:1 + C18:1 or C19:0 + C18:1 or C18:1 + C18:0

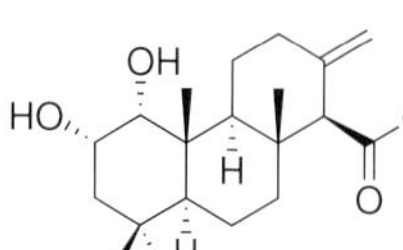

3

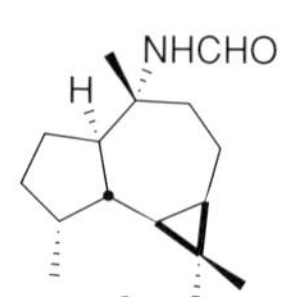

4

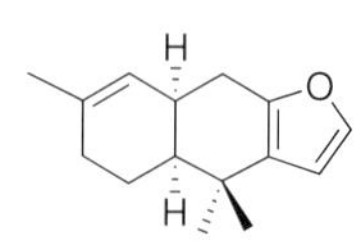

5

Jorunna funebris (Kelaart 1858) is distributed in the tropical Indo-West Pacific Ocean. Sri Lankan *J. funebris* was reported to sequester a series of isoquinoline–quinone metabolites (e.g. compound **6**) from its prey, the sponge *Xestospongia* sp. (Karuso 1987; De Silva and Gulavita 1988). *J. funebris* from Mandapam (Tamil Nadu, India) afforded the dimeric isoquinoline alkaloid jorumycin (**7**; Fontana et al. 2000), which displays a close resemblance to renieramycin E, a metabolite of the sponge *Reniera* sp. (He and Faulkner 1989). Jorumycin (**7**) showed antimicrobial activity against *Bacillus subtilis* and *Staphylococcus aureus*, and cytotoxic activity at very low concentrations against different tumour cell lines (Cimino et al. 2001b).

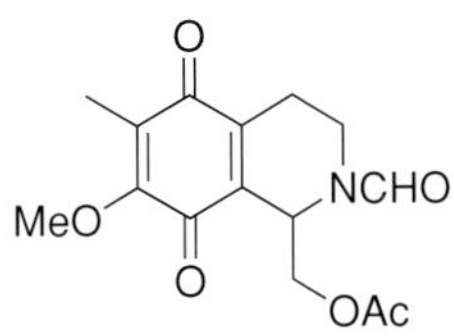

6

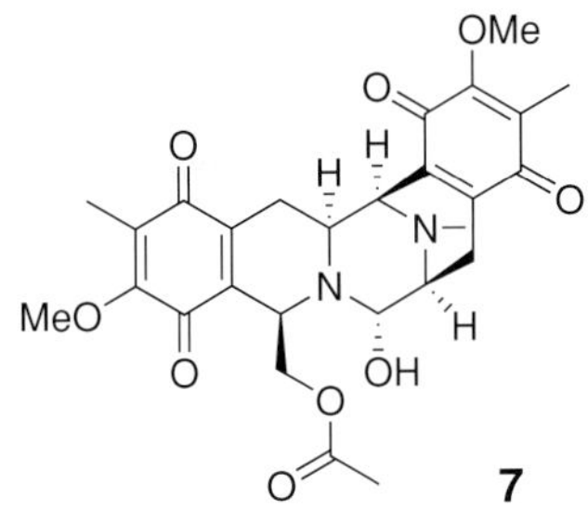

7

Glossodoris atromarginata (Cuvier 1804) is found throughout the tropical and subtropical Pacific and Indian Oceans. This mollusc, collected off Mandapam (Tamil Nadu, India) was found in association with the two different potential preys, the sponge *Hyatella cribriformis* and an unidentified sponge probably belonging to the genus *Spongia* (Fontana et al. 1999a, 2001). The sesterterpene deoxoscalarin (**8**), previously reported by Cimino et al. (1973) from a Mediterranean sponge, was the sole product isolated in *G. atromarginata* individuals collected on *H. cribriformis*. The presence of **8** also in *Hyatella* confirmed its dietary origin. The nudibranchs found on *Spongia* contained the sesterterpene heteronemin (**9**), already known from the sponge *Heteronema erecta* (Kazlauskas et al. 1976; Kashman and Rudi 1977), and the two new scalaranes **10** and **11**. Compound **11** was also identified in the extract of the associated sponge, consistent with its dietary origin. This finding suggested that the mollusc could obtain compound **10** by bio-transformation of dietary **11**. In a preliminary assay, selective cytoxicity against human thyroid carcinoma was reported for compounds **9** and **10** (Fontana et al. 1999a). Interestingly, Sri Lankan *G. atromarginata* (=*Casella*), unlike the Indian specimen, contained spongiane furanoditerpenoids (**12**–**17**; De Silva and Scheuer 1982) previously reported as metabolites of Australian sponges, *Spongia* spp. (Kazlauskas et al. 1979). Although the authors lacked information about the preferred food of the nudibranch, a dietary origin of the diterpenoids from *Spongia* was suggested.

Glossodoris cincta (Bergh 1888), widely distributed in the Indo-West Pacific, seems to have regional colour forms (Rudman 1986). Nine spongian diterpenoids (**18**–**26**) have been isolated from a nudibranch, first identified as *G. atromarginata* and subsequently revised as *G. cincta*, collected along the Egyptian coast of the Red Sea (Fontana et al. 1997, 1999a). Compounds **23**–**26** are known sponge metabolites, some of which display cytotoxic and antiviral properties (Kohmoto et al. 1987; Cambie et al. 1988). A dietary origin from sponges is also inferred for the novel compounds **18**–**22**, on the basis of their structural analogies with the earlier sponge metabolites.

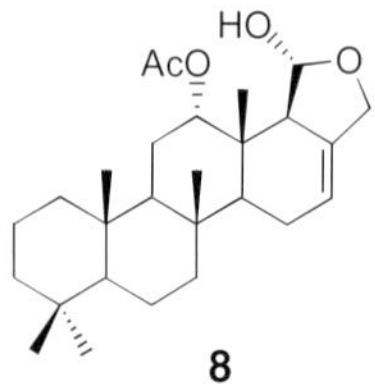

8

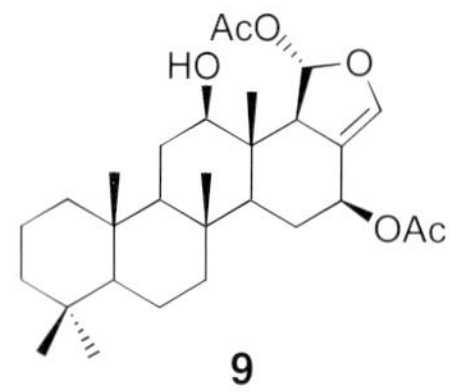

9

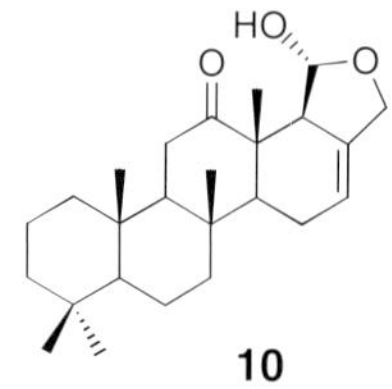

10

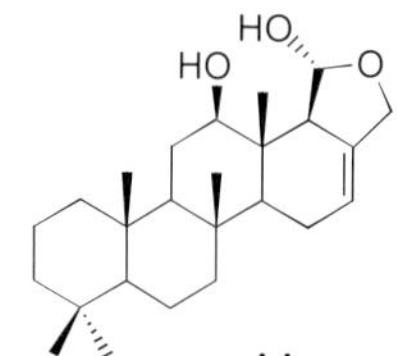

11

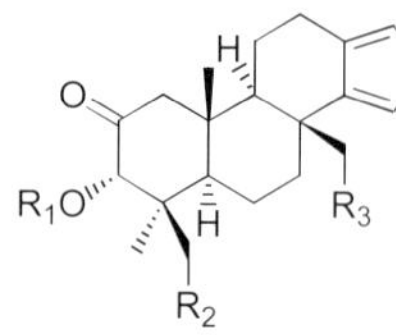

12 R_1=Ac, R_2=R_3=OAc
13 R_1=H, R_2=R_3=OAc
16 R_1=Ac, R_2=OAc, R_3=H
17 R_1=R_3=H, R_2=OAc

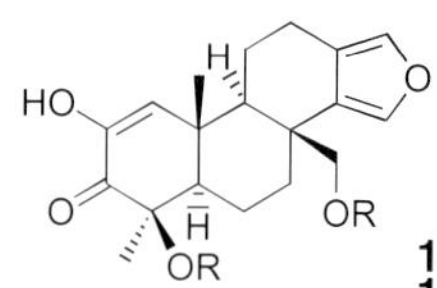

14 R=Ac
15 R=H

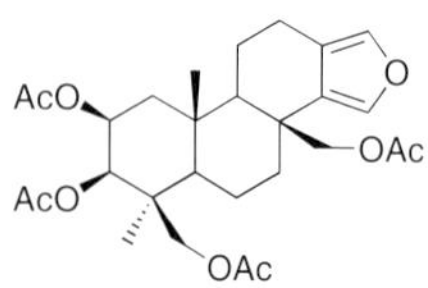

18

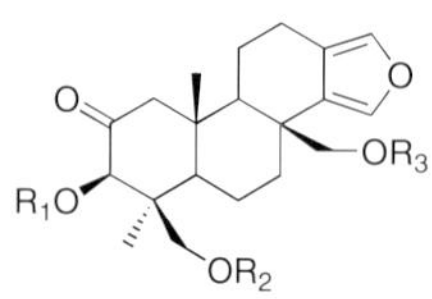

19 R_1=H, R_2=R_3=Ac
20 R_1=R_2=Ac, R_3=H
21 R_1=Ac, R_2=R_3=H
23 R_1=R_2=R_3=Ac
24 R_1=R_2=R_3=H

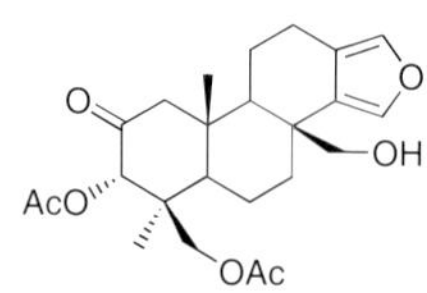

22

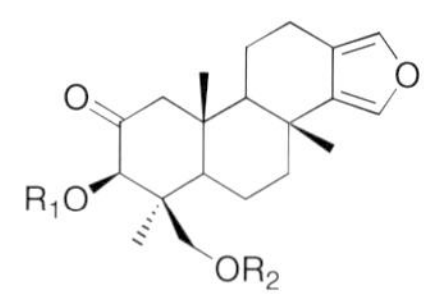

25 R_1=Ac, R_2=H
26 R_1=R_2=H

Glossodoris rufomarginata (Bergh 1890) is distributed in the tropical Indo-West Pacific. Chemical studies on this nudibranch and the sponge on which it was caught off Hainan (China) led to the isolation and identification of seven sesterterpenes (**27–33**) belonging to the class of the scalarane compounds, in structural analogy with the metabolites from the Indian *G. atromarginata*. Scalaranes have been described as an extremely important class of anti-inflammatory compounds (Jacobs et al. 1993), the first member of which is the well-known scalaradial (**27**; Cimino et al. 1974). Scalaradial (**27**) and its 12-deacetyl derivative (**28**) were the main constituents of the associated sponge, whereas scalaranes **8, 10, 28–31** were isolated from *G. rufomarginata*. The trophic relationship between the two organisms was confirmed by the presence of 12-deacetylscalaradial (**28**) in both the mollusc and the sponge, while the absence of scalaradial (**27**) in a nudibranch containing a series of its derivatives suggested the ability of the mollusc to transform dietary toxic **27** into related molecules (Gavagnin et al. 2004). Interestingly, other scalarane metabolites (e.g. 12-epi-scalarin, **32**; and 6-keto-deoxoscalarin, **33**) have been isolated from nudibranchs of different genera and provenance (Cimino et al. 1999; and literature cited therein).

Chromodoris mandapamensis (Valdés, Mollo and Ortea 1999), a new species described from Mandapam (Tamil Nadu, India), was found to contain spongiadiol (**26**), also found in the earlier-mentioned *G. cincta* from Egypt and previously isolated from Australian sponges (Kazlauskas et al. 1979), together with a mixture of related spongiane compounds

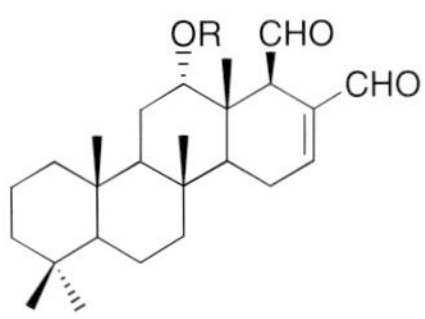

27 R=Ac
28 R=H

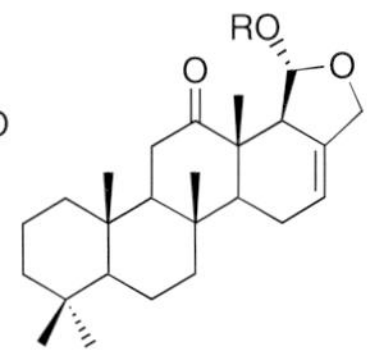

29 R=Ac

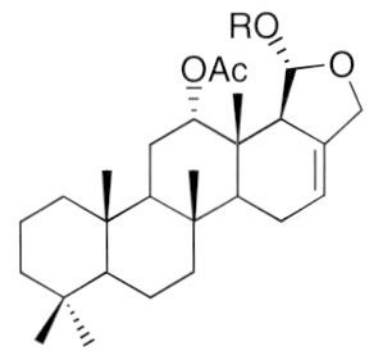

30 R=Ac

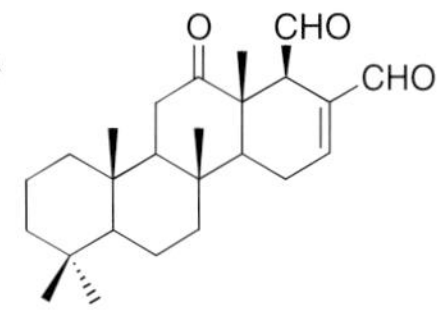

31

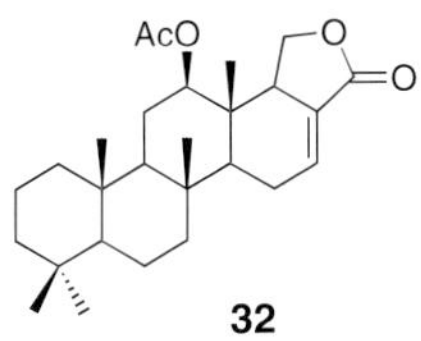

32

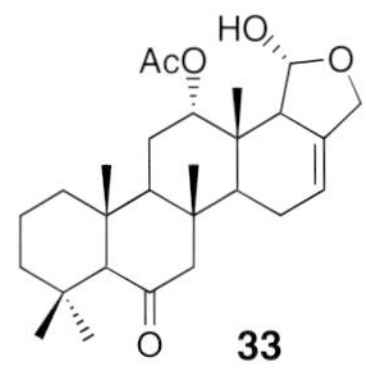

33

(Fontana et al. 2001). The unidentified sponge on which the nudibranch was collected also yielded spongiadiol (**26**), supporting its dietary origin in the mollusc. Compound **26** has been reported to have antiviral activity against *Herpes simplex* virus (Kohmoto et al. 1987). Here, mention must be made that chemical studies on the skin extract of four *Chromodoris* nudibranchs from neighbouring sites in Sri Lanka led to the identification of rearranged spongiane diterpenes (De Silva et al. 1991). In particular, shahamin K (**34**), along with dendrillolide A (**35**) and 12-desacetoxyshahamin C (**36**), found in the sponge *Dendrilla* sp. (Bobzin and Faulkner 1989a), were reported from *C. gleneii*; also, two new acetoxyspongianes (**37**, **38**) together with 12-epi-aplysillin (**39**), previously isolated from the Mediterranean *C. luteorosea* (Cimino et al. 1990), were present in the extracts of *C. geminus*, whereas *C. annulata* contained shahamin F (**40**; Carmely et al. 1988); and finally, the known sponge metabolites aplyroseol 2 (**41**; Karuso et al. 1986), spongian-16-one (**42**; Kernan et al. 1990) and its acetoxy derivative (**43**; Karuso et al. 1986; Bobzin and Faulkner 1989b) occurred in *C. inopinata*. The last two compounds are also known from the Japanese *C. obsoleta*, showing cytotoxic activity (Miyamoto et al. 1996).

Chromodoris reticulata (Quoy and Gaimard 1832) is one of a complex of Indo-Pacific species. Chemical studies on one individual found along the coast of Hainan (China) yielded, in analogy with the earlier Sri Lankan *C. inopinata*, the known cytotoxic furano diterpenoids **41** and **43** (Gavagnin et al., unpublished data), in addition to the related compound **44** reported earlier as a constituent of sponges (Schmitz et al. 1985). Such chemical analogy supported the idea that the two species should be considered a different colour form to *C. tinctoria* (Ruppell and Leuckart 1828), as proposed by Rudman (1973).

Chromodoris africana (Eliot 1904) is a nudibranch common in the Red Sea. Chemical studies on individuals collected along Egyptian coasts led to the identification of the furanoterpene kurospongin (**45**) and the 14-membered macrolide with an attached 2-thiazolidinone unit, latrunculin B (**46**; Guo 1997). Compound **45** was previously isolated from the sponge

Spongia sp. in Okinawan waters, showing ichthyotoxic and feeding-deterrent activities (Tanaka and Higa 1988), while compound **46**, described as being ichthyotoxic and capable of inducing changes in cell morphology, is a member of the group of latrunculins which have been found in the red-coloured sponge *Latrunculia magnifica* from the Gulf of Eilat (Kashman et al. 1980, 1985). It is noteworthy that there are similarities between *C. africana* and *C. quadricolor* from the Gulf of Aqaba (Mebs 1985; reported as *Glossodoris quadricolor*), in that both contain latrunculin B (**46**).

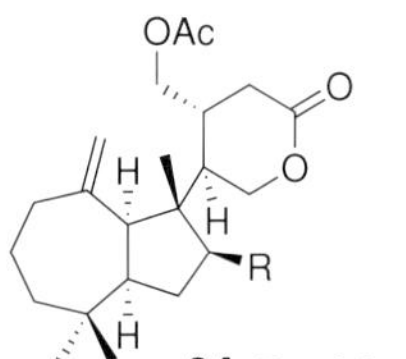
34 R = OAc
35 R = H

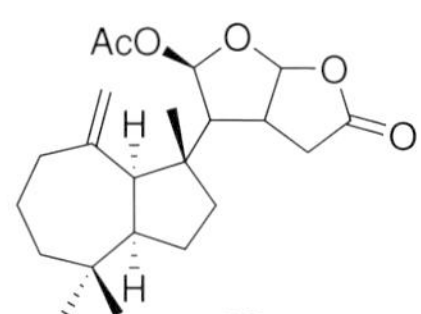
36

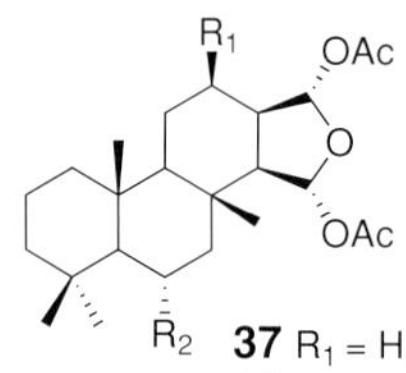
37 R_1 = H, R_2 = OAc
38 R_1 = OAc, R_2 = OAc
39 R_1 = OAc, R_2 = H

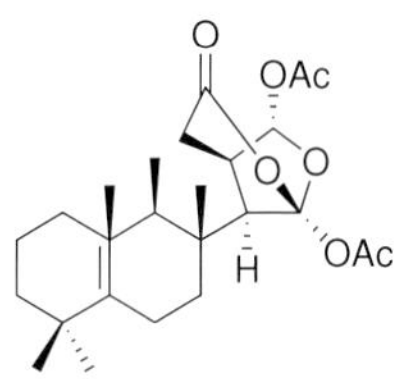
40

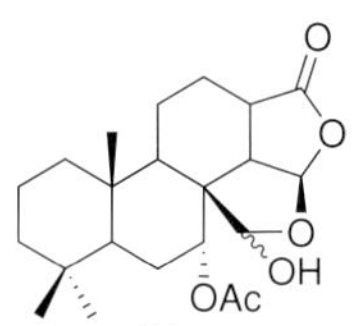
41

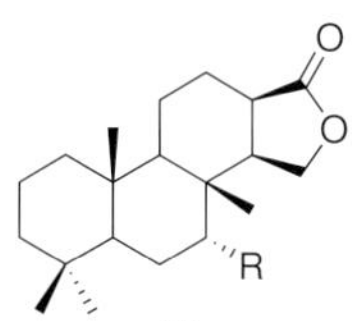
42 R = H
43 R = OAc

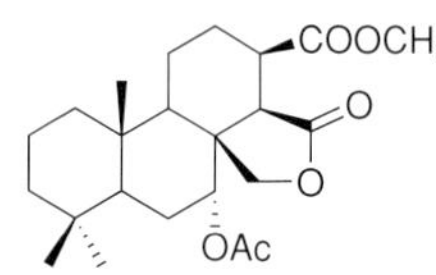
44

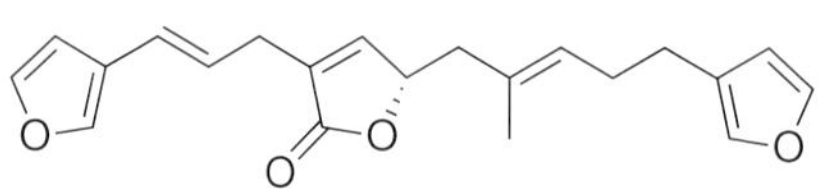
45

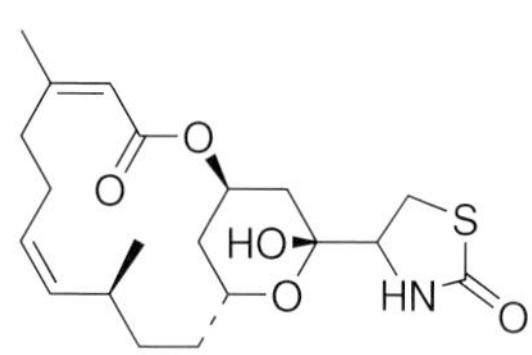
46

Hypselodoris kanga (Rudman 1977) is distributed from the Western Indian Ocean to Hong Kong. Specimens of this mollusc from Mandapam (Tamil Nadu, India) and a sample of the associated sponge *Dysidea* sp. were both found to contain the sesquiterpenoid furodysinin (**47**), already known from a *Dysidea* sponge (Kazlauskas et al. 1978), suggesting a trophic relationship (Fontana et al. 2001). The presence of compound **47** as the major metabolite in the mantle gland of *H. kanga* suggested its selective transfer into defensive structures. Here, mention may be made of a furodysinin lactone derivative, also isolated from a *Dysidea* sponge, as a potent agonist to human leukotriene B4 receptors (Mong et al. 1990).

Ceratosoma trilobatum (Gray 1827) and *C. gracillimum* Semper in (Bergh 1876) are tropical Indo-West Pacific nudibranchs. Both species collected along the South China Sea coast were found to contain the four furanosesquiterpenoids **48–50** and **5** (Mollo et al. 2005). These compounds, previously isolated from *Dysidea* sponges (Cimino et al. 1975; Charles et al.

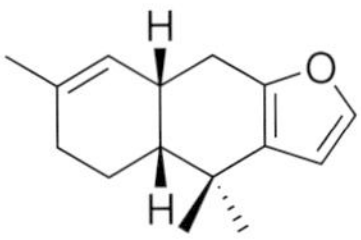
47
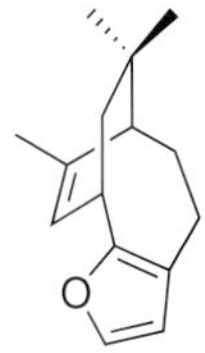
48
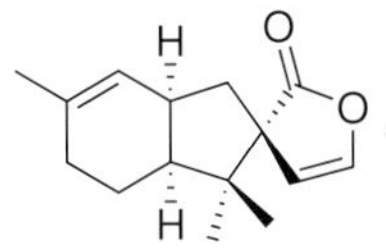
49
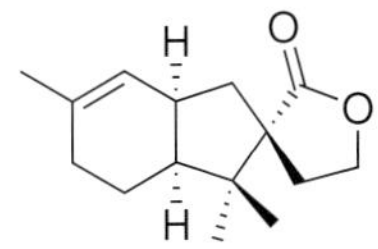
50

1978; Guella et al. 1985; Cameron et al. 2000), most probably have a dietary origin. The almost exclusive presence of (–)-furodysinin (**5**), which showed significant feeding-deterrent and ichthyotoxic properties, in the mantle glands concentrated in the dorsal horn of *C. gracillimum* suggested a defensive role of that dorsal protuberance.

Phyllidiella pustulosa (Cuvier 1804) is one of the most common nudibranchs throughout the tropical Indo-West Pacific. This nudibranch, analogously with other phyllidiids, is known to contain sesquiterpene isocyanides (Karuso 1987; Cimino et al. 1999; Garson and Simpson 2004). Antifouling, antibiotic, antifungal and antitumour properties of these molecules have been extensively investigated (Chang et al. 1987; Fusetani et al. 1990; Okino et al. 1995, 1996; Hirota et al. 1996; Ciavatta et al. 1999). The recent chemical investigation on a Chinese population of *P. pustulosa* from Hainan (China) resulted in the first finding of isocyanide diterpenoids (**51–54**) in a phyllidiid nudibranch, along with the expected isocyanide sesquiterpenoids (**55–58**; Manzo et al. 2004). Compounds **55**, **57** and **58** were novel compounds, while **51–53** were previously identified from *Acanthella* sponges (Chang et al. 1984, 1987; Patra et al. 1984; Fusetani et al. 1990; Shimomura et al. 1999), confirming the prey-predator relationship between the studied nudibranch and sponges of the genus *Acanthella*, proposed by Dumdei et al. (1997). In contrast, amphilectene **54** was previously isolated from *Hymeniacidon* and *Cribochalina* sponges (Wratten et al. 1978; Ciavatta et al. 1999), while 3-isocyano-theonellin **56**, previously found in *Phyllidia* sp. from Sri Lanka (Gulavita et al. 1986), is closely related to a cyanide reported from an *Axinyssa* sponge (Iwashima et al. 2002).

Dermatobranchus ornatus (Bergh 1874) is distributed in the tropical Indo-West Pacific. Chemical studies on *D. ornatus* from Hainan (China) led to the isolation of four known diterpenoids of the eunicellin class (**59–62**) from its mantle extract (Zhang et al., unpublished data). Compounds **60** and **61** were also present in the cnidarian *Muricella sinensis* (Verrill), collected off the same place where the nudibranch was found (Yan 2004), while **62** was isolated from an unidentified Pacific soft coral (Hochlowski and Faulkner 1980). Ophirin (**59**), previously found in *Muricella* spp. from the Red Sea (Kashman 1980) and from Korea (Seo et al. 1997), has been described to have inhibitory activities against both the growth of silkworm *Bombyx mori* (Ochi et al. 1991) and the cell division of fertilized

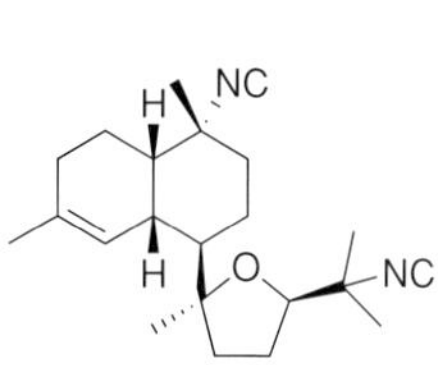

51

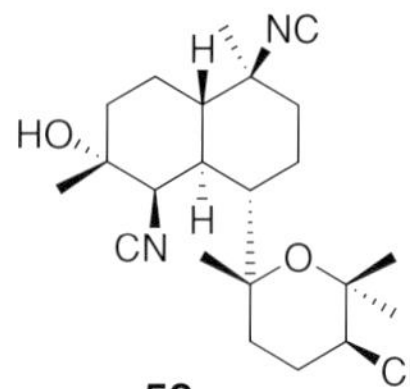

52

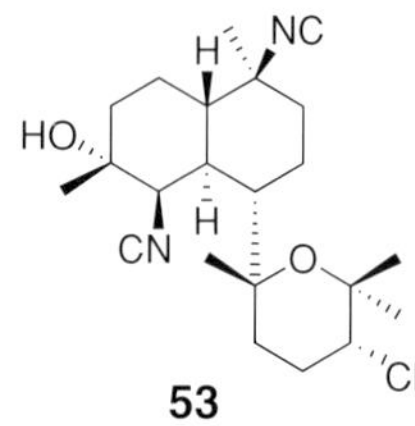

53

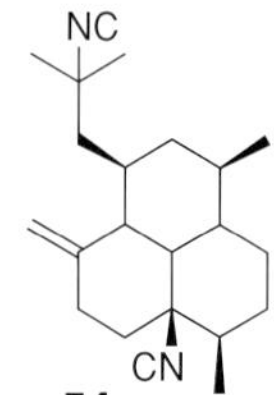

54

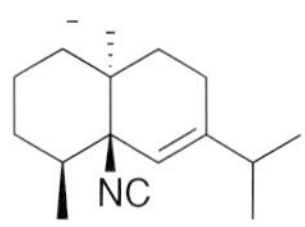

55

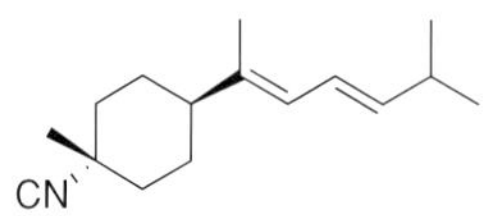

56

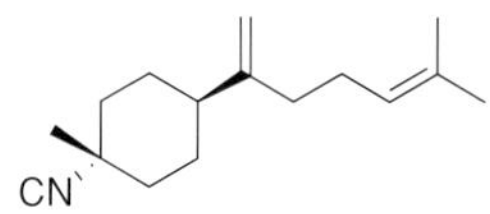

57

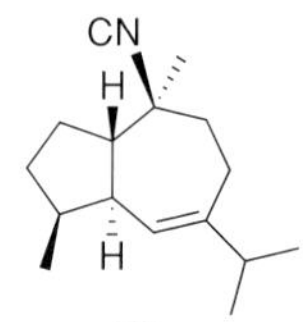

58

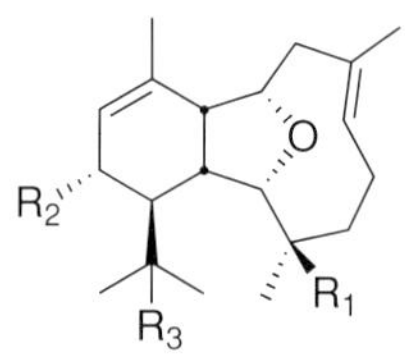

59 $R_1 = R_2 = R_3 = OAc$
60 $R_1 = R_2 = OAc$, $R_3 = H$
61 $R_1 = OAc$, $R_2 = R_3 = H$
62 $R_1 = OH$, $R_2 = R_3 = H$

starfish eggs (Fusetani et al. 1989). These compounds also exhibited significant brine shrimp lethality and moderate cytotoxicity (Seo et al. 1997).

8.3 Studies on Sacoglossans

Sacoglossans display a complete evolutionary series from shelled molluscs (Oxynoacea) to shell-less species (Placobranchacea; Jensen 1996). They feed upon algae and their evolution has been closely linked to their specialized suctorial herbivorous habits. Chemical studies have shown that these molluscs have the ability to either accumulate sesquiterpenoids, diterpenoids and depsipeptides from their prey or modify such molecules or to biosynthesize polypropionates (Cimino and Ghiselin 1998, 2001; Cimino et al. 1999). A parallel has been suggested between the evolution from shelled to shell-less species and the chemical trend from bio-accumulation to bio-synthesis of defensive compounds (Cimino et al. 1999).

Volvatella sp. (genus: Pease 1860), a shelled sacoglossan collected off Mandapam (Tamil Nadu, India) while grazing on siphonalean algae of the genus *Caulerpa*, was found to contain the linear sesquiterpenoid caulerpenyne (**63**), a typical caulerpalean metabolite (Amico et al. 1978).

Compound **63** was present both in the animal and in the alga, while the mollusc and its mucous secretion contained, in addition, a more polar and highly unstable unprecedented caulerpenyne derivative characterized as volvatellin (**64**; Fontana et al. 1999b).

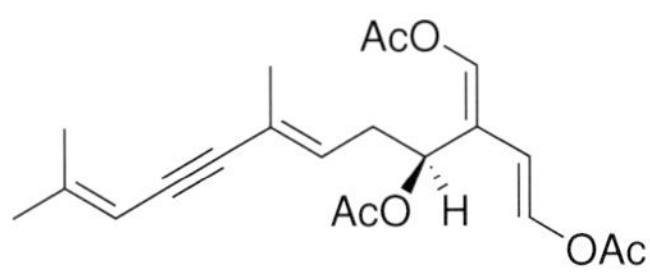

63

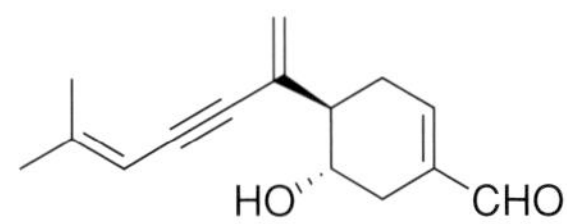

64

Plakobranchus ocellatus (van Hasselt 1824) is a shell-less sacoglossan found throughout the tropical Indo-West Pacific. *P. ocellatus* from Kurusadai Island (Tamil Nadu, India) was found to contain two γ-pyrone propionates (**65**, **66**), possessing a new bi-cyclic carbon skeleton (Manzo et al. 2005a). The reported structure **65** is that recently revised (Moses et al. 2005; Miller and Trauner 2005). In addition, in this animal were also present the known 9, 10-deoxy-tridachione (**67**) previously isolated from the Pacific sacoglossan *Tridachiella diomedea* (Ireland et al. 1978), photodeoxy-tridachione (**68**) probably originated by a photoconversion of **67** (Ireland and Scheuer 1979), tridachiahydropyrone B (**69**) and C (**70**), which are known metabolites of *P. ocellatus* from Philippine (Fu et al. 2000), and *iso*-9, 10-deoxy-tridachione (**71**), previously reported from the Mediterranean sacoglossan *Elysia timida* (Gavagnin et al. 1994).

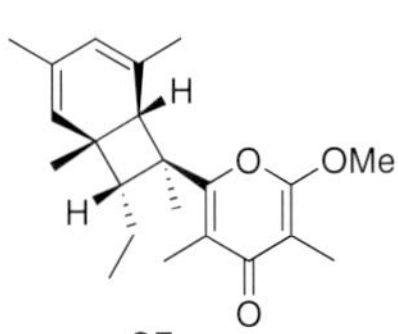

65

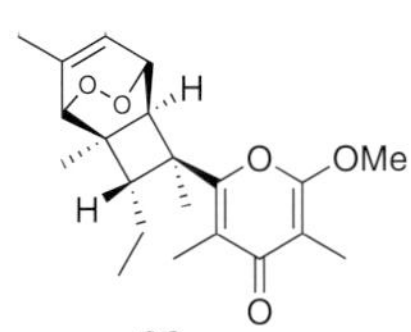

66

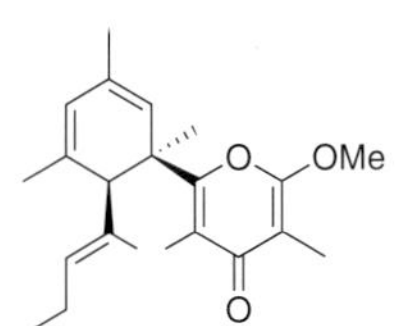

67

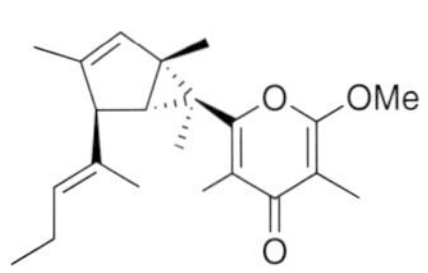

68

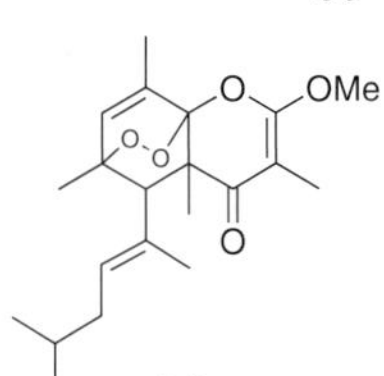

69

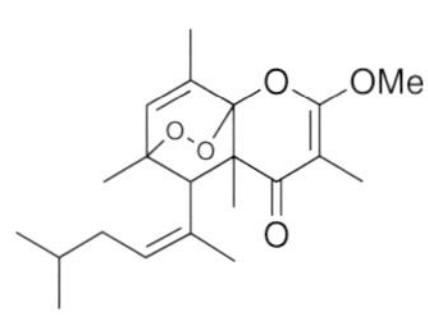

70

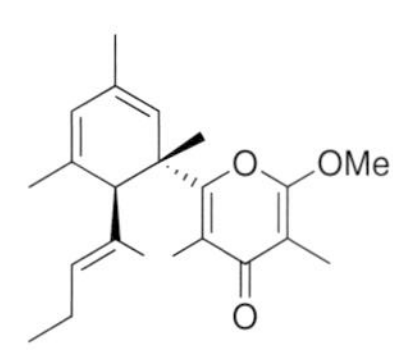

71

Elysia grandifolia (Kelaart 1858) is a shell-less sacoglossan originally described from Sri Lanka. Indian specimens of the mollusc were found to contain cyclic toxic depsipeptides (Fontana et al. 2001; as *Elysia* sp.) previously described in *E. rufescens*, *E. ornata* and in green algae of the genus *Bryopsis* (Hamann and Scheuer 1993; Hamann et al. 1996; Horgen et al. 2000). Kahalalide A (**72**) and F (**73**) have been identified as common constituents of both the mollusc and its host, *Bryopsis plumosa*, consistent with a dietary relationship. ESI-MS data of the peptide mixture from the sacoglossan also indicated the presence of two new depsipeptides tentatively identified as kahalalide P and kahalalide Q (Naik, unpublished data). Significant antifungal activity against food spoilage *Aspergillus* strains is reported for the crude extract of the mollusc (Bhosale et al. 1999). Padmakumar (1997) also describes the preference of an unidentified *Elysia* sp. for *B. plumosa*. The extracts from the alga, the mollusc and its mucus showed feeding-deterrent and broad-spectrum antimicrobial properties as well as moderate cytotoxicity against murine cancer cell lines, Dalton's lymphoma and Ehrlich ascites. Kahalalides are suspected of originating from epibiotic cyanobacteria rather than from algae (Kan et al. 1999).

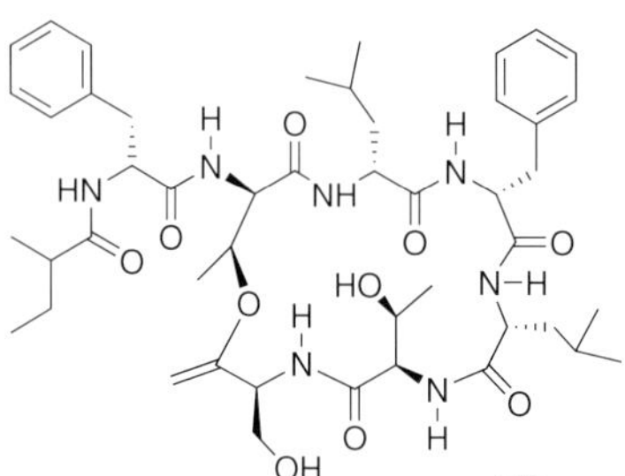
72

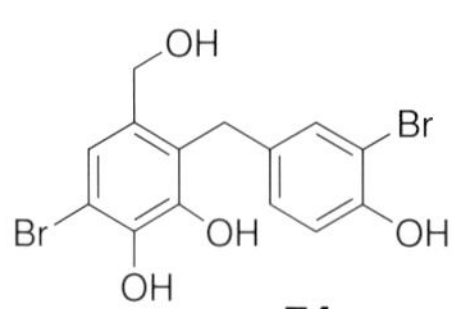
74

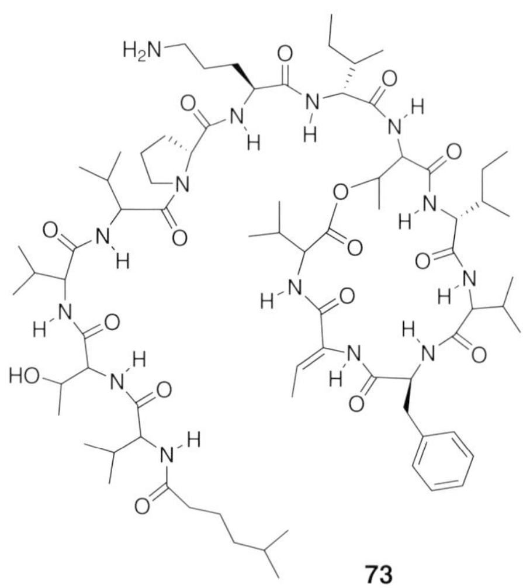
73

Costasiella sp. (genus: Pruvot-Fol 1951), a sacoglossan collected in the Gulf of Mannar showed a feeding preference for *Avrainvillea erecta* though it had access to other green algae: *Udotea flabellum*, *Halimeda macroloba*, *Caulerpa scalpelliformis*, *Enhalus acroides* and *Ulva lactuca* (Padmakumar and Lali 1996). Extracts of the alga and the animal deterred both herbivorous and carnivorous fishes as well as the sea star *Protoreaster lineti*. In analogy, earlier reports on *Costasiella ocellifera* from the Caribbean feeding upon the green alga *A. longicaulis* showed that the sacoglossan obtains from the alga both functional chloroplasts and a brominated diphenylmethane derivative avrainvilleol (74) that has shown to be an effective deterrent to fish (Hay et al. 1990).

8.4
Studies on Cephalaspideans

Cephalaspidean opisthobranchs are usually shelled, though the shell is somewhat reduced. They are recorded as sources of various metabolites, including polypropionates, alkaloids, terpenes and peptides (Cimino and Ghiselin 2001; Cimino et al. 2001a). A unique chemical work has been reported so far on cephalaspideans from the geographical areas considered here.

Haminoea cymbalum (Quoy and Gaimard 1833) is a cephalaspidean found throughout the Indo-West Pacific. Chemical investigation of the digestive gland, the mantle and the mucus of *H. cymbalum* from the shallow waters of Mandapam (Tamil Nadu, India), yielded the known brominated tetrahydropyran (**75**; Fontana et al. 2001). This compound, which was previously described from the Australian sponge *Haliclona* sp. (Capon et al. 1982), is structurally related to the rearranged trisnorsesquiterpene kumepaloxane (**76**), a metabolite of *H. cymbalum* from Guam with deterrent properties toward carnivorous fish (Poiner et al. 1989). In addition, the mantle and mucus of the Indian specimen also yielded a linear ketoalcohol (**77**) as a minor constituent. Given the absence of **77** in the digestive gland and the analogy with the antifeedant properties of similar compounds such as kumepaloxane (**76**), it was hypothesized that the Indian cephalaspidean may be able to biosynthesize its own defensive chemicals *de novo*. It is noteworthy that, differently from *H. cymbalum*, Mediterranean *Haminoea* species are characterized by alkylpyridines (Cimino et al. 1991, 2001a; Spinella et al. 1993; Marin et al. 1999).

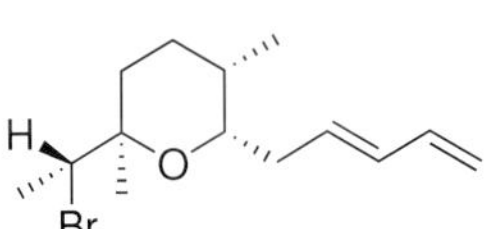

75

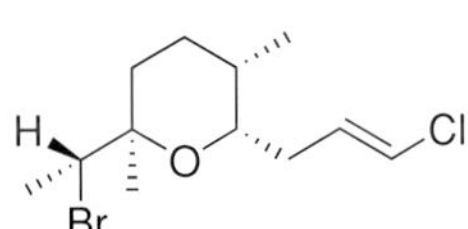

76

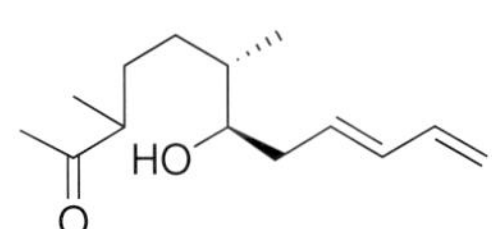

77

8.5
Studies on Anaspideans

Since anaspideans, commonly known as sea hares, are exclusively herbivorous, they contain typical algal metabolites even though some peculiar molecules isolated from the skin, including degraded sterols, oxylipins and polypropionates, have been suggested to be biosynthesized *de novo*. Also, a cyanobacterial origin of some anaspidean metabolites, e.g. antitumour dolastatins in *Dolabella auricularia*, has been proposed (Cimino et al. 2001a; and literature cited therein).

Bursatella leachii (de Blainville 1817) is an anaspidean commonly found in warm temperate and tropical waters throughout the world. Seven or eight geographical subspecies are recognized by some authorities. Studied from different geographical areas, it is known to contain very unusual metabolites (Gopichand and Schmitz 1980; Appleton et al. 2002). There are recent reports on the identification of four antitumour substances (**78**–**81**) in the extracts of a *Bursatella* subspecies from the South China Sea, reported as *Notarchus* (*Bursatella*) *leachii cirrosus* (Stimpson 1855) (Lin et al. 2001, 2002). The cytotoxic polyhalogenated monoterpenes **78** and **79**, previously isolated from the sea hare *Aplysia punctata* (Ortega et al. 1997), confirmed significant antineoplastic activities in vitro (Lin et al. 2001). Steroids **80** and **81** also showed cytotoxic activities (Lin et al. 2002). Here, mention must be made that the purple fluid from an unidentified subspecies of the sea hare *B. leachii* from Indian coasts yielded an anti-HIV protein, bursatellanin P, with a molecular mass of 60 kDa (Rajaganapathi et al. 2002), the molecular structure of which has not yet been reported. Heparinase activity exhibited by the same fluid has also been observed (Rajaganapathi and Kathiresan 2002).

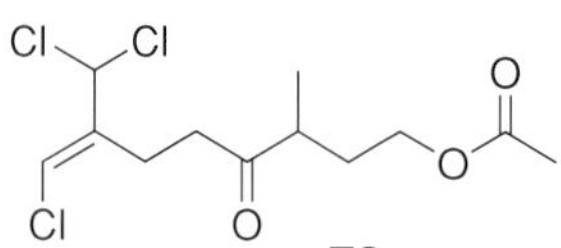

78

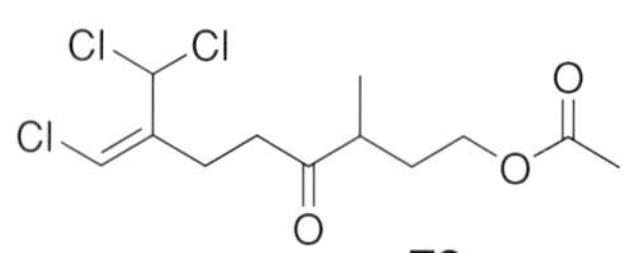

79

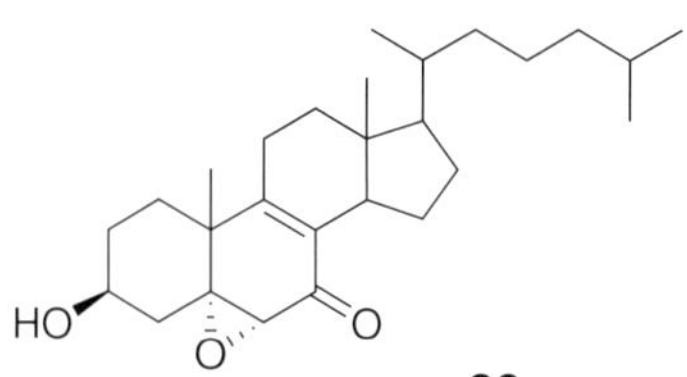

80

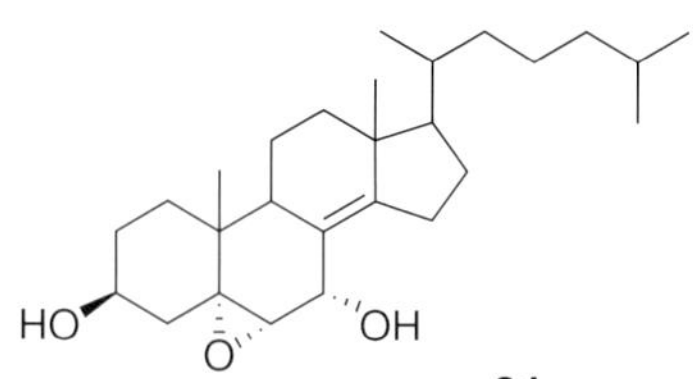

81

Aplysia dactylomela (Rang 1828) is distributed worldwide in tropical to warm temperate waters. A plethora of secondary metabolites have been isolated from *Aplysia* sea hares, including *A. dactylomela* (Yamada and Kigoshi 1997; Cimino et al. 2001a). *A. dactylomela* from the Gulf of Mannar (India) has yielded compounds **82**–**91** (Rao et al. 1989). The origin of these metabolites has been attributed to the algal diet of the mollusc. A dietary origin is substantiated because the bromoindoles **86** and **87** from this anaspidean are identical to those reported from the alga *Laurencia brongniartii* (Carter et al. 1978). Rao et al. (1994) also report the isolation of the new chamigrane derivative **92** from the digestive glands of *A. dactylomela*, besides the sterol **93**. Chemical studies on specimens of *A. dactylomela* from Hainan (China) led to the characterization of three novel halogenated acetogenins (**94**–**96**; Manzo et al. 2005b). These metabolites were isolated along with known related cyclic ethers **97** and

98, and the sesquiterpenoid **99**, all already known from *Laurencia* algae (Stallard et al. 1978; Falshaw et al. 1980; Gonzalez et al. 1982; Amico et al. 1991).

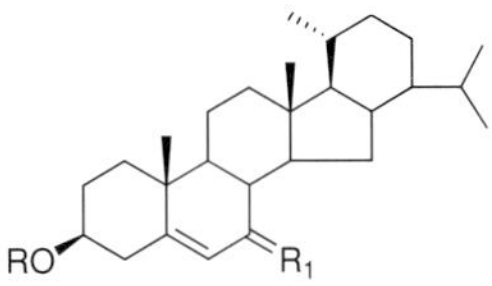

82 R = H, $R_1 = H_2$
83 R = H, R_1 = O

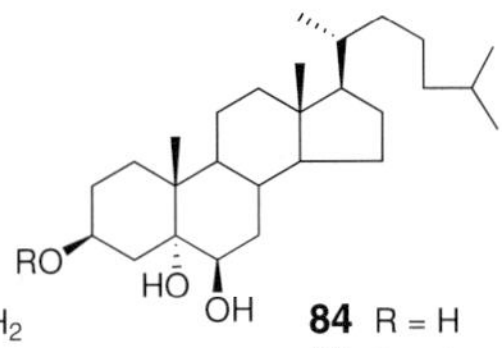

84 R = H
85 R = Ac

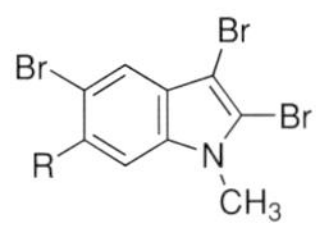

86 R = H
87 R = Br

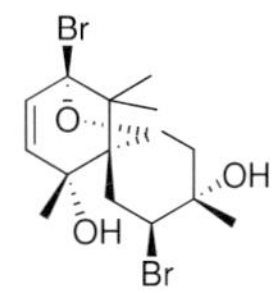

88

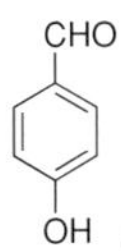

89

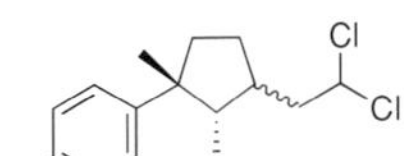

90

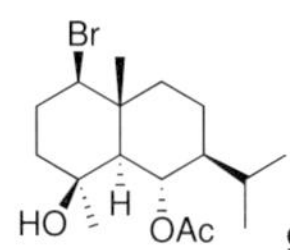

91

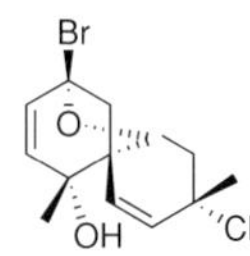

92

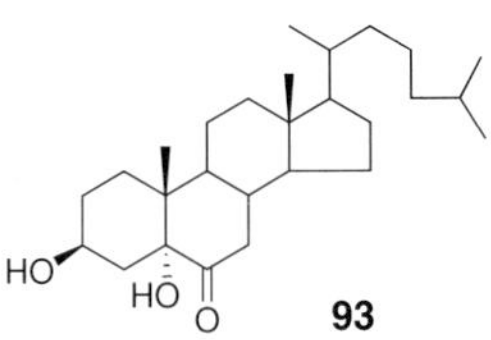

93

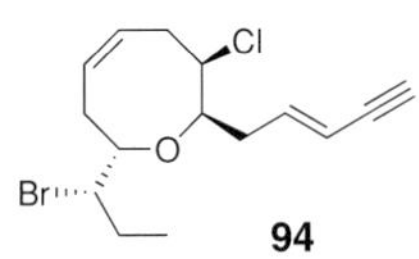

94

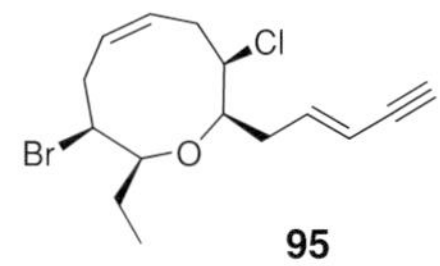

95

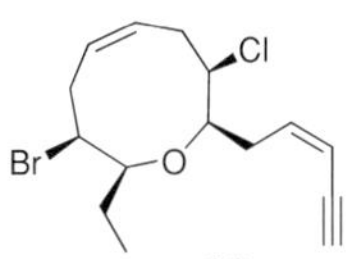

96

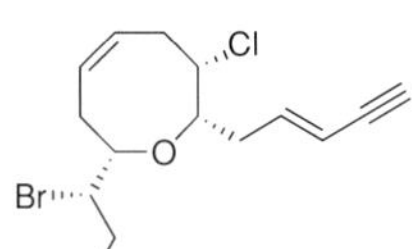

97

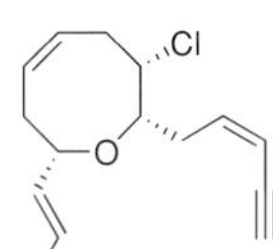

98

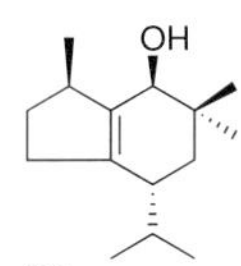

99

8.6 Comparative Discussion

Even though the chemicals derived from opisthobranchs may not be good taxonomic markers when they have a dietary origin (Karuso 1987), the comparative analysis of the chemical data reviewed in this chapter, taking into account alimentary habits and geographical chemical variations, suggests some interesting generalizations.

8.6.1 Nudibranchs

The intra-specific geographic variations observed in the chemical composition of *H. sanguineus* from China and Egypt significantly extend the range of chemicals isolated from the nudibranch, previously known to contain macrolides. The findings strongly support the dietary origin of the metabolites in *Hexabranchus*, which appear to be a generalist sponge feeder genus (Francis 1980), excluding specialized feeding habits on macrolide-containing sponges.

Even though the dimeric compound jorumycin isolated in the Indian *J. funebris* was not detected in the same nudibranch from Sri Lanka, isoquinoline alkaloids seem to characterize both populations. In spite of their dietary origin, these metabolites could represent good chemotaxonomical markers for *Jorunna*, which appear to have a specialized diet.

The results on Indian *G. atromarginata* and Chinese *G. rufomarginata* confirm that most of the nudibranchs belonging to the genus *Glossodoris* contain sesterterpenoids, as suggested by Cimino et al. (1999). In contrast, the Egyptian *G. cincta* and the Sri Lankan *G. atromarginata*, which have been found to contain diterpenoids, seem to be particular cases diverging from the above generalization. Analogously, the chemical composition of the Indian *C. mandapamensis* and the Chinese *C. reticulata* supports the generalization that *Chromodoris* nudibranchs prefer sponges characterized by diterpene metabolites (Cimino et al. 1999), while the macrolide isolated in *C. africana* and *C. quadricolor* from the Red Sea seems to indicate different alimentary habits. Even though the specialization of *Chromodoris* on diterpenoids, like that of *Glossodoris* on sesterterpenoids, is a real phenomenon, it seems to be only an evolutionary tendency; and it would appear that the various species are not equally specialized. In addition, the chemical composition of a given species (e.g. *G. atromarginata* from India and Sri Lanka) is contingent upon environmental circumstances. For such reasons, the emphasis upon diterpenoids or sesterterpenoids tells us a great deal about the evolutionary ecology of these organisms, but it does not furnish us with "taxonomic markers" or an "essence" for the groups.

The chemical composition of the Indian *H. kanga* is in agreement with the typical furanosesquiterpenoidic pattern proposed by Cimino et al. (1999) for the genus *Hypselodoris*, while the observed chemical similarity between *Ceratosoma* and *Hypselodoris*, both characterized by sole sesquiterpenoids, suggests a close chemotaxonomical relationship between the two genera.

The *P. pustulosa* chemical pattern confirms the chemotaxonomic features of the family Phyllidiidae, based on the presence of dietary sesquiterpene-isocyanides and suggests that the nudibranch also sequesters diterpene-isocyanides from sponges belonging to different genera, with a range of alimentary habits broader than previously described.

According to previous chemical studies on the little investigated family Arminidae, the isolation of four known diterpenoids of eunicellin class from the Chinese *D. ornatus*, also found in gorgonians, suggests that they feed selectively on cnidarians and accumulate their metabolites. In fact, similar diterpenoids have been analogously isolated in the Mediterranean *Armina maculata* and its prey, the pennatulacean octocoral *Veretillum cynomorium* (Guerriero et al. 1987).

8.6.2 Sacoglossans

The chemistry of the Indian *Volvatella* sp. supports the generalization that shelled sacoglossans feed exclusively on green algae of the genus *Caulerpa*, also confirming their ability to modify the algal sesquiterpenoid caulerpenyne into more toxic defensive metabolites.

In contrast, the Indian shell-less sacoglossans *E. grandifolia* and *Costasiella* sp. derive their protective compounds from different macroalgae. *E. grandifolia* shares with some co-generic species the depsipeptide pattern as well as a trophic relationship with algae of the genus *Bryopsis*. *Costasiella* sp., however, feeds on *Avrainvillea* algae, from which it most probably sequesters algal brominated diphenylmethane derivatives with feeding-deterrent properties, in agreement with previous results obtained by Hay et al. (1990) on the Caribbean *C. ocellifera*.

Finally, the chemical studies on *P. ocellatus* from India were in agreement with previous results on populations of the same mollusc from the Philippines. The presence of pyrone polypropionates, most likely biosynthesized *de novo*, supports including *P. ocellatus* in the heterogeneous group of shell-less sacoglossans belonging to different families which are able to produce their own defensive compounds.

8.6.3 Cephalaspideans

The chemical results on *H. cymbalum* from India, which were in accordance with those previously obtained on individuals of the same species from Guam, could have interesting biogeographical implications. In fact, the chemistry of this Indo-West Pacific cephalaspidean is surprisingly very different from that of the Mediterranean *Haminoea* species, all of which are characterized by the presence of alkylpyridines.

8.6.4 Anaspideans

The chemical analysis on *A. dactylomela* from India and China has confirmed the general rule that, with a very few exceptions, *Aplysia* molluscs contain typical halogenated red algae metabolites (Cimino et al. 2001a).

However, polyhalogenated monoterpenes found in *B. leachii* from the South China Sea have never been obtained from macroalgae. Analogously, the dietary source from which Spanish populations of *A. punctata* could obtain the same compounds is still unknown. A cyanobacterial origin could be suggested for these metabolites by analogy with *B. leachii* from New Zealand that is known to feed on cyanobacteria (Willan 1979) and to contain their typical metabolites (Appleton et al. 2002).

8.7 Conclusions

With both reviewed data and generalizations, we hope to provide a further informational base for a better understanding of the phyletic distribution of natural products in the opisthobranch molluscs, trying to overcome partial points of view derived from local phenomena. However, it would be valuable to have a larger sample of opisthobranch metabolites from the studied areas and from other less-studied regions, to confirm the chemotaxonomic generalizations and to provide chemical data for poorly represented taxa. Future chemical work there should be conducted with this need in mind.

References

Amico V, Oriente G, Piattelli M, Tringali C, Fattorusso E, Magno S, Mayol L (1978) Caulerpenyne, an unusual sequiterpenoid from the green alga *Caulerpa prolifera*. Tedrahedron Lett 38:3593–3596

Amico V, Caccamese S, Neri P, Russo G, Foti M (1991) Brasiliane-type sesquiterpenoids from the Mediterranean red alga *Laurencia obtusa*. Phytochemistry 30:1921–1927

Appleton DR, Sewell MA, Berridge MV, Copp BR (2002) A new active malyngamide from a New Zealand collection of the sea hare *Bursatella leachii*. J Nat Prod 65:630–631

Bhosale SH, Jagtap TG, Naik CG (1999) Antifungal activity of some marine organisms from India against food spoilage *Aspergillus* strains. Mycopathologia 147:133–138

Bobzin SC, Faulkner DJ (1989a) Novel rearranged spongian diterpenes from the Palauan sponge *Dendrilla* sp.: reassessment of the structures of dendrillolide A and dendrillolide B. J Org Chem 54:5727–5731

Bobzin SC, Faulkner DJ (1989b) Diterpenes from the marine sponge *Aplysilla polyrhaphis* and the dorid nudibranch *Chromodoris norrisi*. J Org Chem 54:3902–3907

Cambie RC, Craw PA, Stone MJ, Bergquist PR (1988) Chemistry of sponges, IV. Spongian diterpenes from *Hyatella intestinalis*. J Nat Prod 51:293–297

Cameron GM, Stapleton BL, Simonsen SM, Brecknell DJ, Garson MJ (2000) New sesquiterpene and brominated metabolites from the tropical marine sponge *Dysidea* sp. Tetrahedron 56:5247–5252

Capon RJ, Ghisalberti EL, Jefferies PR (1982) New tetrahydropyrans from a marine sponge. Tetrahedron 38:1699–1703

Carmely S, Cojocaru M, Loya Y, Kashman Y (1988) Two new rearranged spongian diterpenes from two *Dysidea* species. J Org Chem 53:4801–4807

Carter TE, Rinehart LK Jr, Lin LH, Kuentzl SL, Connor JL (1978) Brominated metabolites from *Laurencia brongniartii*. Tetrahedron Lett 19:4479–4482

Chang CWJ, Patra A, Deborah MR, Scheuer PJ (1984) Kalihinol – a, a highly functionalized diisocyanoditerpenoid antibiotic from a sponge. J Am Chem Soc 106:4644–4646

Chang CWJ, Patra JA, Baker JA, Scheuer PJ (1987) Kalihinols, multifunctional diterpenoid antibiotics from marine sponges *Acanthella* spp. J Am Chem Soc 109:6119–6123

Charles C, Braekman JC, Daloze D, Tursch B, Declercq JP, Germain G, Van Meerssche M (1978). Chemical studies of marine invertebrates. XXXIV. Herbadysidolide and herbasolide, two unusual sesquiterpenoids from the sponge *Dysidea herbacea*. Bull Soc Chim Belg 87:481–486

Ciavatta ML, Fontana A, Puliti R, Scognamiglio G, Cimino G (1999) Structures and absolute stereochemistry of isocyanide and isothiocyanate amphilectenes from the Caribbean sponge *Cribochalina* sp. Tetrahedron 55:12629–12636

Cimino G, Ghiselin MT (1998) Chemical defense and evolution in the Sacoglossa (Mollusca: Gastropoda: Opisthobranchia). Chemoecology 8:51–60

Cimino G, Ghiselin MT (1999) Chemical defense and evolutionary trends in biosynthetic capacity among dorid nudibranch (Mollusca: Gastropoda: Opisthobranchia). Chemoecology 9:187–207

Cimino G, Ghiselin MT (2001) Marine natural products chemistry as an evolutionary narrative. In: McClintock JB, Baker BJ (eds) Marine chemical ecology. CRC Press, Washington, D.C., pp. 115–156

Cimino G, De Stefano S, Minale L (1973) Deoxoscalarin, a further sesterterpene with unusual tetracyclic carbon skeleton of scalarin, from *Spongia officinalis*. Experientia 29:934–936

Cimino G, De Stefano S, Minale L (1974) Scalaradial, a third sesterterpene with a tetracarbocyclic skeleton from the sponge *Cacospongia mollior*. Experientia 30:846–847

Cimino G, De Stefano S, Guerriero A, Minale L (1975) Furanosesquiterpenoids in sponges – III. Pallescensins A–D from *Disidea pallescens*: new skeletal types. Tetrahedron Lett 17:1425–1428

Cimino G, Crispino A, Gavagnin M, Sodano G (1990). Diterpenes from the nudibranch *Chromodoris luteorosea*. J Nat Prod 53:102–106

Cimino G, Passeggio A, Sodano G, Spinella A, Villani G (1991) Alarm pheromones from the Mediterranean opisthobranch *Haminoea navicula*. Experientia 47:61–63

Cimino G, Fontana A, Gavagnin M (1999) Marine opisthobranch molluscs: chemistry and ecology in sacoglossan and dorids. Curr Org Chem 3:327–372

Cimino G, Ciavatta ML, Fontana A, Gavagnin M (2001a) Metabolites of marine opisthobranchs: chemistry and biological activity. In: Tringali C (ed) Bioactive compounds from natural sources – isolation, characterization and biological properties. Taylor & Francis, London, pp. 579–637

Cimino G, Fontana A, Garcia Gravalos D, Wahidulla S (2001b) New antitumor marine alkaloids WO 2001019824 A2 2001032 WO 2000-GB3489 20000911. PCT Int Appl 2001:1–9

Cimino G, Fontana A, Cutignano A, Gavagnin M (2004) Biosynthesis in opisthobranch molluscs: general outline in the light of recent use of stable isotopes. Phytochem Rev 3:285–307

De Silva ED, Gulavita NK (1988) Isoquinolinequinones from a marine sponge *Xestospongia* sp. and the nudibranch *Jorunna funebris*. IUPAC Int Symp Chem Nat Prod 16:610

De Silva ED, Scheuer PJ (1982) Furanoditerpenoids from the dorid nudibranch *Casella atromarginata*. Heterocycles 17:167–170

De Silva ED, Morris SA, Miao S, Dumdei E, Andersen R (1991) Terpenoid metabolites from the skin extracts of four Sri Lankan nudibranchs in the genus *Chromodoris*. J Nat Prod 54:993–997

Dumdei EJ, Flowers AE, Garson MJ, Moore CJ (1997) The biosynthesis of sesquiterpene isocyanides and isothiocyanates in the marine sponge *Acanthella cavernosa* (Dendy): evidence for dietary transfer to the dorid nudibranch *Phyllidiella pustulosa*. Comp Biochem Physiol 118:1385–1392

Falshaw CP, King TJ, Imre S, Islimayeli S, Thomson RH (1980) Laurenyne, a new acetylene from *Laurencia obtusa*: crystal structure and absolute configuration. Tetrahedron Lett 21:4951–4954

Fattorusso E, Magno S, Mayol L, Santacroce C, Sica D (1975) New sesquiterpenoids from the sponge *Axinella cannabina*. Tetrahedron 31:269–270

Faulkner DJ, Ghiselin MT (1983) Chemical defense and evolutionary ecology of dorid nudibranch and some other opisthobranch gastropods. Mar Ecol Prog Ser 13:295–301
Faulkner DJ, Molinski TF, Andersen RJ, Dumdei EJ, De Silva ED (1990) Geographical variation in defensive chemicals from Pacific coast dorid nudibranchs and some related marine molluscs. Comp Biochem Physiol 97C:233–240
Fontana A, Mollo E, Ricciardi D, Fakhr I, Cimino G (1997) Chemical studies of Egyptian opisthobranchs: spongian diterpenoids from *Glossodoris atromarginata*. J Nat Prod 60:444–448
Fontana A, Cavaliere P, Ungur N, D'Souza L, Parameswaran PS, Cimino G (1999a) New scalaranes from the nudibranch *Glossodoris atromarginata* and its sponge prey. J Nat Prod 62:1367–1370
Fontana A, Ciavatta ML, Mollo E, Naik CG, Wahidulla S, D'Souza L, Cimino G (1999b) Volvatellin, caulerpenyne-related product from the sacoglossan *Volvatella* sp. J Nat Prod 62:931–933
Fontana A, Cavaliere P, Wahidullah S, Naik CG, Cimino G (2000) A new antitumor isoquinoline alkaloid from the marine nudibranch *Jorunna funebris*. Tetrahedron 56:7305–7308
Fontana A, Ciavatta ML, D'Souza L, Mollo E, Naik CG, Parameswaran PS, Wahidulla S, Cimino G (2001) Selected chemo-ecological studies of marine opisthbranchs from Indian coasts. J Ind Inst Sci 81:403–415
Francis MP (1980) Habitat, food and reproductive activity of the nudibranch *Hexabranchus sanguineus* on Tongatapu Island. Veliger 22:252–258
Fu X, Hong EP, Schmitz FJ (2000) New polypropionates pyrones from the Philippine Sacoglossan *Placobranchus ocellatus*. Tetrahedron 56:8989–8993
Fusetani N, Nagata H, Hirota H, Tsuyuki T (1989) Astrogorgiadiol and astrogorgin, inhibitors of cell division in fertilized starfish eggs, from a gorgonian *Asterogorgia* sp. Tetrahedron Lett 30:7079–7082
Fusetani N, Yasumuro K, Kawai H, Natori T, Brinen L, Clardy J (1990) Kalihinene and isokalihinol B, cytotoxic diterpenes isonitriles from the marine sponge *Acanthella klethra*. Tetrahedron Lett 31:3599–3602
Garson MJ, Simpson JS (2004) Marine isocyanides and related natural products – structure, biosynthesis and ecology. Nat Prod Rep 21:164–179
Gavagnin M, Spinella A, Cimino G, Marin A (1994) Polypropionates from the Mediterranean Mollusc *Elysia timida*. J Nat Prod 57:298–304
Gavagnin M, Mollo E, Docimo T, Guo Y-W, Cimino G (2004) Scalarane metabolites of the nudibranch *Glossodoris rufomarginata* and its dietary sponge from the South China Sea. J Nat Prod 67:2104–2107
Gonzalez AG, Martin JD, Martin VS, Norte M, Perez R, Ruano JZ (1982) Non-terpenoid C-15 metabolites from the red seaweed *Laurencia pinnatifida*. Tetrahedron 38:1009–1014
Gopichand Y, Schmitz FJ (1980) Bursatellin, a new diol dinitrile from the sea hare *Bursatella leachii pleii*. J Org Chem 45:5383–5385
Guella G, Mancini I, Guerriero A, Pietra F (1985) New furanosesquiterpenoids from Mediterranean sponges. Helv Chim Acta 68:1276–1282
Guerriero A, D'Ambrosio M, Pietra F (1987) Verecynarmin A, a novel briarane diterpenoid isolated from both the Mediterranean nudibranch mollusc *Armina maculata* and its prey, the pennatulacean octocoral *Veretillum cynomorium*. Helv Chim Acta 70:984–991
Gulavita NK, De Silva ED, Hagadone MR, Karuso P, Scheuer PJ (1986) Nitrogenous bisabolene sesquiterpenes from marine invertebrates. J Org Chem 51:5136–5139
Guo Y-W (1997) Chemical studies of the novel bioactive secondary metabolites from the benthic invertebrates: isolation and structure characterization. Ph.D. thesis, University of Naples, Naples
Guo Y-W, Gavagnin M, Mollo E, Trivellone E, Cimino G (1998) Structure of the pigment of the Red Sea nudibranch *Hexabranchus sanguineus*. Tetrahedron Lett 39:2635–2638

Hamann MT, Scheuer PJ (1993) Kahalalide F: a bioactive depsipeptide from the sacoglossan mollusk *Elysia rufescens* and the green alga *Bryopsis* sp. J Am Chem Soc 45:5825–5826
Hamann MT, Otto CS, Scheuer PJ, Dumbar C (1996) Kahalalides: bioactive peptides from marine mollusk *Elysia rufescens* and its algal diet *Bryopsis* sp. J Org Chem 61:6594–6600
Hay ME, Duffy JE, Paul VJ, Renauld PE, Fenical W (1990) Specialist herbivores reduce their susceptibility to predation by feeding on the chemically defended seaweed *Avrainvillea longicaulis*. Limnol Oceanogr 35:1734–1743
He H, Faulkner DJ (1989) Renieramycins E and F from the sponge *Reniera* sp. reassignment of the stereochemistry of the renieramycins. J Org Chem 54:5822–5824
Hertzberg S, Liaaen-Jensen S, Enzell CR, Francis GW (1969) Animal carotenoids. Acta Chem Scand 23:3290–3312
Hirota H, Tomono Y, Fusetani N (1996) Terpenoids with antifouling activity against barnacle larvae from the marine sponge *Acanthella cavernosa*. Tetrahedron 52:2359–2368
Hochlowski JE, Faulkner DJ (1980) A diterpene related to cladiellin from a Pacific soft coral. Tetrahedron Lett 21:4055–4056
Horgen FD, delos Santos DB, Goetz G, Sakamoto B, Kan Y, Nagai H, Scheuer PJ (2000) A new depsipeptide from the sacoglossan mollusk, *Elysia ornata* and the green alga, *Bryopsis* sp. J Nat Prod 63:152–154
Ireland C, Scheuer PJ (1979) Photosynthetic marine mollusks: in vivo ^{14}C incorporation into metabolites of the sacoglossan *Placobranchus ocellatus*. Science 205:922–923
Ireland C, Faulkner DJ, Solheim BA, Clardy J (1978) Tridachione, a propionate derived metabolites of the opisthobranch mollusk *Tridachiella diomedea*. J Am Chem Soc 100:1002–1003
Iwashima M, Terada I, Iguchi K, Yamori T (2002) New biologically active marine sesquiterpenoid and steroid from the Okinawan sponge of the genus *Axinyssa*. Chem Pharm Bull 50:1286–1289
Jacobs RS, Bober MA, Pinto I, Williams AB, Jacobson PB, de Carvalho MS (1993). In: Attaway DH, Zaborsky OR (eds) Marine biotechnology, vol 1. Plenum, New York, pp. 77–99
Jensen KR (1996) Phylogenetic systematics and classification of the Sacoglossa (Mollusca, Gastropoda, Opisthobranchia). Philos Trans R Soc Lond B 351:91–122
Kan Y, Fujita T, Sakamoto B, Hokama Y, Nagai H (1999) Kahalalide K: a new cyclic depsipeptide from the Hawaiian green alga *Bryopsis* species. J Nat Prod 62:1169–1172
Karuso P (1987) Chemical ecology of the nudibranchs. In: Schueur PJ (ed) Bioorganic marine chemistry, vol 1. Springer, Berlin Heidelberg New York, pp. 31–60
Karuso P, Taylor WC, White AH (1986) The constituents of marine sponges 11. Isolation of 8 diterpenes from *Aplysilla rosea*. Aust J Chem 39:1629–1641
Kashman Y (1980) A new diterpenoid related to eunicellin and cladiellin from a *Muricella* sp. Tetrahedron Lett 21:879–880
Kashman Y, Rudi A (1977) The C-NMR spectrum and stereochemistry of heteronemin from sponge *Heteronema erecta*; genus *Cacospongia*. Tetrahedron 33:2997–2998
Kashman Y, Croweiss A, Shmueli U (1980) Latrunculin, a new 2-thiazolidinone macrolide from the marine sponge *Latrunculia magnifica*. Tetrahedron Lett 21:3629–3632
Kashman Y, Croweiss A, Kidor R, Blasberger D, Carmely S (1985) Latrunculins: NMR study, two new toxins and a synthetic approach. Tetrahedron 41:1905–1914
Kazlauskas R, Murphy PT, Quinn RJ, Wells RJ (1976) Heteronemin, a scalarin type sesterterpene from the sponge *Heteronema erecta*. Tetrahedron Lett 30:2631–2634
Kazlauskas R, Murphy PT, Wells RJ, Daly JJ, Schonholzer P (1978) Two sesquiterpene furans with new carbocyclic ring systems and related thiol acetates from a species of the sponge genus *Dysidea*. Tetrahedron Lett 49:4951–4954
Kazlauskas R, Murphy PT, Wells RJ, Noack K, Oberhansli WE, Schonholzer P (1979) A new series of diterpenes from Australian *Spongia* species. Aust J Chem 32:867–880

Kernan MR, Molinski TF, Faulkner DJ (1988) Macrocyclic antifungal metabolites from the Spanish dancer nudibranch *Hexabranchus sanguineus* and sponges of the genus *Halicondria*. J Org Chem 53:5014–5020
Kernan MR, Cambie CR, Bergquist RP (1990) New diterpenes from the marine sponge *Dictyodendrilla cavernosa*. J Nat Prod 53:724–727
Kohmoto S, McConnell OJ, Wrigth A, Cross S (1987) Isospongiadiol, a cytotoxic and antiviral diterpene from a Caribbean deep water marine sponge, *Spongia* sp. Chem Lett 1687–1690
Lin GY (1975) Opisthobranchia from the Intertidal zone of Xisha Islands, Guangdong Province, China. Stud Mar Sin 10:141–154
Lin GY (1981) A study of the family Arminidae (Opisthobranchia) of the China coast. Stud Mar Sin 18:181–206
Lin GY (1986) Additions to the opisthobranch fauna of the Hainan and Xisha Islands of China. Stud Mar Sin 26:117–128
Lin H-W, Zhang C, Yi Y-Y, Shen Y, Guo C, Shao Y-F (2001) Studies on antineoplastic constituents from Chinese marine mollusc sea hare (I). Chin J Mar Drugs 20:3–6
Lin H-W, Tang H-F, Liu G-L, Zhang C, Shen Y, Hua Y-Y, Wu H-M (2002) Studies on antineoplastic constituents from Chinese marine mollusc sea hare (II). Acad J Sec Mil Med Univ 23:243–245
Manzo E, Ciavatta ML, Gavagnin M, Mollo E, Guo Y-W, Cimino G (2004) Isocyanide terpene metabolites of *Phyllidiella pustulosa*, a nudibranch (Mollusca: Gasteropoda) from the South China Sea. J Nat Prod 67:1701–1704
Manzo E, Ciavatta ML, Gavagnin M, Mollo E, Wahidulla S, Cimino G (2005a) New γ-pyrone propionates from the Indian Ocean sacoglossan *Placobranchus ocellatus*. Tetrahedron Lett 46:465–468
Manzo E, Ciavatta ML, Gavagnin M, Puliti M, Mollo E, Guo Y-W, Mattia CA, Mazzarella L, Cimino G (2005b) Structure and absolute stereochemistry of novel C_{15}-halogenated acetogenins from anaspidean mollusc *Aplysia dactylomela*. Tetrahedron 61:7456–7460
Marin A, Alvarez LA, Cimino G, Spinella A (1999) Chemical defence in cephalaspidean gastropods: origin, anatomical location and ecological roles. J Mol Stud 65:121–131
Matsunaga S, Fusetani N, Hashimoto K, Koseki K, Noma M (1986) Kabiramide C, a novel antifungal macrolide from nudibranch eggmasses. J Am Chem Soc 108:847–849
Matsunaga S, Fusetani N, Hashimoto K, Koseki K, Noma M, Noguchi H, Sankawa U (1989) Further kabiramides and halicondramides, cytotoxic macrolides embracing trisoxazole, from the *Hexabranchus* egg masses. J Org Chem 54:1360–1363
Mebs D (1985) Chemical defense of a dorid nudibranch, *Glossodoris quadricolor*, from the Red Sea. J Chem Ecol 11:713–716
Miller AK, Trauner D (2005) Mining the tetraene manifold: total synthesis of complex pyrones from *Placobranchus ocellatus*. Angew Chem Int Ed 44:4602–4606
Miyamoto T, Sakamoto K, Arao K, Komori T, Higuchi R, Sasaki T (1996) Dorisenones, cytotoxic spongian diterpenoids, from the nudibranch *Chromodoris obsoleta*. Tetrahedron 52:8187–8189
Mollo E, Gavagnin M, Carbone M, Guo Y-W, Cimino G (2005) Chemical studies on Indopacific *Ceratosoma* nudibranchs illuminate the protective role of their dorsal horn. Chemoecology 15:31–36
Mong S, Votta B, Sarau HM, Foley JJ, Schmidt D, Carte BK, Poehland B, Westley J (1990) 15-Acetylthioxyfurodysinin lactone, isolated from a marine sponge *Dysidea* sp. is a potent agonist to human leukotriene B4 receptor. Prostaglandins 39:89–97
Moses JE, Adlington RM, Rodriguez R, Eade SJ, Baldwin JE (2005) Biomimetic synthesis of (±)- 9,10-deoxytridachione. Chem Commun 1687–1689
Ochi M, Yamada K, Shirase K, Kotsuki H (1991) Calicophirins A and B, two new insect growth inhibitory diterpenoids from a gorgonian coral *Calicogorgia* sp. Heterocycles 32:19–21
Okino T, Yoshimura E, Hirota H, Fusetani N (1995) Antifouling kalihinenes from the marine sponge *Acanthella cavernosa*. Tetrahedron Lett 36:8637–8640

Okino T, Yoshimura E, Hirota H, Fusetani N (1996) New antifouling kalihipyrans from the marine sponge *Acanthella cavernosa*. J Nat Prod 59:1081–1083

Ortega MJ, Zubía E, Salvá J (1997) New polyhalogenated monoterpenes from the sea hare *Aplysia punctata*. J Nat Prod 60:482–484

Padmakumar K (1997) Bioactivity and chemical ecology of an opisthobranch *Elysia* sp. and the dietary alga *Bryopsis plumosa*. Phuket Mar Biol Cent Spec Publ 18:187

Padmakumar K, Lali G (1996) Chemical defence of the specialist ascoglossan *Costasiella sp*. from the Indian Coast. Phuket Mar Biol Cent Spec Publ 17:61–68

Patra A, Chang CWJ, Scheuer PJ, Van Duyne GD, Matsumoto GK, Clardy J (1984) An unprecedented triisocyano diterpenoid antibiotic from a sponge. J Am Chem Soc 106:7981–7983

Poiner A, Paul VJ, Scheuer PJ (1989) Kumepaloxane, a rearranged trisnorsesquiterpene from the bubble shell *Haminoea cymbalum*. Tetrahedron 45:617–622

Rajaganapathi J, Kathiresan K (2002) Heparinase in purple fluid of the sea hare, *Bursatella leachii*. Curr Sci 82:264–266

Rajaganapathi J, Kathiresan K, Singh TP (2002) Purification of anti-HIV protein from purple fluid of the sea hare *Bursatella leachii* de Blainville. Mar Biotechnol 4:447–453

Rao CB, Satyanarayana C, Rao DV, Fahy E, Faulkner DJ (1989) Metabolites of *Aplysia dactylomela* from Indian Ocean. Ind J Chem 28B:322–325

Rao CB, Satyanarayana C, Rao DV, Thompson MF, Nagabhushanam R, Sarojini R, Fingerman M (1994) Recent development in biofouling control. Oxford/IBH, New Delhi, pp. 259–261

Roesener JA, Scheuer PJ (1986) Ulapualide A and B: extraordinary antitumor macrolides from nudibranch eggmasses. J Am Chem Soc 108:846–847

Rudman WB (1973) Chromodorid opisthobranch Mollusca from the Indo-West Pacific. Zool J Linn Soc 52:175–199

Rudman WB (1986) The Chromodorididae (Opisthobranchia: Mollusca) of the Indo-West Pacific: the genus *Glossodoris* Ehrenbergh (=*Casella*, H.& A. Adams). Zool J Linn Soc 86:101–184

Rudman WB, Darvell BW (1990) Opisthobranch molluscs of Hong Kong: part 1. Goniodorididae, Onchidorididae, Triophidae, Gymnodorididae, Chromodorididae (Nudibranchia). Asian Mar Biol 7:31–79

Schmitz JF, Chang SJ, Hossain BM, van der Helm D(1985) Marine natural products: spongiane derivatives from the Sponge *Irgenella notabilis*. J Org Chem 50:2862–2865

Seo Y, Rho JR, Cho KW, Shin J (1997) Isolation of diterpenoids of the cladiellane class from gorgonians of the genus *Muricella*. J Nat Prod 60:171–174

Shimomura M, Miyaoka H, Yamada Y (1999) Absolute configuration of marine diterpenoid kalihinol A. Tetrahedron Lett 40:8015–8017

Spinella A, Alvarez LA, Passeggio A, Cimino G (1993) New 3-alkylpyridines from three Mediterranean cephalaspidean molluscs: structure, ecological role and taxonomic relevance. Tetrahedron 49:1307–1314

Stallard MO, Fenical W, Kittredge JS (1978) The brasilenols, rearranged sesquiterpene alcohols isolated from the marine opisthobranch *Aplysia brasiliana*. Tetrahedron 34:2077–2081

Subba Rao NV (2003) Indian seashells: part I: Polyplacophora and Gastropoda. Zoological Survey of India, Calcutta, 416 pp

Subba Rao NV, Dey A (2000) Catalogue of marine molluscs of Andaman and Nicobar Islands. Zoological Survey of India, Calcutta, 323 pp

Tanaka J, Higa T (1988) The absolute configuration of kurospongin a new furanoterpene from a marine sponge, *Spongia* sp. Tetrahedron 44:2805–2810

Todd CD, Lambert WD, Davies J (2001) Some perspectives on the biology and ecology of nudibranch molluscs: generalisations and variations on the theme that prove the rule. Boll Malacol 37:105–120

Willan RC (1979) The ecology of two New Zealand opisthobranch molluscs. Ph.D. thesis, The University of Auckland, Auckland

Wratten SJ, Faulkner DJ, Hirotsu K, Clardy J (1978) Diterpenoids isocyanides from the marine sponge *Hymeniacidon amphilecta*. Tetrahedron Lett 19:4345–4348

Yamada K, Kigoshi H (1997) Bioactive compounds from the sea hares of two genera: *Aplysia* and *Dolabella*. Bull Chem Soc Jpn 70:1479–1489

Yan X (2004) Study on chemical constituents and bioactivities of seven Chinese marine invertebrates. Ph.D. thesis, Shanghai Institute of Materia Medica, Shanghai, pp. 145–154

Yonow N (1990) Red Sea Opisthobranchia 3: the orders Sacoglossa, Cephalaspidea, and Nudibranchia: Doridacea (Mollusca, Opisthobranchia). Fauna Saudi Arabia 11:286–299

Yonow N (2000) Red Sea Opisthobranchia 4: the orders Cephalaspidea, Anaspidea, Notaspidea and Nudibranchia: Dendronotacea and Aeolidacea. Fauna Saudi Arabia 18:87–131

Selected Bioactive Compounds from Japanese Anaspideans and Nudibranchs

T. Miyamoto

Abstract. Marine opisthobranchs are mollusks scarcely protected by a shell which can be either reduced or completely absent. It is generally accepted that they possess chemical defenses (allelochemicals) against predators. These protective molecules can derive from dietary habits (algae, sponges, tunicates, soft corals, etc.) and can be used either in their **intact** form or as derived metabolites. Added to this, other compounds are biogenetically obtained. These defensive compounds are very often concentrated in the external parts of the animal and are also released into the mucous secretion to deter predators. The chemical defenses of sea hares (order Anaspidea) and sea slugs (order Nudibranchia) are well studied. This review highlights the allelochemicals reported from Japanese anaspideans and nudibranchs, with particular attention to the finding of our group.

9.1 Introduction

"Anaspidea" means "no-shield" or "shield-less." The mollusks belonging to this order do not have shields but possess a clear thin shell like a leaf in their bodies. Four species of anaspidean mollusks are frequently observed in Japan. The most widespread, *Aplysia kurodai*, generally eats red algae of the genera *Plocamium* and *Laurencia*. An analogous diet is adopted by both the large *A. dactylomela* and the small *A. parvula*. Finally, the small *A. juliana* eats green algae of the genus *Ulvales* (Hirano 2000).

Most anaspidean allelochemicals are derived from algal metabolites. Chemical diversity is due to the location where the mollusks live and the kind of alga that grows there.

In contrast, the order "Nudibranchia," with a name which means "bare-gill" or "naked-gill," comprises shell-less mollusks dressed in a colorful coat. More than 300 species of nudibranchs have been observed around Japan (Ono 1999; Suzuki 2000); and many chemical studies on chromodoridid and phyllidid nudibranchs have been made. Nudibranch allelochemicals, which often derive from the sponges they prey on, are classified according to the terpenoid skeleton exhibited. Some dietary metabolites are transformed by nudibranchs in order to obtain derivatives more active against predators.

Tomofumi Mtyamoto
Graduate School of Pharmaceutical Sciences, Kyushu University, 3-1-1 Maidashi
812-8582 Higashio-ku, Fukuoka, Japan

Progress in Molecular and Subcellular Biology
Subseries Marine Molecular Biotechnology
G. Cimino, M. Gavagnin (Eds.): Molluscs

Cytotoxic substances, including aplyronins, dolastatins, and aurilide from Japanese sea hares, are referred to in the chapter by Kamiya et al. in this book and in previous reviews by Yamada and Kigoshi (1997) and Yamada et al. (2000), whereas trisoxazole macrolides including kabiramides and halichondramides from Japanese *Hexabranchus* spp. have been dealt with in Matsunaga's chapter in this volume.

9.2 Metabolites of Anaspideans

9.2.1 Halogenated Compounds

Acetogenins

Bromoallenyl acetogenins are typical metabolites of red algae of the genus *Laurencia*, but they are also found in some anaspideans. The aromatic feeding-deterrent panacene (**1**) was first isolated from the sea hare *A. brasiliana*, collected in Florida (Kinnel et al. 1977). Dactylallene (**2**) was isolated from the Atlantic *A. dactylomela* as a defensive product (Ciavatta et al. 1997). Two bromoallenyl acetogenins have been reported from Japanese mollusks: aplypervunin (**3**), isolated from *A. parvula*, showed potent ichthyotoxicity against mosquito fish at the LC_{100} value of 3 ppm (Miyamoto et al. 1995), and aplysiallene (**4**), isolated from the same animal, inhibited Na,K-ATPase at the IC_{50} value of 0.7 mM (Okamoto et al. 2001).

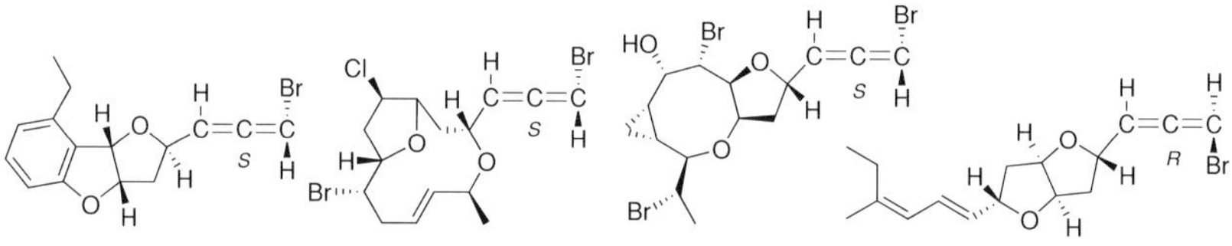

panacene (**1**) dactylallene (**2**) aplyparvunin (**3**) aplysiallene (**4**)

Cyclic Monoterpenoids

A number of polyhalogenated monoterpenes have been isolated from both red algae of the family Plocaminacea and sea hares of the genus *Aplysia*. Aplysiaterpenoids A (**5**) and B (**6**), isolated from the Japanese sea hare *A. kurodai* displayed cytotoxic and ichtyotoxic properties (Miyamoto et al. 1988). Some additional biological properties, such as insecticidal activity and feeding-inhibition, were reported for aplysiaterpenoid A (**5**) isolated from *Plocamium telfiriae* (Watanabe et al. 1990) and *P. leptophyllum* (Sakata et al. 1991), respectively.

As shown in Scheme 9.1, 13 specimens of *A. kurodai* were dissected into mantles and midgut glands. The mantle was extracted with MeOH and the concentrated MeOH extract was partitioned between $CHCl_3$ and H_2O. Also, the midgut glands were lyophilized and extracted with *n*-hexane. The *n*-hexane extracts were chromatographed on silica gel and reversed phase HPLC to give compounds **5** and **6**, and other halogenated terpenes.

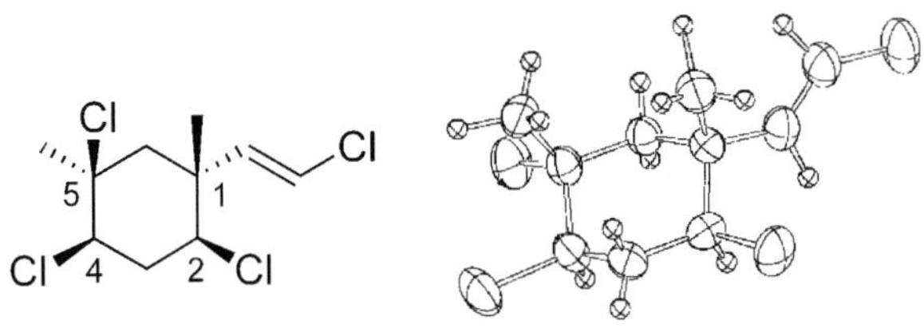

aplysiaterpenoid A (**5**)

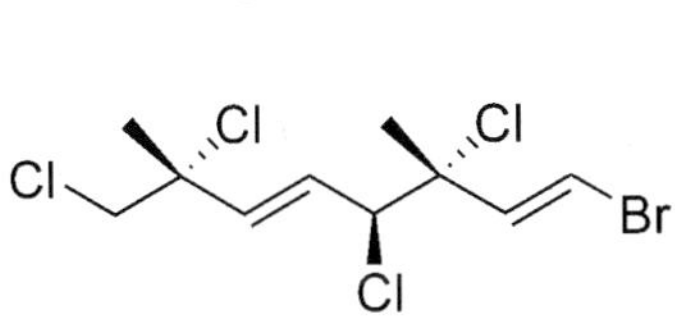

aplysiaterpenoid B (**6**)

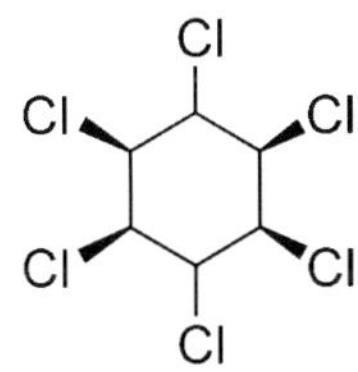

γ-BHC

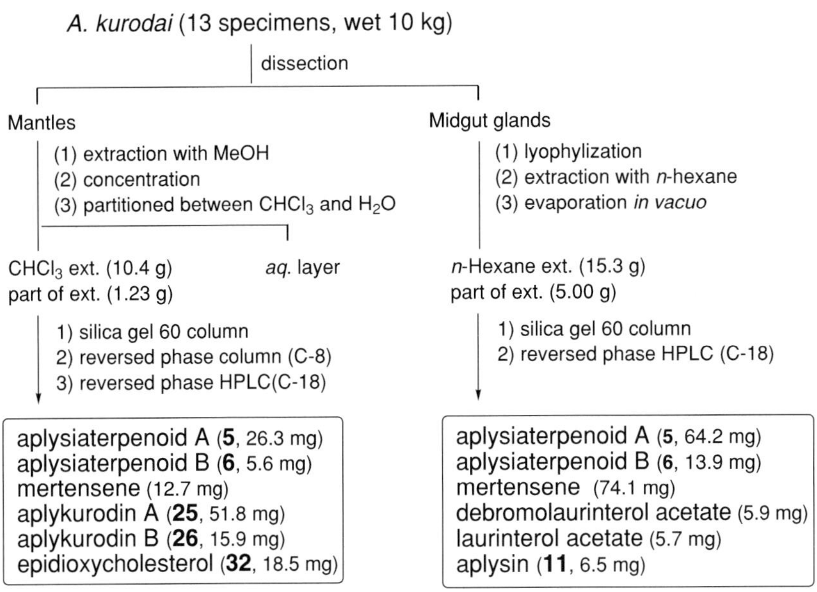

Scheme 9.1. Isolation of various allelochemicals from *A. kurodai*

The planar structure of **5** was determined by NMR and mass spectral data. The relative stereochemistry was elucidated by analysis of $^3J_{H-H}$ and nOe. correlations, whereas the absolute stereochemistry was determined to be (1*R*,2*S*,4*R*,5*R*)-2,4,5-trichloro-1-*E*-chlorovinyl-1,5-dimethylcyclo-hexane by comparison between the Bijvoet pairs in the X-ray analysis of a single crystal of **5** (Miyamoto et al. 1988).

Aplysiaterpenoid A (**5**) showed mild cytotoxicity against various tumor cell lines (IC_{50} values were: 10 μg ml^{-1} for L1210 mouse lymphoma, 15.3 μg ml^{-1} for QG-90 human lung carcinoma, 30.2 μg ml^{-1} for MCF-7 human breast cancer). It exhibited significant ichtyotoxicity against mosquito fish (*Oryzias latipes*; LC_{100} value 10 ppm, within 3 h). A strong insecticidal activity was also demonstrated against both the German cockroach (*Blatella germanica*; 60% and 80% mortality against the German cockroach, by injection of 6 and 8 μg per male, respectively, into the abdomen) and mosquito larvae (*Anopheles gambiae*; LC_{50} values of 0.1 ppm for a susceptible strain, 0.24 ppm for a dieldrin-resistant strain). On the basis of molecular topographical studies, it has been proposed that aplysiaterpenoid A (**5**) could interact with the picrotoxinin receptor on the presynaptic membrane of the central nervous system in the same manner as the known insecticide, lindane (1α,2α,3β,4α,5α,6β-hexachlorocyclohexane, γ-BHC). In the "Avicel plate method," a small amount of **5** (40 μg) is enough to inhibit feeding of four marine herbivores, the gastropod *Turbo cornutus*, the top shell *Omphalius pfeifferi*, the abalone *Haliotis discus*, and the sea urchin *Strongylocentrotus intermedicus*, used on the sample zone together with 40 μg of digalactosyldiacylglycerol (feeding stimulant).

Pyrano-Monoterpenoids

Polyhalogenated pyranoid monoterpenes, aplysiapyranoids A (**7**) and B (**8**), were isolated from the midgut gland of *Aplysia kurodai*. Their structures were determined by X-ray crystal analysis of compound **8**. These molecules adopt two conformations in solution (Inoue et al. 1987; see formula **8**). Further investigation led to the isolation of related aplysiapyranoids C (**9**) and D (**10**), which have fixed conformations. Aplysiapyranoids exhibited moderate cytotoxicity against monkey kidney Vero epithelial-like, MDCK dog kidney epithelial, and B16 mouse melanoma cell lines at IC_{50} values of 19–96 μg ml^{-1} (Kusumi et al. 1987).

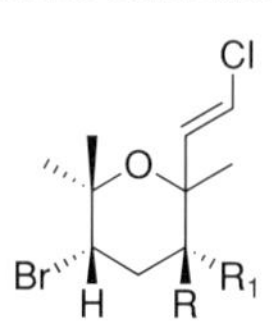

R=Br, R_1=H, α-chlorovinyl: aplysiapyranoid A (**7**)
R=Br, R_1=H, β-chlorovinyl: aplysiapyranoid B (**8**)
R=H, R_1=Cl, α-chlorovinyl: aplysiapyranoid C (**9**)
R=H, R_1=Cl, β-chlorovinyl: aplysiapyranoid D (**10**)

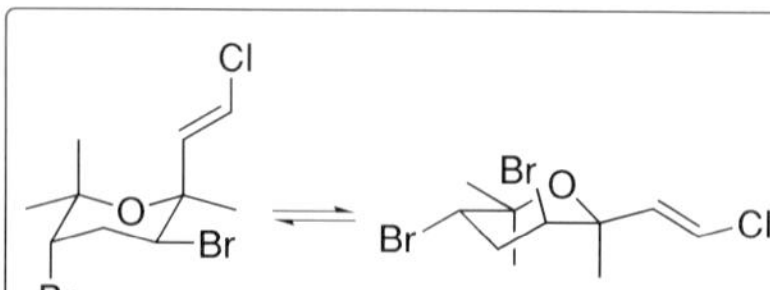

8

Sesquiterpenoids

More than 100 halogenated compounds have been isolated from the Japanese *Laurencia* species (Suzuki et al. 2002). Also, halogenated chamigrene-, cuparene-, and bisabolene-type sesquiterpenes have been found in some sea hares that prey on *Laurencia*. Aplysin (**11**), debromoaplysin (**12**), and aplysinol (**13**) were first isolated from the Japanese sea hare *A. kurodai* (Yamamura and Hirata 1963). Laurinterol (**14**) and debromolaurinterol (**15**) were first reported from the red alga *L. intermedia* (Irie et al. 1970), since when they have been identified as constituents of *A. kurodai* and *A. dactylomela*.

Higa and coworkers reported the isolation and structure elucidation of four chamigrenes: (*E*)-and (*Z*)-9-(bromomethylidene)-1,2,5-trimethylspiro [5.5]undeca-1,7-dien-3-ones (**16**, **17**), (8*R*,9*R*)-8-bromo-9-chloro-5,5,9-trimethyl-1-methylidenspiro[5.5]undec-3-en-2-one (**18**), and its (11*R*)-11-acetoxy derivative (**19**; Sakai et al. 1986), and five cuparene-derived sesquiterpenes: cyclolaurene (**20**), cyclolaurenol (**21**), cyclolaurenol acetate (**22**), cupalaurenol (**23**), and cupalairenol acetate (**24**) from the Okinawan sea hare *A. dactylomela*. These cuparene-derived sesquiterpenes showed antibacterial activity against *Bacillus subtilis* and also showed antifungal activity against *Aspergillus* sp. In addition, these sesquiterpenes, with the exception of **20**, were ichthyotoxic against guppies at a concentration of 5–10 ppm (Ichiba and Higa 1986).

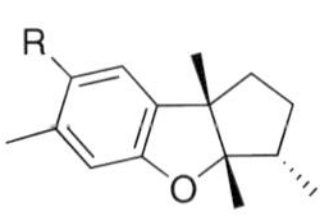

R=Br aplysin (**11**)
R=H debromoaplysin (**12**)

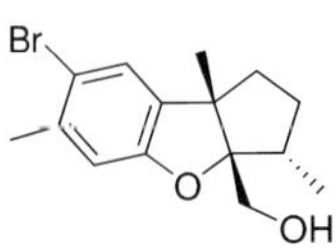

aplysinol (**13**)

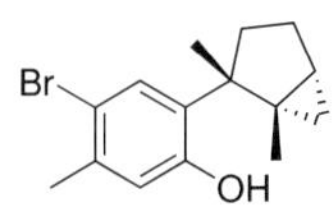

laurinterol (**14**)

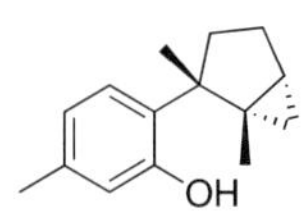

debromolaurinterol (**15**)

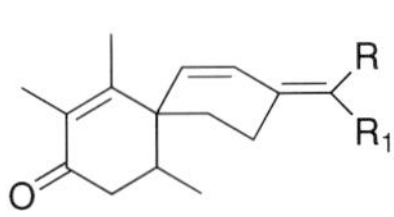

R=H, R_1=Br **16**
R=Br, R_1=H **17**

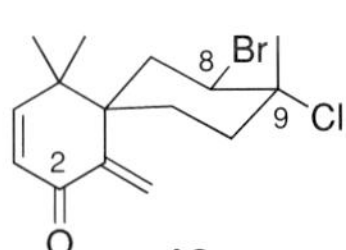

18

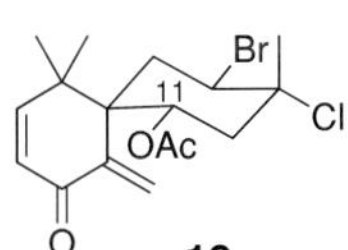

19

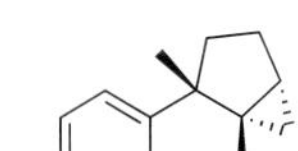

cyclolaurene (**20**)

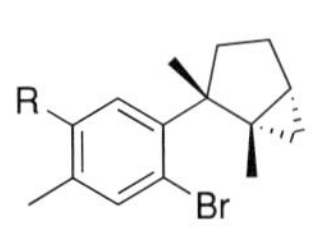

R=OH cyclolaurenol (**21**)
R=OAc cyclolaurenol acetate (**22**)

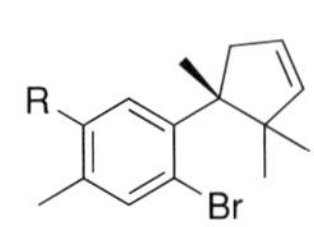

R=OH cupalaurenol (**23**)
R=OAc cupalaurenol acetate (**24**)

9.2.2 Degraded Sterols

Aplykurodins belong to the rare class of highly degraded marine steroids and have been isolated only from the mantle of some anaspideans: aplykurodins A (**25**) and B (**26**) from Japanese *Aplysia kurodai* (Miyamoto et al. 1986), 4-acetylaplykurodin B (**27**) and aplykurodinone B (**28**) from Mediterranean *A. fasciata* (Spinella et al. 1992), 3-*epi*-aplykurodinone B (**29**) from Atlantic *A. fasciata* (Ortega et al. 1997), and aplykurodinone-1 (**30**) and -2 (**31**) from *Syphonota geographica*, collected along Greek coasts (Gavagnin et al. 2005).

The $CHCl_3$-soluble part of the MeOH extract of the mantle of *A. kurodai* was separated with silica gel, reversed phase column chromatography and reversed phase HPLC to give **25** and **26**, together with halogenated monoterpenes and epidioxycholesterol (**32**), as shown in Scheme 9.1.

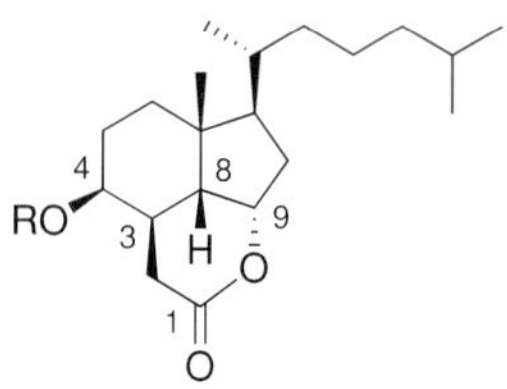

R=H aplykurodin A (**25**)
R=Ac **34**

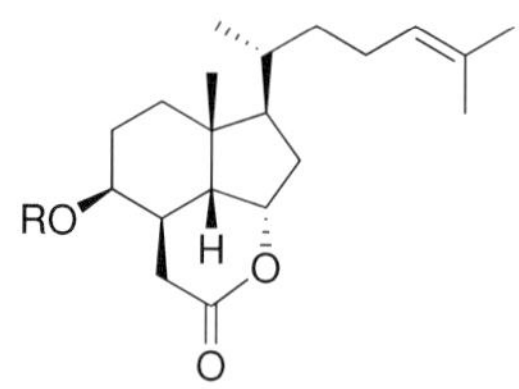

R=H aplykurodin B (**26**)
R=Ac **27**

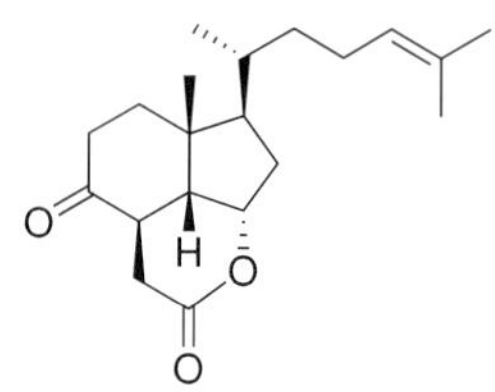

aplykurodinone B (**28**)

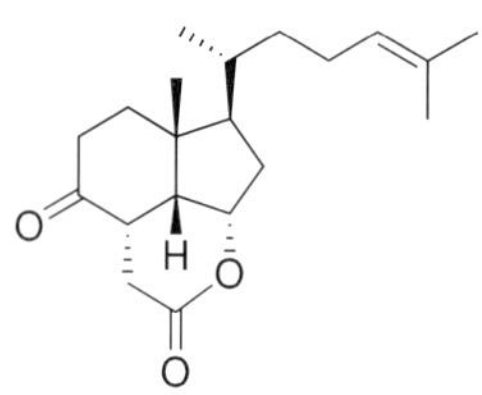

3-*epi*-aplykurodinone B (**29**)

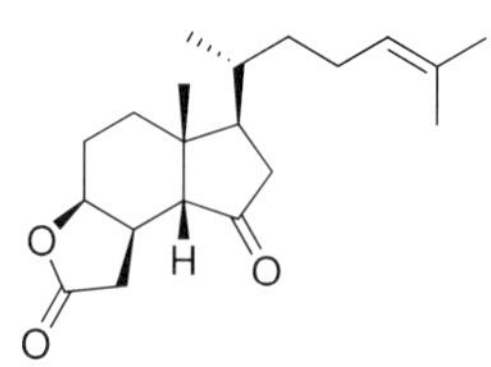

aplykurodinone-1 (**30**)

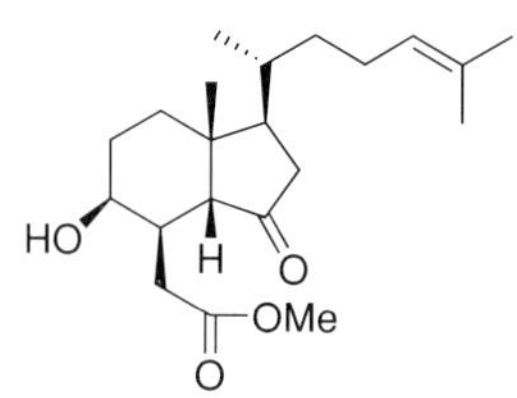

aplykurodinone-2 (**31**)

The planar structure of **25** and of its γ-lactone isomer (**33**) was supported by NMR analysis. The relative stereochemistry was established by single crystal X-ray analysis for aplykurodin A acetate (**34**), whereas the absolute stereochemistry was determined by applying the octant rule to the 3-oxo derivative, aplykurodinone A (**35**). The interconversion of the two lactone isomers was investigated by Spinella et al. (1992). Neutralization of the carboxylate form led to the formation of the δ-lactone as a kinetic product while, when this latter was put in acidic solution, the γ-lactone was obtained as a thermodynamic product (Fig. 9.1).

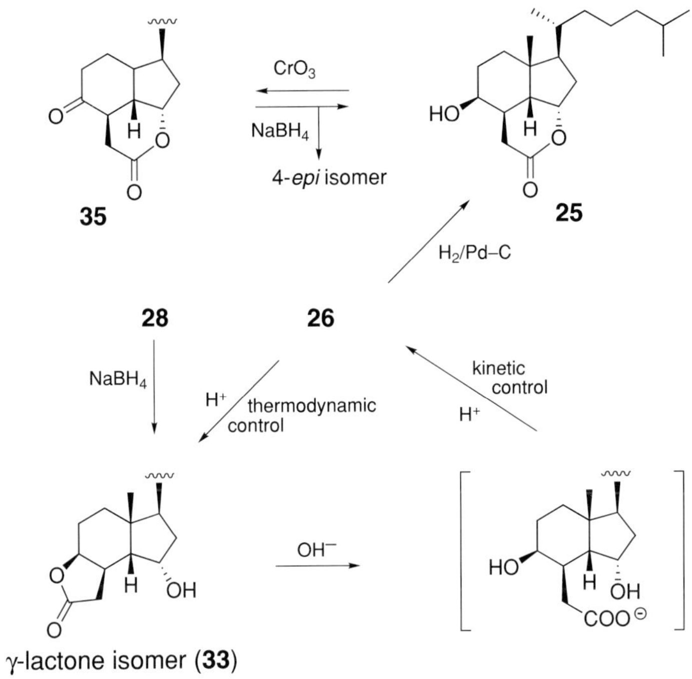

Fig. 9.1. Chemical interconversions of aplykurodins

Synthesis of non-natural 3,9-di-*epi*-aplykurodinone A, via the Grundmann ketone from vitamin D_3, was reported by Sodano's group (Izzo et al. 2000).

Aplykurodin A (**25**) showed cytotoxicity against the L1210 cell line, with an IC_{50} value of 5.5 µg ml^{-1}; and 3-*epi*-aplykurodinone B (**29**) exhibited cytotoxicity against P-388 mouse lymphoma, A-549 human lung carcinoma, HT-29 human colon carcinoma, and MEL-28 human melanoma cell lines, with IC_{50} values of 2.5 µg ml^{-1} in all cases. However, 4-acetylaplykurodin B (**27**) and aplykurodinone B (**28**) showed ichthyotoxicity against mosquito fish at 10 ppm and exhibited antifeedant activity against the fish *Carassius auratus* at the concentration of 60 µg cm^{-2} of food pellets. Compound **27** also showed toxicity against brine shrimp at the LC_{50} value of 29.1 ppm.

Aplykurodins could derive biogenetically from a parent sterol by oxidative pathways involving the loss of A-ring carbon atoms (Fig. 9.2). 5α,8α-Epidioxycholesterol (**32**), which is the peroxidative product of 5,7-dehydrocholesterol, has been identified in the mantle extract of *A. kurodai* (Scheme 9.1). Ultraviolet irradiation could lead to an oxidative cleavage of the C_9–C_{11} bond; and the resulting cholecalciferol (**36**) could undergo a further oxidative cleavage of the Δ5 bond to afford the carbon skeleton of aplykurodins. Degraded sterols related to **36**, astrogorgiadiol (**37**) and calicoferol B (**38**) have been found in gorgonians (Fusetani et al. 1989; Ochi et al. 1991). It has been speculated that epimerization at C_{14} of the sterol skeleton (C_8 in aplykurodins) could occur during the degradation process. Feeding experiments with appropriately labeled precursors could give useful information on the biosynthetic pathway of aplykurodins.

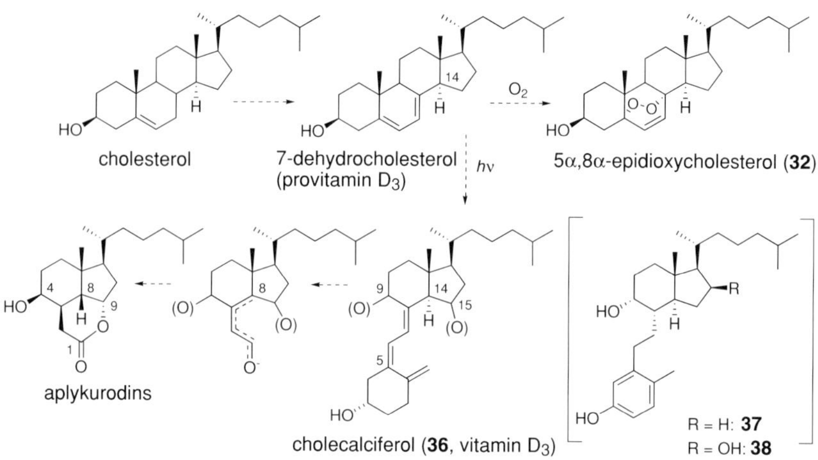

Fig. 9.2. Plausible biogenetic pathway of aplykurodins

9.3 Metabolites of Nudibranchs

9.3.1 Sesquiterpenes

Dendrodoris carbunculosa (Kelaart, 1858; family Dendrodorididae) is an exceptionally large nudibranch. This species usually lives in tropical areas and is rarely found in Japan. Drimane sesquiterpenes, dendocarbins A–N (**39–52**), were isolated from a single individual of *D. carbunculosa*, together with two known sesquiterpenes, isodrimeninol (**53**) and 11-epivaldiviolide (**54**; Sakio et al. 2001). The cytotoxic activities of these sesquiterpenes against P388 mouse leukemia were evaluated. The acetyl derivative of dendocarbin A, compound **55**, was also tested. The IC_{50} values against adriamycin (ADR)- and vincristin (VCR)-resistant P388 cells as well as their sensitive cells are shown in Table 9.1. Although all sesquiterpenes did not reverse multidrug resistance, dendocarbin J (**48**)

Table 9.1. Cytotoxic activity of the sesquiterpenes isolated from *D. carbunculosa*.

cell lines	IC_{50} (μg ml^{-1})															
	39	**40**	**41**	**42**	**43**	**44**	**45**	**46**	**47**	**48**	**49**	**50/51**	**52**	**53**	**54**	**55**
P388/S	>25.0	11.5	>25.0	15.0	>25.0	>25.0	>25.0	10.8	10.0	17.0	10.5	>25	>25	13.0	3.2	10.2
P388/VCR(–)	>25.0	10.2	>25.0	16.0	22.5	>25.0	>25.0	10.1	10.2	4.0	10.5	>25	>25	22.0	2.5	10.2
P388/VCR(+)	>25.0	9.0	>25.0	14.0	>25.0	>25.0	>25.0	8.0	10.2	4.0	9.0	>25	>25	>25.0	2.5	9.0
P388/ADR(–)	>25.0	7.0	>25.0	16.0	19.0	>25.0	>25.0	10.1	8.0	11.0	9.0	>25	>25	18.0	2.5	8.0
P388/ADR(+)	>25.0	10.5	>25.0	16.0	>25.0	22.0	>25.0	10.5	10.0	8.0	10.5	>25	>25	18.0	2.5	10.1

P388/S is a drug-sensitive P388 cell line; P388/ADR and P388/VCR are adriamycin- and vincristine-resistant P388 cell lines (Sakio et al. 2001)

and 11-epivaldiviolide (54) exhibited moderate cytotoxicity against both sensitive and resistant cell strains (Sakio et al. 2001).

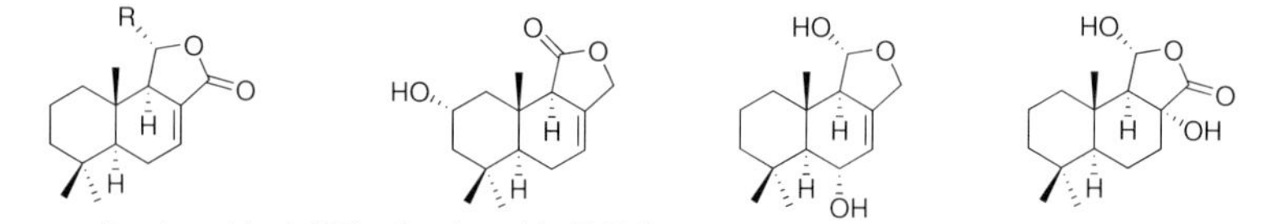

R=OH: dendocarbin A (**39**) dendocarbin B (**40**) dendocarbin C (**41**) dendocarbin D (**42**)
R=Ac: **55**

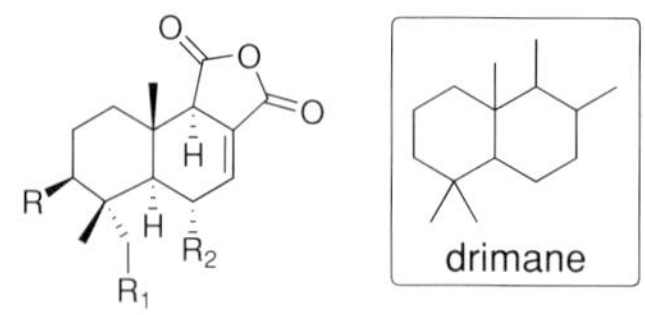

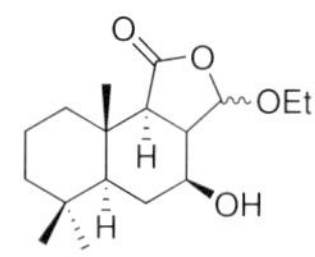

dendocarbin I (**47**)

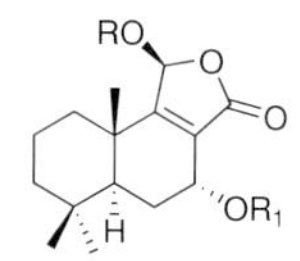

R=Et, R_1=H: dendocarbin J (**48**)
R=R_1=H: dendocarbin L (**50**)

R=OH, R_1=R_2=H: dendocarbin E (**43**)
R=H, R_1=OH, R_2=H: dendocarbin F (**44**)
R=R_1=H, R_2=OH: dendocarbin G (**45**)
R=R_1=R_2=H: dendocarbin H (**46**)

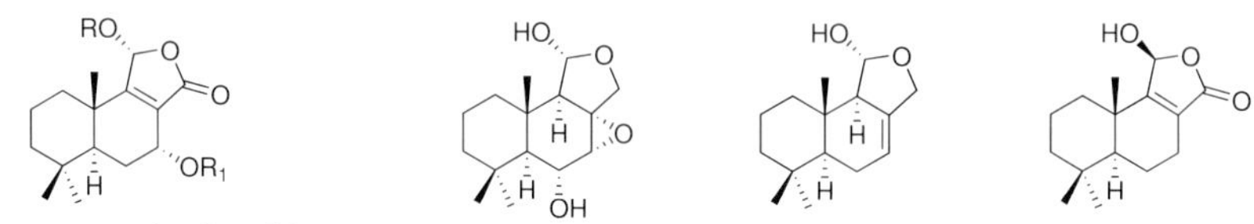

R=Et, R_1=H: dendocarbin K (**49**) dendocarbin N (**52**) isodrimeninol (**53**) 11-epivaldiviolide (**54**)
R=R_1=H: dendocarbin M (**51**)

Sesquiterpene carbonimidic chlorides, reticulidins, were isolated from the phyllidiid nudibranch *Reticulidia fungi*, together with two known carbonimidic chlorides (Tanaka and Higa 1999). Reticulidins A (56) and B (57) exhibit moderate cytotoxicity, with IC_{50} values of 0.41 and 0.42 μg ml^{-1} against KB cells, and 0.59 and 0.11 μg ml^{-1} against L1210 cells, respectively (Tanaka and Higa 1999).

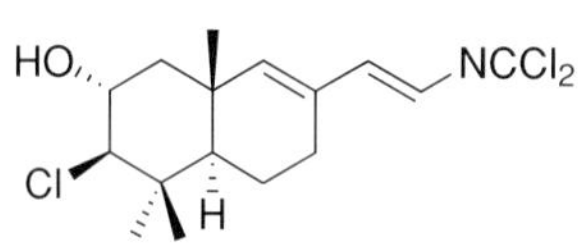

reticulidin A (**56**)

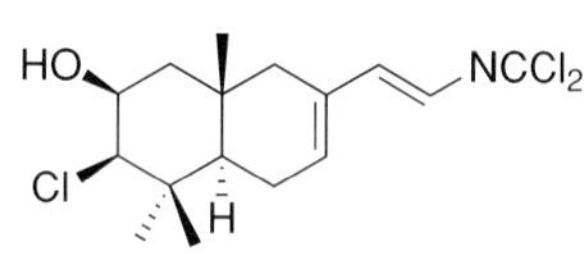

reticulidin B (**57**)

Nudibranchs of the family Phyllidiidae are well known to sequester sesquiterpene isocyanides and thiocyanates from their sponge prey to protect themselves against predators. 9-Isocyanopupukeanane (58) is the first reported allomone isolated from both the Hawaiian nudibranch *Phyllidia varicosa* and its prey, the sponge *Hymeniacidon* sp. (Burreson et al. 1975; Hagadone et al. 1979). Fusetani and coworkers have investigated the chemical defense of phillidid nudibranchs in Japan; and they isolated 9-*epi*-9-isocyanopupukeanane (59) from *P. bourguini*

(Fusetani et al. 1990), cavernothiocyanate (**60**), and isocyano-4-amorphene (**61**) from *P. ocellata* (Fusetani et al. 1992). Furthermore, they investigated the antifouling substances from nudibranchs and their sponge preys and found 10-*epi*-axionitrile-3 (**62**), 10-isocyano-4-cadinene (**63**), and 2-isocyanotrachyopsane (**64**), along with the known sesquiterpenes, axisonitrile-3 (**65**), (–)-10-isothiocyanato-4-amorphene (**66**), 2-thiocyanatoneopupukeanane (**67**), 4-thiocyanatoneopupukeanane (**68**), 3-isocyanotheonellin (**69**), 10-isocyano-4-amorphene (**70**), and 1,7-epidioxy-5-cadinene (**71**) in *P. pustulosa*, *P. varicosa*, and *P. krempfi* (Okino et al. 1996), and an isocyanosesquiterpene alcohol (**72**) in *P. pustulosa* (Hirota et al. 1998).

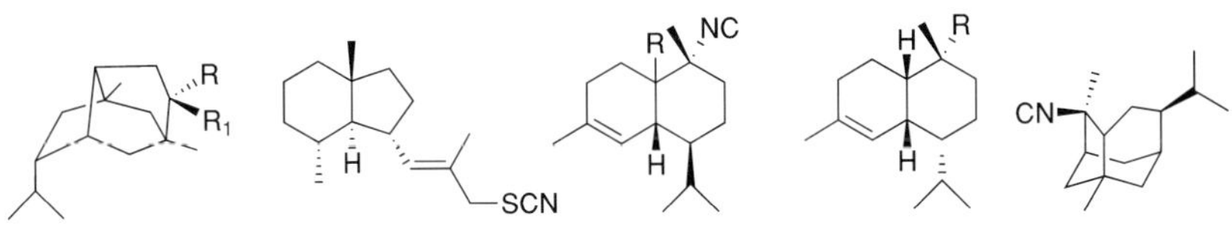

R=NC, R_1=H: **58**
R=H, R_1=NC: **59**
60
R=β-H: **61**
R=α-H: **63**
R=NCS: **66**
R=NC: **70**
64

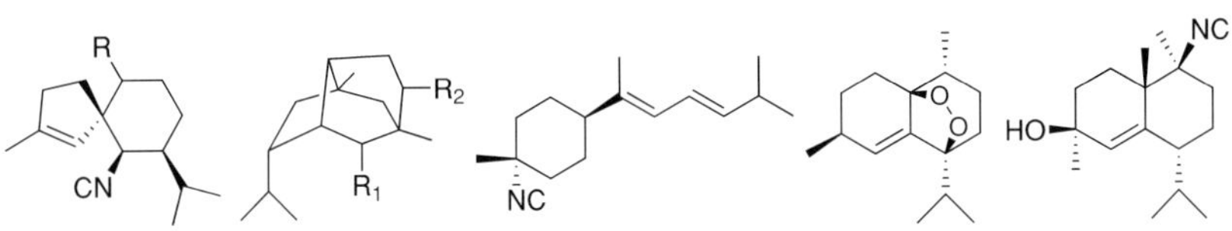

R=α-CH_3: **65**
R=β-CH_3: **62**
R_1=SCN, R_2=H: **67**
R_1=H, R_1=SCN: **68**
69
71
72

9-Isocyanopupukeanane (**58**) and its 9-epimer **59** showed strong ichthyotoxicity with, respectively, LC_{50} values of 2.0 μg ml^{-1} (2 ppm) and 1.0 μg ml^{-1} (1 ppm) in a 6.5-h assay. There is the possibility that the nudibranch partially metabolizes compounds **58** into **59** to reduce the total toxicity (Fusetani et al. 1990). The antifouling activity of the compounds tested was established by evaluating their inhibitory effect on the settlement and metamorphosis of cyprid larvae of the barnacle *Balanus amphitrite*. Compounds **63**, **69**, and **72** showed potent antifouling activity at EC_{50} values of 0.14, 0.13, and 0.17 μg ml^{-1}, respectively, with very weak lethality (Okino et al. 1996; Hirota et al. 1998).

9.3.2 Diterpenes

Dorisenones A–D (**73**–**76**) are cytotoxic spongiane diterpenes that were isolated from the Japanese chromodorid nudibranch *Chromodoris obsoleta* (the nomenclature of *C. obsoleta* has been revised to *C. tinctoria*; Miyamoto et al. 1996). Three additional unprecedented diterpenes,

7α-hydroxyspongian-16-one (**77**), 15α16α-diacetoxy-11,12β-epoxyspongian (**78**), and 7α-acetoxydendrillol-3 (**79**), together with four known spongianes (compounds **80–83**) were also found in the same nudibranch. The cytotoxic activity of all these diterpenes was examined against L1210 and KB human epidermoid carcinoma cell lines. Compounds **73**, **78**, and **79** showed strong activities, as shown in Table 9.2 (Miyamoto et al. 1996).

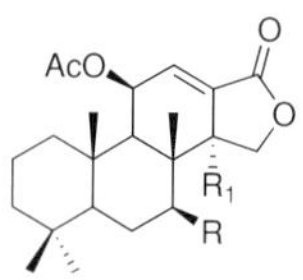

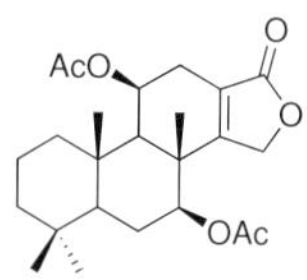

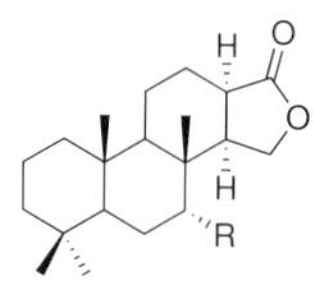

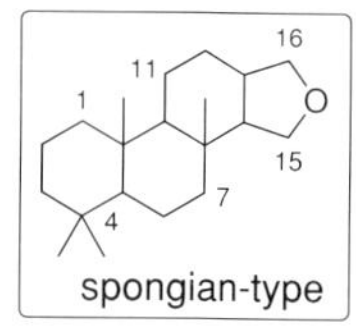

R=OAc, R_1=OH: dorisenone A (**73**)
R=H, R_1=OH: dorisenone B (**74**)
R=OAc, R_1=H: dorisenone D (**76**)
R=H, R_1=H: 11β-hydroxyspongi-12-en-16-one (**81**)

dorisenone C (**75**)

R=OH: **77**
R=H: spongian-16-one (**82**)
R=OAc: 7α-acetoxyspongian-16-one (**83**)

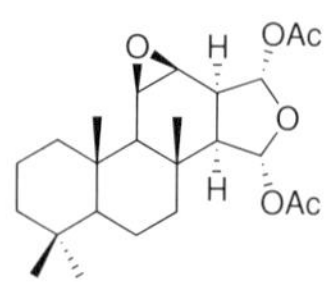

78

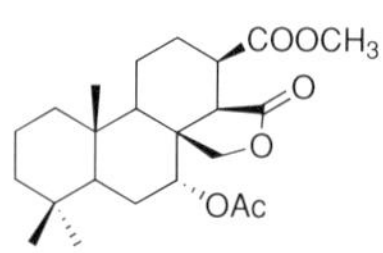

79

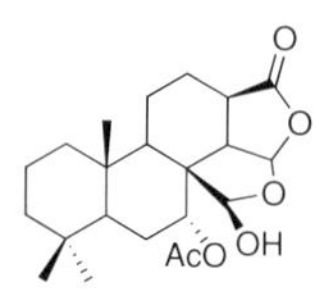

7α-acetoxy-17β-hydroxy-15, 17-oxidospongian-16-one (**80**)

Table 9.2. Cytotoxic activity of the diterpenes isolated from *C. obsoleta*

cell lines	IC_{50} (μg/ml)										
	73	74	75	76	77	78	79	80	81	82	83
L1210	0.21	1.0	7.5	0.18	7.5	0.18	4.8	1.9	1.0	5.0	2.2
KB	0.22	1.5	19.0	1.4	10.2	0.98	15.0	2.5	1.9	9.2	16.0

9.3.3 Sesterterpenes

Inorolides A (**84**), B (**85**), and C (**86**) are cytotoxic sesterterpenes, which were isolated from a Japanese chromodorid nudibranch *C. inornata* (the nomenclature of *C. inornata* has been revised to *C. orientalis*). Four new scalaranes, 12-epideoxyscalarin-3-one (**87**), deoxoscalarin-3-one (**88**), 21-hydroxydeoxoscalarin (**89**), 21-acetoxydeoxoscalarin (**90**), and one scalarolbutenolide-type, 12-*O*-acetyl-16-*O*-deacetyl-12, 16-episcalarolbutenolide (**91**) were also found, along with the known 12-epideoxoscalarin (**92**) and deoxoscalarin (**93**), in the extract of this nudibranch (Miyamoto et al. 1992).

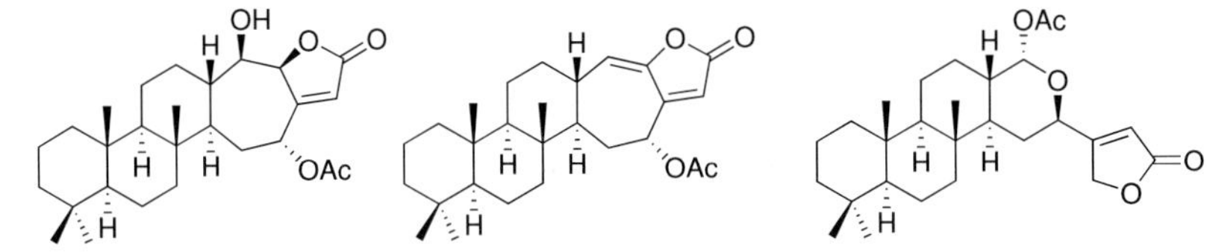

inorolide A (**84**) inorolide B (**85**) inorolide C (**86**)

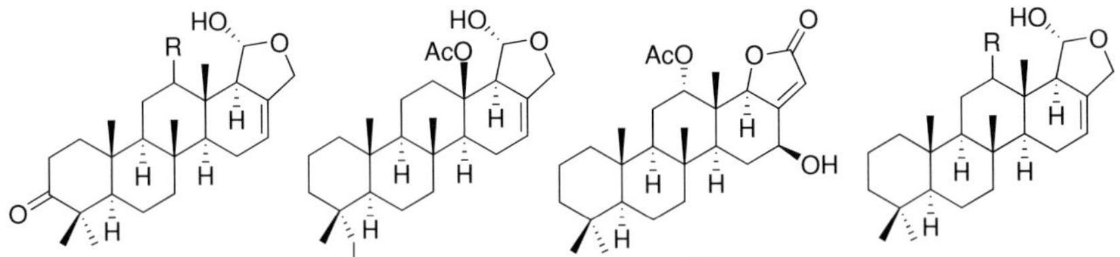

R=β-OAc: **87**
R=α-OAc: **88**

R=OH: **89**
R=OAc: **90**

91

R=β-OH: **92**
R=α-OH: **93**

It could be suggested that inorolide-type sesterterpenes are biogenetically derivable by ring closure between C_{24} and C_{25} of the cheilanthane skeleton, whereas scalarane and scalarolbutenolide-type sesterterpenes are derivable from the same skeleton by ring closure between C_{13} and C_{18}, and between C_{13} and C_{25}, respectively (Fig. 9.3).

The cytotoxic activity of these sesterterpenes against L1210 and KB human epidermoid carcinoma cell lines was examined. Inorolide B (**85**) and 21-acetoxydeoxoscalarin (**90**) showed potent activity, as shown in Table 9.3. Furthermore, neuritogenic activity against the PC-12 rat pheochromocytoma cell line was evaluated. Activity was observed only in deoxoscalarin-3-one (**88**) at a concentration above 10 μg ml^{-1} (Miyamoto et al. 1992).

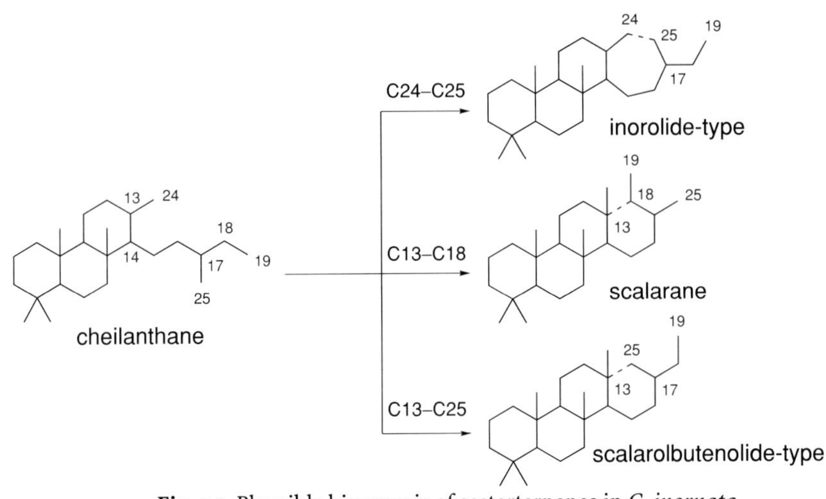

Fig. 9.3. Plausible biogenesis of sesterterpenes in *C. inornata*

Table 9.3. Cytotoxic activity of the sesterterpenes isolated from *C. inornata*

cell lines	IC_{50} (μg ml^{-1})									
	84	85	86	87	88	89	90	91	92	93
L1210	1.9	0.72	1.9	6.6	0.95	4.1	0.35	2.4	8.2	1.4
KB	3.4	2.2	6.4	22.8	5.2	21.0	3.1	7.6	>30.0	6.4

9.4 Conclusions

Over the past four decades, research into allelochemicals from Japanese marine mollusks has yielded a variety of bioactive substances. Anaspidean mollusks are herbivorous and nudibranchs are carnivorous; and both utilize these substances for chemical defenses against predators. Aplykurodins are absent in any possible diet of the mollusk. The sea hare *A. kurodai* can probably biosynthesize aplykurodins for chemical defense and store them in its mantle. Allelochemics are defined as interactions of attack, defense, and behavioral response involving not physical force but chemical agents between different species. These chemical agents are classified into "allomones," which give an adaptive advantage to the producing organism, and "kairomones," which give an adaptive advantage to the receiving organism (Whittaker and Fenny 1971). Aplysiaterpenoid A (**5**) is a versatile allelochemical. It acts as a "kairomone" for anaspidean mollusks, but it acts as "allomone" for the red alga against gastropodeans. Furthermore, anaspideans use compound **5** as an "allomone" against fish, but it acts as a "kairomone" for humans; and humans intend to use aplysiaterpenoid A as an "allomone" against insects.

In the long history of development, marine mollusks belonging to the Opisthobranchia have undergone modification in their defense systems against predators. When marine mollusks got rid of their heavy armor, they protected themselves with invisible weapons, so-called allelochemicals, against predators. It is interesting to report that new strategies against dangerous human pathologies can be discovered by studying these simple animals.

References

Burreson JB, Scheuer PJ, Finer J, Clardy J (1975) 9-Isocyanopupukeanane, a marine invertebrate allomone with a new sesquiterpene skeleton. J Am Chem Soc 97:4763–4764

Ciavatta ML, Gavagnin M, Puliti R, Cimino G (1997) Dactylallene: a novel dietary C_{15} bromoallene from the Atlantic anaspidean mollusk *Aplysia dactylomela*. Tetrahedron 53:17343–17350

Fusetani N, Nagata H, Hirota H, Tsuyuki T (1989) Astrogorgiadiol and astrogorgin, inhibitors of cell division in Fertilized starfish eggs, from a gorgonian *Astrogorgia* sp. Tetrahedron Lett 30:7079–7082

Fusetani N, Wolstenholme HJ, Matsunaga S (1990) Co-occurrence of 9-isocyano-pupukeanane and its C-9 epimer in the nudibranch *Phyllidia bourguini*. Tetrahedron Lett 31:5623–5624

Fusetani N, Wolstenholme HJ, Shinoda K, Asai N, Matsunaga S, Onuki H, Hirota H (1992) Two sesquiterpene isocyanides and a sesquiterpene thiocyanate from the marine sponge *Acanthella cf. cavernosa* and the nudibranch *Phyllidia ocellata*. Tetrahedron Lett 33:6823–6826

Gavagnin M, Carbone M, Nappo M, Mollo E, Roussis V, Cimino G (2005) First chemical study of anaspidean *Syphonota geographica*: structure of degraded sterols aplykurodinone-1 and -2. Tetrahedron 61:617–621

Hagadone MR, Burreson JB, Scheuer PJ, Finer J, Clardy J (1979) Defense allomones of the nudibranch *Phyllidia varicosa* Lamarck 1801. Helv Chim Acta 62:2484–2494

Hirano Y (2000) Umiushi-Gaku. Tokai Daigaku Shuppankai, Tokyo

Hirota H, Okino T, Yoshimura E, Fusetani N (1998) Five new antifouling sesquiterpenes from two marine sponges of the genus *Axinyssa* and the nudibranch *Phyllidia pustulosa*. Tetrahedron 54:13971–13980

Ichiba T, Higa T (1986) New cuparene-derived sesquiterpenes with unprecedented oxygenation patterns from the sea hare *Aplysia dactylomela*. J Org Chem 51:3364–3366

Inoue Y, Uchida H, Kusumi T, Kakisawa H (1987) Structure and absolute configuration of aplysiapyranoid B, a new polyhalogenated pyranoid monoterpene from *Aplysia kurodai*. J Chem Soc Chem Commun 5:346–347

Irie T, Suzuki M, Kurosawa E, Masamune T (1970) Laurinterol, debromolaurinterol and isolaurinterol, constituents of *Laurencia intermedia* Yamada. Tetrahedron 26:3271–3277

Izzo I, Meduri G, Avallone E, De Riccardis F, Sodano G (2000) Studies towards the synthesis of aplykurodins – synthesis of 17,18-dihydro-3,9-di-*epi*-aplykurodinone B–. Eur J Org Chem:439–448

Kinnel R, Duggan A J, Eisner T, Meinwald J (1977) Panacene: an aromatic bromoallene from a sea hare (*Aplysia brasilina*). Tetrahedron Lett 44:3913–3916

Kusumi T, Uchida H, Inoue Y, Ishitsuka M, Yamamoto H, Kakisawa H (1987) Novel cytotoxic monoterpenes having a halogenated tetrahydropyran from *Aplysia kurodai*. J Org Chem 52:4597–4600

Miyamoto T, Higuchi R, Komori T, Fujioka T, Mihashi K (1986) Isolation and structures of aplykurodins A and B, two new isoprenoids from the marine mollusk *Aplysia kurodai*. Tetrahedron Lett 27:1153–1156

Miyamoto T, Higuchi R, Marubayashi N, Komori T (1988) Two new polyhalogenated monoterpenes from the sea hare *Aplysia kurodai*. Liebigs Ann Chem 1988:1191–1193

Miyamoto T, Sakamoto K, Amano H, Higuchi R, Komori T, Sasaki T (1992) Three new cytotoxic sesterterpenoids, inorolide A, B, and C from the nudibranchi *Chromodoris inornata*.Tetrahedron Lett 33:5811–5814

Miyamoto T, Ebisawa Y, Higuchi R (1995) Aplyparvunin, a bioactive acetogenin from the sea hare *Aplysia parvula*. Tetrahedron Lett 36:6073–6074

Miyamoto T, Sakamoto K, Arao K, Komori T, Higuchi R, Sasaki T (1996) Dorisenones, cytotoxic spongian diterpenoids, from the nudibranch *Chromodoris obsoleta*. Tetrahedron 52:8187–8198

Ochi M, Yamada K, Kotsuki H, Shibata K (1991) Calicoferols A and B, two novel secosterols possessing brine-shrimp lethality from the gorgonian *Calicogorgia* sp. Chem Lett :427–430

Okamoto Y, Nitanda N, Ojika M, Sakagami Y (2001) Aplysiallene, a new bromoallene as an Na, K-ATPase inhibitor from the sea hare, *Aplysia kurodai*. Biosci Biotechnol Biochem 65:474–476

Okino T, Yoshimura E, Hirota H, Fusetani N (1996) New antifouling sesquiterpenes from four nudibranchs of the family *Phyllididae*. Tetrahedron 52:9447–9454

Ono A (1999) Opisthobranchs of Kerama Islands. TBS/Britannica, Tokyo
Ortega JM, Zubia E, Salva J (1997) 3-*epi*-aplykurodinone B, a new degraded sterol from *Aplysia fasciata*. J Nat Prod 60:488–489
Sakai R, Higa T, Jefford CW, Bernardinelli G (1986) The absolute configurations and biogenesis of some new halogenated chamigrenes from the sea hare *Aplysia dactylomela*. Helv Chim Acta 69:91–105
Sakata K, Iwase Y, Ina K, Fujita D (1991) Halogenated terpenes isolated from the red alga *Plocamium leptophyllum* as feeding inhibitors for marine herbivores. Nippon Suisan Gakkaishi 57:743–746
Sakio Y, Hirano YJ, Hayashi M, Komiyama K, Ishibashi M (2001) Dendocarbins A-N, new drimane sesquiterpenes from the nudibranch *Dendrodoris carbunculosa*. J Nat Prod 64:726–731
Spinella A, Gavagnin M, Crispino A, Cimino G (1992) 4-Acetylaplykurodin B and aplykurodinone B, two ichthyotoxic degraded sterols from the Mediterranean mollusk *Aplysia fasciata*. J Nat Prod 55:989–993
Suzuki K (2000) Opisthobranchs of Izu Peninsula. TBS/Britanica, Tokyo
Suzuki M, Takahashi Y, Mitome Y, Itoh T, Abe T, Masuda M (2002) Brominated meabolites from an Okinawan *Laurencia intricata*. Phytochemistry 60:861–867
Tanaka J, Higa T (1999) Two new cytotoxic carbonimidic dichlorides from the nudibranch *Reticulidia fungia*. J Nat Prod 62:1339–1340
Watanabe K, Umeda K, Kurita Y, Takayama C, Miyakado M (1990) Two insecticidal monoterpenes, telfairine and aplysiaterpenoid A, from the red alga *Plocamium telfairiae*: structure elucidation, biological activity, and molecular topographical consideration by a semiempirical molecular orbital study. Pest Biochem Physiol 37:275–286
Whittaker RH, Fenny PP (1971) Allelochemics: chemical interactions between species, chemical agents are of major significance in the adaptation of species and organization of communities. Science 171:757–770
Yamada K, Kigoshi H (1997) Bioactive compounds from the sea hares of two genera: *Aplysia* and *Dolabella*. Bull Chem Soc 70:1479–1489
Yamada K, Ojika M, Kigoshi H, Suenaga K (2000) Cytotoxic substances from opisthobranch mollusks. In: Fusetani N (ed) Drugs from the sea. Karger, Basel, pp. 59–73
Yamamura S, Hirata Y (1963) Structure of aplysin and apysinol, naturally occurring bromo-compounds. Tetrahedron 19:1485–1496

Bioactive Molecules from Sea Hares

H. Kamiya, R. Sakai, M. Jimbo

Abstract. Sea hares, belonging to the order Opisthobranchia, subclass Gastropoda, are mollusks that have attracted many researchers who are interested in the chemical defense mechanisms of these soft and "shell-less" snails. Numbers of small molecules of dietary origin have been isolated from sea hares and some have ecologically relevant activities, such as fish deterrent activity or toxicity. Recently, however, greater attention has been paid to biomedically interesting sea hare isolates such as dolastatins, a series of antitumor peptide/macrolides isolated from *Dolabella auricularia*. Another series of bioactive peptide/macrolides, as represented by aplyronines, have been isolated from sea hares in Japanese waters. Although earlier studies indicated the potent antitumor activity of aplyronines, their clinical development has never been conducted because of the minute amount of compound available from the natural source. Recent synthetic studies, however, have made it possible to prepare these compounds and analogs for a structure–activity relationship study, and started to uncover their unique action mechanism towards their putative targets, microfilaments. Here, recent findings of small antitumor molecules isolated from Japanese sea hares are reviewed. Sea hares are also known to produce cytotoxic and antimicrobial proteins. In contrast to the small molecules of dietary origin, proteins are the genetic products of sea hares and they are likely to have some primary physiological functions in addition to ecological roles in the sea hare. Based on the biochemical properties and phylogenetic analysis of these proteins, we propose that they belong to one family of molecule, the "Aplysianin A family," although their molecular weights are apparently divided into two groups. Interestingly, the active principles in *Aplysia* species and *Dolabella auricularia* were shown to be L-amino acid oxidase (LAAO), a flavin enzyme that oxidizes an α-amino group of the substrate with molecular oxygen and liberates hydrogen peroxide, with a sequence similar to other known LAAOs, including snake venom. Possible antibacterial activity and cytotoxic activity mechanisms of these proteins are also discussed.

Hisao Kamiya
School of Fisheries Sciences, Kitasato University, Sanrikucho
Ofunato-shi, 022-0101 Iwate, Japan

Progress in Molecular and Subcellular Biology
Subseries Marine Molecular Biotechnology
G. Cimino, M. Gavagnin (Eds.): Molluscs

10.1 Introduction

According to *Poisonous and venomous marine animals of the world* by Halstead and Courville (1965), references to the toxicity of sea hares date back to the first century AD, in ancient Rome. Dioscorides, the originator of the medical profession, recognized the sea hare *Aplysia* as being poisonous. The poisonous qualities of sea hares seemed to be received as truth in the European Middle Ages, probably because of their odd appearance and fetid nauseating odor. In contrast, peoples in tropical areas and some Asian countries have used sea hares as food. In Japan, for example, people on Okinoshima Island eat cooked local species of *Aplysia* as a delicacy after removing the viscera and washing the body surface with common salt. Toxic cases of sea hares, however, are also known in those areas. In the southern coasts of China, it was warned that overeating of *Stylocheilus leachii* could cause gastrointestinal disorders, loss of sight, and death in severe cases (Tchang and Lin 1964). These symptoms could be attributed to photosensitizing toxin pheophorbides, degraded chlorophylls, since pyropheophorbides a and b were identified from *A. juliana* collected in Iwate, Japan (Kobayashi et al. 1991).

The dried egg mass of *A. kurodai* has been used as a traditional folk remedy for more than 100 years (Tchang and Lin 1964). However, sporadic occurrences of food poisoning due to sea hares or their eggs are also reported. Ingestion of *Dolabella auricularia* causes neurological symptoms such as ataxia and muscular itching (Sorokin 1988). Hino et al. (1994) described nausea, acute liver damage, and hyperbilirubinemia but no neurological disturbance caused by boiled Aplysiidae eggs. Acute liver damage with apoptosis of numerous hepatocytes after ingestion of boiled mantle and albumen gland of *A. kurodai* have also been reported (Sakamoto et al. 1998).

The Aplysiacea order comprises the Akeridae and Aplysiidae families. Genera of the latter family are generally referred to as "sea hare" because of their appearance. Like most of the Opisthobranchia, sea hares are regarded as "soft and defenseless animals" since they are "shell-less" in their appearance. However, they do actually have a small degenerated shell inside their mantle cavity, but of course, this shell is too small to serve as a shelter.

This seemingly defenseless animal thrives in coastal areas of temperate, subtropical, and tropical waters all over the world. They are hermaphroditic herbivores and feed mainly on macroalgae, but in some cases, on colonies of cyanobacteria. In Japanese waters, they breed between spring and autumn and deposit egg masses on the rock or algal substrate.

Not only their unique appearance, but also their odd behavior characterizes these animals. Most Aplysiidae discharge purple "ink" or white fluid when disturbed. This sudden reaction might be surprising enough for the attacking predators (or people) to retract their interest in or threat to the individual further; and thus this behavior can be considered as one of a few (or the only) "physical" defense mechanisms of sea hare, although the effectiveness or biological role of purple secretions aimed at predators is still a matter of controversy (Nolen et al. 1995; Johnson and Willows 1999; Nolen and Johnson 2001). Numbers of chemists who believe that organisms poor in physical defense mechanisms might have a high possibility of having a chemical defense mechanism, and therefore contain some biologically interesting compounds, have been attracted to this animal. This belief has been shown to be true because extracts of sea hares have afforded numbers of structurally and biologically interesting compounds. Most of them, however, are believed to have their origins in their algal or microalgal diet. Modern structural research of sea hare metabolites started with the isolation of aplysin, a brominated sesquiterpene, in 1963 (Yamamura and Hirata 1963). Literature dealing with natural products, secretions, and active extracts of opisthobranch mollusks available up to May 1994 is listed in a review by Avila (1995). All the chemical studies on opisthobranchs are summarized in two recent reviews (Cimino et al. 1999, 2001). Bioactive compounds from sea hares were reviewed, focusing especially on chemistry of cytotoxic compounds, by Yamada and Kigoshi (1997) and Yamada et al. (2000). More recent cases dealing specifically with the sea hare isolates, most likely originated from cyanobacteria, have also been reviewed (Luesch et al. 2002).

Water-soluble compounds – particularly biopolymers of sea hares – seem to have received little attention in the area of marine natural product chemistry, although there is considerable evidence that the macromolecular compounds are as important in the defense process as small molecules. Numbers of proteins have been isolated from sea hares and they display a variety of bioactivities. Lectins, carbohydrate-binding proteins, are interesting functional biopolymers in sea hares. Lectins specific to D-galacturonic acid/D-galactose have been isolated from laid eggs, gonads, and skin mucus of *Aplysia* species (Kamiya and Shimizu 1981; Gilboa-Garber et al. 1985; Ozeki 1998; Wu et al. 2000). As for physiological roles, animal lectins display interesting cellular activities, such as induction of cell proliferation, differentiation, and controlled cell death, as well as foreign material recognition (Vasta 1991). Often, functions of lectins are also discussed in connection with ecological events such as symbiosis (Müller et al. 1981). *Aplysia* gonad lectin shows the modulation of growth of *Aplysia* neurons (Wilson et al. 1992). The physiological role of *Aplysia* gonad lectin in the sea hare, however, remains unknown.

Aplysia attractins, water-borne pheromones of *Aplysia* species, are also an interesting subject from the chemical–ecological and behavioral–biological points of view. Although this topic is not discussed in this review, recent articles cover the progress and newer aspects of *Aplysia* attractin research (Cummins et al. 2004a,b; Painter et al. 2004). Although numbers of neuropeptides have been identified from *Aplysia*, they are excluded from this review since most neuropeptides are hormonal and few semiochemical functions are expected. Reviews are available in literature (Weiss et al. 1993; Whim et al. 1993).

A series of antimicrobial and cytotoxic proteins, which may have a primary role in the chemical defense of sea hares, have been identified. Biochemical properties and phylogenetic analysis of these proteins suggest that they belong to one protein family, although apparent molecular weights divide them into two groups: 300 and 60 kDa. Here, we call these proteins the "aplysianin A (ApA) family," since ApA is the first protein for which the complete amino acid sequence is determined in the family. Earlier perspectives of chemical and pharmacological properties of these proteins have been reviewed by Yamazaki (1993).

The present review focuses on recent literature which deals with biomedically interesting small molecules from Aplysiidae, then moves on to recent chemical and biological aspects of bioactive proteins of the ApA family.

10.2 Bioactive Small Molecules from Aplysiidae

Two classes of metabolites, dolastatins and aplyronines, and related peptides and macrolides isolated from sea hares are of significant biomedical interest.

Dolastatins are a series of cytotoxic compounds isolated from *Dolabella auricularia* by Pettit and coworkers who have reported 19 different dolastatins since 1974. Chemical and biomedical aspects of dolastatins have been extensively reviewed (Pettit 1997; Poncet 1999; Schwartsmann et al. 2001; Amador et al. 2003). Further, Yamada has reviewed cytotoxic metabolites isolated from sea hares inhabiting Japanese waters (Yamada and Kigoshi 1997; Yamada et al. 2000). In this section, we focus mainly on recent progress in research on antitumor macrolides and peptides from Japanese sea hares.

10.2.1 Cytotoxic Metabolites form *Dolabella auricularia*

A series of dolastatins reported by Pettit were isolated from *D. auricularia* collected from the Indian Ocean (Mauritius), Papua New Guinea, and the Gulf of California (Pettit et al. 2004). Alternatively, Yamada and coworkers collected the sea hare in Mie Prefecture, Japan, and investigated its cytotoxic peptidyl constituents to afford seven new dolastatins. They were also very minor components. Structures were determined for dolastatins C (Sone et al. 1993a), D (Sone et al. 1993b), E (Ojika et al. 1995), G and nordorasrtatin G (Mutou et al. 1996), dolastatin H, isodolastatin H (Sone et al. 1996a), and dolastatin I (Sone et al. 1997). All of these compounds were again similar to cyanobacterial products. In fact, dolastatin C, G, and H are close analogues of dolastatin 10, which has structural similarity to cyanobacterial products. However, dolastatin H and I possess striking resemblance to terrestrial cyanobacterial products dendroamide A and teruecyclamide A (Yamada and Kigoshi 1997; Luesch et al. 2002).

Yamada also found other new cytotoxic depsipeptides: doliculide (Ishiwata et al. 1994), aurilide (Suenaga et al. 1996), and dolabellin (Sone et al. 1995). In addition to these nitrogenous compounds, Japanese *D. auricularia* afforded several new polyketides, including aurisides A and B (Sone et al. 1996b), which have a skeletal structure similar to dolastatine 19, a polyketide dolastatin from the Gulf of California specimen, and to the cyanobacterial isolate, lyngbyaloside (Pettit et al. 2004).

Recently, doliculide was shown to promote actin hyperpolymerization and displace FITC-labeled phalloidin form actin polymer (Bai et al. 2002). This action is analogous to phalloidin (mushroom toxin), chondramide (mixobacterial product) and jasplakinolide (sponge isolate), whose structure shares some similarity to doliculide (Fig. 10.1). Interestingly, dolastatin 11, which was recently shown to have actin hyperpolymerization

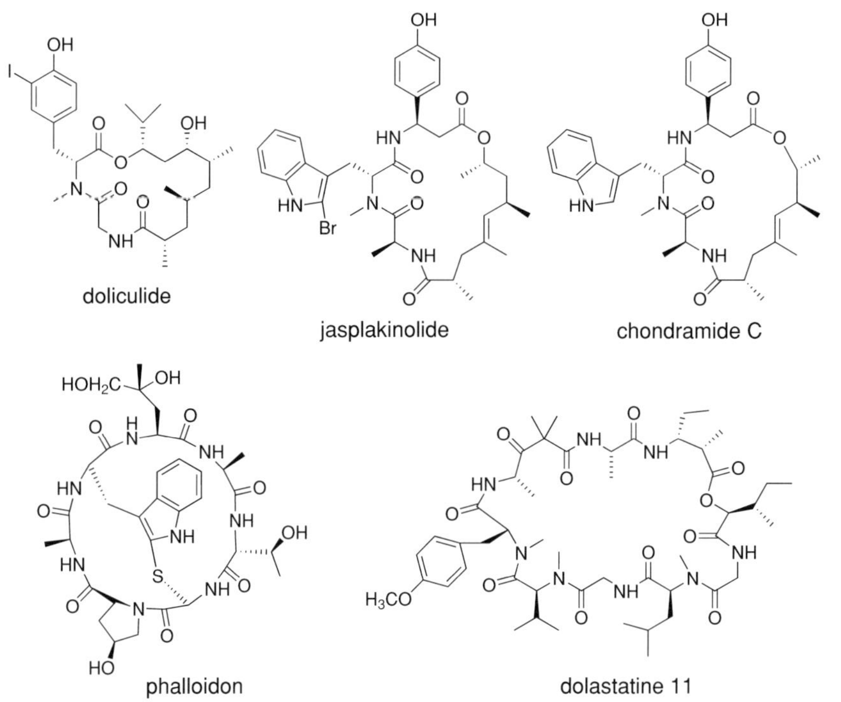

Fig. 10.1. Structures of doliculide and other actin hyperpolymerizing agents

activity like doliculide, does not displace the fluorescent phalloidin derivative. This clearly indicates that doliculide binds to the phalloidin site of actin, but dolastatin 11 binds to some other site, to display a similar activity to actin. A computer modeling study indicated that phalloidin, chondramide, and jasplakinolide share the pharmacophore with doliculide but not with dolastatin 11, as expected (Bai et al. 2001).

Aurilide (Fig. 10.2) was recently shown to be a strong microtubule stabilizer, but the stabilization mechanism was different to that of taxol. Although aurilide showed in vivo activity in NCI's hollow fiber assay, it was not active in a human tumor xenograft model because of its narrow therapeutic window (Suenaga et al. 2004). Recently, kurokekahilide-2, a close analogue of aurilide was isolated from the Hawaiian carnivorous mollusk *Philinopsis speciosa*. Despite the structural similarity to aurilide, kurokekahilide-2 did not interact with tubulin, intermediate filaments, or actin (Nakao et al. 2004).

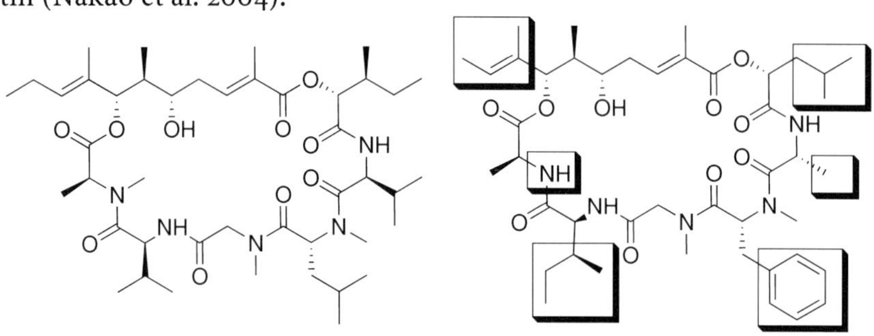

Fig. 10.2. Aurilide and kurokekahilide-2. Side-chain structures of kurokekahilide-2 differing from those of aulilide are shown in *boxes*

10.2.2 Cytotoxic Metabolites from *Aplysia* sp.

Yamada et al. (1993, 2000) have also explored cytotoxic metabolites contained in another genus of sea hare, *Aplysia*. A large amount (300 kg) of *A. kurodai* was collected in Mie Prefecture, Japan, of which the methanol extract showed strong cytotoxicity. Cytotoxicity-guided separation of the extract of the whole animal afforded a trace amount of macrolide aplyronine A–H (Yamada et al. 2000; Fig. 10.3). Aplyronines were strongly cytotoxic against HeLa S3 cells, with IC_{50} values of each aplyronine, A, B, and C, being 0.48, 3.11, and 21.2 ng mL^{-1}, respectively. The most remarkable feature of aplyronine is the in vivo efficacy of aplyronine A in mice. Aplyronine A showed life elongation of 200–560% in five different tumor-grafted mice models (Yamada et al. 1993, 2000; Yamada and Kigoshi 1997). Aplyronine A was shown to depolymerize F-actin and inhibit polymerization of G-actin

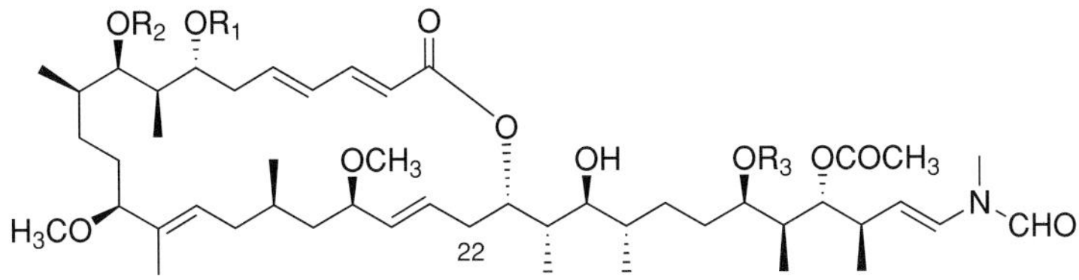

aplyronines

A: R_1=*N,N,O*-trimethylserinyl (TMSer), R_2=H, R_3= *N,N*-dimethylalanyl (DMAla)
B: R_1=H, R_2=TMSer, R_3=DMAla
C: R_1=H, R_2=H, R_3=DMAla
D: R_1=TMSer, R_2=H, R_3=*N,N*-dimethylglycinyl
E: 22-methyl aplyronine A
F: R_1=*N,O*-dimethylserinyl (DMSer), R_2=H, R_3=*N*-methylalanyl
G: R_1=DMSer, R_2=H, R_3=DMAla
H: R_1=H, R_2=DMSer, R_3=DMAla

Fig. 10.3. Structures of aplyronines

(Saito et al. 1996). Several natural toxins including mushroom toxin phalloidin and cytochalasins have shown to bind to actin. Recently growing numbers of marine-derived macrolides have been shown to target actin, such as latrunculin A (Spector et al. 1983), swinholide A (Bubb et al. 1995), misakinolide A (bistheonellide A; Terry et al. 1997), dolastatin 11 (Bai et al. 2001), doliculide (Bai et al. 2002), goniodomin (Furukawa et al. 1993), jasplakinolide (jaspamide; Bubb et al. 1994), a series of trioxazole macrolides including mycalolide (Saito et al. 1994) and kabiramide (Klenchin et al. 2003; Tanaka et al. 2003); and pectenotoxin (Zhou et al. 1994), hectochlorin (Marquez et al. 2002), sphinxolide (Zhang et al. 1997), and amphidinolide H (Saito et al. 2004). Note that the skeletal structure of the side-chain portion of aplyronines shares significant similarity to trioxazole macrolides, in which this portion plays a key role in binding to actin molecules (Klenchin et al. 2003; Tanaka et al. 2003). Among them, however, only a few examples of in vivo antitumor activities were reported (Kobayashi et al. 1994; Takeuchi et al. 1998). Moreover, no actin-targeted anticancer drugs have been developed so far, unlike their tubulin-binding counterparts, probably because actin-targeting compounds have rather been regarded the "toxin" as represented by a deadly mushroom toxin phalloidin (Bonnet and Basson 2002). Moreover, larger variations of structure- and actin-modulating mechanisms of actin-targeting compounds become available only recently. In any event, the in vivo efficacy of aplyronine A is exceptional among the known actin-disrupting drugs, and thus this might encourage the further study of this class of compounds, as well as compounds to regulate actin dynamics, such as antineoplastic drugs (Johnson and Willows 1999; Rao and Li 2004).

Because actin is one of the most abundant proteins in cells, regardless of cell type, and aplyronine A shows its in vivo activity at low dosage, the actin-disrupting activity of aplyronines A alone may not be sufficient to explain its in vivo antitumor activity. Interestingly, structure–activity relationship studies of aplyronines show that the actin-depolymerizing activity of aplyronine analogues does not necessarily coincide with cytotoxicity: that is, some analogs of aplyronines show strong actin depolymerizing activity although cytotoxicity is attenuated significantly. For example, an analog with $R_1 = R_3 =$ *N,N*-dimethylglycyl, $R_2 = H$ (cf. Fig. 10.3) showed actin depolymerizing activity around 63% of that of aplyronine A, while its cytotoxicity was attenuated more than 4,000-fold compared to aplyronine A. In contrast, all the analogs which displayed strong cytotoxicity also showed potent actin disrupting activity (Suenaga et al. 1997; Kigoshi et al. 2002). These results indicated that actin-disrupting activity and some other factors are necessary for aplyronines to display their full cytotoxicity. If in vivo activities of these analogues are tested, more discussions regarding the actin disrupting activity and in vivo efficacy of aplyronines will become possible. Hopefully, this information could uncover the "secret" of aplyronines as such an effective compound in vivo and help actin-targeting compounds to be developed as anticancer drugs with new action mechanisms.

The structure of aplyronine, which is related to the cyanobacterial products scytophycin C and tolytoxin, implies its cyanobacterial origin (Luesch et al. 2002). The functional importance of aplyronines in the sea hare is not known. Since aplyronine was isolated from the whole animal, the localization of these compounds was not determined. If the compounds are from algae they have fed on, the digestive gland is the organ where these compounds are likely to be concentrated. From the chemical/ecological point of view, this is an interesting problem to solve, since many macroalgal metabolites are accumulated in the digestive gland of the sea hare, and in this case, it would not be feasible for these compounds to be considered as functional feeding deterrents for the animal. If potent toxic substances are accumulated or secreted near the surface of the animal, however, they could conceivably serve as "defense" molecules. Further ecologically relevant studies are necessary to clarify these points.

10.3
Bioactive Proteins: Aplysianin A Family

Most bioactive small molecules found in sea hares are of dietary origin; and their amount and content may vary significantly. One may find a sea hare extract that has no measurable "biological activity," but nevertheless sea hares still seem to be able to resist being eaten by predators or infection by microorganisms or parasites. Collective evidence indicates that lectins or hemagglutinins, antimicrobial proteins (Millet and Ewbank 2004), and phenoloxidases (Rolff and Siva-Jothy 2004) are well known to function as defense molecule in invertebrates. Thus, an innate immune system was also postulated in sea hares (Pauley et al. 1971), and some bioactive biomolecules, not secondary metabolites, could be responsible for the defense mechanism of this class of animals.

In search of such bioactive proteins, we surveyed the presence of antibacterial or hemagglutinating activities in the saline extract of various organs from Japanese *A. kurodai*; and we found that the extract of egg mass and the albumen gland showed strong antibacterial activity against both gram-positive and gram-negative bacteria (Kamiya et al. 1984). These observations led to the discovery of series of novel antimicrobial and antitumor proteins; and sea hares have begun to be recognized as rich sources not only of secondary metabolites, but also of proteins with novel bioactivities.

10.3.1 Isolation and Amino Acid Sequence of Aplysianin A

The antimicrobial protein aplysianin A (ApA) was first isolated from the albumen gland of *A. kurodai*. The albumen gland is generally recognized as an organ responsible for the production of perivitelline fluid in mollusks (Beeman 1977). The albumen gland of sea hares is covered with the mucous gland and forms a convoluted portion, generally referred to as the genital mass. The extraction with saline of the genital mass gave a highly viscous material with which it was almost impossible to conduct column chromatographic separations, but freezing the genital mass at −70°C enabled us to separate the albumen gland from other organs. The active principle, ApA, in 40% saturated ammonium sulfate was effectively adsorbed onto hydrophobic Butyl-Toyopearl 650S gel, was eluted with 20% saturated ammonium sulfate, and was finally purified by gel filtration. Intact ApA showed a molecular mass of 320 kDa in gel filtration and showed a single band at 85 kDa with or without reducing agent. Its sugar content was determined to be 9.8% by the phenol–sulfuric acid method (Kamiya et al. 1986). These results suggested that ApA is a glycoprotein composed of four non-covalently bonded identical subunits.

The amino acid sequence of ApA was not known for years, but cDNA clones were finally isolated from an albumen gland cDNA library and sequenced (Takamatsu et al. 1995). ApA was found to be produced as a precursor protein of 556 amino acid residues with a signal peptide of 19 amino acid residues and six potential *N*-glycosylation sites. ApA mRNA was expressed tissue-specifically in the albumen gland and its size was approximately 2.5 kb; but it was not expressed in the purple gland, ovotestis, midgut gland, ctenidium, and body wall. Homology search reveals that ApA has an overall homology (48% identity, 59% similarity) to achacin, an antibacterial glycoprotein of the giant African snail, *Achatina fulica* (Obara et al. 1992; Otsuka-Fuchino et al. 1992), while no significant similarities to other known proteins have been pointed out. Recently, however, Vallon (2000) suggested that ApA is a flavin enzyme, due to a highly conserved GG motif (RxGGRxxS/T) and βαβ dinucleotide-binding motif present in its deduced amino acid sequence. Proteins with the above structural motif are often descrived as L-amino acid oxidase (LAAO).

10.3.2 Related Sea Hare-Derived Proteins of Aplysianin A

To date, a number of antimicrobial or cytotoxic proteins have been reported from sea hares, as listed in Table 10.1. A total of 13 proteins have been reported from four *Aplysia* species, *Dolabella auricularia*, and *Bursatella leachii*. The apparent molecular weights of these molecules are largely divided into two groups, a 300 and a 60-kDa group. The larger proteins are composed of glycosylated polypeptide subunit of 70–100 kDa, while the smaller proteins are claimed to be single polypeptides. Of those proteins, complete amino acid sequences have only been reported for five molecules (accessions Q17043, AAN78211, AAR14185, AAT12273, CAC19362). A comparison of the amino acid sequences of these proteins with that of ApA revealed the possible structural similarity with ApA; and thus we denote them as the ApA family in this review. It is likely that other sequence-unknown molecules may belong to this family. It should be noted that, in our recent examination, an N-terminal amino acid sequence of up to 15 residues of ApA, and aplysianin E (ApE), a protein isolated from the egg mass of *A. kurodai*, were found to be identical. Moreover, ApE in the egg mass extract was lost as the egg cleavage proceeded. These observations strongly suggested that these proteins are actually identical and might be metabolized gradually during their development (unpublished data).

Phylogenetic analysis of these proteins with representative LAAO indicated that the phylogenetic tree structure was separated into gastropods, bacteria, and vertebrates (Fig. 10.4). In the gastropods, three clades were identified: the first consisted of ApA, cyplasin, and an aplysianin A homologue from *A. californica*, and they were all cloned from albumen gland cDNAs. In contrast, the second clade was formed by APIT and escapin, an antimicrobial protein from *A. californica*. They were obtained from purple fluid. These data suggested that the Aplysiidae would have at least two paralogs of genes for LAAO: one of albumen gland origin and the other of purple gland origin.

Table 10.1. Properties of antibacterial and antitumor proteins from sea hares

species	molecular weight (kDa)	number of subunit (kDa)	sugar (%)	activities	references
Aplysia kurodai					
egg (aplysianin E)	250	3 (76, 88, 102)	8	antibacterial, antifungal, tumor cell-lytic	Yamazaki et al. (1985)
albumen gland (aplysianin A)	320	4 (85)	9.8	antibacterial, tumor cell-lytic, LAAO active	Kamiya et al. (1986), Jimbo et al. (2003)
purple fluid (aplysianin P)	60	1	18	antibacterial, tumor cell-lytic	Yamazaki et al. (1986)
Aplysia juliana					
egg (julianin E-1)	375	4 (78)	15.9	antibacterial, tumor cell-lytic	Kamiya et al. (1988)
genital mass (julianin G-1)	375	4 (78)	9.9	antibacterial, tumor cell-lytic	Kamiya et al. (1988)
white secretion (julianin S)	67	1	nd[a]	antibacterial, tumor cell-lytic	Kamiya et al. (1989)
Aplysia punctata					
mucus (cyplasin)	nd	? (56)	nd	tumor cell-toxic	Petzelt et al. (2002)
purple fluid (APIT)	60	1	nd	tumor cell-toxic, LAAO active	Butzke et al. (2004, 2005)
Aplysia dactylomela					
purple fluid	60	1	nd	antibacterial, hemagglutinating	Melo et al. (2000)
Bursatella leachii					
purple fluid	60	1	nd	anti-HIV	Rajanagapathi et al. (2002)
Dolabella auricularia					
albumen gland (dolabellanin A)	250	4 (70)	10.9	antibacterial, tumor cell-lytic, LAAO active	Yamazaki et al. (1989b) and Iijima et al. (2003a)
purple fluid (dolabellanin P)	60	1	0.8	tumor cell-lytic	Yamazaki et al. (1989c)
hemolymph (dolabellanin C)	215	3 (70)	10.9	tumor cell-lytic	Kisugi et al. (1989)
skin and mucus (dolabellanin B2)	0.39	33 amino acid residues	0	antibacterial, antifungal	Iijima et al. (2003b)

[a] not determined

10.3.3
Aplysianin A as LAAO

Since the amino acid sequence of ApA also showed similarity to LAAO, we re-characterized ApA as LAAO, a flavin enzyme that catalyzes a reaction to oxidize the α-amino group of an L-amino acid into an α-keto group by molecular oxygen as electron acceptor (Jimbo et al. 2003). Oxidation of 1 mol of amino acid results in the production of 1 mol of H_2O_2. All known LAAO contains some flavin-type cofactor, which is essential for their enzymatic action. The absorbance spectrum for ApA shows the absorption maxima at $\lambda = 377$ and 458 nm, indicating the presence of flavin. The content of flavin was deduced to be 1 mol of cofactor per 1 mol subunit, on the basis of the ε value. All other LAAO were reported to have one cofactor per subunit in the molecule. Only the basic amino acids L-arginine or L-lysine were consumed as substrates of ApA. The LAAO activity of ApA, as K_{cat}/K_m and value of ApA, is the highest of all the LAAOs reported (Table 10.2).

It is interesting to note that AIP, an LAAO with apoptosis-inducing activity purified from fish, also has strict preference towards the basic amino acid L-lysine, although the sequence similarity is only about 20% to ApA. Despite the close structural similarity (50%), however, a spectrum of amino acid specificity of achacin is much broader than that of ApA and it prefers both basic and hydrophobic amino acids. Terrestrial LAAOs, including snake venom, achacin, or mouse milk LAAO, generally seem to show a broader amino acid preference regardless of their amino acid sequence.

10.3.4
Biological Activities of the ApA Family Proteins

Sea hare-derived bioactive proteins of the ApA family have shown a wide variety of biological activities, including antimicrobial activity, cytotoxicity, induction of apoptosis, and anti-HIV activities. The mechanisms of these activities, however, especially antimicrobial activity and cytotoxicity, have been elusive. Yamazaki et al. (1989a) have claimed that inhibition of macromolecule synthesis is the key mechanism for these proteins to exhibit cytotoxicities. LAAO activities established for ApA suggested that all the above proteins share, at least in part, a common mechanism of action which is ascribable to the production of H_2O_2 through their LAAO action.

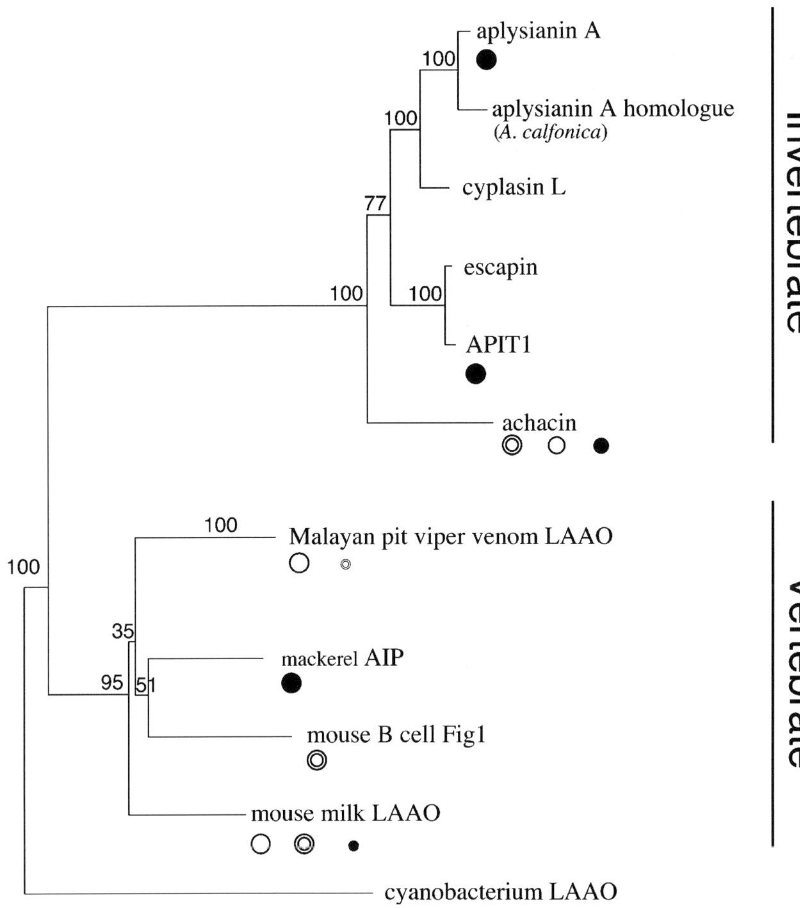

Fig. 10.4. Phylogenetic tree and substrate specificity of various L-amino acid oxidases. The phylogenetic tree was constructed by a neighbor-joining method, using the gene sequences of LAAOs of various animals listed. *Shynechococcus* was used as an outgroup. Values above the line represent bootstrap support values of nodes from the neighbor-joining analysis. Bootstrap support values are percentages of 100 resamplings. Substrate specificity is indicated by: *closed circles* basic amino acids, *open circles* aromatic amino acids, and *double circles* aliphatic amino acids. Sizes represent relative activity. Data used include: aplysianin A (accession no. Q17043; Jimbo et al. 2003), *A. californica* (AAN78211), cyplasin L (CAC19362; Petzelt et al. 2002), escapin (AAT12273), APIT1 (AAR14185; Butzke et al. 2004, 2005), achacin (CAA45871), snake *Gloydius halys* venom LAAO (AR20248), *Calloselasma rhodostoma* (P81382; Ponnudurai et al. 1994), AIP, a mackerel *Scomber japonicus* LAAO (CAC00499; Jung et al. 2000), Fig1 (AAO65453; Mason et al. 2004), mouse milk LAAO (NP_598653; Sun et al. 2002)

Table 10.2. Properties of LAAO from various origins

species	name	subunits (kDa)	cytotoxicity (minimum conc.)	specificity[b]	K_{cat}/K_m (Ms^{-1})	references
Aplysia kurodai	aplysianin A (E)	85 (homotetramer)	yes[a]	R, K	1.64×10^7 (Arg)	Jimbo et al. (2003)
Dolabella auricularia	dolabellanin A	70 (homotetramer)	yes[a](10 ng/mL)	R, K	nd[c]	Iijima et al. (2003a)
A. punctata	APIT	60	yes (10 ng/mL)	R, K	nd[c]	Butzke et al. (2004, 2005)
A. punctata	cyplasin	56	yes (2 nM)	nd[c]	nd[c]	Petzelt et al. (2002)
Scomber japanicus	AIP	64 (homodimer)	apoptosis (10 ng/mL)	K	nd[c]	Jung et al. (2000)
*Achatina fulica*Férussac	achacin	62 (homodimer)	yes[a](1.5 μg/mL)	M, L, W, K, R	nd[c]	Kanzawa et al. (2004)
Agkistrodon halys pallas	AHP-LAAO	60.7	apoptosis (0.5 μg/mL)	nd[c]	5.08×10^4 (Leu)	Zhang et al. (2004)
Crotalus atrox	apoxin I	55 (homodimer)	apoptosis (2.5 μg/mL)	nd[c]	nd[c]	Torii et al. (1997)
Naja naja kaouthia		57.4 (homodimer)	nd[c]	F, W, M, L, R	1.79×10^5 (Phe)	Ponnudurai et al. (1994)
Ophiophagus hannah		65 (homodimer)	yes (1 μg/mL)	K, F, L, W, R	5.41×10^5 (Phe)	Tan and Saifuddin (1991), Ahn et al. (1997)

[a]Apoptosis was occurred when catalase was added.
[b]Representative amino acid (one-letter abbreviations) substrates.
[c]not determined.

Antimicrobial Activity

To test this hypothesis, the relation between LAAO action and antimicrobial activity of ApA was first examined. The process by which H_2O_2 kills bacteria or halts bacterial growth is complicated and concentration-dependent (Brandi et al. 1989; Hyslop et al. 1995). We therefore designed a series of experiments to determine the mechanism of antimicrobial action with respect to its LAAO actions and found that an addition of catalase inhibited the bactricidal action of ApA (Jimbo et al. 2003). This result indicated that most of the antimicrobial action of ApA results from H_2O_2 generated by the reactions catalyzed by ApA. We observed that treatment of bacterial culture with H_2O_2 did not cause cell lysis, unlike treatment with ampicillin. In our experiment, ApA was able to generate about 800 μM H_2O_2 which is higher than the bactericidal concentration of H_2O_2 reported earlier (Hyslop et al. 1995; Fig. 10.5A). Of notable interest, achacin was also evidenced to kill bacteria through the generation of H_2O_2, but its antibacterial action should be facilitated by binding of the protein to bacterial cells for effective delivery of the bactericidal agent, because the amount of H_2O_2 produced by the enzyme was not sufficient to reach the concentration needed to kill bacteria in the media (Ehara et al. 2002; Kanzawa et al. 2004).

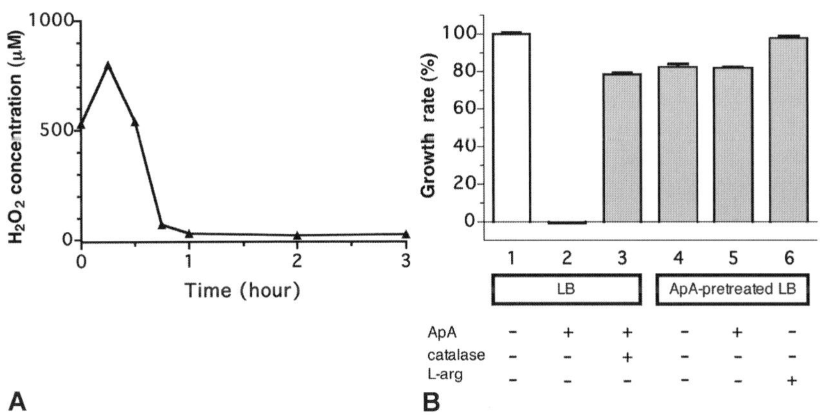

Fig. 10.5. H_2O_2 production and antimicrobial activity of ApA. (**A**) 12.5 μg ml^{-1} ApA was incubated with *Bacillus subtilis* cultured in LB medium. The concentration of hydrogen peroxide was measured against time. (**B**) *B. subtilis* was cultured with LB medium or ApA-pretreated LB medium at 37°C for 3 h. The growth rate was indicated as the increase of absorption at 655 nm, relative to the control. To prepare ApA-pretreated LB medium, 1 ml LB medium was incubated with 10 μg ApA and 10 μg catalase at 8°C overnight and then boiled for 5 min to inactivate ApA. *Lane 1* LB medium (control), *lane 2* LB medium with 16 μg ml^{-1} ApA, *lane 3* LB medium with 16 μg ml^{-1} ApA plus 10 μg ml^{-1} catalase, *lane 4* ApA-pretreated LB, *lane 5* ApA pretreated LB with 16 μg ml^{-1} ApA, *lane 6* ApA pretreated LB with 5 mM L-arginine

We even found that, in the presence of excessive catalase, viability of bacteria was only about 70% of the control, while catalase itself did not attenuate bacterial growth. Thus, an additional mechanism of ApA to suppress bacterial growth was suspected. However, this growth suppression might be eventually attributed to the quick consumption of basic amino acid in the culture medium by ApA, since the growth of bacteria was apparently suppressed in media pretreated with ApA, while their growth was recovered by addition of L-arginine (Fig. 10.5B, bar 6). ApA alone did not show antimicrobial activity in the amino acid-deficient media as expected (Fig. 10.5).

This result may explain the earlier observation that prolonged incubation resumed bacterial growth: in that case, consumption of substrate in the medium by the enzyme limited the generation of H_2O_2,whereby surviving bacteria could re-grow.

A similar phenomenon has been reported for the apoptosis-inducing activity of AIP, in which consumption of amino acid from the culture media by the enzyme could induce apoptosis (Murakawa et al. 2001).

Cytotoxicity

Proteins of the putative ApA family exhibited cytotoxicity to various cultured murine and human tumor cells, killing all the tumor cells at an order of nanograms of protein per milliliter. Inhibition of macromolecule synthesis has been proposed as an apparent mechanism of action for aplysianins. However, their primary interaction with the cells was not well understood. Moreover, some peculiar observations made elucidation of the activity more complicated. For example, *N*-acetylneuraminic acid (NANA) inhibited the cytotoxicity of ApE (Kisugi et al. 1987). ApE was highly cytotoxic against cultured mouse and human tumor cells but not against sheep spleen cells and erythrocytes (Yamazaki et al. 1985). Recent work, however, has started to provide evidence that the activities of ApA family proteins on cells can be ascribed to their LAAO actions.

Iijima et al. (2003a) first reported that the cytotoxicity of dolabellanin A was mediated by its LAAO activity. Upon the addition of dolabellanin A, 10 ng ml^{-1}, a cytotoxic concentration of H_2O_2 was detected in the culture medium of EL-4 mouse lymphoma cells. The cytotoxicity was suppressed by antioxidants, especially by catalase. The H_2O_2 produced by the LAAO activity was recognized as being involved in the cytotoxicity; and the cell death seemed to be due to the direct toxicity of H_2O_2. However, the cytotoxicity of dolabellanin A at higher concentration was only partially suppressed by catalase, even when an excess amount of the enzyme was used. Biochemical analysis of the dolabellanin A-treated cells showed features of apoptosis, such as increased caspase 3 activity and DNA fragmentation. This indicated that dolabellanin A induced apoptosis independently of the direct toxicity of H_2O_2.

It was reported earlier that tumor lysis by ApE and Aplysianin P (ApP) from *A. kurodai* purple fluid was inhibited by NANA, but not by other simple sugars such as D-galactose, D-mannnose, and *N*-acetyl-D-glucosamine. These results have been interpreted as recognizing that NANA present on the surface of tumor cells is related to the cytolysis induced by aplysianins (Kisugi et al. 1987; Yamazaki et al. 1989a). Recently, however, Iijima et al. (2004) proposed the role of NANA in the inhibition of antitumor activity, linking it to the LAAO activity of aplysianins. They found that NANA consumes toxic H_2O_2 under physiological conditions and was oxidized by an equimolar amount of H_2O_2, to provide its decarboxylated product, 4-(acetylamino)-2,4-dideoxy-D-glycero-D-galacto-octonic acid (Fig. 10.6).

In cultured cells, the cell death caused by H_2O_2 was suppressed by NANA in a dose-dependent manner, suggesting a novel role for NANA as a reactive oxygen scavenger. This novel chemical reaction of NANA explains the earlier observation.

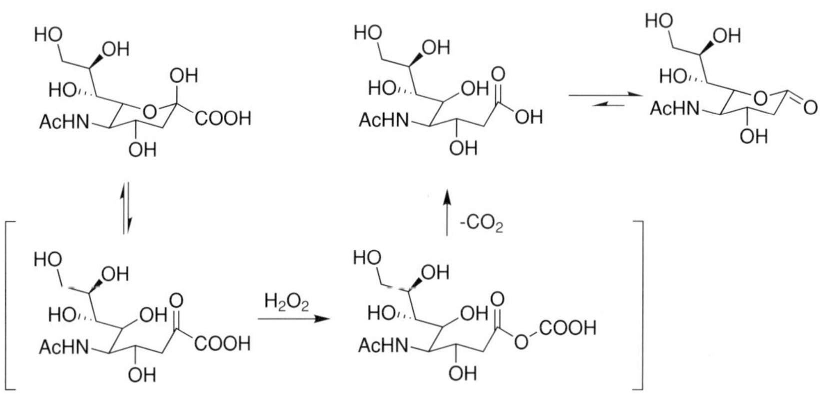

Fig. 10.6. A proposed mechanism for the oxidative transformation of *N*-acetylneuraminic acid (NANA) into 4-(acetylamino)-2,4-dideoxy-D-glycero-D-galacto-octonic acid by H_2O_2

Petzelt et al. (2002) proposed that cell death induced by cyplasin was neither apoptosis nor necrosis. Cyplasin was toxic preferentially to autonomously growing transformed mammalian cells at nanomolar concentrations. In PtK cells (kangaroo rat cell line), the minimum cytotoxic cyplasin concentration was in the order of 2 nM. In contrast, cyplasin was nontoxic to BA2 mice when administered by iv or sc injection (10 μM). The toxicity of cyplasin was cell cycle-dependent. Cyplasin-treated cells exhibited abnormal morphology in a cell cycle-dependent fashion. PtK cells completed anaphase with apparently normal morphology; but entering the interphase, they started to show typical cyplasin-induced changes, including retraction of cell membrane, partial detaching, and vacuole formation. These morphological changes are reminiscent of apoptotic cells, with the exception of the absence of typical indicators like nuclear fragmentation. They also reported the morphological changes of cyplasin-induced cell death are different from those observed in typical apoptosis induced by staurosporine.

In the case of *A. punctata* ink toxin, however, Butzke et al. (2004, 2005) concluded that the mechanism of cell death by APIT is due solely to the production of H_2O_2. In the experiment with Jurkat cells, they found that tumor cell death induced by APIT was independent of apoptosis but was characterized by a rapid loss of metabolic activity, membrane permeabilization, and shrinkage of nuclei. Interestingly, two-dimensional electrophoresis analysis of APIT-treated tumor cells revealed the pI shift of peroxiredoxin I, a cytoplasmic peroxidase involved in the detoxification of peroxides, indicating a modification of the peroxidase by APIT. Knockdown of peroxiredoxin I by RNA interference sensitized cells for APIT-induced cell death. The same cellular modification was observed after treatment with H_2O_2, and catalase completely blocked the activity of APIT.

As to the cytotoxic mechanism of achacin, a giant snail protein with high homology to ApA, Kanzawa et al. (2004) proposed a cellular pathway for cytotoxicity other than that mediated by H_2O_2 production. When HeLa cells were incubated with achacin in the presence of catalase, some features typical of apoptosis, including morphological changes, DNA fragmentation, and poly-ADP-ribose polymerase (PARP) cleavage were observed. Moreover, apoptosis was inhibited by Z-VAD-fmk, a broad-spectrum caspase inhibitor. On the basis of these results as well as former observations, it has been discussed that cytotoxicity of achacin is mediated primarily by generation of H_2O_2. However, the additional mechanism, the caspase-mediated apoptotic pathway, would be driven when L-amino acids were depleted by achacin.

The above observation can be summarized by noting that ApA family proteins induce cell death through: (1) direct toxicity of H_2O_2, or (2) apoptosis. These observations seem consistent with the idea that LAAO activity, i.e. toxicity of H_2O_2, is responsible primarily for the cellular actions observed for all the sea hare-derived bioactive proteins mentioned above and giant snail-derived achacin, although induction of apoptosis by dolabellanin A and achacin was claimed to be independent of H_2O_2. In these cases, however, prolonged treatments of the tumor cells in the presence of LAAO and catalase may result in depletion of amino acids in the media, which in turn lead to the apoptosis of the cells. It should be noted that cell death caused by H_2O_2 is not well defined and the concentration of H_2O_2 induces different paths to cell death. The amount of H_2O_2 produced by LAAO or that actually reaches the cell is not easy to determine, as enzyme kinetics vary from protein to protein. It would become even more complicated if the possibility of the protein binding to the cell surface is taken into account. In fact, the LAAO of snake venom was shown to bind to the cell surface (Suhr and Kim 1999). A similar mechanism was proposed for the antimicrobial activity of achacin (Kanzawa et al. 2004).

10.4 Conclusions

A number of flavin enzymes related to LAAO have been reported from various vertebrate animals. Among the vertebrate LAAO, snake venoms are the most well established class of the enzyme (Du and Clemetson 2002). Besides snake venoms, however, several functionally interesting LAAOs have been discovered. For example, *mFig1* (interleukin-4 induced gene 1) was detected specifically in IL-4-induced mouse B-cells (Chu and Paul 1997). Recently, a *Fig1*-like gene in human (*hFig1*) was characterized to have a large internal amino acid sequence homologous (43%) to LAAO. These observations suggest that Fig1 most likely possesses LAAO activity, although it has never been characterized enzymologically. Recombinant Fig1 (note that Fig1 was renamed to IL4I1) indeed shows LAAO activity (Mason et al. 2004). An LAAO was recently isolated from mouse milk and its putative function was speculated to be anti-infection of the mammary gland (Sun et al. 2002). An apoptosis-inducing protein (AIP) is also a LAAO purified from Chub mackerel which is localized to form capsules surrounding *Anisakis simplex* larvae (Jung et al. 2000). It should be noted that apoptosis induced by AIP is also due to its ability to produce H_2O_2 and deplete L-lysine from the culture media (Murakawa et al. 2001). It is interesting that putative functions of mouse milk-derived LAAO and ApA, chemical defenses against microbes, and those of AIP, chemical defense against parasites, are analogous. It can also be speculated that, as in vertebrates, proteins which possess LAAO activity can be present in a wide variety of invertebrates and play important roles in their chemical defenses.

Studies of bioactive molecules in Aplysiidae reviewed here illustrate that these "defenseless" animals actually have their own way and variety of putative "defenses," mostly by chemical means. Small bioactive molecules, which in most cases have been accumulated from their food, can benefit humans in the areas of cancer research and cell biology. Bioactive proteins which they produce by themselves share the mechanism similar to the immune system or chemical defense of mammals, fish, or snake venoms, in that H_2O_2 produced by the LAAO action of the proteins plays a key role in their bioactivity. This implies that reactive oxygen species (ROS)-based chemical defense might be a common defense mechanism not only in mollusks, but also in other invertebrates. Further studies of bioactive molecules in the sea hare would facilitate an understanding of the ecological and physiological importance of these proteins in the animal and the discovery of useful molecules in biological sciences.

References

Ahn MY, Lee BM, Kim YS (1997) Characterization and cytotoxicity of L-amino acid oxidase from the venom of king cobra (*Ophiophagus hannah*). Int J Biochem Cell Biol 29:911–919

Amador ML, Jimeno J, Paz-Ares L, Cortes-Funes H, Hidalgo M (2003) Progress in the development and acquisition of anticancer agents from marine sources. Ann Oncol 14:1607–1615

Avila C (1995) Natural products of opisthobranch molluscs: a biological review. Oceanogr Mar Biol Annu Rev 33:487–559

Bai R, Verdier-Pinard P, Gangwar S, Stessman CC, McClure KJ, Sausville EA, Pettit GR, Bates RB, Hamel E (2001) Dolastatin 11, a marine depsipeptide, arrests cells at cytokinesis and induces hyperpolymerization of purified actin. Mol Pharmacol 59:462–469

Bai R, Covell DG, Liu C, Ghosh AK, Hamel E (2002) (–)-Doliculide, a new macrocyclic depsipeptide enhancer of actin assembly. J Biol Chem 277:32165–32171

Beeman (1977) Gastropoda:Opisthobranchia. In: Giese AC, Pearse JS (eds) Reproduction of marine invertebrates. Vol IV. Molluscs: gastropods and cephalopods. Academic, New York, pp 115–179

Bonnet MS, Basson PW (2002) The toxicology of *Amanita phalloides*. Homeopathy 91:249–254

Brandi G, Fiorani M, Pierotti C, Albano A, Cattabeni F, Cantoni O (1989) Morphological changes in *Escherichia coli* cells exposed to low or high concentrations of hydrogen peroxide. Microbiol Immunol 33:991–1000

Bubb MR, Senderowicz AM, Sausville EA, Duncan KL, Korn ED (1994) Jasplakinolide, a cytotoxic natural product, induces actin polymerization and competitively inhibits the binding of phalloidin to F-actin. J Biol Chem 269:14869–14871

Bubb MR, Spector I, Bershadsky AD, Korn ED (1995) Swinholide A is a microfilament disrupting marine toxin that stabilizes actin dimers and severs actin filaments. J Biol Chem 270:3463–3466

Butzke D, Machuy N, Thiede B, Hurwitz R, Goedert S, Rudel T (2004) Hydrogen peroxide produced by *Aplysia* ink toxin kills tumor cells independent of apoptosis via peroxiredoxin I sensitive pathways. Cell Death Differ 11:608–617

Butzke D, Hurwitz R, Thiede B, Goedert S, Rudel T (2005) Cloning and biochemical characterization of APIT, a new L-amino acid oxidase from *Aplysia punctata*. Toxicon 46:479 489

Chu CC, Paul WE (1997) Fig1, an interleukin 4-induced mouse B cell gene isolated by cDNA representational difference analysis. Proc Natl Acad Sci USA 94:2507–2512

Cimino G, Fontana A, Gavagnin M (1999) Marine opisthobranch molluscs: chemistry and ecology in sacoglossans and dorids. Curr Org Chem 3:327–372

Cimino G, Ciavatta ML, Fontana A, Gavagnin M (2001) Metabolites of marine opisthobranchs: chemistry and biological activity. In: Tingali C (ed) Bioactive compounds from natural sources. Taylor and Francis, London, pp 577–637

Cummins SF, Nichols AE, Amare A, Hummon AB, Sweedler JV, Nagle GT (2004a) Characterization of *Aplysia* enticin and temptin, two novel water-borne protein pheromones that act in concert with attractin to stimulate mate attraction. J Biol Chem 279:25614–25622

Cummins SF, Nichols AE, Rajarathnam K, Nagle GT (2004b) A conserved heptapeptide sequence in the waterborne attractin pheromone stimulates mate attraction in *Aplysia*. Peptides 25:185–189

Du XY, Clemetson KJ (2002) Snake venom L-amino acid oxidases. Toxicon 40:659–665

Ehara T, Kitajima S, Kanzawa N, Tamiya T, Tsuchiya T (2002) Antimicrobial action of achacin is mediated by L-amino acid oxidase activity. FEBS Lett 531:509–512

Furukawa K, Sakai K, Watanabe S, Maruyama K, Murakami M, Yamaguchi K, Ohizumi Y (1993) Goniodomin A induces modulation of actomyosin ATPase activity mediated through conformational change of actin. J Biol Chem 268:26026–26031

Gilboa-Garber N, Susswein AJ, Mizrahi L, Avichezer D (1985) Purification and characterization of the gonad lectin of Aplysia depilans. FEBS Lett 181:267–270

Halstead BW, Courville DA (1965) Poisonous sea hares. In: USGPO (ed) Poisonous and venomous marine animals of the world, Vol 1. United States Government Printing Office, Washington, DC, pp 709–715

Hino K, Mitsui Y, Hirano Y (1994) Four cases of acute liver damage following the ingestion of a sea hare egg. J Gastroenterol 29:679

Hyslop PA, Hinshaw DB, Scraufstatter IU, Cochrane CG, Kunz S, Vosbeck K (1995) Hydrogen peroxide as a potent bacteriostatic antibiotic: implications for host defense. Free Radic Biol Med 19:31–37

Iijima R, Kisugi J, Yamazaki M (2003a) L-Amino acid oxidase activity of an antineoplastic factor of a marine mollusk and its relationship to cytotoxicity. Dev Comp Immunol 27:505–512

Iijima R, Kisugi J, Yamazaki M (2003b) A novel antimicrobial peptide from the sea hare *Dolabella auricularia*. Dev Comp Immunol 27:305–311

Iijima R, Takahashi H, Namme R, Ikegami S, Yamazaki M (2004) Novel biological function of sialic acid (*N*-acetylneuraminic acid) as a hydrogen peroxide scavenger. FEBS Lett 561:163–166

Ishiwata H, Nemoto T, Ojika M, Yamada K (1994) Isolation and structure of doliculide, a cytotoxic cyclodepsipeptide from Japanese sea hare *Dolabella auricularia*. J Org Chem 59:4712–4713

Jimbo M, Nakanishi F, Sakai R, Muramoto K, Kamiya H (2003) Characterization of L-amino acid oxidase and antimicrobial activity of aplysianin A, a sea hare-derived antitumor-antimicrobial protein. Fish Sci 69:1240–1246

Johnson P, Willows A (1999) Defensive sea hare (Gastropoda, Opisthobranchia, Anaspidea): multiple layers of protection from egg to adult. Mar Freshwater Behav Physiol 32:147–180

Jung SK, Mai A, Iwamoto M, Arizono N, Fujimoto D, Sakamaki K, Yonehara S (2000) Purification and cloning of an apoptosis-inducing protein derived from fish infected with *Anisakis simplex*, a causative nematode of human anisakiasis. J Immunol 165:1491–1497

Kamiya H, Shimizu Y (1981) A natural agglutinin inhibitable by D-galacturonic acid in sea hare *Aplysia* eggs: characterization and purification. Bull Jpn Soc Fish 47:255–259

Kamiya H, Muramoto K, Ogata K (1984) Antibacterian activity in the egg mass of a sea hare. Experientia 40:947–949

Kamiya H, Muramoto K, Yamazaki M (1986) Aplysianin-A, an antibacterial and antineoplastic glycoprotein in the albumen gland of a sea hare, *Aplysia kurodai*. Experientia 42:1065–1067

Kamiya H, Muramoto K, Goto R, Yamazaki M (1988) Characterization of the antibacterial and antineoplastic glycoproteins in a sea hare *Aplysia juliana*. Nippon Suisan Gakkaishi 54:773–777

Kamiya H, Muramoto K, Goto R, Sakai M, Endo Y, Yamazaki M (1989) Purification and characterization of an antibacterial and antineoplastic protein secretion of a sea hare, *Aplysia juliana*. Toxicon 27:1269–1277

Kanzawa N, Shintani S, Ohta K, Kitajima S, Ehara T, Kobayashi H, Kizaki H, Tsuchiya T (2004) Achacin induces cell death in HeLa cells through two different mechanisms. Arch Biochem Biophys 422:103–109

Kigoshi H, Suenaga K, Takagi M, Akao A, Kanematsu K, Kamei N, Okugawa Y, Yamada K (2002) Cytotoxicity and actin-depolymerizing activity of aplyronine A, a potent antitumor macrolide of marine origin, and its analogs. Tetrahedron 58:1075–1102

Kisugi J, Kamiya H, Yamazaki M (1987) Purification and characterization of aplysianin E, an antitumor factor from sea hare eggs. Cancer Res 47:5649–5653

Kisugi J, Kamiya H, Yamazaki M (1989) Purification of dolabellanin-C an antineoplastic glycoprotein in the body fluid of a sea hare, *Dolabella auricularia*. Dev Comp Immunol 13:3–8

Klenchin VA, Allingham JS, King R, Tanaka J, Marriott G, Rayment I (2003) Trisoxazole macrolide toxins mimic the binding of actin-capping proteins to actin. Nat Struct Biol 10:1058–1063

Kobayashi M, Kanda F, Kamiya H (1991) Occurrence of pyropheophorbides a and b in the viscera of the sea hare *Aplysia juliana*. Nippon Suisan Gakkaishi 57:1983

Kobayashi M, Kawazoe K, Okamoto T, Sasaki T, Kitagawa I (1994) Marine natural products. XXXI. Structure–activity correlation of a potent cytotoxic dimeric macrolide swinholide A, from the Okinawan marine sponge *Theonella swinhoei*, and its isomers. Chem Pharm Bull 42:19–26

Luesch H, Harrigan G, Goetz G, Horgen F (2002) The cyanobacterial origin of potent anticancer agents originally isolated from sea hares. Curr Med Chem 9:1791–1806

Marquez BL, Watts KS, Yokochi A, Roberts MA, Verdier-Pinard P, Jimenez JI, Hamel E, Scheuer PJ, Gerwick WH (2002) Structure and absolute stereochemistry of hectochlorin, a potent stimulator of actin assembly. J Nat Prod 65:866–871

Mason JM, Naidu MD, Barcia M, Porti D, Chavan SS, Chu CC (2004) IL-4-induced gene-1 is a leukocyte L-amino acid oxidase with an unusual acidic pH preference and lysosomal localization. J Immunol 173:4561–4567

Melo VM, Duarte AB, Carvalho AF, Siebra EA, Vasconcelos IM (2000) Purification of a novel antibacterial and haemagglutinating protein from the purple gland of the sea hare, *Aplysia dactylomela* Rang, 1828. Toxicon 38:1415–1427

Millet AC, Ewbank JJ (2004) Immunity in *Caenorhabditis elegans*. Curr Opin Immunol 16:4–9

Müller WE, Zahn RK, Kurelec B, Lucu C, Müller I, Uhlenbruck G (1981) Lectin, a possible basis for symbiosis between bacteria and sponges. J Bacteriol 145:548–558

Murakawa M, Jung SK, Iijima K, Yonehara S (2001) Apoptosis-inducing protein, AIP, from parasite-infected fish induces apoptosis in mammalian cells by two different molecular mechanisms. Cell Death Differ 8:298–307

Mutou T, Kondo T, Ojika M, Yamada K (1996) Isolation and stereostructures of dolastatin G and nordolastatin G, cytotoxic 35-membered cyclodepsipeptides from the Japanese sea hare *Dolabella auricularia*. J Org Chem 61:6340–6345

Nakao Y, Yoshida WY, Takada Y, Kimura J, Yang L, Mooberry SL, Scheuer PJ (2004) Kulokekahilide-2, a cytotoxic depsipeptide from a cephalaspidean mollusk *Philinopsis speciosa*. J Nat Prod 67:1332–1340

Nolen T, Johnson P (2001) Defensive inking in *Aplysia* spp: multiple episodes of ink secretion and the adaptive use of a limited chemical resource. J Exp Biol 204:1257–1268

Nolen T, Johnson P, Kicklighter C, Capo T (1995) Ink secretion by the marine snail *Aplysia californica* enhances its ability to escape from a natural predator. J Comp Physiol 176A:239–254

Obara K, Otsuka-Fuchino H, Sattayasai N, Nonomura Y, Tsuchiya T, Tamiya T (1992) Molecular cloning of the antibacterial protein of the giant African snail, *Achatina fulica* Ferussac. Eur J Biochem 209:1–6

Ojika M, Nemoto T, Nakamura M, Yamada K (1995) Dolastatin E, a new cyclic hexapeptide isolated from the sea hare *Dolabella auricularia*. Tetrahedron Lett 36:5057–5058

Otsuka-Fuchino H, Watanabe Y, Hirakawa C, Tamiya T, Matsumoto J, Tsuchiya T (1992) Bactericidal action of a glycoprotein from the body surface mucus of giant African snail. Comp Biochem Physiol C 101:607–613

Ozeki Y (1998) Purification and cell attachment activity of a D-galactose-binding lectin from the skin of sea hare, *Aplysia kurodai*. Biochem Mol Biol Int 45:989–995

Painter SD, Cummins SF, Nichols AE, Akalal DB, Schein CH, Braun W, Smith JS, Susswein AJ, Levy M, de Boer PA, ter Maat A, Miller MW, Scanlan C, Milberg RM, Sweedler JV, Nagle GT (2004) Structural and functional analysis of *Aplysia* attractins, a family of water-borne protein pheromones with interspecific attractiveness. Proc Natl Acad Sci USA 101:6929–6933

Pauley GB, Krassner SM, Chapman FA (1971) Bacterial clearance in the California sea hare, *Aplysia californica*. J Invertebr Pathol 18:227–239

Pettit GR (1997) The dolastatins. Fortschr Chem Org Naturst 70:1–79

Pettit GR, Xu JP, Doubek DL, Chapuis JC, Schmidt JM (2004) Antineoplastic agents. 510. Isolation and structure of dolastatin 19 from the Gulf of California sea hare *Dolabella auricularia*. J Nat Prod 67:1252–1255

Petzelt C, Joswig G, Stammer H, Werner D (2002) Cytotoxic cyplasin of the sea hare, *Aplysia punctata*, cDNA cloning, and expression of bioactive recombinants in insect cells. Neoplasia 4:49–59

Poncet J (1999) The dolastatins, a family of promising antineoplastic agents. Curr Pharm Des 5:139–162

Ponnudurai G, Chung MC, Tan NH (1994) Purification and properties of the L-amino acid oxidase from Malayan pit viper (*Calloselasma rhodostoma*) venom. Arch Biochem Biophys 313:373–378

Rajaganapathi J, Kathiresan K, Singh TP (2002) Purification of anti-HIV protein from purple fluid of the sea hare *Bursatella leachii* de Blainville. Mar Biotechnol 4:447–453

Rao J, Li N (2004) Microfilament actin remodeling as a potential target for cancer drug development. Curr Cancer Drug Targets 4:345–354

Rolff J, Siva-Jothy MT (2004) Selection on insect immunity in the wild. Proc R Soc Lond B Biol Sci 271:2157–2160

Saito S, Watabe S, Ozaki H, Fusetani N, Karaki H (1994) Mycalolide B, a novel actin depolymerizing agent. J Biol Chem 269:29710–29714

Saito S, Watabe S, Ozaki H, Kigoshi H, Yamada K, Fusetani N, Karaki H (1996) Novel actin depolymerizing macrolide aplyronine A. J Biochem 120:552–555

Saito SY, Feng J, Kira A, Kobayashi J, Ohizumi Y (2004) Amphidinolide H, a novel type of actin-stabilizing agent isolated from dinoflagellate. Biochem Biophys Res Commun 320:961–965

Sakamoto Y, Nakajima T, Misawa S, Ishikawa H, Itoh Y, Nakashima T, Okanoue T, Kashima K, Tsuji T (1998) Acute liver damage with characteristic apoptotic hepatocytes by ingestion of *Aplysia kurodai*, a sea hare. Intern Med 37:927–929

Schwartsmann G, Brondani da Rocha A, Berlinck RG, Jimeno J (2001) Marine organisms as a source of new anticancer agents. Lancet Oncol 2:221–225

Sone H, Nemoto T, Ojika M, Yamada K (1993a) Isolation, structure and synthesis of dolastatin C, a new depsipeptide from the sea hare *Dolabella auricularia*. Tetrahedron Lett 34:8445–8448

Sone H, Nemoto T, Ishiwata H, Ojika M, Yamada K (1993b) Isolation, structure and synthesis of dolastatin D, a cytotoxic cyclic depsipeptide from the sea hare *Dolabella auricularia*. Tetrahedron Lett 34:8499–8452

Sone H, Kondo T, Kiryu M, Ishiwata H, Ojika M, Yamada K (1995) Dolabellin, a cytotoxic bisthiazole isolated from the sea hare *Dolabella auricuralia*: structure determination and synthesis. J Org Chem 60:4774–4781

Sone H, Shibata T, Fujita T, Ojika M, Yamada K (1996a) Dolastatin H and isodolastatin H, potent cytotoxic peptides from sea hare *Dolabella auricularia*: isolation, stereostructure and synthesis. J Am Chem Soc 118:1874–1880

Sone H, Kigoshi H, Yamada K (1996b) Auriside A and B, cytotoxic macrolide glycosides from the Japanese sea hare *Dolabella auricularia*. J Org Chem 61:8956–8960

Sone H, Kigoshi H, Yamada K (1997) Isolation and stereostructure of dolastatin I, a cytotoxic cyclic hexapeptide from the Japanese sea hare *Dolabella auricularia*. Tetrahedron 53:8149–8154

Sorokin M (1988) Human poisoning by ingestion of a sea hare (*Dolabella auricularia*). Toxicon 26:1095–1097

Spector I, Shochet NR, Kashman Y, Groweiss A (1983) Latrunculins: novel marine toxins that disrupt microfilament organization in cultured cells. Science 219:493–495

Suenaga K, Mutou T, Shibata T, Itoh T, Kigoshi H, Yamada K (1996) Isolation and stereostructure of aurilide, a novel cyclodepsipeptide from the Japanese sea hare *Dolabella auricularia*. Tetrahedron 53:8149–8154

Suenaga K, Kamei N, Okugawa Y, Takagi M, Akao A, Kigoshi H, Yamada K (1997) Cytotoxicity and actin depolymerizing activity of aplyronine A, a potent antitumor macrolide of marine origin, and the natural and artificial analogs. Bioorg Med Chem Lett 7:269–274

Suenaga K, Mutou T, Shibata T, Itoh T, Fujita T, Takada N, Hayamizu K, Takagi M, Irifune T, Kigoshi H, Yamada K (2004) Aurilide, a cytotoxic depsipeptide from the sea hare *Dolabella auricularia*: isolation, structure determination, synthesis, and biological activity. Tetrahedron 60:8509–8527

Suhr SM, Kim DS (1999) Comparison of the apoptotic pathways induced by L-amino acid oxidase and hydrogen peroxide. J Biochem 125:305–309

Sun Y, Nonobe E, Kobayashi Y, Kuraishi T, Aoki F, Yamamoto K, Sakai S (2002) Characterization and expression of L-amino acid oxidase of mouse milk. J Biol Chem 277:19080–19086

Takamatsu N, Shiba T, Muramoto K, Kamiya H (1995) Molecular cloning of the defense factor in the albumen gland of the sea hare *Aplysia kurodai*. FEBS Lett 377:373–376

Takeuchi H, Ara G, Sausville EA, Teicher B (1998) Jasplakinolide: interaction with radiation and hyperthermia in human prostate carcinoma and Lewis lung carcinoma. Cancer Chemother Pharmacol 42:491–496

Tan NH, Saifuddin MN (1991) Substrate specificity of king cobra (*Ophiophagus hannah*) venom L-amino acid oxidase. Int J Biochem 23:323–327

Tanaka J, Yan Y, Choi J, Bai J, Klenchin VA, Rayment I, Marriott G (2003) Biomolecular mimicry in the actin cytoskeleton: mechanisms underlying the cytotoxicity of kabiramide C and related macrolides. Proc Natl Acad Sci USA 100:13851–13856

Tchang S, Lin G (1964) A study on Aplysiidae from China coast (in Chinese). Stud Mar Sin 5:1–25

Terry DR, Spector I, Higa T, Bubb MR (1997) Misakinolide A is a marine macrolide that caps but does not sever filamentous actin. J Biol Chem 272:7841–7845

Torii S, Naito M, Tsuruo T (1997) Apoxin I, a novel apoptosis-inducing factor with L-amino acid oxidase activity purified from Western diamondback rattlesnake venom. J Biol Chem 272:9539–9542

Vallon O (2000) New sequence motifs in flavoproteins: evidence for common ancestry and tools to predict structure. Proteins 38:95–114

Vasta R (1991) The multiple biological roles of invertebrate lectins: their participation in nonself recognition mechanism. In: Warr G, Cohen N (eds) Phylogenesis of immune functions. CRC, Boston, pp 73–101

Weiss KR, Brezina V, Cropper EC, Heierhorst J, Hooper SL, Probst WC, Rosen SC, Vilim FS, Kupfermann I (1993) Physiology and biochemistry of peptidergic cotransmission in *Aplysia*. J Physiol 87:141–151

Whim MD, Church PJ, Lloyd PE (1993) Functional roles of peptide cotransmitters at neuromuscular synapses in *Aplysia*. Mol Neurobiol 7:335–347

Wilson MP, Carrow GM, Levitan IB (1992) Modulation of growth of *Aplysia* neurons by an endogenous lectin. J Neurobiol 23:739–750

Wu AM, Song SC, Chen YY, Gilboa-Garber N (2000) Defining the carbohydrate specificities of aplysia gonad lectin exhibiting a peculiar D-galacturonic acid affinity. J Biol Chem 275:14017–14024

Yamada K, Kigoshi H (1997) Bioactive compounds from the sea hares of two genera *Aplysia* and *Dolabella*. Bull Chem Soc Jpn 70:1479–1489

Yamada K, Ojika M, Ishigaki T, Yoshida Y, Ekimoto H, Arakawa M (1993) Aplyronine A, a potent antitumor substance, and the congeners aplyronine B and C isolated from the sea hare *Aplysia kirodai*. J Am Chem Soc 115:11020–11021

Yamada K, Ojika M, Kigoshi H, Suenaga K (2000) Cytotoxic substances from opisthobranch mollusks. In: Fusetani N (ed) Drugs from the sea. Kargar, Basel, pp 59–73

Yamamura S, Hirata Y (1963) Structures of aplysin and aplysinol, naturally occurring bromo-compounds. Tetrahedron 19:1485–1496

Yamazaki M (1993) Antitumor and antimicrobial glycoproteins from sea hares. Comp Biochem Physiol C 105:141–146

Yamazaki M, Kisugi J, Kimura K, Kamiya H, Mizuno D (1985) Purification of antineoplastic factor from eggs of a sea hare. FEBS Lett 185:295–298

Yamazaki M, Kimura K, Kisugi J, Kamiya H (1986) Purification of a cytolytic facter from purple fluid of a sea hare. FEBS Lett 198:25–28

Yamazaki M, Kimura K, Kisugi J, Muramoto K, Kamiya H (1989a) Isolation and characterization of a novel cytolytic factor in purple fluid of the sea hare, *Aplysia kurodai*. Cancer Res 49:3834–3838

Yamazaki M, Kisugi J, Kamiya H (1989b) Biopolymers from marine invertebrates. XI. Characterization of an antineoplastic glycoprotein, dolabellanin A, from the albumen gland of a sea hare, *Dolabella auricularia*. Chem Pharm Bull 37:3343–3346

Yamazaki M, Tansho S, Kisugi J, Muramoto K, Kamiya H (1989c) Purification and characterization of a cytolytic protein from purple fluid of the sea hare, *Dolabella auricularia*. Chem Pharm Bull 37:2179–2182

Zhang X, Minale L, Zampella A, Smith CD (1997) Microfilament depletion and circumvention of multiple drug resistance by sphinxolides. Cancer Res 57:3751–3758

Zhang H, Teng M, Niu L, Wang Y, Wang Y, Liu Q, Huang Q, Hao Q, Dong Y, Liu P (2004) Purification, partial characterization, crystallization and structural determination of AHP-LAAO, a novel L-amino-acid oxidase with cell apoptosis-inducing activity from *Agkistrodon halys pallas* venom. Acta Crystallogr D Biol Crystallogr 60:974–977

Zhou ZH, Komiyama M, Terao K, Shimada Y (1994) Effects of pectenotoxin-1 on liver cells in vitro. Nat Toxins 2:132–135

Trisoxazole Macrolides from *Hexabranchus* Nudibranchs and Other Marine Invertebrates

S. Matsunaga

Abstract. Trisoxazole macrolides are cytotoxic and antifungal metabolites initially isolated from the egg-ribbons of the *Hexabranchus* nudibranch and later found in other marine invertebrates. They possess a characteristic macrolide portion, in which three contiguous oxazole units are integrated, and a side-chain with an *N*-methyl-vinylformamide terminus. The planar structures of the first members of this group, ulapualides and kabiramide C, were determined by interpretation of spectral data in conjunction with chemical degradation. Following these studies, the structures of approximately 35 congeners have been reported, including mycalolides from a marine sponge *Mycale* sp. The absolute stereochemistry of mycalolides was determined by chemical methods. Trisoxazole macrolides depolymerize F-actin and form a 1:1 complex with G-actin, thereby exhibiting potent toxicity toward eukaryotic cells. X-ray crystallography established the mode of binding of some of the members to G-actin and their absolute stereochemistry.

11.1 Introduction

The observations that shell-less gastropods are immune to predation by their surrounding carnivores led to a hypothesis that they are chemically defended. Succeeding studies disclosed diverse protection strategies: acid secretion, displaying cnidocytes obtained from the cnidarian prey, and protection by repugnant organic compounds which are acquired from their prey, such as algae or sponges, or biosynthesized de novo (Faulkner and Ghiselin 1983; Faulkner 1988; Avila 1995; Cimino and Ghiselin 1999; Cimino et al. 1999, 2001). Dorid nudibranchs procure defensive chemicals from sponges on which they prey. *Hexabranchus sanguineus* is a dorid nudibranch distributed commonly in the Indo-Pacific area and well known for its large size (up to 60 cm), brilliant color, and swimming behavior (Gohar and Soliman 1963; Thompson 1972; Francis 1980; Wägele and Willan 2000; Valdés 2002). More conspicuous in the reef are their bright red egg-ribbons, prominently deposited on rocks or coral debris.

Shigeki Matsunaga
Laboratory of Aquatic Natural Products Chemistry
Graduate School of Agricultural and Life Sciences, The University of Tokyo, 1-1-1 Yayoi
Bankyo-ku, 113-8657 Tokyo, Japan

Progress in Molecular and Subcellular Biology
Subseries Marine Molecular Biotechnology
G. Cimino, M. Gavagnin (Eds.): Molluscs

In the antimicrobial screening of marine invertebrates, we found that the extract of the egg-ribbons deposited by *Hexabranchus* and collected at Kabira Bay on Ishigaki Island in the Ryukyu Islands exhibited prominent antifungal but no antibacterial activity; and we started to study the active constituents (Matsunaga et al. 1986). Simultaneously, studies on the biologically active constituents from the *Hexabranchus* nudibranch and its egg-ribbons from Oahu and those from the Kwajalein Atoll were in progress (Roesner and Scheuer 1986; Kernan et al. 1988). Also at the same time, *Halichondria* sponges from Kwajalein and Palau were shown to contain metabolites found in *H. sanguineus* (Kernan and Faulkner 1987; Andersen et al. 2004), which was followed by the isolation of further congeners from sponges (Fusetani et al. 1989). The biologically active constituents contained in *H. sanguineus* were a series of trisoxazole macrolides with an unprecedented structure with no counterpart in the natural products from terrestrial organisms.

11.2 Isolation and Structure Elucidation of Trisoxazole Macrolides from Marine Invertebrates

At the time of the discovery, the structures of trisoxazole macrolides were unique, especially the presence of three contiguous oxazoles in a macrolide ring and the side-chain terminating with *N*-methyl-vinylformamide. In order to confirm this novel structure, we had to detect small ^{1}H–^{13}C long-range couplings through time-consuming ^{13}C NMR spectroscopy (Matsunaga et al. 1986) and Professor Scheuer's group had to isolate a small amount of an oxidation product (Roesner and Scheuer 1986). The side-chain portions of this class of molecules were similar to the one in tolytoxin, which had been isolated from a marine blue-green alga and for which only a partial structure was reported in a review (Moore 1981). Because of the similarity, a close relationship between trisoxazole macrolides and the tolytoxin class of metabolites was speculated (Ishibashi et al. 1986). After the structure elucidation of ulapualides and kabiramides, important NMR techniques such as inverse-detection and pulse field gradient were introduced. These techniques and other advances in instrumental analysis significantly decreased the sample size and time-frame required for structure elucidation and made the structure elucidation of complex molecules less laborious.

In this section, the structure elucidation of trisoxazole macrolides is summarized.

11.2.1 Ulapualides

Ulapualides were isolated from the egg-ribbons of *H. sanguineus* collected at Pupukea, Oahu (Roesner and Scheuer 1986). The extract of the egg-ribbons was subjected to solvent partitioning, followed by reversed-phase

HPLC and silica HPLC to furnish ulapualides A (1) and B (2; Fig. 11.1). The structural study was mainly conducted on ulapualide B. The molecular formula was determined by high-resolution fast atom bombardment mass spectrometry, together with elemental analysis. Partial structures V, W, X, and Y were assigned on the basis of 2D-NMR data such as COSY, CH-COSY, and long-range CH-COSY together with difference decoupling experiments in ^{1}H NMR. Restricted rotation about the amide bond in the *N*-methyl-vinylformamide moiety gave rise to doubled ^{1}H and ^{13}C NMR signals in the vicinity of this group. The most difficult part of the structure elucidation was the assignment of the trisoxazole moiety, whose presence was suggested by the ^{13}C NMR and ^{15}N NMR data (δ_N231.9, 222.5, 213.7). Because the orientation of the central oxazole ring was not assignable by spectroscopic methods, ulapualide B was subjected to ozonolysis, followed by a reductive work-up and acetylation. An aromatic compound (3; Fig. 11.1), analyzed for $C_{10}H_9N_3O_5$ and structurally assigned on the basis of ^{1}H NMR data, was isolated from a complex mixture of reaction products. The structure implied that the compound arose from cleavage of ring C and the Δ^{19}-olefin. Compound 3 was converted into a nonsymmetrical bis-oxazole (4). If the orientation of the central oxazole ring is reversed, the product should be symmetrical. Because 4 exhibited a pair of oxazole protons, the orientation of the central oxazole unit was determined as shown. Ulapualide A was a C30-ketone analogue.

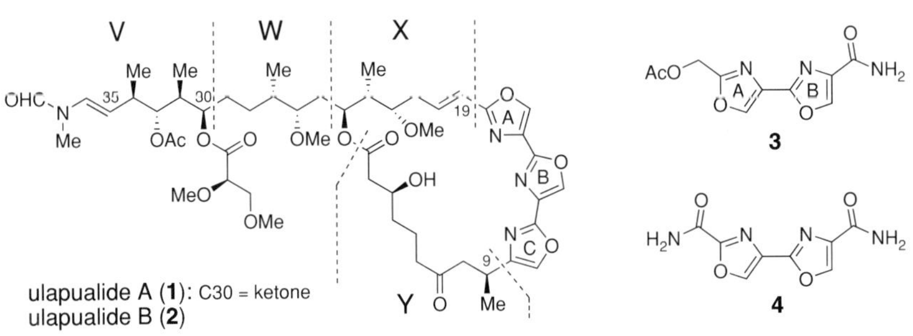

Fig. 11.1. Structures of ulapualides and the degradation products

The stereochemistry of all the stereogenic centers in ulapualide A (1) was predicted by hypothesizing that ulapualide A should form a complex with metals and that the compound possessing the stereochemistry of the natural product should form the lowest energy chelate (Maddock et al. 1993). The proposed structure later turned out to be incorrect by a total synthesis (Chattopadhyay and Pattenden 2000). Although it is of interest, the biological activity of the synthetic compound was not reported. In this connection, the cation-binding property of dihydrohalichondramide (vide infra) was shown not to be associated with its antifungal activity: a trisoxazole fragment which exhibited the same degree of cation-binding property as that of dihydrohalichondramide was not antifungal (James et al. 1993).

The absolute stereochemistry of the C-19 to C-35 portion of ulapualide B (2) was shown to be identical with that of mycalolide B, by converting to the same degradation product (vide infra; Matsunaga et al. 1999) and the total stereochemistry of ulapualide A was solved by X-ray crystallography of a complex with actin (Allingham et al. 2004). Interestingly, the side-chain beyond C-31 was disordered in the crystal structure, demonstrating a weak contact of this portion with actin. This is in agreement with a highly specific binding between actin and swinholide A which differs significantly in the structure of the relevant portion (Bubb et al. 1995; Klenchin et at. 2005).

11.2.2 Kabiramides

Kabiramide C (5; Fig. 11.2) was isolated from the egg-ribbons deposited at Kabira Bay on Ishigaki Island by an unidentified nudibranch (Matsunaga et al. 1986), which was later identified as *Hexabranchus* sp. (Matsunaga et al. 1989). The major HPLC peak of the antifungal fraction of the extract of the egg-ribbons was named kabiramide C. Kabiramide C (5) was very abundant in the egg-ribbons and isolated in a yield of 0.25% wet weight. Several NMR signals were doubled, as was the case with ulapualides. Interpretation of the COSY spectrum together with the CH-COSY spectrum, which assisted the assignments of the crowded aliphatic region,

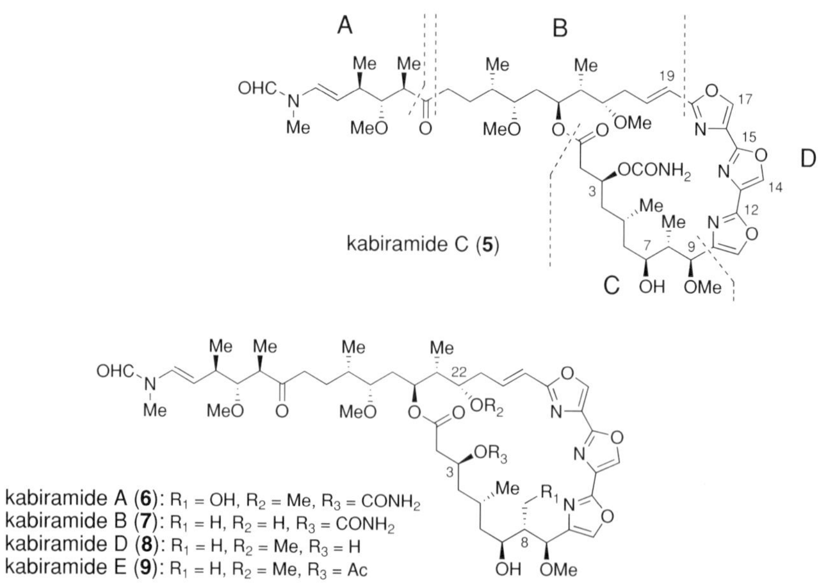

Fig. 11.2. Structures of kabiramides

allowed us to assign partial structures A–C, in which four methoxy groups were placed on the basis of NOE difference experiments. The location of a free hydroxyl group at C-7 was confirmed by a 1.3-ppm shift of the oxymethine proton in the acetate. Units A and B were joined through a doubled ketone (δ 214.0, 214.1) on the basis of NOE difference experiments and chemical shift arguments. A connection between the partial structures B and C was accomplished by a long-range selective proton decoupling (LSPD) experiment. One proton was irradiated overnight and a ^{13}C NMR spectrum without broad-band decoupling was acquired. We had to rely on such time-consuming experiments before the era of inverse-detected 2D-NMR experiments. After the assignment of these units, ten sp^2 carbons, three low-field singlet aromatic protons, and a broad exchangeable signal integrating for 2H remained unassigned. It took some time before we recognized the presence of three iterative units by inspecting the distribution of carbon chemical shifts. At the time of the structure elucidation, heteroaromatic rings such as oxazole and thiazole were not familiar among marine natural products, except for peptides from ascidians (Ireland and Scheuer 1980). Being encouraged by the precedence of two contiguous thiazoles in bleomycins, antitumor nonribosomal peptides of microbial origin (Takita et al. 1978), we presumed the presence of three contiguous oxazole rings (partial structure D), which was unprecedented, and a carbamate ester (δ_C 157.3).

In order to confirm the idea, we again relied on LSPD experiments, which gave not only coupling partners but also magnitudes of coupling constants. We determined long-range ^1H–^{13}C coupling constants within five-membered rings, which matched well with those reported for oxazoles. Unfortunately, long-range couplings across oxazole rings were not obvious from the LSPD experiments, because they were very small. However, careful inspection of carbon signals indicated a clear narrowing due to perturbations of long-range couplings: irradiation of H-14 and H-17 sharpened C-12 and C-15, respectively. This observation allowed the determination of the orientation of the central oxazole ring. By placing the carbamate ester at C-3 on the basis of an LSPD experiment, we completed the structure elucidation of kabiramide C.

Kabiramides A, B, D, E (**6**, **7**, **8**, **9**, respectively; Fig. 11.2) were isolated as minor metabolites from the egg-ribbons collected at Ishigaki Island (Matsunaga et al. 1989). Their structures were determined by comparing the NMR data with those of kabiramide C. In kabiramide A, the C-8 methyl group was oxidized to a hydroxymethyl group; kabiramide B had a hydroxyl group at C-22 instead of a methoxy group in kabiramide C; kabiramide D was the decarbamoyl derivative of kabiramide C; and kabiramide E was the C-3 acetate of kabiramide D. Kabiramides B and C were succeedingly isolated from a sponge *Halichondria* sp., strongly suggesting the sponge origin of kabiramides (Kernan et al. 1988). It is not known whether kabiramides A, D, and E are sponge metabolites or are converted within the body of the nudibranch.

Due to the difficulty in preparing crystals suitable for X-ray crystallography and conformational heterogeneity (cf. Sect. 11.2.5), the stereochemistry of kabiramides remained unknown for almost two decades. However, by immersing a solution of kabiramide C in preformed crystals of actin, the relative and absolute stereochemistry of kabiramide C was determined by X-ray crystallography of the complex (Klenchin et al. 2003).

11.2.3 Halichondramide and Congeners

The discovery of halichondramide (10; Fig. 11.3) from a marine sponge *Halichondria* sp. indicated that the trisoxazole macrolides are originally produced by sponges (Kernan and Faulkner 1987). The structure of halichondramide was determined by interpretation of NMR data, including COSY, CH-COSY, and long-range CH-COSY (COLOC), which was popular before the introduction of the HMBC technique, and comparison of the NMR data with those of kabiramide C.

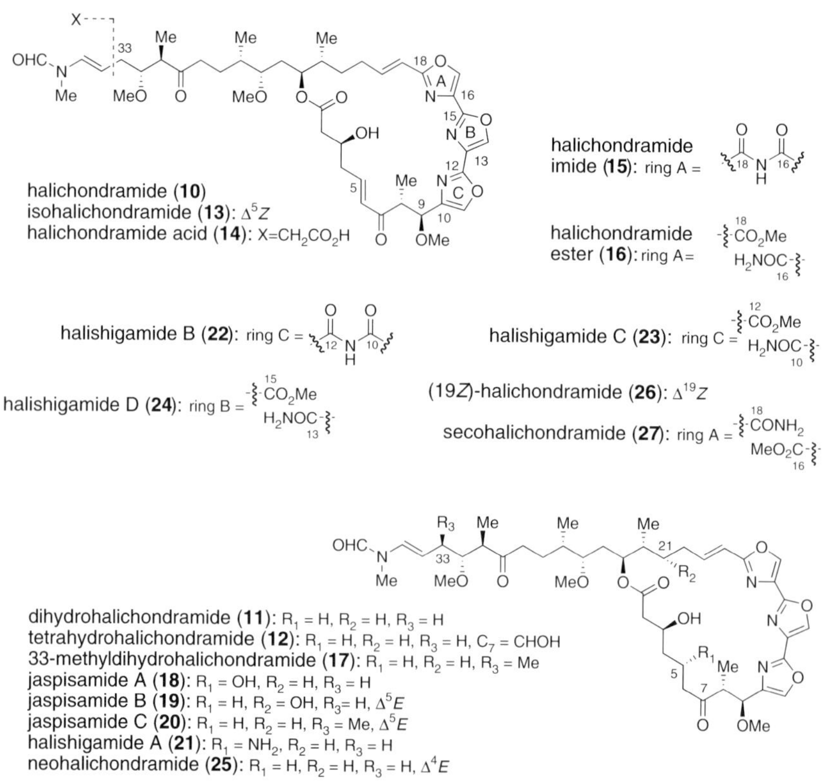

Fig. 11.3. Structures of halichondramide and its congeners

The isolation of halichondramide and its congeners (Fig. 11.3) from *H. sanguineus* which fed on *Halichondria* sp. was reported by Kernan et al. (1988). Dihydrohalichondramide (**11**) was the reduction product of halichondramide at the Δ^5-olefin; and tetrahydrohalichondramide (**12**) was the compound further reduced at the C7-ketone. Tetrahydrohalichondramide was produced by reduction of dihydrohalichondramide with Li(*O*-*t*-Bu)$_3$BH. Isohalichondramide (**13**) was the 5*Z*-isomer of halichondramide and was slowly converted to halichondramide in $CDCl_3$ solution. Halichondramide acid (**14**) had the carboxylic acid group instead of the *N*-methyl-vinylformamide group at the end of the side-chain. Jones' oxidation of halichondramide afforded halichondramide acid through the terminal aldehyde generated by acid-catalyzed removal of the *N*-methylformamide group. The NMR signals of halichondramide acid were no more doubled, due to the lack of the terminal amide moiety. Halichondramide imide (**15**) had an imide group [δ_H9.86 (brs, 1H)] instead of the oxazole proton (H-17) and exhibited infrared absorption at 1,760 cm^{-1}, which was ascribed to the absorption of an imide group. The location of the imide group was determined by an NOE difference experiment. The imide was probably produced by air oxidation of the oxazole ring (Wasserman 1970), followed by hydration and generation of carbon dioxide (Scheme 11.1). Halichondramide ester (**16**) was the methanolysis product of the imide, which was confirmed by a chemical transformation from the imide to the ester. Given the susceptibility of oxazole rings toward oxidation, compounds **15** and **16** may be artifacts. Dihydrohalichondramide (**11**) and a new congener, 33-methyldihydrohalichondramide (**17**), were isolated from the egg-ribbons collected at Hachijo Island, which also contained ulapualides A and B (Matsunaga et al. 1989).

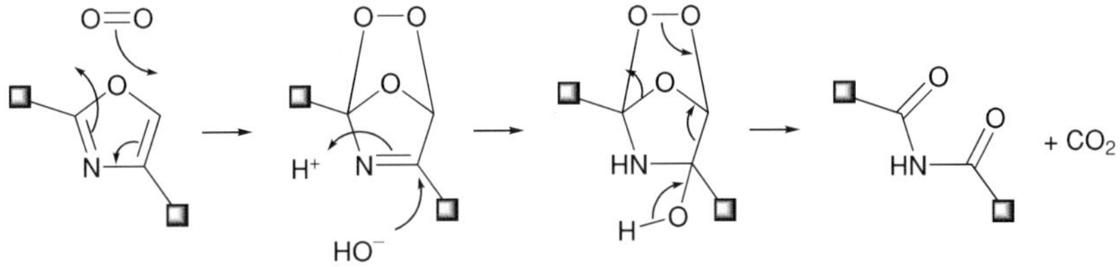

Scheme 11.1. Opening of ring A of halichondramide and formation of imide functional group

Jaspisamides are minor cytotoxic constituents of a marine sponge *Jaspis* sp. collected at Ishigaki Island, halichondramide being the major constituent of the sponge (Kobayashi et al. 1993). Jaspisamide A (**18**) is a hydration product of halichondramide at C-5, jaspisamide B is 21-hydroxyhalichondramide (**19**), and jaspisamide C is 33-methylhalichondramide (**20**). Hydroxylation at C-21 and methylation at C-33 are both observed in ulapualides and kabiramides. Halishigamides are also minor constituents of a marine sponge *Halichondria* sp. collected at Ishigaki Island (Kobayashi et al. 1997). Halishigamide A (**21**) is the ammonia adduct of halichondramide at C-5. The presence of an amino group was confirmed by a positive ninhydrin reaction and the preparation of an *N*-Boc derivative. Halishigamide B (**22**), in which ring C was converted to an imide, is isomeric to halichondramide imide (**15**). The location of the imide group was shown by the HMBC and NOESY correlations. Halishigamide C (**23**) is a methanolysis product of halishigamide B. The locations of the methyl ester and amide groups were demonstrated by a decoupled HMBC experiment. Halishigamide D (**24**) with oxidized ring B was isomeric to halishigamide C. The locations of the methyl ester and amide groups in **24** were determined on the basis of decoupled-HMBC experiment in which several four-bond H–C couplings were observed. Neohalichondramide (**25**), (19*Z*)-halichondramide (**26**), and secohalichondramide (**27**) were isolated from the marine sponge *Chondrosia corticata* collected at Guam, the predominant trisoxazole constituent of the sponge being halichondramide (**10**; Shin et al. 2004). Dihydrohalichondramide (**11**), jaspisamide A (**18**), and halishigamide D (**24**) were also found in the sponge. Neohalichondramide was the Δ^4-isomer of halichondramide, whereas secohalichondramide was the positional isomer of halichondramide ester. Compound **26** was the first 19*Z*-isomer in this class of metabolites.

The absolute stereochemistry of halichondramide was determined by X-ray crystallography of a complex with actin (Klenchin et al. 2003). Actin-catalyzed hydration of halichondramide to jaspisamide A was observed during crystallization.

11.2.4 Mycalolides

Mycalolides A–C (Fig. 11.4) were isolated from a marine sponge *Mycale* sp. (Fusetani et al. 1989). Their structures were assigned by analysis of 2D-NMR data. Mycalolide A (**28**) is a hybrid of halichondramide (C-1 to C-9 portion) and ulapualide A (C-19 to C-35 portion), whereas mycalolide B (**29**) is a hybrid of halichondramide and ulapualide B. Mycalolide C (**30**) is the 38-*O*-des-methyl derivative of mycalolide B.

Mycalolide C and two related macrolides, mycalolides D (**31**) and E (**32**; Fig. 11.4), were isolated from the stony coral *Tubastrea faulkneri* collected at the Great Barrier Reef (Rashid et al. 1995). The structure of mycalolide D, the imide derivative of a putative dehydroulapualide B, was assigned on the basis of HMBC data. Mycalolide E is the ring C oxidation product of a positional isomer of a putative dihydroulapualide A.

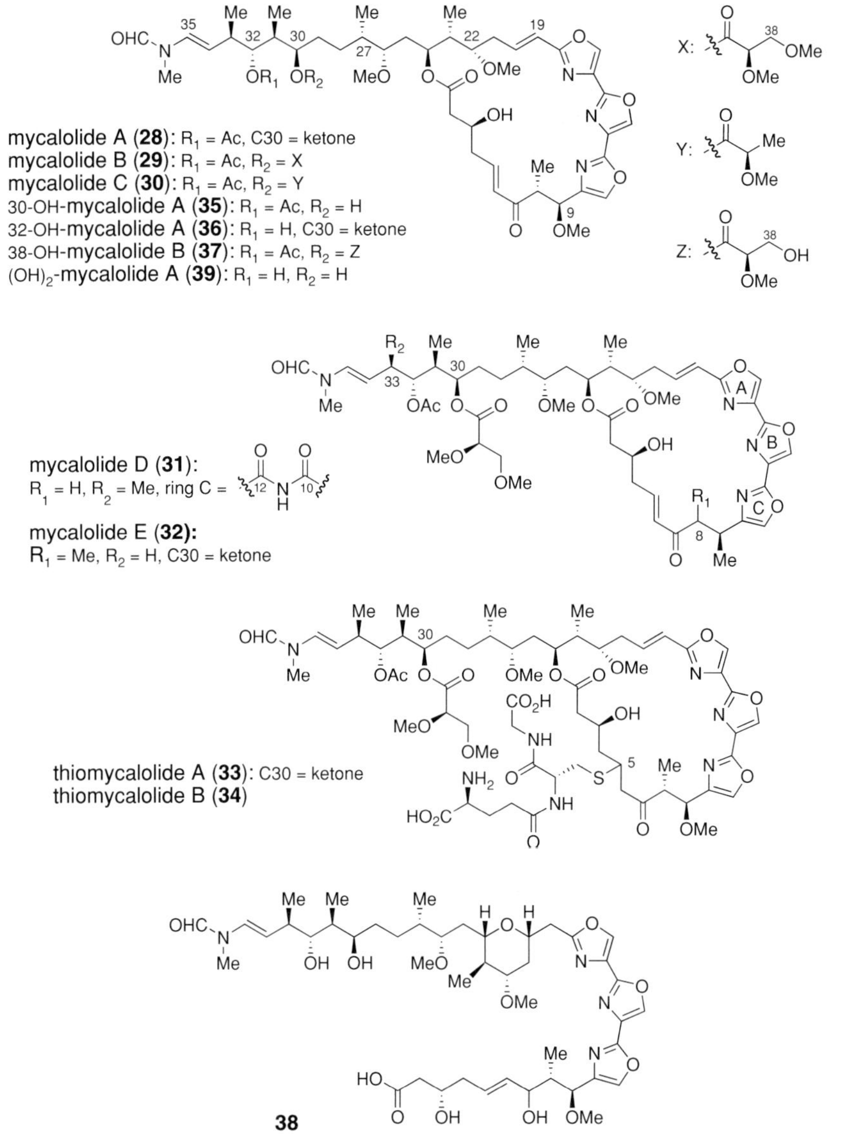

Fig. 11.4. Structures of mycalolides and their degradation products

A polar fraction of the *Mycale* sp. that contained mycalolides A–C showed potent cytotoxic activity. From the fraction were isolated thiomycalolides A, B (33, 34; Fig. 11.4; Matsunaga et al. 1998a). Thiomycalolide B is a glutathione adduct of mycalolide B, whereas thiomycalolide A is that of mycalolide A. Thiomycalolide A was prepared from the reaction between mycalolide A and glutathione.

Three congeners of mycalolides (**35**–**37**; Fig. 11.4), all of which had a hydroxyl group near the terminus of the side-chain, were isolated from *M. magellanica* collected in Suruga Bay (Matsunaga et al. 1998b). Their structures were determined by interpretation of spectral data. In 30-hydroxymycalolide A (**35**), the C30-ketone of mycalolide A was reduced to a secondary alcohol. This was confirmed by converting **28** and **35** to a common reduction product. 32-Hydroxymycalolide A (**36**) was the deacetylation product of mycalolide A, which was confirmed by converting **28** and **36** to the same diacetate. 38-Hydroxymycalolide B (**37**) possessed a hydroxyl group at C-38 instead of a methoxy group in mycalolide B. This was substantiated by converting **29** and **37** to the same tetrahydropyran (**38**) by saponification of the $NaBH_4$ reduction products. Additionally, mycalolide C and 30-hydroxymycalolide A were also converted to **38**. From these experiments, it was demonstrated that the stereochemistries of all stereogenic centers in mycalolides from *Mycale* spp. were identical.

Additionally, 30,32-dihydroxymycalolide A (**39**) was isolated as a minor constituent of *M. izuensis* (Phuwaraisirisan et al. 2002).

11.2.5 Stereochemistry of Mycalolides

Initially, it was envisaged that the stereochemistry of mycalolides could be determined by a combination of conformational analysis and the modified Mosher ester analysis in the ^{1}H NMR spectroscopy. However, it was not possible to correlate the stereochemistry of C-7 with that of C-3 and the assignment of relative stereochemistry of the C-22 to C-27 portion was difficult due to the presence of more than one conformer in this region. Although it is still possible to propose the most probable structure by considering coupling constants as weighted averages of several conformers and NOESY cross-peaks as the sum of each conformer, such analysis carries a high potential of misassignments. Therefore, we elected to carry out chemical degradations in order to prepare conformationally stabilized derivatives and to obtain fragments for direct comparison with authentic samples (Fig.11.5). The degradation started with the cleavage of Δ^{19}- and Δ^{34}-olefinic bonds with RuO_4, followed by methanolysis to remove the esters. The resulting dicarboxylic acid was dehydrated with DCC to form the bis-δ-lactone (**40**). The two δ-lactone rings were both in a boat-like conformation and the relative stereochemistry within the rings was firmly assigned by interpretation of NMR data. Relative stereochemistry at C-26 was correlated to that of C-24 with high credibility because the C-24 to C-26 portion adopted a single conformation with an extended carbon chain. However, a coupling constant of 3.7 Hz between H-26 and H-27 and numerous NOESY cross-peaks observed from these protons suggested that this section of the molecule adopted two or more conformations not only in mycalolides but also in **40**. The relative stereochemistry of C-27 with respect to that of C-26 in mycalolides remained ambiguous, even after analysis of the ^{1}H NMR data of two model compounds (**41**, **42**). Therefore, it appeared necessary to make a direct comparison between a fragment obtained by chemical degradation and synthetic compounds with defined stereochemistry. Before preparing synthetic fragments, we had to reduce the number of possible isomers.

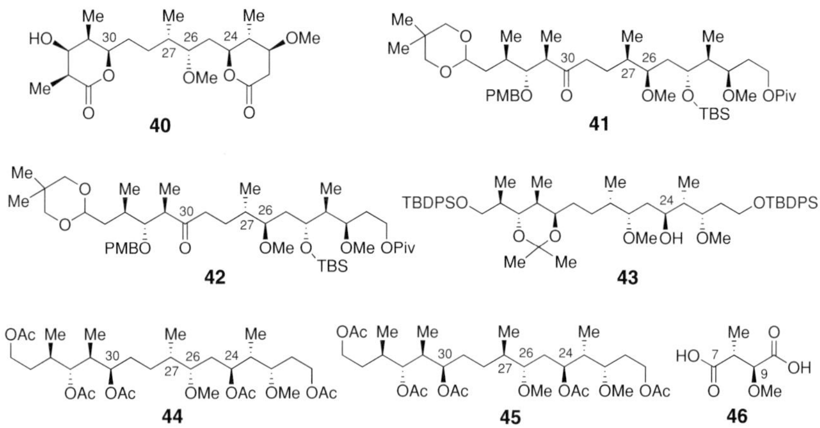

Fig. 11.5. Structures of degradation products and synthetic fragments of mycalolides

It is possible to determine the stereochemistry of compounds containing many hydroxyl groups by preparing a per-MTPA ester and applying the modified Mosher method (Kobayashi et al. 1994). In order to carry out the analysis, it is necessary to elucidate the conformation of the molecule and estimate the additive effects of spatially close MTPA groups, which are associated with the possibility of misassignment. Therefore, we intended to avoid the introduction of MTPA groups in proximate hydroxyl groups. Fortunately, there were natural mycalolides whose hydroxyl groups were differentially unmasked. We used them for the modified Mosher analysis (Ohtani et al. 1991). We used 30-hydroxymycalolide A and 32-hydroxymycalolide A for the analysis of C-30 and C-32, respectively. In order to liberate a hydroxyl group at C-24, we prepared acetonide **43** (Fig. 11.5) by a standard sequence of chemical reactions. By application of the Mosher ester analysis to these compounds, we demonstrated the 24*S*,30*R*,32*R*-stereochemistry. Additionally, the 3*S*-stereochemistry was simultaneously defined by the Mosher ester analysis. With these assignments in hand, there were only two possibilities left for the stereochemistry of the C-19 to C-35 portion, differing in the configuration at C-27. Then, two C-27 epimers of the C-20/C-35 pentaacetate fragment (**44**, **45**) were synthesized and compared with the one prepared from 38-hydroxymycalolide B. The ^{1}H and ^{13}C NMR data of the synthetic fragment with 26,27-*syn* relationship coincided with those prepared from the natural product.

Although there was a strong NOE between H-5 and H-8, it was not possible to establish the spatial relationship between H-3 and H-5, because H-5 coupled equally to methylene protons on C-4. We took advantage of the presence of oxidatively cleavable functional groups besides the C-8/C-9 unit. On oxidation with RuO_4, 38-hydroxymycalolide B afforded the expected dicarboxylic acid (**46**), whose relative stereochemistry was assigned on the basis of the ^{1}H NMR data. Compound **46** was converted to the bis-*p*-bromophenacyl ester and its retention time was compared with those of synthetic standards in HPLC analysis using a chiral stationary phase (Chiralcel OJ), which showed the 8*R*,9*S*-stereochemistry. The last stereogenic centers to be assigned were those in the pendant esters at C-30: 2,3-*O*,*O′*-dimethylglycerate in mycalolide B, *O*-methyllactate in mycalolide C, and 2-*O*-methylglycerate in 38-hydroxymycalolide. Each of the three mycalolides were saponified and the product was transformed to the *p*-bromophenacyl ester. After brief purification by silica gel chromatography, each product was analyzed by HPLC with Chiralcel AD or OJ and the retention time was compared with those of synthetic enantiomers. The stereochemistry at C-37 was assigned as *R* in each case, finalizing the assignment of the relative and absolute stereochemistry of mycalolides (Matsunaga et al. 1999).

The proposed stereochemistry of mycalolide A was confirmed by a total synthesis (Panek and Liu 2000). The stereochemistry of scytophycins, e.g., scytophycin B (**47**), and aplyronines, e.g., aplyronine A (**48**; Fig. 11.6), had been determined before we assigned the stereochemistry of mycalolides (Yeung and Paterson 2002). As for the side-chain, these metabolites shared the same stereochemistry for all stereogenic centers. A notable structural similarity in the side-chain was observed in mycalolides as well. Because initial analysis of the NMR data suggested that the relative stereochemistry of the C-22 to C-26 and C-30 to C-33 portions in trisoxazole macrolides was identical to that in scytophycins, it was tempting to propose that the stereochemistry of the C-22 to C-33 portion for both classes of compounds was identical. Contrary to our expectations, mycalolides and scytophycins were enantiomeric in the C-22 to C-26 portion and identical in the C-27 to C-33 portion. Because mycalolides, aplyronines, and scytophycins share the same cellular target (Patterson et al. 1993; Saito et al. 1994, 1996), the biochemical importance of the terminal portion of the side-chain was emphasized by the structure elucidation of mycalolides (Matsunaga et al. 1999).

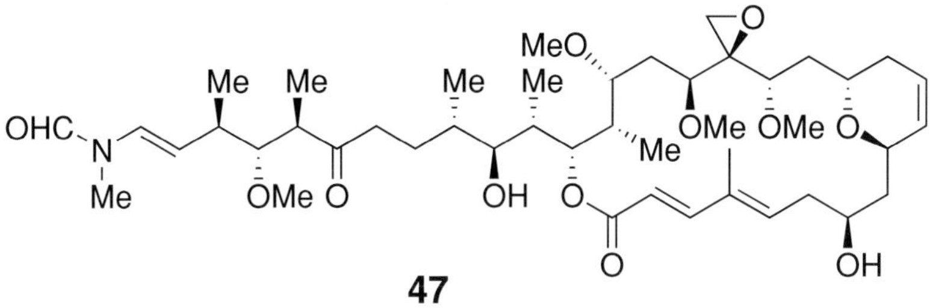

47

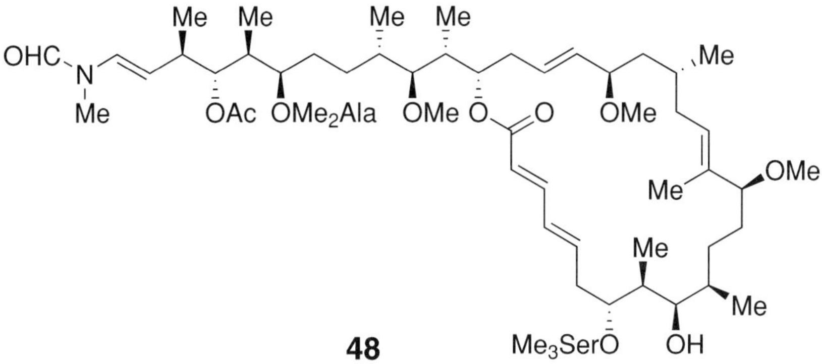

48

Fig. 11.6. Structures of scytophycin B and aplyronine A

11.3 Cellular Target of Trisoxazole Macrolides

Trisoxazole macrolides exhibited potent antifungal and cytotoxic activities but did not show antibacterial activity. Therefore, they were expected to perturb cellular components present specifically in eukaryotes. The effects of mycalolide B on smooth muscle contraction were examined (Hori et al. 1993). The contraction was inhibited by mycalolide B, but the effects were dependent neither on the concentration of Ca^{2+} nor on the phosphorylation state of the myosin light chain, suggesting that one or more of the contractile elements were affected. Mycalolide B inhibited the Mg^{2+}-ATPase activity of actomyosin, but cytochalasin D, the well-known actin-perturbing agent, did not. Therefore, the possibility of mycalolide B affecting actin was excluded and mycalolide B was proposed to be an inhibitor of actin–myosin interaction. Meanwhile, based on microscopic observations of KB and A10 cells and using purified actin, the biological activity of tolytoxin was ascribed to the disruption of microfilament organization and the depolymerization of actin (Patterson et al. 1993). Being inspired by this finding, a further examination of the effects of mycalolide B on purified actin was carried out, which disclosed that mycalolide B severed F-actin and formed a 1:1 complex with G-actin (Saito et al. 1994).

It was shown by the structure–activity relationship study on aplyronines, another actin depolymerizer, that the vinylformamide portion of the side-chain was not essential for activity (Suenaga et al. 1997). Therefore, we envisaged attaching biotin to the terminus of mycalolide B and kabiramide D in order to observe the cellular behavior of the compounds. The *N*-methyl-vinylformamide group was removed with acid to generate an aldehyde to which biocytin was attached by reductive amination (Wada et al. 1998). Many intracellular proteins in cultured rat 3Y1 fibroblast cells bound to biocytinylated mycalolide B (**49**), whereas only actin was detected as the cellular protein that bound to biocytinylated kabiramide D (**50**; Fig. 11.7). The promiscuity of mycalolide B was ascribed to the presence of the enone group, which was susceptible to Michael addition. Biocytinylated kabiramide D bound specifically to stress fibers in the formaldehyde-fixed 3Y1 cells. The binding specificities of other actin-binding molecules were examined using biocytinylated kabiramide D. Purified actin pretreated with either kabiramide D or misakinolide A did not bind to biocytinylated kabiramide D, whereas actin pretreated with either cytochalasin D or latrunculin A bound to biocytinylated kabiramide D. Therefore, the actin-binding site of misakinolide A was identical with or partially overlapped with that of kabiramide D, but those of cytochalasin D and latrunculin A were different. X-ray crystallography of a complex between latrunculin A and actin showed that the binding site of latrunculin was different from that of the trisoxazole macrolides (Morton et al. 2000).

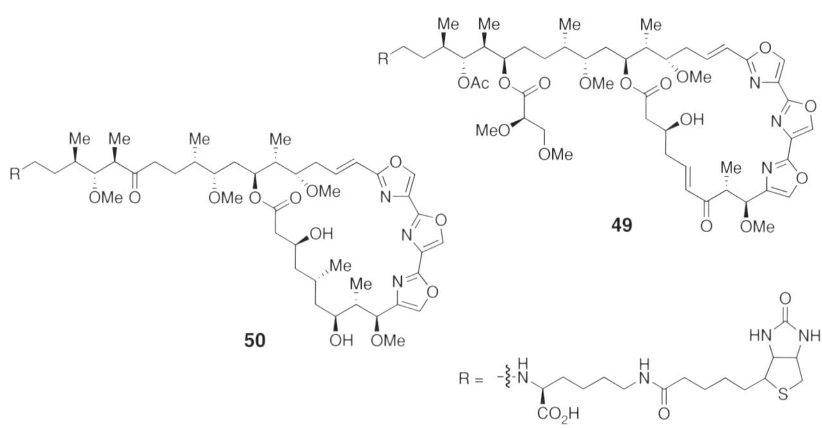

Fig. 11.7. Structures of biocytinylated mycalolide B and kabiramide D

11.4
Modes of Binding of Trisoxazole Macrolides to Actin

The mode of binding of kabiramide C to G-actin was studied by using prodan-actin as a fluorescent probe. The binding of kabiramide C to prodan-G-actin mediated a red-shift in the emission and decreased the quantum yield of the probe. By an examination of the time-course, the binding was shown to consist of two steps. The initial fast step was speculated to be the binding of the macrocyclic portion and the slow step to be that of the side-chain. Fractionation of F-actin treated with substoichiometric amounts of kabiramide C showed that the severing action of kabiramide C was slow (Tanaka et al. 2003). X-ray crystallography of a complex between actin and either kabiramide C, halichondramide, or ulapualides A demonstrated that the side-chain portion of these macrolides is inserted into the hydrophobic cavity which associates with the hydrophobic stretch of another actin molecule in forming F-actin (Klenchin et al. 2003; Allingham et al. 2004).

11.5
Structure–Activity Relationships of Trisoxazole Macrolides

Due to their effects on actin cytoskeletons, trisoxazole macrolides exhibit potent cytotoxic and antifungal activities (Wada et al. 1998; Spector et al. 1999). There is a clear relationship between the chemical structure and cytotoxicity among trisoxazole macrolides (Table 11.1). Compounds with the intact trisoxazole portion possess potent activity. However, when one oxazole ring is oxidized and converted to either imide or methyl ester, the activity of the compound is significantly reduced. Given the close association of the side-chain, as shown by the crystal structures of the complex, we should point out the importance of the side-chain in binding. This was confirmed by the potent actin-depolymerization activity of a synthetic C-22/C-35 fragment of mycalolide B as well as the corresponding portion of aplyronine A (Suenaga et al. 2004). It is of interest to examine whether these fragments possess cytotoxicity or not. Also, the actin-depolymerization activity of the imide and ester derivatives should be examined in order to understand the reason for their diminished cytotoxicity. The low cytotoxicity of the imide and ester derivatives may be attributable to their reluctance to penetrate the cellular membrane.

Table 11.1. Cytotoxicity of trisoxazole macrolides

compound	cytotoxicity (μg/mL)	examined cells
ulapualide A (**1**)	0.01-0.03	L1210
ulapualide B (**2**)	0.01-0.03	L1210
kabiramide C (**5**)	0.01	L1210
kabiramide A (**6**)	0.03	L1210
kabiramide B (**7**)	0.03	L1210
kabiramide D (**8**)	0.02	L1210
kabiramide E (**9**)	0.02	L1210
halichondramide (**10**)	4	sea urchin
dihydrohalichondramide (**11**)	0.03	L1210
tetrahydrohalichondramide (**12**)	1	sea urchin
isohalichondramide (**13**)	4	sea urchin
halichondramide acid (**14**)	>>4	sea urchin
halichondramide imide (**15**)	>>4	sea urchin
halichondramide ester (**16**)	>>4	sea urchin
33-methyldihydrohalichondramide (**17**)	0.05	L1210
jaspisamide A (**18**)	<0.001	L1210
jaspisamide B (**19**)	<0.001	L1210
jaspisamide C (**20**)	<0.001	L1210
halishigamide A (**21**)	0.004	L1210
halishigamide B (**22**)	4.4	L1210
halishigamide C (**23**)	5.2	L1210
halishigamide D (**24**)	1.1	L1210
neohalichondramide (**25**)	0.38	K562
(19*Z*)-halichondramide (**26**)	0.90	K562
secohalichondramide (**27**)	0.90	K562
mycalolide A (**28**)	>500	B-16
mycalolide B (**29**)	<0.001	B-16
mycalolide C (**30**)	<0.001	B-16
mycalolide D (**31**)	<0.001	average[b]
mycalolide E (**32**)	0.6	–
thiomycalolide A (**33**)	NT[a]	P388
thiomycalolide B (**34**)	0.02	P388
30-OH-mycalolide A (**35**)	0.02	L1210
32-OH-mycalolide A (**36**)	0.02	L1210
38-OH-mycalolide B (**37**)	0.01	L1210
$(OH)_2$-mycalolide A (**39**)	0.02	HeLa
	0.003	

For references, see text
[a] Not tested.
[b] Average value for 60 cancer cell lines.

11.6 Biological Significance of Trisoxazole Macrolides in the Nudibranch *H. sanguineus*

The defensive role of the trisoxazole macrolides in *H. sanguineus* was suggested by fish-feeding experiments and anatomical study (Pawlik et al. 1988). Trisoxazole macrolides were the most potent fish-feeding inhibitors studied by the Faulkner team (Pawlik et al. 1988). The nudibranchs were separated into four parts and the macrolide content of each part was determined. This experiment showed that trisoxazole macrolides were concentrated in the dorsal mantle, which is exposed to potential predators, but only small amounts of macrolides were present in the foot tissue. High concentrations of macrolides were detected in a mixture of digestive gland and gonad, because the macrolides came from the diet and a significant amount of the macrolides should be delivered to the egg-ribbons. Egg-ribbons, which were prominently deposited, contained tenfold or greater concentrations of the macrolides than those in the tissues of the nudibranch (Roesner and Scheuer 1986; Pawlik et al. 1988). The aposematic coloration of the animal from the Red Sea was due to a rare carotenoid hurghadin, which was localized in the mantle (Guo et al. 1998).

H. sanguineus fed selectively on the sponge *Halichondria* sp. that contained halichondramide (Pawlik et al. 1988). Because they accumulated dihydrohalichondramide and tetrahydrohalichondramide, which were minor constituents of the sponge, it was speculated that halichondramide was reduced by the nudibranch. Given the susceptibility of halichondramide to nucleophilic attacks at the enone portion, the reduction should take place immediately after the uptake of the molecule. *H. sanguineus* collected in the Red Sea was found to contain swinholide A, another actin-binding molecule (Carmely and Kashman, personal communication) and kabiramide C (Guo et al. 1998).

11.7 Conclusions

H. sanguineus is unique among criptobranch dorids in a morphological sense (Wägele and Willan 2000; Valdés 2002). This uniqueness reflects its peculiar chemistry: it is the only nudibranch that is known to use trisoxazole macrolides for chemical defense. Sponges containing trisoxazole macrolides are widely distributed, because *H. sanguineus* from various geographical areas contain the macrolides. However, no other nudibranchs have so far been reported to contain this class of metabolites. Therefore, it is likely that other nudibranchs are not immune to the trisoxazole macrolides. The mechanism of the tolerance of *H. sanguineus* toward trisoxazole macrolides is an interesting question.

The distribution of trisoxazole macrolides in sponges and hard coral poses a question about the true producer of this class of metabolites. In natural products chemistry, when one class of compounds is found in taxonomically remote organisms, the involvement of symbiotic microorganisms in their production is suspected and, in some cases, verified (Yotsu et al. 1987; Piel et al. 2004). At the time of the discovery of trisoxazole macrolides, the mixed biosynthetic pathway of polyketide and nonribosomal peptide was not common, which led to a proposal of the biosynthesis of oxazole rings through Beckman rearrangement (Ishibashi et al. 1986). Nowadays, many natural products are known to be biosynthesized by the mixed pathway (Shen et al. 2005) and so this is also likely the case with trisoxazole macrolides. The true producer of this class of metabolites and their mechanism of biosynthesis will be clarified in the future.

Actin is involved in several disease states; and compounds interacting with actin are anticipated to be candidates for anticancer agents (Yeung and Paterson 2002). Because actin has integral functions in eukaryotic cells, its perturbation is often associated with lethality. With knowledge of the precise modes of binding of several natural products to actin (Morton et al. 2000; Klenchin et al. 2003, 2005), the creation of therapeutic compounds that bind actin may be actualized.

References

Allingham JS, Tanaka J, Marriott G, Rayment I (2004) Absolute stereochemistry of ulapualide A. Org Lett 6:597–599

Andersen RJ, Ireland CM, Molinski TF, Bewley CA (2004) Research achievement award – D. John Faulkner. J Nat Prod 67:1239–1251

Avila C (1995) Natural products of opisthobranch molluscs: a biological review. Oceanogr Mar Biol Ann Rev 33:487–559

Bubb MR, Spector I, Bershadsky AD, Korn ED (1995) Swinholide A is a microfilament disrupting marine toxin that stabilize actin dimers and severs actin filaments. J Biol Chem 270:3463–3466

Chattopadhyay SK, Pattenden G (2000) A total synthesis of the unique tris-oxazole macrolide ulapualide A produced by the marine nudibranch *Hexabranchus sanguineus*. J Chem Soc Perkin 1:2429–2454

Cimino G, Ghiselin MT (1999) Chemical defense and evolutionary trends in biosynthetic capacity among dorid nudibranchs (Mollusca: Gastropoda: Opisthobranchia). Chemoecology 9:187–207

Cimino G, Fontana A, Gavagnin M (1999) Marine opisthobranch molluscs: chemistry and ecology in sacoglossans and dorids. Curr Org Chem 3:327–372

Cimino G, Ciavatta ML, Fontana A, Gavagnin M (2001) Metabolites of marine opisthobranchs: chemistry and biological activity. In: Tringali C (ed) Bioactive compounds from natural sources. Taylor and Francis, London, pp 577–637

Faulkner DJ (1988) Feeding deterrents in molluscs. In: Fautin DG (ed) Biomedical importance of marine organisms. California Academy of Science, San Francisco, pp 29–36

Faulkner DJ, Ghiselin MT (1983) Chemical defense and evolutionary ecology of dorid nudibranchs and some other opisthobranch gastropods. Mar Ecol Prog Ser 13:295–301

Francis MP (1980) Habitat, food and reproductive activity of the nudibranch *Hexabranchus sanguineus* on Tongatapu Island. Veliger 22:252–258

Fusetani N, Yasumuro K, Matsunaga S, Hashimoto K (1989) Mycalolides A–C, hybrid macrolides of ulapualides and halichondramide, from a sponge of the genus *Mycale*. Tetrahedron Lett 30:2809–2912

Gohar HAF, Soliman GN (1963) The biology and development of *Hexabranchus sanguineus* (Rüpp and Leuck). Publ Mar Biol Stn Ghardaqa 12:219–247

Guo Y, Gavagnin M, Mollo E, Trivellone E, Cimino G, Fakhr I (1998) Structure of the pigment of the Red Sea nudibranch *Hexabranchus sanguineus*. Tetrahedron Lett 39:2636–2638

Hori M, Saito S, Shin YZ, Ozaki H, Fusetani N, Karaki H (1993) Mycalolide-B, a novel and specific inhibitor of actomyosin ATPase isolated from marine sponge. FEBS Lett 322:151–154

Ireland C, Scheuer PJ (1980) Ulicyclamide and ulithiacyclamide, two new small peptides from a marine tunicate. J Am Chem Soc 102:5688–5691

Ishibashi M, Moore RE, Patterson GML, Xu C, Clardy J (1986) Scytophycins, cytotoxic and antimycotic agents from the cyanophyte *Scytonema pseudohofmanni*. J Org Chem 51:5300–5306

James DM, Winter E, Faulkner DJ, Siegel JS (1993) Metal binding by the trisoxazole portion of the marine natural product dihydrohalichondramide. Heterocycles 35:675–678

Kernan MR, Faulkner DJ (1987) Halichondramide, an antifungal macrolide from the sponge *Halichondria* sp. Tetrahedron Lett 28:2809–2812

Kernan MR, Molinski TA, Faulkner DJ (1988) Macrocyclic antifungal metabolites from the Spanish dancer nudibranch *Hexabranchus sanguineus* and sponges of the genus *Halichondria*. J Org Chem 53:5014–5020

Klenchin VA, Allingham JS, King R, Tanaka J, Marriott G, Rayment I (2003) Trisoxazole macrolide toxins mimic the binding of actin-capping proteins to actin. Nat Struct Biol 10:1058–1063

Klenchin VA, King R, Tanaka J, Marriott G, Rayment I (2005) Structural basis of swinholide A binding to actin. Chem Biol 12:287–291

Kobayashi J, Murata O, Shigemori H, Sasaki T (1993) Jaspisamides A–C, new cytotoxic macrolides from the Okinawan sponge *Jaspis* sp. J Nat Prod 56:787–791

Kobayashi J, Tsuda M, Fuse H, Sasaki T, Mikami Y (1997) Halishigamides A–D, new cytotoxic oxazole-containing metabolites from Okinawan sponge *Halichondria* sp. J Nat Prod 60:150–154

Kobayashi M, Aoki S, Kitagawa I (1994) Absolute stereostructures of altohyrtin A and its congeners, potent cytotoxic macrolides from the Okinawan marine sponge *Hyrtios altum*. Tetrahedron Lett 35:1243–1246

Maddock J, Pattenden G, Wight PG (1993) Stereochemistry of ulapualides, a new family of tris-oxazole-containing macrolide ionophores from marine nudibranchs. A molecular mechanics study. J Comput Aided Mol Des 7:573–586

Matsunaga S, Fusetani N, Hashimoto K, Koseki K, Noma M (1986) Kabiramide C, a novel antifungal macrolide from nudibranch eggmasses. J Am Chem Soc 108:847–849

Matsunaga S, Fusetani N, Hashimoto K, Koseki K, Noma M, Noguchi H, Sankawa U (1989) Further kabiramides and halichondramides, cytotoxic macrolides embracing trisoxazole, from the *Hexabranchus* egg masses. J Org Chem 54:1360–1363

Matsunaga S, Nogata Y, Fusetani N (1998a) Thiomycalolides: new cytotoxic trisoxazole-containing macrolides isolated from a marine sponge *Mycale* sp. J Nat Prod 61:663–666

Matsunaga S, Sugawara T, Fusetani N (1998b) New mycalolides from the marine sponge *Mycale magellanica* and their interconversion. J Nat Prod 61:1164–1167

Matsunaga S, Liu P, Celatka CA, Panek JS, Fusetani N (1999) Relative and absolute stereochemistry of mycalolides, bioactive macrolides from the marine sponge *Mycale magellanica*. J Am Chem Soc 121:5605–5606

Moore RE (1981) Constituents of blue-green algae. In: Scheuer PJ (ed) Marine Natural Products, vol 1. Academic Press, New York, pp 1–52

Morton WM, Ayscough KR, McLaughlin PJ (2000) Latrunculin alters the actin-monomer subunit interface to prevent polymerization. Nat Cell Biol 2:376–378

Ohtani I, Kusumi T, Kashman Y, Kakisawa H (1991) High-field FT NMR application of Mosher method. The absolute-configurations of marine terpenoids. J Am Chem Soc 113:4092–4096

Panek JS, Liu P (2000) Total synthesis of the actin-depolymerizing agent (–)-mycalolide A: application of chiral silane-based construction methodology. J Am Chem Soc 122:11090–11097

Patterson GML, Smith CD, Kimura LH, Britton BA, Carmely S (1993) Action of tolytoxin on cell morphology, cytoskeletal organization, and actin polymerization. Cell Motil Cytoskel 24:39–48

Pawlik JR, Kernan MR, Molinski TF, Harper MK, Faulkner DJ (1988) Defensive chemicals of the Spanish dancer and its eggs: macrolides from a sponge diet. J Exp Mar Biol Ecol 119:99–109

Phuwaraisirisan P, Matsunaga S, van Soest RWM, Fusetani N (2002) Isolation of new mycalolide from the marine sponge *Mycale izuensis*. J Nat Prod 65:942–943

Piel J, Hui D, Wen G, Butzke D, Platzer M, Fusetani N, Matsunaga S (2004) Antitumor polyketide biosynthesis by an uncultivated bacterial symbiont of the marine sponge *Theonella swinhoei*. Proc Natl Acad Sci USA 101:16222–16227.

Rashid MA, Gustafson KR, Cardellina II JH, Boyd MR (1995) Mycalolides D and E, new cytotoxic macrolides from a collection of the stony coral *Tubastrea faulkneri*. J Nat Prod 58:1120–1125

Roesner JA, Scheuer PJ (1986) Ulapualide A and B, extraordinary antitumor macrolides from nudibranch eggmasses. J Am Chem Soc 108:846–847

Saito S, Watabe S, Ozaki H, Fusetani N, Karaki H (1994) Mycalolide B, a novel actin depolymerizing agent. J Biol Chem 269:29710–29714

Saito S, Watabe S, Ozaki H, Kigoshi H, Yamada K, Fusetani N, Karaki H (1996) Novel actin depolymerizing macrolide aplyronine A. J Biochem 120:552–555

Shen B, Chen M, Cheng Y, Du L, Edwards DJ, George NP, Huang Y, Oh T, Sanchez C, Tang G, Wendt-Pienkowski E, Yi F (2005) Prerequisites for combinatorial biosynthesis: evolution of hybrid NRPS/PKS gene clusters. Ernst Schering Res Found Workshop 51:107–126

Shin J, Lee HS, Kim JY, Shin HJ, Ahn JW, Paul VJ (2004) New macrolides from the sponge *Chondrosia corticata*. J Nat Prod 67:1889–1892

Spector I, Braet F, Shochet NR, Bubb MR (1999) New anti-actin drugs in the study of the organization and function of the actin cytoskeleton. Microsc Res Tech 47:18–37

Suenaga K, Kamei N, Okugawa Y, Takagi M, Akao A, Kigoshi H, Yamada K (1997) Cytotoxicity and actin depolymerizing activity of aplyronine A, a potent antitumor macrolide of marine origin, and the natural and artificial analogs. Bioorg Med Chem Lett 7:269–274

Suenaga K, Miya S, Kuroda T, Handa T, Kanematsu K, Sakakura A, Kigoshi H (2004) Synthesis and actin-depolymerizing activity of mycalolide analogs. Tetrahedron Lett 45:5383–5386

Takita T, Muraoka Y, Nakatani T, Fujii A, Umezawa Y, Naganawa H (1978) Chemistry of bleomycin. XIX Revised structures of bleomycin and phleomycin. J Antibiot 31:801–804

Tanaka J, Yan Y, Choi J, Bai J, Klenchin VA, Rayment I, Marriott G (2003) Biomolecular mimicry in the actin cytoskeleton: mechanisms underlying the cytotoxicity of kabiramide C and related macrolides. Proc Natl Acad Sci USA 100:13851–13856

Thompson TE (1972) Observation on *Hexabranchus* from the Australian Great Barrier Reef. Veliger 15:1–5

Valdés Á (2002) A phylogenetic analysis and systematic revision of the cryptobranch dorids (Mollusca, Nudibranchia, Anthobranchia). Zool J Linn Soc 136:535–636

Wada S, Matsunaga S, Saito S, Fusetani N, Watabe S (1998) Actin-binding specificity of marine macrolide toxins, mycalolide B and kabiramide D. J Biochem 123:946–952

Wägele H, Willan RC (2000) Phylogeny of the nudibranchia. Zool J Linn Soc 130:83–181

Wasserman HH (1970) Recent reactions of singlet oxygen with heterocyclic and aromatic substrates. Ann NY Acad Sci 171:108–119

Yeung K-S, Paterson I (2002) Actin-binding marine macrolides: total synthesis and biological importance. Angew Chem Int Ed Engl 41:4632–4653

Yotsu M, Yamazaki T, Meguro Y, Endo A, Murata M, Naoki H, Yasumoto T (1987) Production of tetrodotoxin and its derivatives by *Pseudomonas* sp. isolated from the skin of a pufferfish. Toxicon 25:225–228

Sequestration and Possible Role of Dietary Alkaloids in the Sponge-Feeding Mollusk *Tylodina perversa*

C. Thoms, R. Ebel, P. Proksch

Abstract. Opisthobranchs of the genus *Tylodina* are found at exceedingly distant geographical regions in the marine environment but are always associated with sponges of the order Verongida (e.g., *Aplysina* species) which serve as prey for these gastropods. We investigated the chemical ecology of the Mediterranean species *T. perversa* that commonly feeds on *A. aerophoba.* The gastropod sequesters a set of sponge-derived brominated isoxazoline alkaloids which are accumulated in the mantle and egg masses and are furthermore exuded as part of the mucus when the animal is molested. Based on the documented feeding deterrent properties of the sponge alkaloids against fish, it is speculated that the sequestered sponge alkaloids serve also as a defense for *T. perversa.* Interestingly, specimens of *T. perversa* that were either collected while feeding on *A. aerophoba* or had been kept on these sponges under controlled conditions for several weeks almost always contained the brominated alkaloid aerothionin, which is not detected in *A. aerophoba* but occurs in the sibling species *A. cavernicola* instead. The latter sponge is also accepted as a food source by the gastropod, at least under experimental conditions. The possible origin of aerothionin in *T. perversa* is discussed.

12.1 Introduction

Opisthobranchs of the genus *Tylodina* (Notaspidea, Tylodinoidae, Tylodinidae) inhabit the marine environment at exceedingly distant geographical regions. While *T. corticalis* is found in the South Pacific from Southern Queensland, around the southern Australian coast to south-western Australia, *T. fungina* dwells in the Eastern Pacific from Southern California to the Galapagos Islands (Gabb 1865; Willan 1984, 1987, 1998). The Mediterranean Sea is inhabited by *T. perversa* (Gmelin 1791; Riedl 1983), which occasionally can also be found in the north-eastern Atlantic as far as the British Isles (Gainey and Turk 1997). Interestingly, all *Tylodina* species live in close association with sponges of the order Verongida (Andersen and Faulkner 1972; Willan 1984; Faulkner 1992; Teeyapant et al. 1993). The Mediterranean slug *T. perversa* is usually found

Peter Proksch
Institut für Pharmazeutische Biologie, Heinrich-Heine-Universität
Universitätsstr. 1, 40225 Düsseldorf, Germany

Progress in Molecular and Subcellular Biology
Subseries Marine Molecular Biotechnology
G. Cimino, M. Gavagnin (Eds.): Molluscs

on the yellow demosponge *Aplysina aerophoba* (Riedl 1983; Teeyapant et al. 1993; Doneddu and Manunza 1995). This sponge inhabits mostly sun-exposed hard substrates in shallow depths between less than 1 and up to 15 m (Riedl 1983; Pansini 1997). Often on a single *A. aerophoba* individual, several specimens of *T. perversa* can be found, together with their yellow egg ribbons.

The color of *T. perversa* as well as its mucus and egg ribbons is due to uranidine (Fig. 12.1, structure **5**), the same pigment that is found in the sponge *A. aerophoba* (Cimino et al. 1984; Teeyapant et al. 1993; Cimino and Sodano 1994). This, as well as feeding scars that are usually present concomitantly with the opisthobranchs on the sponge, makes it obvious that *T. perversa* preys on *A. aerophoba*. Often the slugs burrow a cavity into the sponge tissue that is large enough to accommodate the whole gastropod. In this case, their limpet-shaped shell protects them towards the opening of the cavity in the sponge, while outside the sponge the tiny shell is no efficient defense, as it only covers part of the soft body of the mollusc.

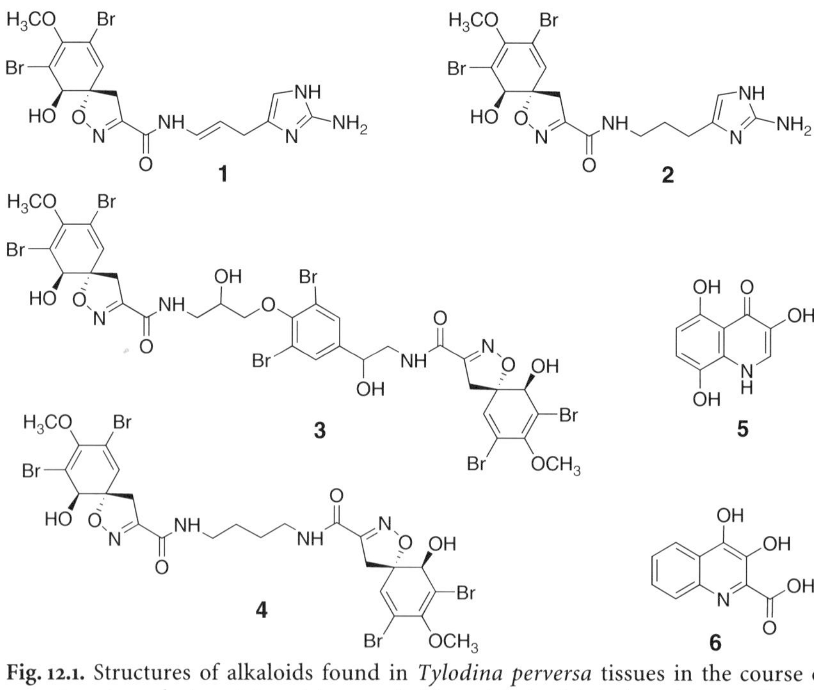

Fig. 12.1. Structures of alkaloids found in *Tylodina perversa* tissues in the course of our experiments: aplysinamisin-1 (**1**), aerophobin-2 (**2**), isofistularin-3 (**3**), aerothionin (**4**), uranidine (**5**), 3,4-dihydroxyquinoline-2-carboxylic acid (**6**)

Sponges of the order Verongida are interesting from a chemical point of view as they accumulate brominated isoxazoline alkaloids (e.g., Fig. 12.1, structures **1–4**) at high concentrations comprising more than 10% of their dry weight (Albrizio et al. 1994; Aiello et al. 1995; Ciminiello et al. 1994a,b, 1995, 1996a,b, 1997, 1999, 2000). These structurally unique secondary metabolites are thought to be biogenetically derived from 3, 5-dibromotyrosine (Tymiak and Rinehart 1981). Whereas the spirocyclohexadienylisoxazoline moiety is identical for most of these alkaloids, the compounds differ by the nature of amine substituents linked to the carbonyl group adjacent to the isoxazoline ring (Fig. 12.1). Several isoxazoline alkaloids from sponges of the genus *Aplysina* (order Verongida) have recently been shown to act as strong feeding deterrents against marine fishes such as *Blennius sphinx* and are thought to play a crucial role in the chemical defense of the sponges against predatory fish (Thoms et al. 2004).

The Mediterranean Sea hosts two species of the order Verongida: *A. aerophoba* and *A. cavernicola*. While the former is abundant in shallow water at locations with high sun exposition, the habitat of the latter is restricted to shadier environments, such as underwater caves or deeper water of about 40 m (Wilkinson and Vacelet 1979; Pansini 1997; Thoms et al. 2003a). Despite these differences in the ecological requirements of the two Mediterranean *Aplysina* species, they share many similarities in their morphology (Vacelet 1959) as well as in their secondary metabolite pattern (Ciminiello et al. 1997; Thoms et al. 2004). There is even controversy as to whether they are really two distinct species or rather two ecotypes belonging to the same species (Voultsiadou-Koukoura 1987; Heim 2003). Nevertheless, some distinctive differences between the two sponges exist with regard both to their chemistry and to their microbiology. Only *A. aerophoba* possesses uranidine (**5**), the characteristic pigment that polymerizes quickly upon air exposure and thereby changes the color of the sponge from yellow to black (Cimino et al. 1984). *A. cavernicola* presents 3,4-dihydroxyquinoline-2-carboxylic acid (**6**) instead, which in comparison to uranidine (**5**) is far more stable (Fattorusso et al. 1971). While many isoxazoline alkaloids such as aerophobin-2 (**2**) and aplysinamisin-1 (**1**) can be found in both species, apparently only *A. cavernicola* accumulates aerothionin (**4**; Ciminiello et al. 1997). Both Mediterranean *Aplysina* species harbor a vast and diverse community of heterotrophic bacteria in their tissues (Friedrich et al. 1999, 2001; Hentschel et al. 2001; Thoms et al. 2003a). *A. aerophoba* is in addition associated with cyanobacteria of the species *Aphanocapsa feldmanii* that thrive in large numbers in its surface tissue (Vacelet 1971; Rützler 1985).

According to Becerro et al. (2003), the Mediterranean opisthobranch *T. perversa* is exclusively found in the zone where *Aplysina aerophoba* is abundant. To our knowledge, it has not yet been reported that *T. perversa* also feeds on the sibling species *A. cavernicola*. Thus, the slug is usually considered to be a specialist predator on *A. aerophoba* (Riedl 1983; Doneddu and Manunza 1995). Becerro et al. (2003) suggested that the cyanobacteria present in the sponge tissue are responsible for the gastropod's feeding choices rather than true sponge compounds, which could explain a preference of *T. perversa* for *A. aerophoba* over the cyanobacteria-free *A. cavernicola* in nature.

Our study on *T. perversa* deals with the secondary metabolite uptake by the opisthobranchs when preying on *Aplysina* sponges. We report on the ecological aspects of this process as well as on the fate of the metabolites in the soft bodies of the slugs. Further aspects examined are the feeding preferences of *T. perversa* and the impact of different prey sponges on the alkaloid pattern in the tissues of the slug.

12.2 Sequestration of Alkaloids from the Prey Sponge by *Tylodina perversa*

As long ago as 1993, when we isolated compounds from *T. perversa* individuals that had been collected from *A. aerophoba* off in the Canary Islands and analyzed the compound structures by ^{1}H NMR, ^{13}C NMR and mass spectrometry, we observed the presence of sponge alkaloids in the opisthobranchs (Teeyapant et al. 1993; Teeyapant 1994). Since the concentrations of these brominated compounds in *T. perversa* appeared too large to account for their presence only in the alimentary tract (30–45%, compared to the total alkaloid concentration in the respective prey sponges), we assumed their accumulation also in organs other than the digestive tissues.

To our surprise, in addition to the secondary metabolites present in *A. aerophoba*, we found the isoxazoline alkaloid aerothionin (Fig. 12.1, structure 4) in crude extracts of *T. perversa* (Teeyapant et al. 1993). Aerothionin is considered a characteristic compound for *A. cavernicola* (Ciminiello et al. 1997), which had not been observed at the collection site of the slugs. This first observation was later confirmed in our subsequent studies on *T. perversa*: irrespective of the fact that the opisthobranchs were always collected from aerothionin-free *A. aerophoba* and irrespective of the absence of *A. cavernicola* at the respective collection sites, nearly all of the *T. perversa* specimens analyzed were found to contain considerable amounts of aerothionin (4; Teeyapant et al. 1993; Ebel et al. 1999; Thoms et al. 2003b).

A subsequent study analyzed in detail the uptake of *A. aerophoba* metabolites by *T. perversa* (Ebel et al. 1999). Opisthobranchs were collected from *A. aerophoba* individuals off the Mediterranean coast at L'Estartit, Spain, and kept along with this sponge in seawater tanks for more than one week. Thereafter, the opisthobranchs were dissected and studied for sequestered sponge-derived alkaloids. The amount of each individual alkaloid in the different tissue samples was determined by HPLC-UV, using calibration curves obtained for the respective isolated compound. The sponge metabolites (the brominated alkaloids as well as the pigment uranidine, 5) were present at high concentrations in the hepatopancreas as well as in the mantle tissue of the *T. perversa* individuals (Table 12.1). Mucus, egg ribbons and feces that had been produced during captivity likewise contained considerable concentrations of these natural products. Of all organs of *T. perversa* investigated in this study, the mantle tissue exhibited the highest total alkaloid concentration (43.3 μmol g^{-1} dry weight compared to a concentration of 62.0 μmol g^{-1} dry weight in *A. aerophoba*), followed by egg ribbons (29.6 μmol g^{-1} dry weight) and the hepatopancreas (24.0 μmol g^{-1} dry weight). Mucus and feces contained brominated alkaloids at a concentration of 21.5 and 16.1 μmol g^{-1} dry weight, respectively.

Table 12.1. Concentrations (μmol g^{-1} dry weight) of alkaloids in various tissues derived from the opisthobranch *Tylodina perversa* and in the sponge *Aplysina aerophoba* collected in L'Estartit (Spain)

	Tylodina perversa					*Aplysina aerophoba*
Compound	mantle	hepatopancreas	egg masses	mucus	feces	
Aplysinamisin-1 (**1**)	2.0 ± 0.6 (4.6 ± 1.4)	4.7 ± 1.9 (19.8 ± 7.8)	1.1 ± 1.1 (3.7 ± 3.9)	0.8 ± 0.3 (3.7 ± 1.3)	1.4 ± 0.1 (8.4 ± 0.5)	16.9 ± 3.5 (27.3 ± 5.6)
Aerophobin-2 (**2**)	34.5 ± 2.9 (79.7 ± 6.6)	9.3 ± 2.5 (38.7 ± 10.6)	26.8 ± 5.3 (90.7 ± 17.9)	16.6 ± 1.6 (77.5 ± 7.5)	6.6 ± 1.0 (41.1 ± 6.1)	32.1 ± 6.6 (51.8 ± 10.6)
Isofistularin-3 (**3**)	0.7 ± 0.7 (1.6 ± 1.5)	9.6 ± 1.4 (40.1 ± 6.0)	0.3 ± 0.2 (0.9 ± 0.7)	0.6 ± 0.3 (2.8 ± 1.6)	7.9 ± 0.6 (48.9 ± 3.5)	13.0 ± 2.1 (20.9 ± 3.3)
Aerothionin (**4**)	6.1 ± 2.0 (14.1 ± 4.6)	0.3 ± 0.3 (1.5 ± 1.1)	1.4 ± 1.1 (4.7 ± 3.7)	3.1 ± 1.1 (16.0 ± 6.7)	0.2 ± 0.3 (1.5 ± 2.2)	n.d. –
Total conc. (mg g^{-1})	24.6 ± 15.2	20.4 ± 6.6	19.2 ± 13.7	12.5 ± 13.6	14.2 ± 13.3	51.2 ± 17.7

Values in brackets represent relative proportions (%) of the respective alkaloids compared to the total brominated alkaloid content in the tissue. Total concentrations comprise additional brominated alkaloids detected in minor amounts and are expressed in mg g^{-1} dry weight to facilitate easier comparison with other chemical studies on opisthobranchs. n.d. - not detected

Comparison of the alkaloid patterns in the different opisthobranch organs analyzed (including mucus, egg ribbons, feces) revealed a distinctive selectivity in the sequestration of the respective isoxazoline alkaloids. Aerophobin-2 (**2**) was the major alkaloid in the mucus, the egg ribbons and in the mantle tissue of *T. perversa*, comprising between 77.5% and 90.7% of all quantified compounds (Table 12.1). In the host sponges, however, this compound made up for only 51.8% of all alkaloids present. In contrast, the isofistularin-3 (**3**) proportion was considerably lower in the opisthobranchs (0.9–2.8%) than in the sponge tissue (20.9%).

Apparently, isofistularin-3 (**3**) was selectively excreted by the opisthobranchs, as in the feces the relative concentration of this compound was 48.9% (total alkaloid content set at 100%) and thus considerably higher than in the sponge tissue. Reasons for a selective excretion of isofistularin-3 (3) versus a selective accumulation of aerophobin-2 (2), however, are yet unclear. Both isoxazoline alkaloids revealed to be equally strong feeding deterrents against the Mediterranean fish species *Blennius sphinx* when tested at their natural concentrations as found in *A. aerophoba* (Thoms et al. 2004). Possibly, the accumulation of isoxazoline alkaloids in exposed tissues such as mantle and egg ribbons as well as in the mucus that is exuded when the opisthobranchs are molested, acts as a chemical defense. Experimental proof for this hypothesis is so far lacking, as the alkaloids have only been tested for feeding deterrent properties at concentrations as found in the sponges, which are approximately twice as high as the alkaloid concentrations in organs (e.g., mantle) of *T. perversa*.

As reported in our first study on *T. perversa* (Teeyapant et al. 1993), our second approach (Ebel et al. 1999) confirmed the presence of aerothionin (4) in *T. perversa*, even though this isoxazoline alkaloid is characteristic for *A. cavernicola* but is not detected in *A. aerophoba*. Separate analysis of the respective opisthobranch tissues revealed the highest concentrations of aerothionin (4) in the mantle (6.1 ± 2.0 µmol g^{-1} dry weight) and in the mucus (3.4 ± 1.4 µmol g^{-1} dry weight), whereas all other tissues analyzed contained this compound at much lower concentrations (Table 12.1). Again no aerothionin (4) was found in *A. aerophoba*, the sponge from which the opisthobranchs had been collected and on which they had preyed under controlled conditions for one week.

12.3 Choice Feeding Experiments with *Tylodina perversa*

To determine whether *A. cavernicola* could serve as a host sponge for *T. perversa* in nature, we performed choice feeding experiments and offered three different sponges including *A. aerophoba*, *A. cavernicola* and (as a control) *Axinella damicornis* to specimens of *T. perversa* held in seawater tanks (Thoms et al. 2003b). Other than the *Aplysina* species, *Axinella damicornis* accumulates alkaloids of the bromopyrrole type. All sponges used for the experiments had a similar yellow color. The opisthobranchs were placed at a starting point equally distanced at about 15 cm from each of the three sponges. We scored a selection when they had crawled completely onto one of the sponges. The sponges were randomly placed in the tanks at each trial to avoid a possible influence of light and flow conditions on our data.

In only 8.3% of a total of 48 individual choice feeding experiments did the opisthobranchs show a preference for *A. damicornis*, whereas in over 90% there was a clear preference for one of the *Aplysina* sponges. However, *T. perversa* apparently did not distinguish between the two *Aplysina* species, as preference for *A. cavernicola* (43.8%) was almost equal to that for *A. aerophoba* (47.9%). This result contradicts the observations of Becerro et al. (2003), who found that *T. perversa* exclusively selects *A. aerophoba* when given the choice between this sponge and *A. cavernicola*.

12.4 Impact of Different *Aplysina* Sponges on the Alkaloid Patterns in *Tylodina perversa*

The acceptance of *A. cavernicola* as a host sponge and food source for *T. perversa* enabled us to perform long-term feeding experiments with the opisthobranchs under controlled conditions and to investigate the impact of the respective prey sponge on the alkaloid profiles in the opisthobranchs. Specimens of *T. perversa* collected in the Mediterranean Sea close to Banyuls-sur-mer, France, were kept for more than two weeks in tanks together with individuals of *A. aerophoba*, followed by an additional feeding period of two weeks on *A. cavernicola* (Thoms et al. 2003b). In a second experiment, we kept *T. perversa* for a period of more than five weeks exclusively on *A. aerophoba*. At the end of each experiment, we dissected the slugs for subsequent HPLC analysis of the mantle and the hepatopancreas. In addition, egg ribbons that had been produced during the experiment were likewise investigated by HPLC. The alkaloids in the respective samples were quantified, based on calibration curves obtained with previously isolated compounds.

After two weeks feeding on *A. cavernicola*, the alkaloid profile of *T. perversa* had clearly shifted towards the profile of this species when compared to the alkaloid patterns of control opisthobranchs that were dissected and analyzed immediately after collection from *A. aerophoba*. The most pronounced similarity of the alkaloid pattern from specimens of *T. perversa* to the pattern of *A. cavernicola* was observed for the hepatopancreas, where high concentrations of aerothionin (Fig. 12.1, structure **4**; $17.1 \pm 6.7\ \mu mol\ g^{-1}$ dry weight) and the *A. cavernicola* pigment 3,4-dihydroxyquinoline-2-carboxylic acid (**6**; $45.6 \pm 14.8\ \mu mol\ g^{-1}$ dry weight) were found (Fig. 12.2).

Moreover, for specimens of *T. perversa* that had been kept on *A. aerophoba* for more than five weeks, the closest similarity of the alkaloid pattern (compared to the host *A. aerophoba*) was also found for the hepatopancreas. Here, the overall concentration of isoxazoline alkaloids in the hepatopancreas amounted for $110.5 \pm 36.5\ \mu mol\ g^{-1}$ dry weight with aplysinamisin-1 (**1**) as the major constituent ($57.1 \pm 20.1\ \mu mol\ g^{-1}$ dry weight). Thus, the total alkaloid content as well as the amount of aplysinamisin-1 in the hepatopancreas of *T. perversa* feeding on *A. aerophoba* from Banyuls-sur-mer (Fig. 12.2) was considerably higher than found in our previous study involving individuals that had been collected at L'Estartit, Spain and kept on *A. aerophoba* for one week (Table 12.1; Ebel et al. 1999). This finding probably reflects intra-specific differences of alkaloid concentrations and patterns of *A. aerophoba* individuals collected at different geographical sites.

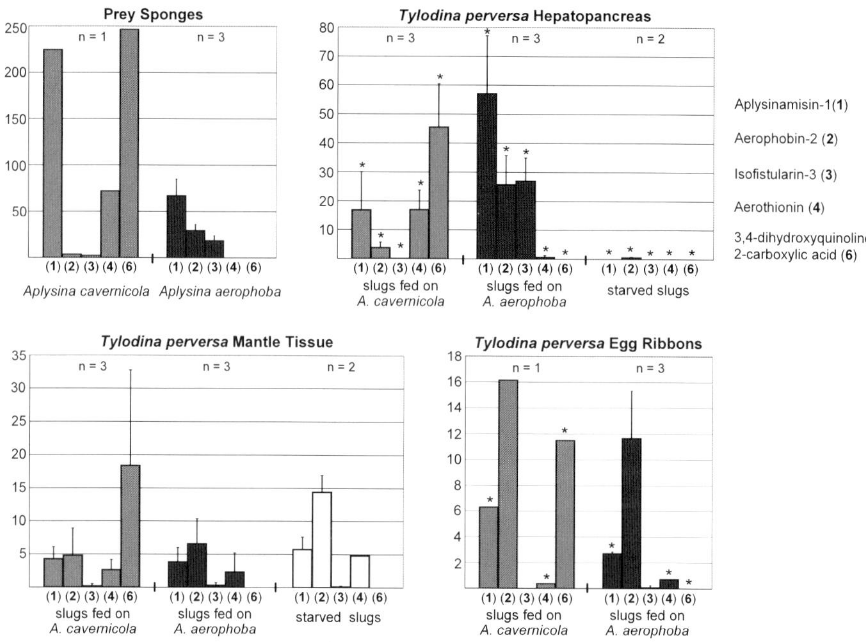

Fig. 12.2. Concentrations (μmol g^{-1} dry weight) of alkaloids in various tissues derived from the opisthobranch *Tylodina perversa* (after two weeks of feeding on *A. cavernicola*, after five weeks of feeding on *A. aerophoba*, after two weeks of starvation) and in their prey sponges. Vertical *bars* represent standard deviation. Numbers of compounds refer to Fig. 12.1. *Asterisks* indicate significant difference in different treatments of the experiment ($P < 0.05$)

The results obtained in the feeding experiments demonstrate a close relationship between the alkaloid profile of the hepatopancreas of *T. perversa* and the alkaloid profile of their most recent host sponge. This close similarity was further exemplified by an additional experiment where specimens of *T. perversa* were starved for two weeks following a five-week feeding period on *A. aerophoba* (Thoms 2004). After starving, the hepatopancreas of the slugs contained almost no alkaloids (total content of isoxazoline alkaloids 0.6 ± 0.2 μmol g^{-1} dry weight), thus demonstrating that dietary alkaloids have to be replenished continuously.

In analogy to the metabolite profiles of the hepatopancreas, the profiles of the mantle tissue of individuals that had fed on *A. cavernicola* for a period of two weeks prior to dissection differed from that of opisthobranchs reared on *A. aerophoba* with regard to the pigments of the respective prey sponges found therein. While the former contained the *A. cavernicola* pigment 3,4-dihydroxyquinoline-2-carboxylic acid (**6**) at concentrations of 18.4 ± 14.3 µmol g^{-1} dry weight, in the latter the *A. aerophoba* pigment uranidine (**5**) was found instead; but due to its instability upon contact with air, it could not be accurately quantified. However, with regard to their isoxazoline alkaloid profiles and contents, the mantle tissues of the two differently fed groups of opisthobranchs were almost identical. This also held true for the aerothionin (**4**) content, which in individuals that had preyed for more than two weeks on aerothionin-yielding *A. cavernicola* amounted to 2.7 ± 1.5 µmol g^{-1} dry weight, whereas in those that had fed on *A. aerophoba* the aerothionin (**4**) concentration amounted to 2.4 ± 2.8 µmol g^{-1} dry weight (Fig. 12.2). Interestingly, no depletion of isoxazoline alkaloids was observable in the mantle tissue of *T. perversa* individuals that were starved for more than two weeks following feeding on *A. aerophoba*. The total isoxazoline alkaloid content detected in starved opisthobranchs was even higher (25.1 ± 4.6 µmol g^{-1} dry weight) than in the mantle tissue of gastropods of the two other groups (13.1 ± 7.7 and 11.9 ± 6.5 µmol g^{-1} dry weight, respectively). The aerothionin (**4**) concentration in starved *T. perversa* individuals amounted to 4.8 ± 0.1 µmol g^{-1} dry weight, compared to 2.4 ± 2.8 µmol g^{-1} dry weight in slugs that had fed on *A. aerophoba* for more than five weeks and 2.7 ± 1.5 µmol g^{-1} dry weight in those that fed for over two weeks exclusively on *A. cavernicola* (Fig. 12.2).

Similar to their mantle tissues, the egg ribbons produced by opisthobranchs following feeding on *A. aerophoba* or *A. cavernicola* under controlled conditions showed no appreciable changes in isoxazoline alkaloid profiles that could reflect differences in alkaloid composition of their host sponges. However, the egg ribbons produced by the slugs after feeding on *A. cavernicola* for more than two weeks contained considerable amounts of the pigment 3,4-dihydroxyquinoline-2-carboxylic acid (**6**), characteristic for this sponge. As the starved opisthobranchs had not produced any egg ribbons during captivity, it was not possible to analyze the metabolite pattern of eggs.

12.5 Conclusions

In the course of our studies on *T. perversa*, we were able to show that the opisthobranchs not only tolerate the defense metabolites of their prey sponges but even sequester them (Teeyapant et al. 1993; Ebel et al. 1999; Thoms et al. 2003b). Unequivocal evidence for this finding was provided by the observation that in all analyzed tissues of the slugs the pigment of the respective sponge on which *T. perversa* had been feeding previously [uranidine (**5**) in the case of *A. aerophoba*, 3,4-dihydroxyquinoline-2-

carboxylic acid (**6**) in the case of *A. cavernicola*] was found (Thoms et al. 2003b). Generally in gastropods, the hepatopancreas produces digestive enzymes and reabsorbs nutrients from the gut (Götting 1996). Chemical analysis of this organ from specimens of *T. perversa* kept in captivity revealed an alkaloid pattern essentially identical to that of the slug's most recent prey sponge (Thoms et al. 2003b). Moreover, an experiment involving starved *T. perversa* individuals demonstrated that the dietary alkaloids have to be replenished continuously in order to keep up their normal levels in the hepatopancreas, as after two weeks of starvation this organ was found to be almost alkaloid free (Thoms 2004).

Sequestration of sponge alkaloids turned out to be far more selective in the mantle tissue and in the egg ribbons of *T. perversa* than in the opisthobranch's hepatopancreas. When specimens of *T. perversa* were kept on *A. cavernicola* after an initial feeding period on *A. aerophoba*, the mantle tissue displayed an enrichment of aerophobin-2 (**2**; present in both *Aplysina* species) and a concomitant depletion of isofistularin-3 (**3**; only present in *A. aerophoba*) when compared to the alkaloid profile of *A. aerophoba* (Ebel et al. 1999; Thoms et al. 2003b). The enrichment of aerophobin-2 (**2**) could possibly be due to a conversion from aplysinamisin-1 (**1**) within the slug, as the structures of these compounds differ only in a double bond. Isofistularin-3 is possibly excreted selectively by the slugs as it is found enriched in the feces (Ebel et al. 1999). It seems likely that the selective sequestration of sponge alkaloids by *T. perversa* serves an ecological purpose as: (a) the alkaloids are preferentially accumulated in exposed tissues such as mantle and egg ribbons and are secreted as part of the mucus when the slugs are molested and (b) the alkaloids were shown to possess pronounced feeding deterrent properties when tested against the Mediterranean fish *Blennius sphinx* (Thoms et al. 2004). However, experimental proof for the feeding deterrent activity of *A. aerophoba* compounds at concentrations as found in the tissues of the slugs is still lacking. Nevertheless, Becerro et al. (2003) were able to show that a crude extract of *T. perversa*, as well as extract of its egg ribbons, had a stronger anti-feeding effect towards the Mediterranean fish *Chromis chromis* than a crude extract of *A. aerophoba*. As the slug tissues hold considerably lower concentrations of the isoxazoline alkaloids than the sponge, the involvement of other so far unknown compounds for their chemical defense is likely. This is also indicated by the study of Becerro et al. (2003), who found that *T. perversa* as well as its egg ribbons yielded a higher percentage of crude extract per dry weight than *A. aerophoba*.

Interesting but still unresolved is the origin of aerothionin (**4**) that is almost always found in *T. perversa* individuals collected from *A. aerophoba* in the wild. According to a number of recent reviews on the chemical ecology of opisthobranchs (e.g., Avila 1995; Cimino et al. 1999; Cimino and Ghiselin 1999, 2001; Gavagnin et al. 2000), three different ecological scenarios can explain the accumulation of secondary metabolites in their tissues:

1. De novo biosynthesis of the compounds by the slug itself (or by its associated microorganisms)
2. Biotransformation of diet-derived compounds
3. Accumulation of diet-derived compounds

Prominent examples for compounds that are products of the secondary metabolism of opisthobranchs formed by de novo biosynthesis are polypropionates found in tissue of the sacoglossans *Cyerce cristallina*, *Elysia viridis* and *Ercolania funerea* (Vardaro et al. 1991; Di Marzo et al. 1991, 1993; Gavagnin et al. 1994). However, in the case of *T. perversa*, de novo synthesis of aerothionin (4) seems rather unlikely, as the isoxazoline alkaloids accumulated in *Aplysina* sponges (e.g., Fig. 12.1, structures 1–4) are structurally unique and almost exclusively found in sponge species of the order Verongida (Albrizio et al. 1994; Aiello et al. 1995; Ciminiello et al. 1994a,b, 1995, 1996a,b, 1997, 1999, 2000).

The opisthobranch *Ascobulla ulla* converts the algal-derived compound caulerpenyne into ascobullin-a and -b during detoxification of these potentially harmful metabolites (Gavagnin et al. 2000). The Mediterranean slug *Hypselodoris orsini* detoxifies scalaradial sequestered from the sponge *Cacospongia mollior* by transforming it into the less toxic deoxoscalarin (Cimino et al. 1993). These examples show that transformation processes of diet-derived metabolites occur in marine opisthobranchs and could possibly explain the occurrence of aerothionin (4) in *T. perversa* after feeding on *Aplysina aerophoba*. However, if aerothionin (4) is formed in *T. perversa* from other isoxazoline alkaloids, such as isofistularin-3 (3) or aerophobin-2 (2), this biotransformation reaction would not only involve cleavage of the putative precursors at the amide bond(s) but also re-assembly of two spirocyclohexadienylisoxazoline moieties and insertion of one putrescine unit, thereby giving rise to aerothionin (4). As there is no experimental evidence for this rather complex bioconversion in the slugs, this hypothesis remains speculative at the moment.

Long-term storage of dietary metabolites in gastropods as observed in this study for starved specimens of *T. perversa* is no isolated case for gastropods. A further example is the terrestrial slug *Chondrina clienta* that sequesters the anthraquinone parietin from its diet, the lichen *Xanrhoria parietina* (Hesbacher et al. 1995). Even after four weeks of feeding on a parietin-free diet, this anthraquinone could be detected in the tissue of slugs that had been collected while feeding on the lichens. This example could indicate that the sponge *A. cavernicola* is the actual source of aerothionin (4) found in *T. perversa*. The observation that, unlike *A. aerophoba*, no specimens of *A. cavernicola* were observed at the sampling sites of *T. perversa* is not necessarily a valid argument against this assumption but rather reflects differences in the ecology of both sponges. Whereas *A. aerophoba* grows exposed and can be found even at low water depths (around 1 m) *A. cavernicola* generally lives hidden in caves or at greater depths and is thus harder to find. Further corroboration for the hypothesis that *A. cavernicola* is the actual source of aerothionin (4) found in *T. perversa* was provided by our choice feeding experiment which revealed that both *A. aerophoba* and *A. cavernicola* are similarly attractive to the gastropods (Thoms et al. 2003b). Thus, *A. cavernicola* could also be a host for *T. perversa* in nature, even though the gastropods have so far not been observed feeding on this particular sponge.

However, some inconsistencies remain if *A. cavernicola* is regarded as the source for aerothionin (4) found in *T. perversa*. First, in the long-term feeding experiment with *T. perversa* on *A. cavernicola*, we demonstrated that, in addition to the isoxazoline alkaloids, the pigment (6) is sequestered by the gastropods. However, the *A. cavernicola* pigment (6) was not found in any of the individuals that had been collected in their natural environment and that contained aerothionin (4). Second, while the amount of sequestered parietin in the above mentioned example *Chondrina clienta* decreased over time on a parietin-free diet (Hesbacher et al. 1995), in specimens of *T. perversa* that had preyed on *A. aerophoba* for more than five weeks under controlled conditions, the aerothionin (4) content in mantle tissue remained largely unchanged (Thoms et al. 2003b). This result is even more surprising, as in the course of the feeding experiment the slugs had produced large amounts of aerothionin-containing egg ribbons, which would have been expected to result in a measurable loss of aerothionin (4). The situation is even more complex, as recently some *Aplysina* specimens were collected from Croatia that yielded aerothionin (4) in addition to the *A. aerophoba* pigment uranidine (5; Heim 2003; Thoms 2004). The true taxonomic status (chemical race, new species?) of these chemically most unusual *Aplysina* specimens is unclear, as is the question whether similar sponges might also occur elsewhere in the Mediterranean Sea where they could be accessible for *T. perversa*. Thus, the origin of aerothionin in *T. perversa* is still an unsolved question and warrants further studies on the fascinating prey–predator relationship of *Tylodina* gastropods and *Aplysina* sponges.

Acknowledgments. We would like to thank PD Dr. Ute Hentschel (Universität Würzburg, Germany), Prof. Franz Brümmer (Universität Stuttgart, Germany) and Prof. Werner E.G. Müller (Universität Mainz) for their cooperation and stimulating discussions. This study was supported by a grant of the BMBF (project: Center of Excellence BIOTECmarin). C.T. gratefully acknowledges a scholarship (Feodor Lynen Fellowship) from the Alexander von Humboldt-Stiftung, Bonn.

References

Aiello A, Fattorusso E, Menna M, Pansini M (1995) Chemistry of Verongida sponges 5. Brominated metabolites from the Caribbean sponge *Pseudoceratina* sp. Biochem Syst Ecol 23:377–381

Albrizio S, Ciminiello P, Fattorusso E, Magno S, Pansini M (1994) Chemistry of Verongida sponges 1. Constituents of the Caribbean sponge *Pseudoceratina crassa*. Tetrahedron 50:783–788

Andersen RJ, Faulkner DJ (1972) Antibiotic from marine organisms of the Gulf of California. Book of Abstracts of the Conference "Food and Drugs from the Sea", Rhode Island, pp. 1111–1115.

Avila C (1995) Natural products from opisthobranch molluscs: a biological review. Oceanogr Mar Biol Annu Rev 33:487–559

Becerro MA, Turon X, Uriz MJ, Templado J (2003) Can a sponge feeder be a herbivore? *Tylodina perversa* (Gastropoda) feeding on *Aplysina aerophoba* (Demospongiae). Biol J Linn Soc 78:429–438

Ciminiello P, Costantino V, Fattorusso E, Magno S, Mangoni A, Pansini M (1994a) Chemistry of Verongida sponges 2. Constituents of the Caribbean sponge *Aplysina fistularis* forma *fulva*. J Nat Prod 57:705–712

Ciminiello P, Fattorusso E, Magno S, Pansini M (1994b) Chemistry of Verongida sponges 3. Constituents of a Caribbean *Verongula* sp. J Nat Prod 57:1564–1569

Ciminiello P, Fattorusso E, Magno S, Pansini M (1995) Chemistry of Verongida sponges 4. Comparison of the secondary metabolite composition of several specimens of *Pseudoceratina crassa*. J Nat Prod 58:689–696

Ciminiello P, Fattorusso E, Magno S, Pansini M (1996a) Chemistry of Verongida sponges 6. Comparison of the secondary metabolic composition of *Aplysina insularis* and *Aplysina fulva*. Biochem Syst Ecol 24:105–107

Ciminiello P, dell'Aversano C, Fattorusso E, Magno S, Carrano L, Pansini M (1996b) Chemistry of Verongida sponges 7. Bromocompounds from the Caribbean sponge *Aplysina archeri*. Tetrahedron 52:9863–9868

Ciminiello P, Fattorusso E, Forino M, Magno S, Pansini M (1997) Chemistry of Verongida sponges 8. Bromocompounds from the Mediterranean sponges *Aplysina aerophoba* and *Aplysina cavernicola*. Tetrahedron 53:6565–6572

Ciminiello P, dell'Aversano C, Fattorusso E, Magno S, Pansini M (1999) Chemistry of Verongida sponges 9. Secondary metabolite composition of the Caribbean sponge *Aplysina cauliformis*. J Nat Prod 62:590–593

Ciminiello P, dell'Aversano C, Fattorusso E, Magno S, Pansini M (2000) Chemistry of Verongida sponges 10. Secondary metabolite composition of the Caribbean sponge *Verongula gigantea*. J Nat Prod 63:263–266

Cimino G, Ghiselin MT (1999) Chemical defense and evolutionary trends in biosynthetic capacity among dorid nudibranchs (Mollusca: Gastropoda: Opisthobranchia). Chemoecology 9:187–207

Cimino G, Ghiselin MT (2001) Marine natural products chemistry as an evolutionary narrative. In: McClintock JB, Baker BJ (eds) Marine chemical ecology. CRC, Washington, DC, pp 115–156

Cimino G, Sodano G (1994) Transfer of sponge secondary metabolites to predators. In: van Soest RWM, van Kempen TMG, Braekman JC (eds) Sponges in space and time. Balkema, Rotterdam, pp 459–472

Cimino G, de Rosa S, de Stefano S, Spinella A, Sodano G (1984) The zoochrome of the sponge *Verongia aerophoba* ("Uranidine"). Tetrahedron Lett 25:2925–2928

Cimino G, Fontana A, Gimenez F, Marin A, Mollo E, Trivellone E, Zubia E (1993) Biotransformation of a dietary sesterterpenoid in the Mediterranean nudibranch *Hypselodoris orsini*. Experientia 49:582–586

Cimino G, Fontana A, Gavagnin M (1999) Marine opisthobranch mollusks: chemistry and ecology in sacoglossans and dorids. Curr Org Chem 3:327–372

Di Marzo V, Vardaro RR, Depetrocellis L, Villani G, Minei R, Cimino G (1991) Cyercenes, novel pyrones from the Ascoglossan mollusk *Cyerce cristallina* – tissue distribution, biosynthesis and possible involvement in defense and regenerative processes. Experientia 47:1221–1227

Di Marzo V, Marin A, Vardaro RR, Depetrocellis L, Villani G, Cimino G (1993) Histological and biochemical bases of defense mechanisms in 4 species of Polybranchioidea ascoglossan mollusks. Mar Biol 117:367–380

Doneddu M, Manunza B (1995) La molluschi gasteropodi rinvenuti in associazione con *Verongia aerophoba* (Schmidt). Conchiglia 275:4–6

Ebel R, Marin A, Proksch P (1999) Organ-specific distribution of dietary alkaloids in the marine opisthobranch *Tylodina perversa*. Biochem Syst Ecol 27:769–777

Fattorusso E, Forenza S, Minale L, Sodano G (1971) Isolation of 3,4-dihydroxyquinoline-2-carboxylic acid from the sponge *Aplysina aerophoba*. Gazz Chim Ital 101:104–105

Faulkner DJ (1992) Chemical defenses of marine mollusks. In: Paul VJ (ed) Ecological roles of marine natural products. Comstock, Ithaca, pp 119–163

Friedrich AB, Merkert H, Fendert T, Hacker J, Proksch P, Hentschel U (1999) Microbial diversity in the marine sponge *Aplysina cavernicola* (formerly *Verongia cavernicola*) analyzed by fluorescence in situ hybridization (FISH). Mar Biol 134:461–470

Friedrich AB, Fischer I, Proksch P, Hacker J, Hentschel U (2001) Temporal variation of the microbial community associated with the Mediterranean sponge *Aplysina aerophoba*. FEMS Microbiol Ecol 38:105–113

Gabb WM (1865) Description of new species of marine shells from the coast of California. Proc Calif Acad Sci 3:182–190

Gainey PA, Turk SM (1997) More records of *Charonia lampas* (L.), a recent record of Pteria *hirundo* (L.) and a marine slug *Tylodina perversa* Gmelin new to Britain. J Conchol 36:78–79

Gavagnin M, Marin A, Mollo E, Crispino A, Villani G, Cimino G (1994) Secondary metabolites from Mediterranean Elysioidea – origin and biological role. Comp Biochem Physiol B Biochem Mol Biol 108:107–115

Gavagnin M, Mollo E, Montanaro D, Ortea J, Cimino G (2000) Chemical studies of Caribbean sacoglossans: dietary relationships with green algae and ecological implications. J Chem Ecol 26:1563–1578

Gmelin JRV (1791) Caroli A Linne systema naturae per regna tria naturae, vol 1, pt 6, cl 6. Vermes, Leipzig, pp 3021–3910

Götting (1996) Mollusca. In: Westheide W, Rieger RM (eds) Einzeller und Wirbellose. Spektrum, Heidelberg, pp 276–330

Heim I (2003) MSc thesis, University of Stuttgart, Stuttgart

Hentschel U, Schmid M, Wagner M, Fieseler L, Gernert C, Hacker J (2001) Isolation and phylogenetic analysis of bacteria with antimicrobial activities from the Mediterranean sponges *Aplysina aerophoba* and *Aplysina cavernicola*. FEMS Microbiol Ecol 35:305–312

Hesbacher S, Baur B, Baur A, Proksch P (1995) Sequestration of lichen compounds by three species of terrestrial snails. J Chem Ecol 21:233–246

Pansini M (1997) Effects of light on the morphology, distribution and ecology of some Mediterranean sponges. Biol Mar Mediterr 4:74–80

Riedl H (1983) Fauna und Flora des Mittelmeeres. Parey, Hamburg

Rützler K (1985) Associations between Caribbean sponges and photosynthetic organisms. In: Rützler K (ed) New perspectives in sponge biology. Smithsonian Institution, Washington, DC, pp 455–466

Teeyapant R (1994) Brominated secondary metabolites of the marine sponge *Verongia aerophoba* Schmidt and the sponge feeding gastropod *Tylodina perversa* Gmelin: identification, biological activities and biotransformation. Ph.D. thesis, University of Würzburg, Würzburg

Teeyapant R, Kreis P, Wray V, Witte L, Proksch P (1993) Brominated secondary compounds from the marine sponge *Verongia aerophoba* and the sponge feeding gastropod *Tylodina perversa*. Z Naturforsch 48c:640–644

Thoms C (2004) Alkaloide in Schwämmen der Gattung *Aplysina* – chemische Ökologie und assoziierte Mikroorganismen. Ph.D. thesis, University of Düsseldorf, Düsseldorf

Thoms C, Horn M, Wagner M, Hentschel U, Proksch P (2003a) Monitoring microbial diversity and natural product profiles of the sponge *Aplysina cavernicola* following transplantation. Mar Biol 142:685–692

Thoms C, Ebel R, Hentschel U, Proksch P (2003b) Sequestration of dietary alkaloids by the spongivorous marine mollusc *Tylodina perversa*. Z Naturforsch 58c:426–432

Thoms C, Wolff M, Padmakumar K, Ebel R, Proksch P (2004) Chemical defense of Mediterranean sponges *Aplysina cavernicola* and *Aplysina aerophoba*. Z Naturforsch 59c:113–122

Tymiak AA, Rinehart KL Jr (1981) Biosynthesis of dibromotyrosine-derived antimicrobial compounds by the marine sponge *Aplysina fistularis*. J Am Chem Soc 103:6763–6765

Vacelet J (1959) Répartition générale des eponges cornees de la region de Marseille et de quelques stations méditerranées. Rec Trav Stn Mar Endoume 26:39–101

Vacelet J (1971) Electron-microscopic study on association between blue-green-algae (Chroococcales) and sponge of genus *Verongia*. J Microsc 12:363–380

Vardaro RR, Dimarzo V, Crispino A, Cimino G (1991) Cyercenes, novel polypropionate pyrones from the autotomizing Mediterranean mollusk *Cyerce cristallina*. Tetrahedron 47:5569–5576

Voultsiadou-Koukoura E (1987) Some remarks on the Mediterranean species of the genus *Aplysina* (Demospongia, Verongida). In: Vacelet J, Boury-Esnault N (eds) Taxonomy of Porifera. (NATO ASI Ecological Series G, vol 13) Springer, Berlin Heidelberg New York, pp 275–279

Wilkinson CR, Vacelet J (1979) Transplantation of marine sponges to different conditions of light and current. J Exp Mar Biol Ecol 37:91–104

Willan RC (1984) A review of diets in the Notaspidea (Mollusca: Opisthobranchia). J Malacol Soc Aust 6:125–142

Willan RC (1987) Phylogenetic systematics of the Notaspidea (Opisthobranchia) with reappraisal of families and genera. Am Malacol Bull 5:215–241

Willan RC (1998) Order Notaspidea. In: Beesley PL, Ross GJB, Wells A (eds) Mollusca: the southern synthesis (Fauna of Australia, vol 5, part B) CSIRO, Melbourne, pp 977–980

Skin Chemistry of Nudibranchs from the West Coast of North America

R.J. Andersen, K. Desjardine, K. Woods

Abstract. The Pacific coastline of North America extends from Alaska in the north to Panama in the south. Chemical studies of skin extracts from nudibranchs collected along this coast have resulted in the isolation of over 100 chemically diverse secondary metabolites. The majority of the compounds are terpenoids, but polyketides, steroids, and alkaloids have also been found. Observations of geographic variation in metabolite content and stable isotope-feeding experiments have provided information about the de novo biosynthetic or dietary sequestration origins of the skin extract metabolites.

13.1 Introduction

Nudibranchs are some of the most beautiful animals found in the oceans. Their delicate bodies are usually brightly colored, often in striking multihued geometric patterns (Behrens 1980). Nudibranchs are missing the hard shell used by most of their molluscan relatives for protection, which leaves their sensory rhinophores and oxygen-gathering branchial plumes decoratively exposed on their dorsums. They have a large foot that provides locomotion, but only at speeds that are slow relative to potential predators. Nudibranchs tend to be found in shallow water habitats, where they frequently sit out in the open blatantly advertising their vulnerability. Despite their apparent lack of physical attributes and behavioral patterns suited for defensive purposes, nudibranchs have few documented predators (Thompson 1960a,b; Edmunds 1966, 1968). Astute field observations and some simple antifeedant experiments led marine biologists to propose that chemicals provide an invisible protective armor for these soft-bodied molluscs. In his book *Between Pacific Tides* (Ricketts et al. 1968), the naturalist Ricketts wrote that "Many nudibranchs, but especially the dorids, have a penetrating fruity odor that is pleasant when mild but nauseating when concentrated. Undoubtedly, this odor is one of the reasons why nudibranchs seem to be let strictly alone by predatory animals."

Raymond Andersen
Departments of Chemistry and Earth & Ocean Sciences, University of British Columbia
V6T 1Z1 Vancouver, B C., Canada

Progress in Molecular and Subcellular Biology
Subseries Marine Molecular Biotechnology
G. Cimino, M. Gavagnin (Eds.): Molluscs

In 1963, Johannes reported that the dorid nudibranch *Phyllidia varicosa* secretes a strong and unusual-smelling, heat-stable substance, lethal to fish and crustaceans (Johannes 1963). Scheuer's group at the University of Hawaii, inspired by Johannes' observation, undertook a chemical study of the exudates of *P. varicosa*, resulting in the identification of the novel sesquiterpenoid 9-isocyanopupukeanane (**1**; Burreson et al. 1975). They also showed that *P. varicosa* sequesters 9-isocyanopupukeanane from the sponge *Hymeniacidon* sp., which is part of the nudibranch's diet. This pioneering investigation provided the first example of a secondary metabolite defensive allomone from a nudibranch and revealed that nudibranchs can acquire their chemical defenses from sponges. Following Scheuer's report on *P. varicosa*, other groups started to investigate the secondary metabolites present in nudibranch skin extracts and their origins. A second milestone in the study of nudibranch chemical defenses came in 1983 when Cimino's group demonstrated with radioisotope-feeding experiments that the dorid *Dendrodoris limbata* elaborates the drimane sesquiterpenoid polygodial (**2**) and structural analogs via de novo biosynthesis (Cimino et al. 1983).

The Pacific coastline of North America extends from Alaska in the north to Panama in the south. It spans arctic, temperate, and tropical climatic zones encompassing a wide range of near-shore water temperatures and marine habitats. There are more than 100 described species and many more undescribed species of nudibranchs from these waters (Behrens 1980, 2004). Chemical studies of nudibranchs from this coast started in the early 1980s at two locations. Faulkner's group at the Scripps Institution of Oceanography examined species collected in La Jolla and the Gulf of California, while our group at the University of British Columbia investigated species collected from Vancouver Island (Barkley Sound) and north along the mainland British Columbia (BC) coast from Vancouver to southern Alaska. Many of the species studied by the two groups have ranges that extend from BC to southern California, which facilitated comparisons of their skin chemistry at the two locales and provided data on geographic variation in metabolite content and its relationship to the origin of the compounds.

This chapter will provide a summary of the chemical structures and origins of the secondary metabolites reported to date from skin extracts of nudibranchs collected on the west coast of North America. Nudibranchs are in the Phylum Mollusca, Class Gastropoda, Subclass Opisthobranchiata, Order Nudibranchia, as shown in Fig. 13.1. Secondary metabolites have been reported from northeastern Pacific nudibranchs in the two suborders Doridina and Dendronotina, with the majority of the compounds isolated from the Doridina.

Phylum:	Mollusca
Class:	Gastropoda
Subclass:	Opisthobranchia
Order:	Nudibranchia

Suborder	family	subfamily	genus	species
Doridina	Onchidorididae		*Acanthodoris*	*nanaimoensis; hudsoni*
			Adalaria	
	Dendrodorididae		*Doriopsilla*	*albopunctata; janaina*
			Dendrodoris	*krebsii*
	Dorididae		*Aldisa*	*sanguinea cooperi*
			Anisodoris	*nobilis*
			Archidoris	*ohdneri; montereyensis*
			Diaulula	*sandiegensis*
	Chromodorididae		*Cadilina*	*luteomarginata*
			Chromodoris	*marislae; norrisi; sedna; macfarlandi*
			Hypselodoris	*agassizi; ghiselini; califoriensis; porterae*
			Sclerodoris	*tanya*
	Polyceridae	Polycerinae	*Polycera*	*tricolor*
		Triophinae	*Triopha*	*catalinae*
		Nembrothinae	*Roboastra*	*tigris*
			Tambja	*elitra; abdere*
Dendronotina	Tritoniidae		*Tochuina*	*tetraquetra*
	Tethydidae		*Melibe*	*leonina*

Fig. 13.1. Taxonomy of northeastern Pacific nudibranchs (Behrens 1980, 2004)

13.2 Secondary Metabolites Reported from Northeastern Pacific Nudibranchs

Acanthodoris nanaimoensis is a small pleasant-smelling nudibranch that is found in rocky subtidal habitats from Vancouver Island to Shell Beach, California. Three sesquiterpenoid aldehydes, nanaimoal (**3**; Ayer et al. 1984a), acanthodoral (**4**), and isoacanthodoral (**5**; Ayer et al. 1984b), representing the first examples of the nanaimoane, acanthodorane, and isoacanthodorane carbon skeletons, were isolated from *A. nanaimoensis* skin extracts. The structure of nanaimoal (**3**) was proposed on the basis of spectroscopic analysis and confirmed by synthesis of the *p*-bromophenylurethane (**6**), which was identical by NMR and MS comparison with the same derivative prepared from the reduced natural product. Single crystal X-ray diffraction analysis of the *p*-bromophenylurethane derivatives of the alcohols obtained by reducing the natural products was used to elucidate the structures of acanthodoral (**4**) and isoacanthodoral (**5**). *A. hudsoni*, which has a similar distribution range but is much less abundant than *A. nanaimoensis* in BC waters, has a similar fruity smell; and analysis of its skin extracts revealed the presence of the same three sesquiterpenoid aldehydes **3**, **4**, and **5** (Kubanek 1998). Interestingly, the relative abundance of nanaimoal (**3**), isoacanthodoral (**5**), and acanthodoral (**4**) in *A. nanaimoensis* extracts is 40:10:1, whereas the ratio in *A. hudsoni* extracts is 1:10:40, suggesting that the nudibranchs might send species-specific signals by varying the ratio of the same three chemical components.

Skin extracts of an undescribed species of *Adalaria* collected from the surface of shallow water kelp fronds in Copper Cove BC were found to contain a mixture of known sterol peroxides and the new compound **7** (Stonard et al. 1980). As with other sterol peroxides reported from marine invertebrates, the *Adalaria* compounds were isolated as mixtures of the 5α,8α and 5β,8β cyclic peroxides. Specimens of *Aldisa sanguinea cooperi* found in Barkley Sound, BC are normally deeply embedded in the sponge *Anthoarcuata graceae*, from which they obtain cryptic pigmentation and nutrition. The steroids **8** and **9**, having bile-acid side-chains, were isolated from skin extracts of Barkley Sound specimens of *A. sanguinea cooperi*, along with cholestenone (**10**) and several cholestenone analogs with methylated and unsaturated side-chains (Ayer and Andersen 1982). The sponge *A. graceae* contained the same cholestenone mixture found in *A. sanguinea cooperi*, but the sponge did not contain steroids **8** and **9**. A standard goldfish bioassay was used to test **9** and **10** for antifeedant properties. The acid **9** effectively inhibited feeding, while cholestenone (**10**) was inactive. It was proposed that the nudibranch is obtaining an inactive metabolite **10** from its diet and chemically modifying it to produce an active antifeedant **9**.

The large nudibranch *Anisodoris nobilis* is commonly called the "speckled sea lemon" because of its bright yellow-orange color and its fruity odor. Skin extracts obtained from specimens of *A. nobilis* collected in Barkley Sound were shown to contain the degraded sesquiterpenoid **11** as the only odoriferous metabolite (Gustafson and Andersen 1985). At the time of its isolation from *A. nobilis*, aldehyde **11** had not been previously reported as a natural product but a synthetic version had been used in the perfume industry.

There are two species of nudibranchs in the genus *Archidoris* that have ranges extending from Alaska to San Diego. *A. odhneri* is a large (10–20 cm in length) pure white nudibranch, while *A. montereyensis* is somewhat smaller (up to 15 cm in length) and has a light yellow color with black markings. A preliminary chemical examination of *A. odhneri* yielded the farnesic acid glycerides **12**, **13**, and **14**, which represented the first terpenoic acid glycerides reported from nudibranchs (Andersen and Sum 1980). Subsequently, many related terpenoic acid glycerides have been isolated from dorids collected worldwide (Gavagnin and Fontana 2000). These include the cyclized diterpenoic acid glyceride **15**, isolated from *A. montereyensis* collected in both La Jolla and Barkley Sound, along with the monoacetates **16** and **17**, the drimane **18**, and the monocyclofarnesane **19**, which were isolated only from Barkley Sound specimens (Gustafson et al. 1984; Gustafson and Andersen 1985).

The gross structure and relative configuration of **15** were elucidated via single crystal X-ray diffraction analysis and its absolute configuration was determined by DIBAL reduction to the known sponge metabolite **20** (Cimino et al. 1974). In a similar manner, the absolute configuration of the drimane fragment of **18** was established by reduction to the known drimanol **21**. Interestingly, the isocopalane diterpenoid skeleton found in the diacylglycerols **22** and **23**, isolated from the Patagonian nudibranch *Anisodoris fontaini*, is the enantiomer of the *ent*isocopalane skeleton found in **15** (Gavagnin et al. 1999). Both series of compounds have the *S* configuration at C-2 of the glycerol fragment. Subsequent analysis of *A. odhneri* extracts revealed the presence of minor amounts of **15**, **18**, and **19** along with the major farnesic acid glycerides **12**, **13**, and **14** (Gustafson and Andersen 1985). The drimane glyceride **18** was found to be active, while the glycerides **12** and **15** were inactive, in fish antifeedant assays using the tidepool sculpin *Oligocottus maculosus* (Gustafson and Andersen 1985).

Cadlina luteomarginata, which has a range extending from Alaska to Mexico, is one of the most common nudibranchs in the rocky intertidal and subtidal habitats of coastal BC and southern California. Skin extracts of *C. luteomarginata* have been an extremely rich source of terpenoid metabolites. Many of the compounds have also been found in the sponges that form the diet of *C. luteomarginata*, and as a consequence, the skin chemistry of this nudibranch varies significantly from collecting site to collecting site as its sponge diet changes (Dumdei et al. 1997). Specimens of *C. luteomarginata* collected in Scripps Canyon, off La Jolla, and at Point Loma, San Diego, California yielded the furanosesquiterpenoids dendrolasin (**24**), pallescensin-A (**25**), pleraplysillin-1 (**26**), furodysinin (**27**), and dihydropallescensin-2 (**28**), along with the sesterterpenoid idiadione (**29**), the isonitriles **30**, **32**, and **34**, and the corresponding isothiocyanates **31**, **33**, and **35** (Thompson et al. 1982). Several of the terpenoids were also found in sponges eaten by *C. luteomarginata*. Idiadione (**29**) was found in *Leiosella idia*, furodysinin (**27**) and pallescensin-A (**25**) were found in *Dysidea amblia*, and the isonitrile **30** was present in an *Axinella* sp. It was noted that many of the major terpenoids identified in these same sponges were never encountered in *C. luteomarginata* skin extracts, suggesting that the nudibranchs are selectively sequestering compounds from their diets.

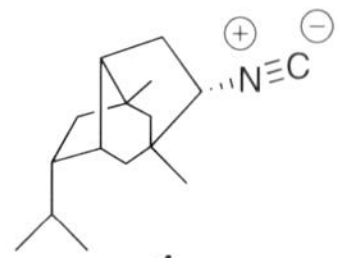

1

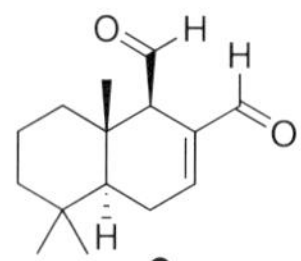

2

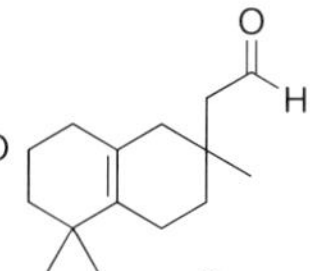

3

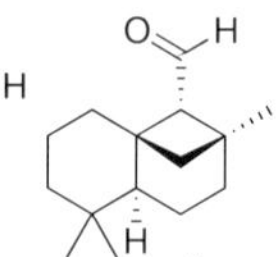

4

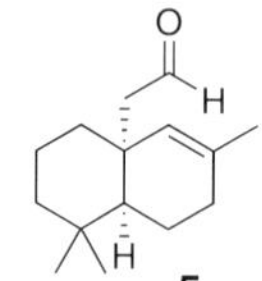

5

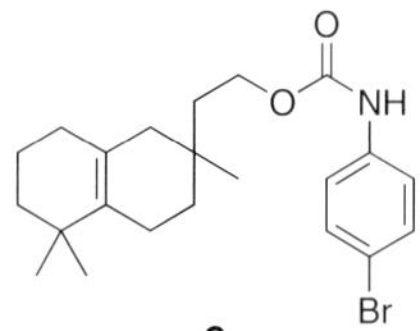

6

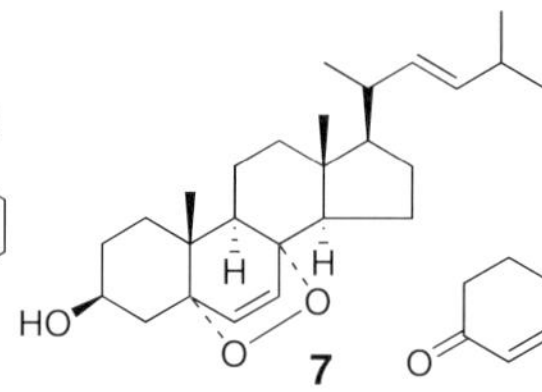

7

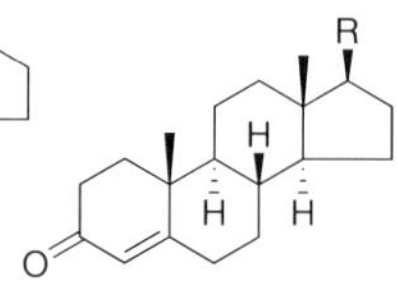

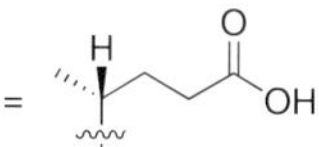

8 R =

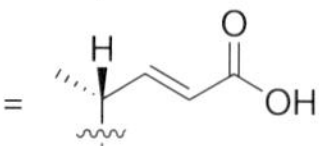

9 R =

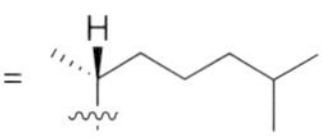

10 R =

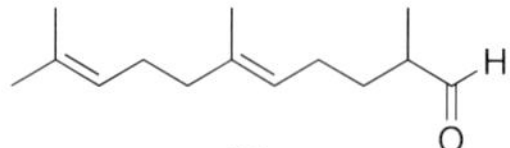

11

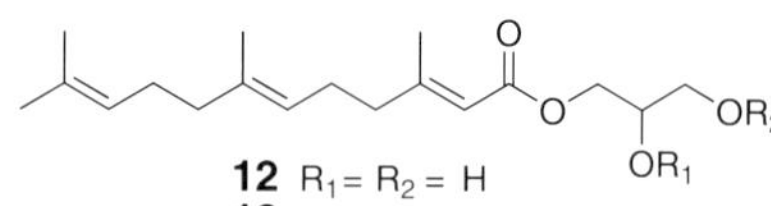

12 $R_1 = R_2 = H$
13 $R_1 = H$, $R_2 = Ac$
14 $R_1 = Ac$, $R_2 = H$

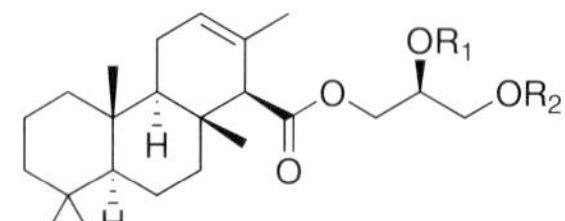

15 $R_1 = R_2 = H$
16 $R_1 = H$, $R_2 = Ac$
17 $R_1 = Ac$, $R_2 = H$

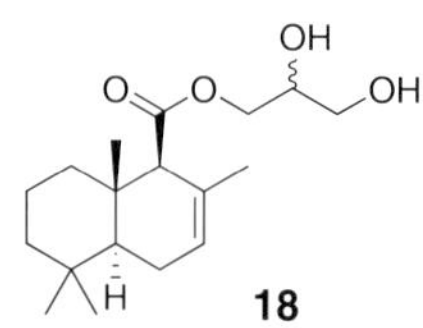

18

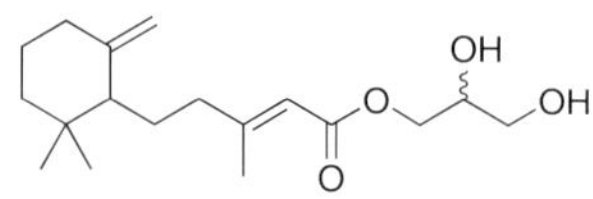

19

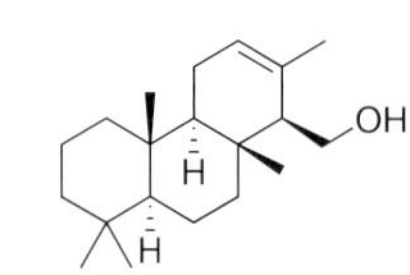

20

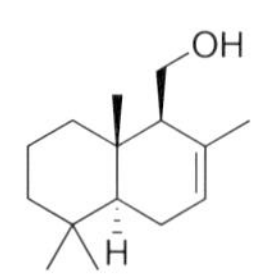

21

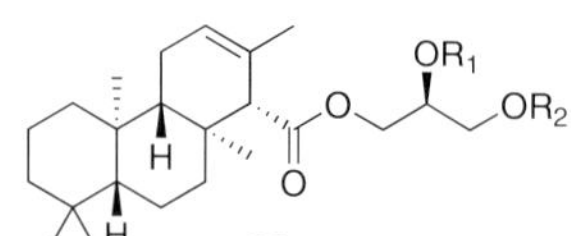

22 $R_1 = H$, $R_2 = Ac$
23 $R_1 = Ac$, $R_2 = H$

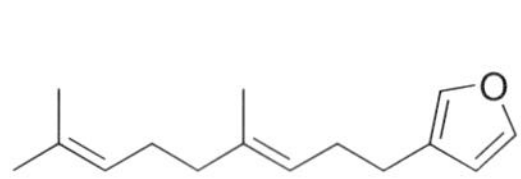

24

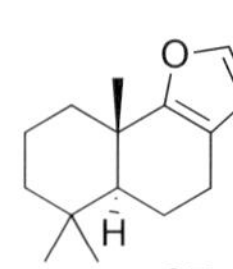

25

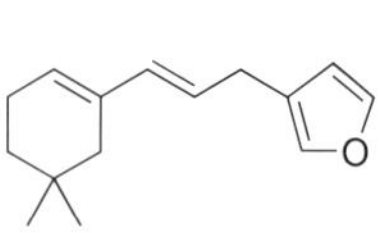

26

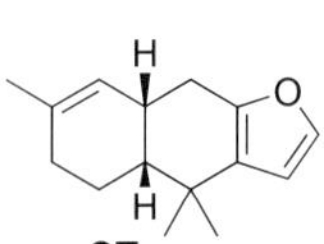

27

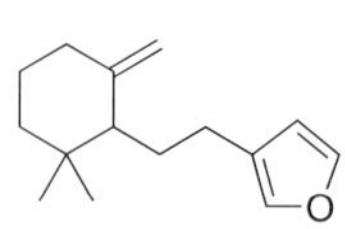

28

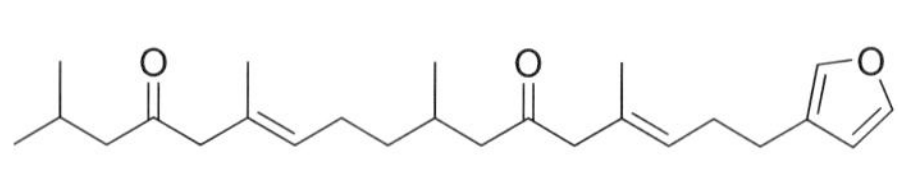

29

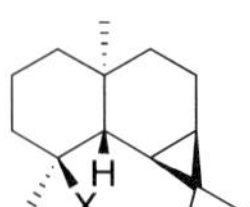

30 X = NC
31 X = NCS
43 X = NHCHO

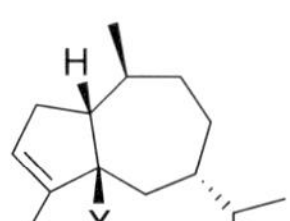

32 X = NC
33 X = NCS

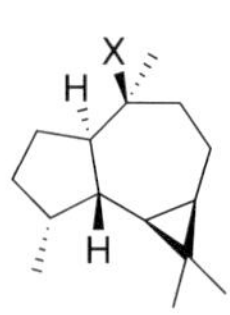

34 X = NC
35 X = NCS

Skin extracts of specimens of *C. luteomarginata* collected along the BC coast yielded the sesquiterpenoids albicanyl acetate (**36**; Hellou et al. 1982), albicanol (**37**), furodysinin (**27**), furodysin (**38**), microcionin-2 (**39**), the isonitriles **30** and **40**, the isothiocyanates **41** and **42**, and the formamides **43** and **44** (Dumdei et al. 1997). A number of diterpenoids, degraded diterpenoids, and degraded sesterterpenoids have also been isolated from these nudibranchs. Included in this group are marginatafuran (**45**; Gustafson et al. 1985), glaciolide (**46**; Tischler and Andersen 1989), cadlinolide A (**47**; Tischler et al. 1991), tetrahydroaplysulfurin (**48**), 9,11-dihydrogracillin A (**49**), 20-acetoxymarginatone (**50**), lutenolide (**51**), the spongiane diterpenoids **52**, **53**, and **54**, and the two degraded sesterterpenoids, luteone (**55**; Hellou et al. 1981) and cadlinaldehyde (**56**; Dumdei et al. 1997). Extracts from an egg mass of *C. luteomarginata* collected in Rennell Sound BC yielded two acetoxy derivatives of albicanyl acetate (**36**). Only 1α,2α-diacetoxyalbicanyl-acetate (**57**) was obtained in sufficient quantity for complete structural characterization.

Many of the terpenoids isolated from BC specimens of *C. luteomarginata* have also been traced to the nudibranch's diet. Glaciolide (**46**) and cadlinolide A (**47**) have been isolated from *Aplysilla glacialis*, marginatafuran (**45**) and 9,11-dihydrogracillin A (**49**) came from a second unidentified species of *Aplysilla*, and furodysinin (**27**) and furodysin (**38**) have been found in a *Pleraplysilla* sp. One collection of *C. luteomarginata* from the Queen Charlotte Islands in BC contained the halogenated monoterpene violacene (**58**) that is a known metabolite of red algae in the genus *Plocamium* (van Engen et al. 1978). Violacene (**58**), the isonitriles **30** and **40**, the isocyanates **41** and **42**, and the formamides **43** and **44** were also isolated from the sponge *Axinella* sp. collected in the Queen Charlotte Islands (Burgoyne et al. 1993). The presence of violacene (**58**) in *Axinella* sp. can best be explained by the sponge concentrating exudates of specimens of a *Plocamium* sp. living near to the sponge. If this scenario is correct, it represents an interesting example of metabolite transfer from a red alga to a sponge and then to a nudibranch.

Chemical investigations of *C. luteomarginata* have resulted in the identification of a truly remarkable collection of terpenoid metabolites. To date, a total of 35 terpenoids representing 21 different carbon skeletons (Fig. 13.2) have been isolated from *C. luteomarginata* skin and egg mass extracts (Dumdei et al. 1997). The 21 carbon skeletons include monoterpenoid, sesquiterpenoid, diterpenoid, sesterterpenoid, degraded diterpenoid, and degraded sesterterpenoid representatives. Four of these, the marginatane (**xv**; Gustafson et al. 1985), glaciane (**xvi**; Tischler and Andersen 1989), cadlinalane (**xx**; Dumdei et al. 1997), and luteane (**xxi**; Hellou et al. 1981) skeletons, were first encountered in *C. luteomarginata* metabolites. The *C. luteomarginata* terpenoids represent a range of biosynthetic diversity of secondary metabolites rarely

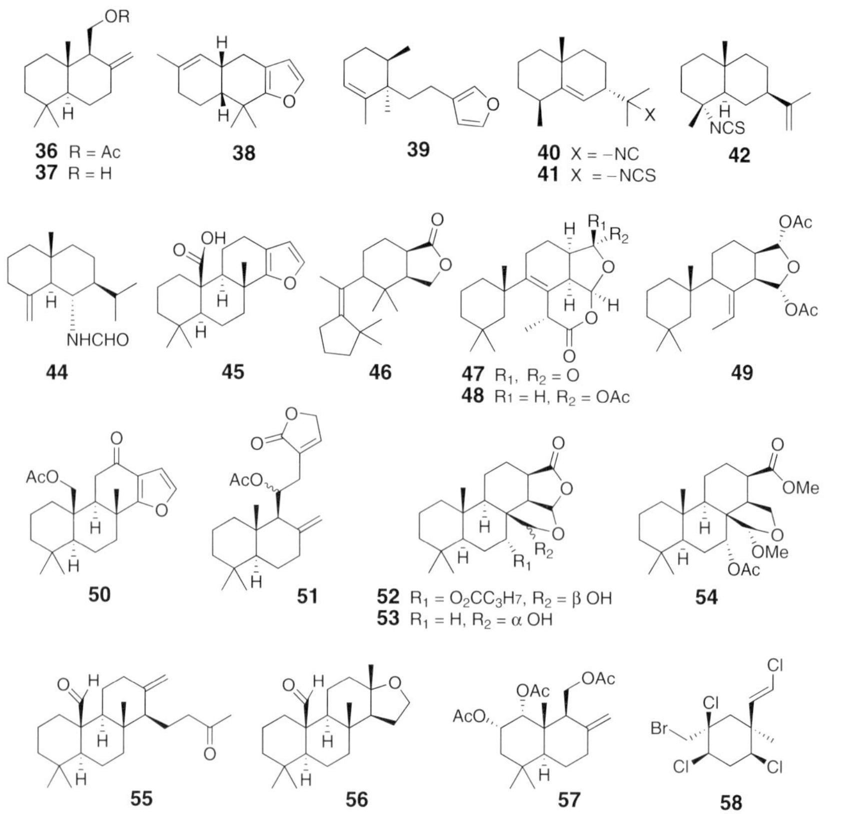

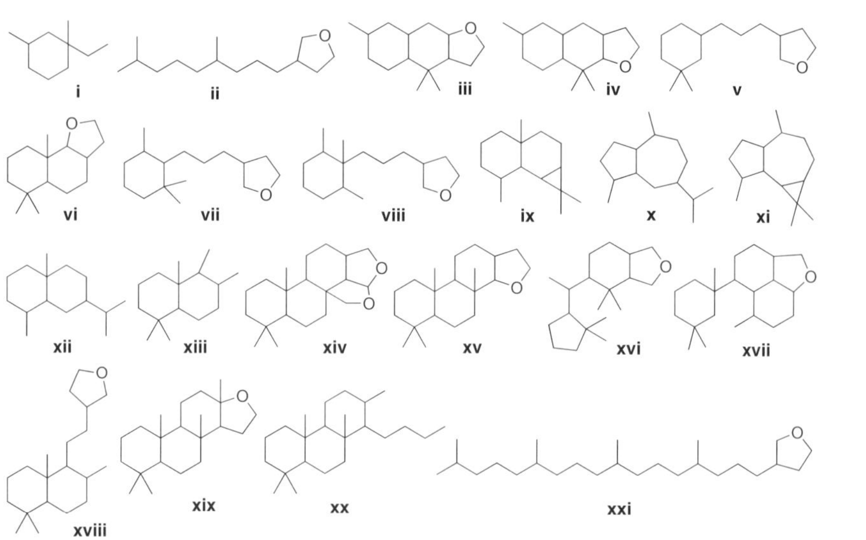

Fig. 13.2. Terpenoid carbon skeletons represented in metabolites isolated from *Cadlina luteomarginata* skin extracts

encountered in a single living organism. This great metabolite diversity results primarily from the highly developed capabilities of *C. luteomarginata* for sequestering terpenoids from sponges in its diet. Thus, *C. luteomarginata* did not need to develop the enzymes required to make this wide range of compounds de novo. Antifeedant assays conducted by Thompson et al. (1982) using the goldfish *Carassius auratus* and the tidepool sculpin *Clinocottus analis* showed significant activity in a mixture of isonitriles (**30**, **32**, **34**), a mixture of isothiocynates (**31**, **33**, **35**), furodysin (**27**), pallescensin-A (**25**), and idiadione (**29**). Similar assays carried out by Hellou et al. (1982) using the goldfish *Carassius auratus* showed albicanyl acetate (**36**) and furodysinin (**38**) to be active antifeedants.

Faulkner's group reported the isolation of terpenoids from four species of nudibranchs in the genus *Chromodoris*. *C. marislae* collected in the Gulf of California yielded pure marislin (**59**) and the minor compounds **60a**, **b** and **61a**, **b** as inseparable mixtures (Hochlowski and Faulkner 1981). The structure of marislin (**59**) was confirmed by converting it into the known sponge metabolite pleraplysillin-2 (**62**) by treatment with boron trifluoride etherate. Norrisolide (**63**), a rearranged spongian diterpenoid, was isolated from *C. norrisi* also collected in the Gulf of California (Hochlowski et al. 1983a). Skin extracts of *C. sedna*, collected at Bahia de Concepcion in the Gulf of California, contained the sesterterpenoids sedenolide (**64**), 12-deacetyl-20-methyldeoxyscalarin (**65**), 23-hydroxy-20-methyldeoxoscalarin (**66**), 23-hydroxy-20-methylscalarolide (**67**), and sednolide 23-acetate (**68**; Hochlowski et al. 1983b). Two aromatic diterpenoids, macfarlandins-A (**69**) and -B (**70**), and the rearranged spongian diterpenoids macfarlandins C (**71**), D (**72**), and E (**73**), were isolated from *C. macfarlandi* collected in Scripps Canyon, La Jolla (Molinski and Faulkner 1986; Molinski et al. 1986).

The porostome nudibranchs are notable for their absence of a radula. Skin extracts of the porostomes *Dendrodoris krebsii* collected at Bahia de los Angeles, Baja California, Mexico, *Doriopsilla albopunctata* collected at Bahia de los Angeles and La Jolla, and *D. janaina* collected at Bahia de los Angeles all contained drimane sesquiterpenoids (Okuda et al. 1983). *D. krebsii* contained polygodial (**2**) and olepupuane (**74**), while *D. albopunctata* and *D. janaina* contained olepupuane (**74**) and the mixed esters **75**, previously found in *Dendrodoris limbata* collected in the Mediterranean. Cimino's finding that polygodial (**2**) is made de novo by *D. limbata* (Cimino et al. 1983) suggests that the drimanes found in the northeastern Pacific *Dendrodoris* and *Doriopsilla* species are also made by the nudibranchs.

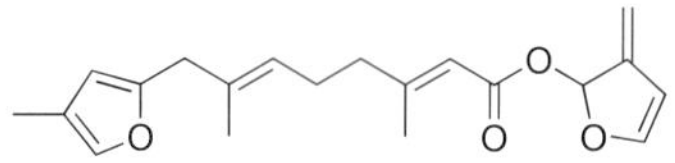
59

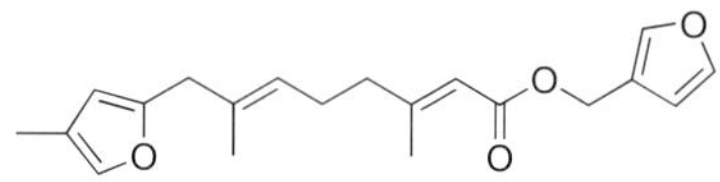
62

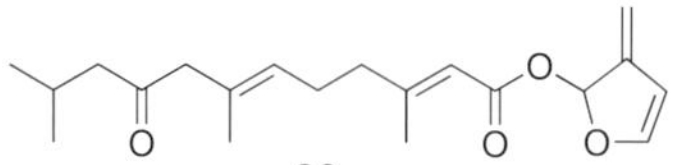
60a

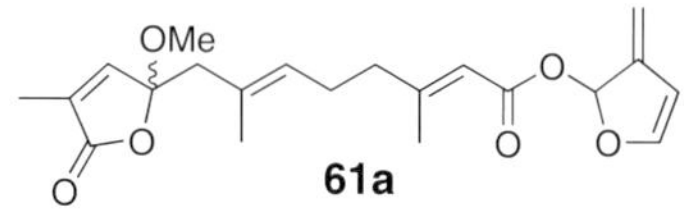
61a

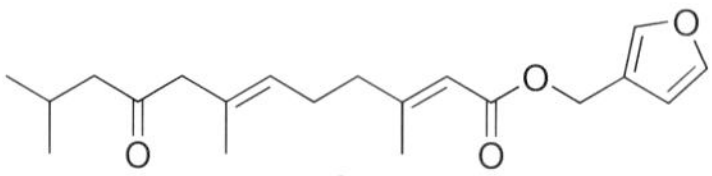
60b

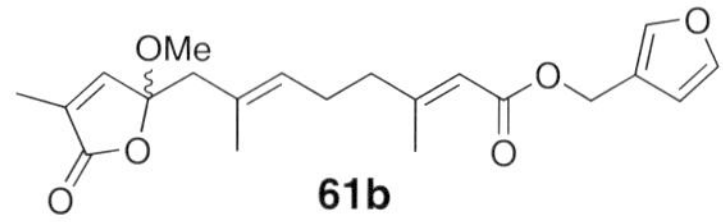
61b

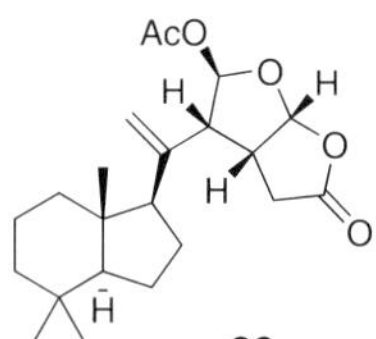
63

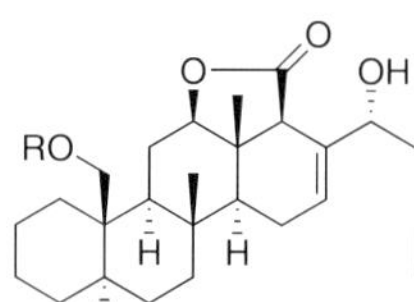
64 R = H
68 R = Ac

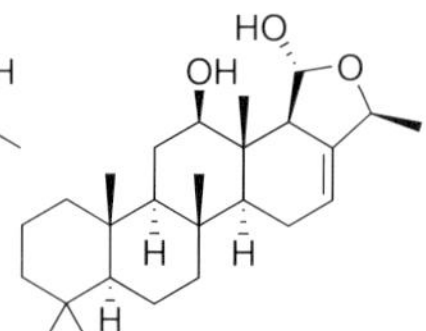
65

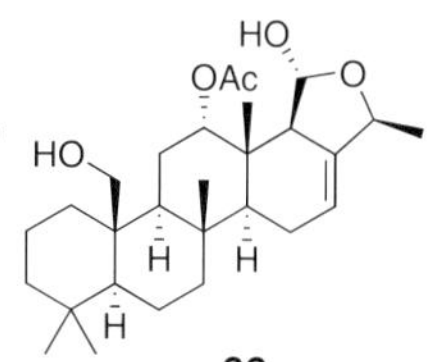
66

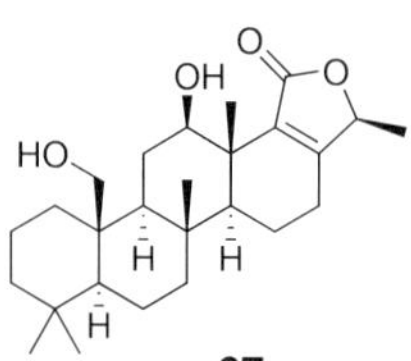
67

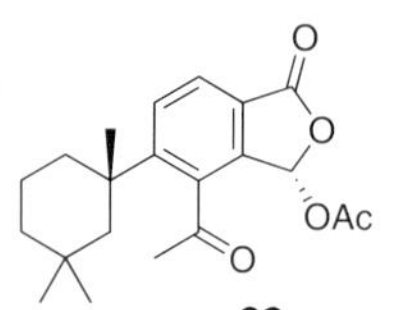
69

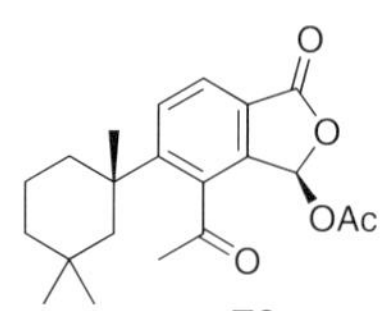
70

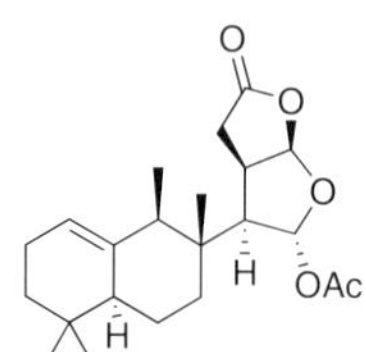
71

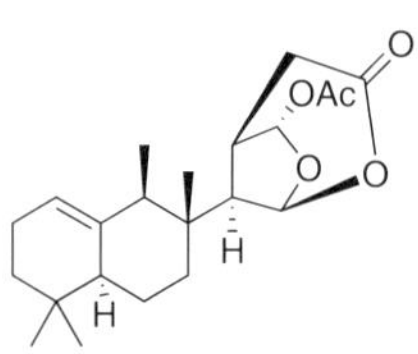
72

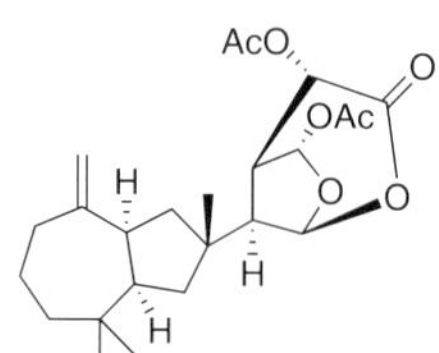
73

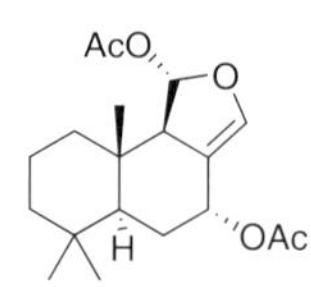
74

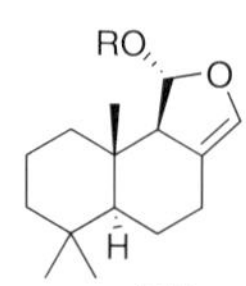
75

Skin extracts of *Diaulula sandiegensis* collected at Point Loma, San Diego, contained the nine chlorinated acetylenes **76–84** (Walker and Faulkner 1981). Diaulusterols A (**85**) and B (**86**), but no chlorinated acetylenes, were obtained from skin extracts of *D. sandiegensis* specimens collected in Barkley Sound BC (Williams et al. 1986). Four northeastern Pacific nudibranchs in the genus *Hypselodoris* have been examined and they all contained furanoterpenoids (Hochlowski et al. 1982). *H. agassizi* collected at Cruz de Juanacaxtle, Nayarit, Mexico, contained agassizin (**87**) and *H. ghiselini* collected at Isla Danzante in the Gulf of California contained nakafuran 9 (**88**), ghiselinin (**89**), the butenolide **90**, and dendrolasin (**24**). *H. californiensis* specimens collected at Isla San Jose in the Gulf of California contained dendrolasin (**24**) and nakafuran 9 (**88**), while specimens collected at Casa Cove in La Jolla contained furodysinin (**27**) and euryfuran (**91**). Six specimens of *H. porterae* collected at Point

Loma were found sitting on the sponge *Dysidea amblia*. These nudibranchs contained furodysinin (**27**) and euryfuran (**91**), but there was no evidence of the sponge containing the same compounds. However, euryfuran (**91**) was found in *Euryspongia* sp. collected at Casa Cove, La Jolla.

Sclerodoris tanya collected at Casa Cove in La Jolla was found to contain the terpenoic acid glycerides tanyolides A (**92**) and B (**93**), which have sesquiterpenoid residue linked to the 2-position of glycerol (Krug et al. 1995). The structures of **92** and **93**, including the absolute configuration of **93**, were confirmed by synthesis of the natural products. *Triopha catalinae* is a common BC nudibranch that feeds exclusively on bryozoans. Skin extracts of *T. catalinae* were found to contain a single secondary metabolite, the diacylguanidine triophamine (**94**; Gustafson and Andersen 1982). The structure of triophamine (**94**), including the configuration of the trisubstituted alkenes in the two identical acyl residues, was confirmed by total synthesis (Piers et al. 1984). A closely related bryozoan-eating nudibranch found in BC waters, *Polycera tricolor*, also contains triophamine (**94**; Gustafson and Andersen 1985).

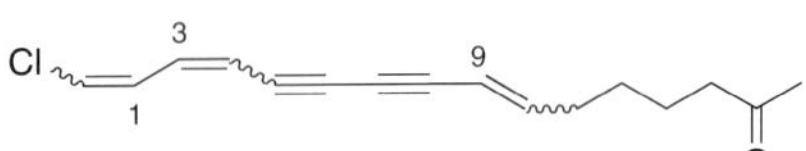

76 (1*Z*,3*E*,9*Z*)
77 (1*Z*,3*Z*,9*Z*)

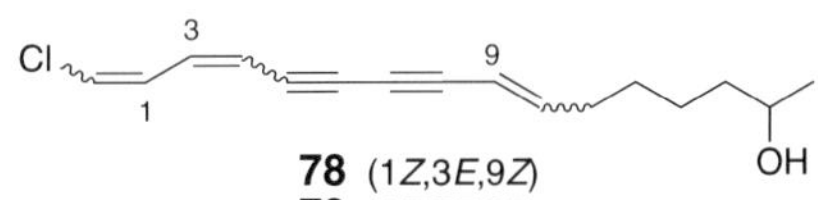

78 (1*Z*,3*E*,9*Z*)
79 (1*Z*,3*Z*,9*Z*)
80 (1*E*,3*E*,9*Z*)

81 (1*Z*,3*E*,9*Z*)
82 (1*Z*,3*Z*,9*Z*)
83 (1*E*,3*E*,9*Z*)

84 (1*Z*,3*E*)

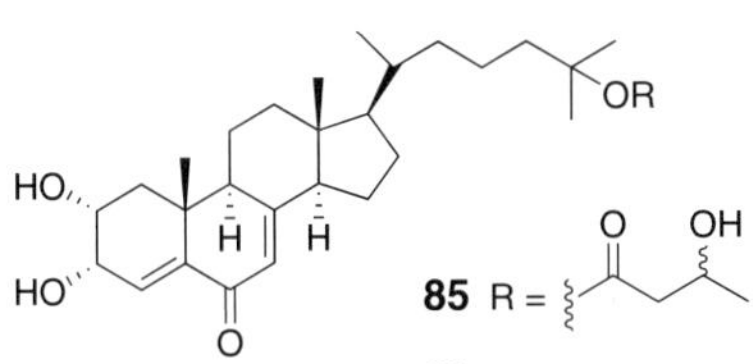

85 R =

86 R = H

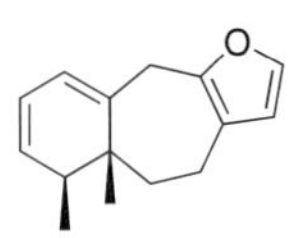

87

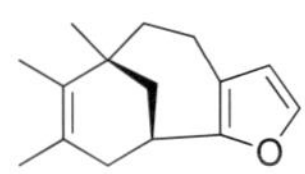

88

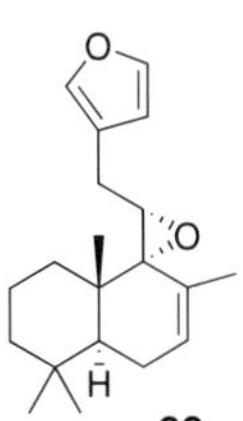

89

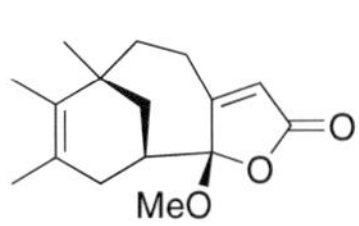

90

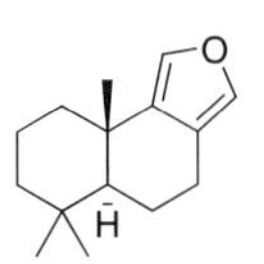

91

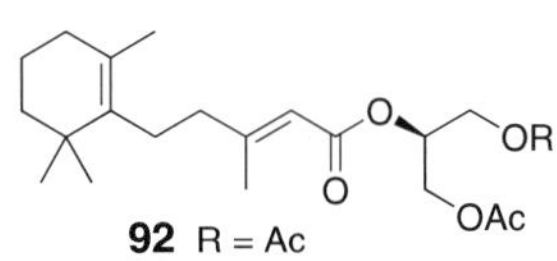

92 R = Ac
93 R = H

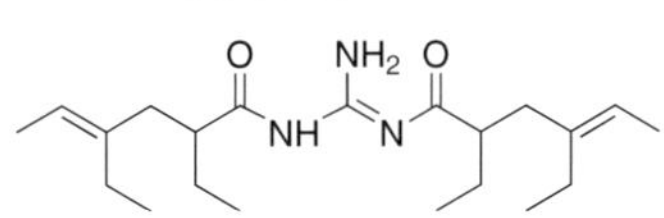

94

Tochuina tetraquetra is a large (up to 30 cm) orange-yellow dendronotid nudibranch that is commonly found from the Kuril Islands, Russia, to the Santa Cruz Islands, California. Specimens collected in Port Hardy, BC, contained tochuinyl acetate (**95**), dihydrotochuinyl acetate (**96**), rubifolide (**97**), and pukalide (**98**; Williams and Andersen 1987). Terpenoids **95**, **96**, and **97** were also isolated from the soft coral *Gersemia rubiformis*, which is a major part of the diet of *T. tetraquetra* at Port Hardy (Williams et al. 1987). *T. tetraquetra* specimens collected in Barkley Sound BC, contained ptilosarcenone (**99**) and the butanoate analogue **100**, which are metabolites of the sea pen *Ptilosarcus gurneyi* (Wratten et al. 1977), an important component in the diet of *T. tetraquetra* in Barkley Sound.

The dendronotid nudibranch *Melibe leonia* has one of the most unusual feeding behaviors of any member of the Phylum Mollusca. Unlike other nudibranchs, *M. leonina* is not a predator of sessile bottom-dwelling animals; rather, it feeds upon zooplankton by majestically sweeping the sea with its large oral hood. Chemical studies on *M. leonina* were prompted by a report that the nudibranch's primary defense is an odoriferous substance that is repugnant to potential predators. Skin extracts of *M. leonina* were found to contain the fragrant aldehyde 2,6-dimethyl-5-heptenal (**101**) and the corresponding acid (**102**; Ayer and Andersen 1983).

Roboastra tigris is a large carnivorous nudibranch that is known to feed on two smaller nudibranchs, *Tambja eliora* and *T. abdere*. Skin extracts of all three molluscs contained the same mixture of tambjamines A (**103**) to D (**106**; Carte and Faulkner 1983). The tambjamines A–D were also found in the bryozoan *Sessibugula translucens*, which is in the diet of *T. eliora* and *T. abdere*. When attacked by *R. tigris*, *T. abdere* produces a yellow mucus containing tambjamines from goblet cells on its skin. The secretion appears to cause *R. tigris* to break off its attack, suggesting a defensive role for the tambjamines (Carte and Faulkner 1986).

13.3 De Novo Biosynthesis by Northeastern Pacific Nudibranchs

The first clear indication of geographical variation in chemical constituents of a nudibranch came from comparing the metabolites found in *C. luteomarginata* collected at various sites in BC and southern California (Hellou et al. 1982; Thompson et al. 1982; Faulkner et al. 1990; Dumdei et al. 1997). Many of the terpenoids obtained from *C. luteomarginata* at both sites were shown to be sequestered unchanged from the sponges that make up the nudibranch's diet (vide supra). It was also found that the terpenoid content of skin extracts of specimens of *C. luteomarginata* can vary dramatically from collecting site to collecting site, reflecting

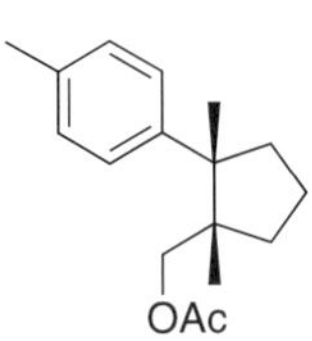

95

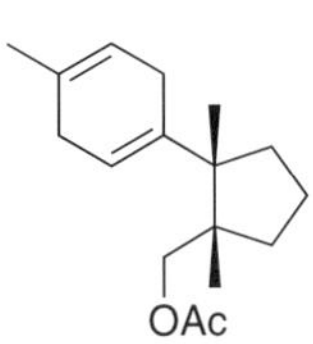

96

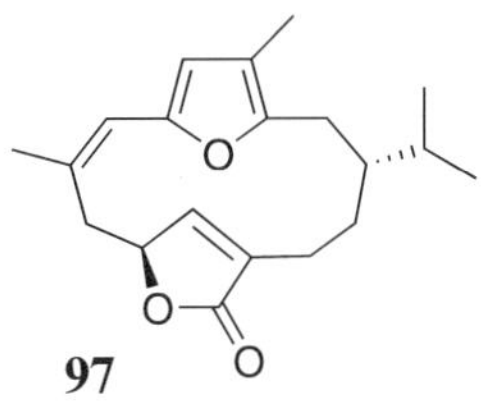

97

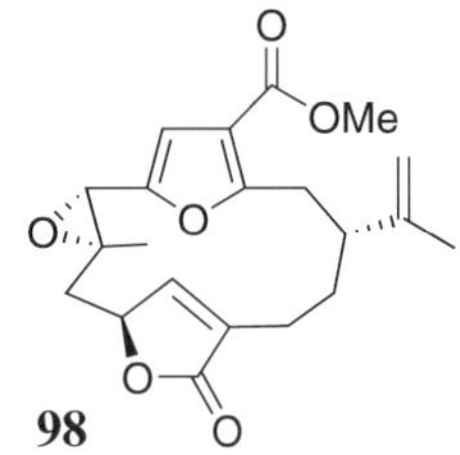

98

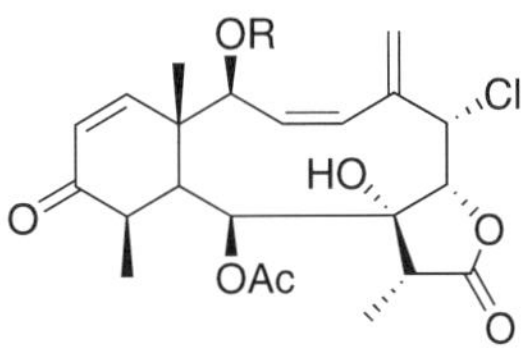

99 R1 = Ac

100 R1 = 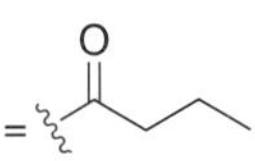

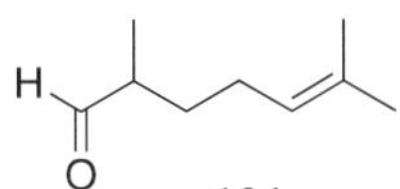

101

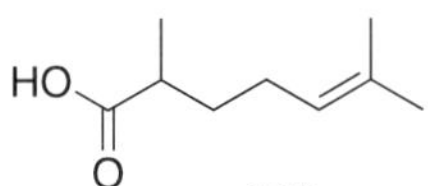

102

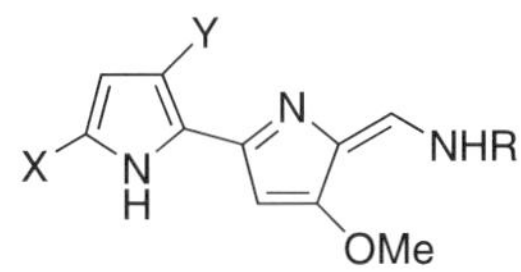

103 X = Y = R = H
104 X = Br, Y = R = H
105 X = Y = H, R = iBu
106 X = H, Y = Br, R = iBu

the different terpenoid constituents present in the dietary sponges available at each site. The discovery of geographic variability in *C. luteromarginata* terpenoid constituents prompted investigations of geographic variation in metabolite content of other species. Based on the results of these investigations, it was proposed that nudibranchs capable of making their defensive chemicals via de novo biosynthesis will produce the same species-specific array of chemicals no matter where they are collected (Faulkner et al. 1990). Conversely, the species that show variable chemical content from site to site must be acquiring their chemicals from their diet. It was suggested that the absence of geographic variation in secondary metabolites for a particular species could be used to identify promising candidates for biosynthetic investigations.

Based on the geographic variation hypothesis, several metabolites isolated from BC nudibranchs, that were present in specimens collected at all locations over a wide geographic range, became the focus of biosynthetic investigations. The terpenoic acid glycerides **12–19** found in *A. odhneri* and *A. montereyensis* were the first compounds examined. Following Cimino's precedent (Cimino et al. 1983), the digestive glands of *A. odhneri* and *A. montereyensis* were injected with ^{14}C-labeled mevalonic acid and, after a period of 24 h, the glyceride **12** was isolated from *A. ohdneri* and the glycerides **15** and **18** were isolated from *A. montereyensis* (Gustafson and Andersen 1985). Compound **12** was reduced to farnesol and converted to the dinitrobenzoyl ester **107** that was recrystallized to constant radioactivity. The glycerides **15** and **18** were reduced to the alcohols **20** and **21**, respectively, which were also recrystallized to constant radioactivity. These experiments clearly demonstrated incorporation of

^{14}C-labeled mevalonic acid into the glycerides **12**, **15**, and **18**, providing the first evidence that terpenoic acid glycerides are biosynthesized de novo by nudibranchs.

Skin extracts of *A. nanaimoensis* collected at many sites along the BC coast always contain nanaimoal (**3**), acanthodoral (**4**), and isoacanthodoral (**5**), which all have unprecedented sesquiterpenoid skeletons. Interest in the biogenesis of the three new terpenoid skeletons represented by **3**, **4**, and **5** prompted isotope incorporation studies with *A. nanaimoensis* (Graziani and Andersen 1996a). The objectives were to demonstrate that **3**, **4**, and **5** were being made de novo by *A. nanaimoensis*, to increase the levels of precursor incorporation to a point where NMR detection of the stable isotope ^{13}C was feasible, and to compare the incorporation patterns with those predicted by initial biogenetic proposals for the formation of the nanaimoane, acanthodorane, and isoacanthodorane skeletons (Ayer et al. 1984a,b).

Prior to the feeding studies with *A. nanaimoensis*, the low levels of precursor incorporation into metabolites of marine invertebrates observed by other investigators had been attributed to a very slow rate of biosynthesis. Nudibranch biology appeared ideally suited to overcoming this impediment. When nudibranchs are collected in the field and transported live back to a laboratory for study, they usually shed most of the secondary metabolites that can be extracted by simply immersing them whole in solvent. The physical act of handling the nudibranchs perhaps simulates an attack by a predator, which causes them to release chemicals that might play a defensive role. Any manipulation that causes a nudibranch to shed a substantial portion of its secondary metabolites should in principle activate existing de novo biosynthetic pathways that are capable of replenishing the metabolites. With the biosynthetic pathway activated and the pool of unlabeled compounds depleted, conditions should be ideally suited to achieving high levels of incorporation of isotopically labeled precursors.

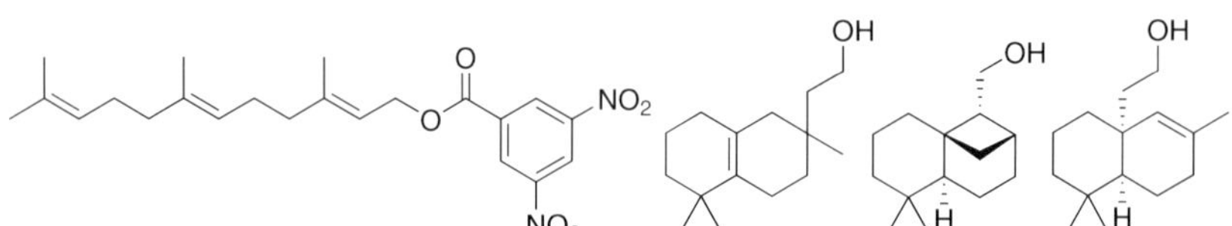

107 **108** **109** **110**

The feeding protocol used in the *A. nanimoensis* biosynthetic experiments involved injecting the digestive gland of individual specimens with 100 μl of a 550 mM solution of [1,2-$^{13}C_2$] acetate every second day for 16 days. Extraction of the nudibranchs at the end of the injection cycle gave a mixture of the aldehydes **3**, **4**, and **5** that were reduced to the corresponding alcohols **108**, **109**, and **110** for NMR analysis. Shown in Fig. 13.3 are the

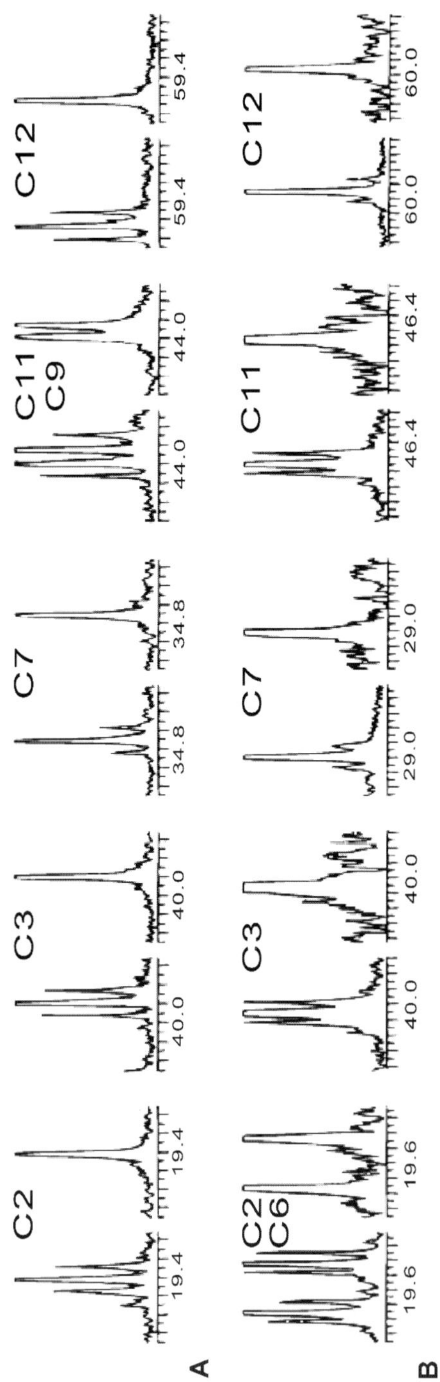

Fig. 13.3. (**a**) Truncated normalized ^{13}C NMR resonances for labeled (*left*) and unlabeled (*right*) samples of nanaimool (**108**). (**b**) Truncated normalized ^{13}C NMR resonances for labeled (*left*) and unlabeled (*right*) samples of isoacanthodorol (**110**)

truncated ^{13}C NMR signals for a representative group of carbon atoms in both a labeled sample and an unlabeled control sample of **108**. All of the resonances in Fig. 13.3, and indeed all of the resonances in the spectrum, showed flanking doublets resulting from incorporation of ^{13}C-labeled acetate units. Analysis of the intensities of the doublets indicated that there was one set of relatively intense doublets (i.e., C-2, C-3, C-12: average specific incorporation 0.35%) and a second set of relatively weak doublets (i.e., C-7: average specific incorporation 0.11%). A complete analysis of the coupling constants for the intense flanking doublets revealed a pattern of intact acetate unit incorporation that was consistent with a biogenesis from mevalonic acid (Scheme 13.1).

If only one labeled acetate had been incorporated per molecule of **108**, then the resonances for C-1, C-7, and C-14 should appear as enriched singlets without flanking doublets resulting from one bond coupling, because these carbons should not be part of intact acetate units. The most likely explanation for the weak flanking doublets observed in the C-1, C-7, and C-14 resonances of labeled **108** is that they originate from incorporation of more than one ^{13}C-labeled acetate unit into a single nanaimoal (**3**) molecule. *A. nanaimoensis* specimens were starved during the 16-day injection period; and this could have led to a highly labeled acetate pool and a reasonable probability of more than one labeled acetate unit being

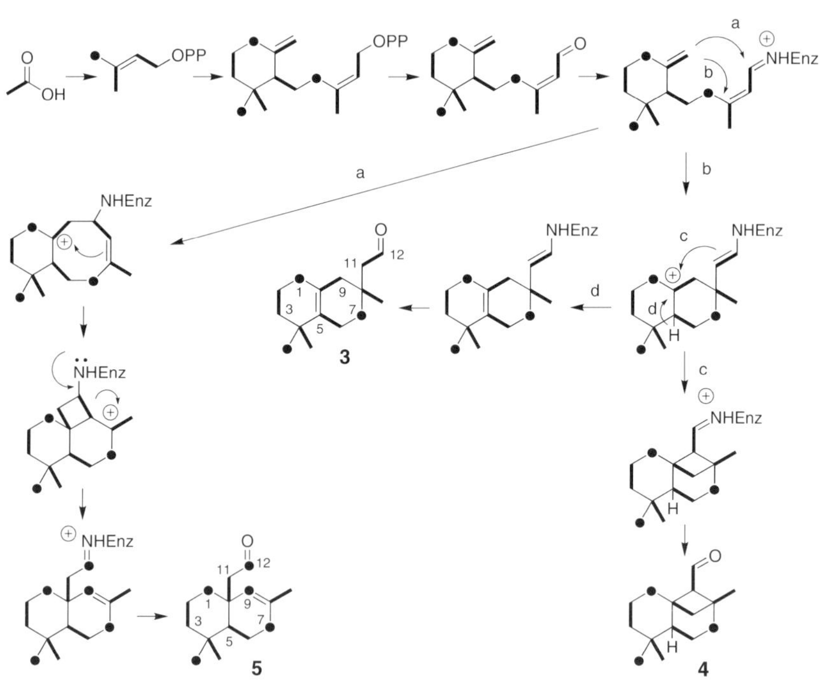

Scheme 13.1. Proposed biogenesis of nanaimoal (**3**), acanthodoral (**4**), and isoacanthodoral (**5**) that accounts for the observed labeling patterns resulting from feeding [1,2-$^{13}C_2$] acetate

incorporated into individual mevalonic acid molecules and/or more than one labeled mevalonic acid unit being incorporated into individual molecules of **3**. As a result, the C-7 doublet would arise from molecules of **108** having a singly enriched carbon at C-7 and an intact acetate unit at either C-5/C-6 or C-8/C-15.

The labeling patterns observed in **108**, **109**, and **110** are consistent with the biosynthetic proposal shown in Scheme 13.1 that suggests an imonium ion might serve as a cationic initiator of the final cyclization reactions. Imine intermediates were invoked to account for the observation that the oxygen atoms in [1,2-$^{13}C_2$, $^{18}O_1$] acetate were not incorporated into the aldehyde functionalities of **3**, **4**, and **5** (Kubanek 1998). Of particular interest is the rearrangement involved in the formation of the isoacanthodorane skeleton (i.e., **5**) that is readily apparent from the ^{13}C NMR experiments but would not have been evident from simple ^{14}C-labeled mevalonic acid incorporation without complex chemical degradation experiments. The [1,2-$^{13}C_2$] acetate-feeding experiments with *A. nanaimoensis* represented the first successful use of stable isotope detection by NMR to study the biosynthesis of terpenoid carbon skeletons by a marine invertebrate.

The *A. nanaimoensis* experiments were also the first to show that it is possible to get high levels of incorporation of isotopically labeled precursors in marine invertebrate feeding experiments. They suggest that it is not the slow rate of biosynthesis by marine invertebrates that limits incorporation. Instead, the terpenoid metabolism of adult marine invertebrates can be viewed as having reached a steady state where the pool of unlabeled compound is at a maximum and, therefore, very little active biosynthesis is taking place. A single injection under these circumstances results in little transformation of precursor to labeled metabolite and, furthermore, the small amount of labeled compound must then be detected in the presence of a large amount of unlabeled material. However, if one can deplete the terpenoid pool, as was done in the *A. nanaimoensis* feeding experiments simply by handling during injection, which simulates attack by a predator causing release of putative defensive terpenoids, this turns on active biosynthesis to replace the compounds that were shed. This active biosynthesis is rapid and the newly formed compounds do not have to be detected in the presence of a large unlabeled pool, since those compounds have been released by the nudibranchs during handling.

Using the protocol developed for *A. nanaimoensis*, it was subsequently confirmed by injecting [1,2-$^{13}C_2$] acetate that the terpenoic acid glycerides **12** and **15** were made de novo from mevalonic acid by *A. ohdneri* and *A. montereyensis*, respectively (Graziani et al. 1996). Similar experiments demonstrated that *M. leonina* makes the degraded monoterpenoids **101** and **102** from mevalonic acid (Barsby et al. 2002).

Individual terpenoids obtained from specimens of *C. luteomarginata* collected in BC showed examples of both geographic variability and lack of geographic variability. In particular, albicanyl acetate (**36**), luteone (**55**), and cadlinaldehyde (**56**) were present in extracts of all specimens studied, suggesting that this subset of *C. luteomarginata* terpenoids were being made via de novo biosynthesis. Feeding studies with [1,2-$^{13}C_2$] acetate and [2-^{13}C] mevalonolactone gave the incorporation patterns for **36**, **55**, and **56** shown in Scheme 13.2, which were consistent with de novo biosynthesis via mevalonic acid (Kubanek et al. 1997). Luteone (**55**) and cadlinaldehyde (**56**) are degraded terpenoids, probably degraded sesterterpenoids.

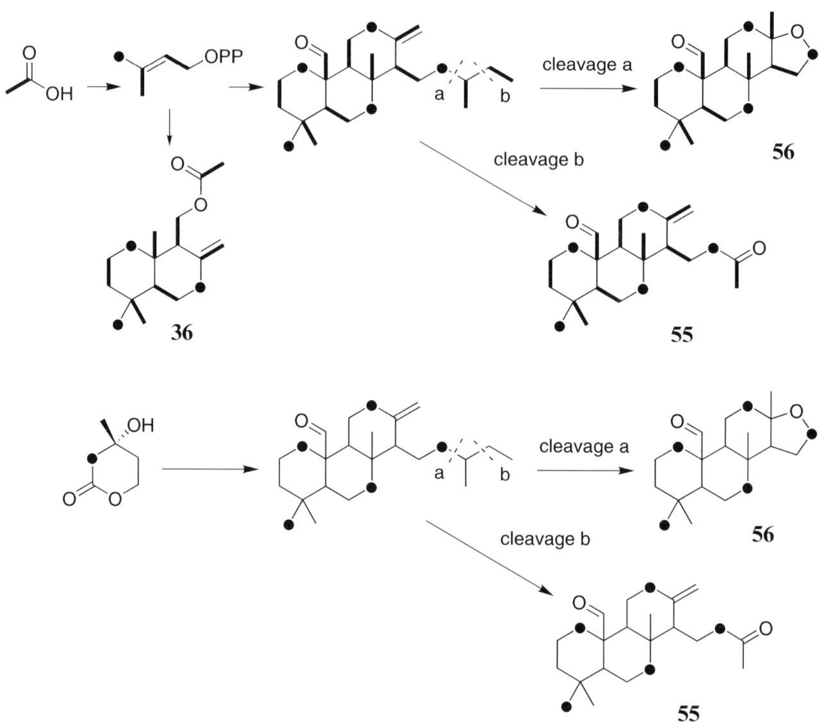

Scheme 13.2. Proposed biogenesis of albicanyl acetate (**36**), luteone (**55**), and cadlinaldehyde (**56**) that accounts for the observed labeling patterns resulting from feeding [1,2-$^{13}C_2$] acetate and [2-^{13}C] mevalonolactone

Initially, the terpenoids **36**, **55**, and **56**, which are made de novo by *C. luteomarginata*, appeared to contradict the geographic variability hypothesis because they had not been found in specimens collected in California. Subsequently, a detailed GC–MS analysis of extracts from California specimens did show the presence of all three compounds, although **55** and **56** were present at much lower concentrations, consistent with the geographic variability hypothesis. The disparity in concentration of **55** and **56** in BC and California specimens prompted a detailed GC–MS analysis of skin extracts from individual animals collected in BC and California (Kubanek et al. 2000). Two distinct patterns emerged (1) individuals possessed high concentrations of **36**, **55**, and **56** and almost no other GC detectable constituents or (2) individuals showed very low concentrations of **36**, **55**, and **56** and equal or higher concentrations of many other GC detectable constituents. These observations led to the proposal that the de novo biosynthesis of the putative defensive allomones **36**, **55**, and **56** might be regulated according to need. When there is an abundant supply of dietary compounds that could fulfill the same ecological requirements as the endogenous terpenoids, the de novo biosynthesis of **36**, **55**, and **56** would be down-regulated, and if suitable dietary compounds were scarce, the de novo biosynthesis would be up-regulated.

Skin extracts of *Triopha catalinae* contained the unique diacylguanidine metabolite triophamine (**94**) at all collecting sites between Alaska and southern California. *T. catalinae* feeds exclusively on bryozoans, but triophamine could never be found in extracts of the obvious dietary sources. These two observations suggested that triophamine was being made via de novo biosynthesis. Feeding experiments, first with [1,2-$^{13}C_2$] acetate (Graziani and Andersen 1996b) and subsequently with [2,3-$^{13}C_2$] ethylmalonate and [2,3-$^{13}C_2$] butyrate (Kubanek and Andersen 1997), demonstrated that the ten carbon acyl residues in triophamine are made from two units of butyrate and one unit of acetate, as shown in Scheme 13.3.

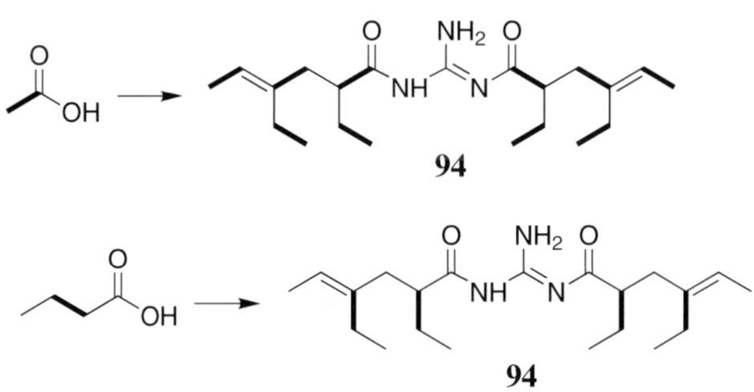

Scheme 13.3. Proposed biogenesis of triophamine (**94**) that accounts for the observed labeling patterns resulting from feeding [1,2-$^{13}C_2$] acetate and [2,3-$^{13}C_2$] butyrate

Diaulula sandiegensis is common in rocky subtidal habitats from Alaska to Cabo San Lucas, Mexico. Extracts of specimens collected in Alaska and BC always contained significant amounts of diaulusterols A (**85**) and B (**86**), but no chlorinated acetylenes, while specimens collected in California contained only trace amounts of diaulusterol B (**86**), but large quantities of the chlorinated acetylenes **76–84**. The distribution pattern of **85** and **86** in *D. sandiegensis* mimicked the pattern observed for albicanyl acetate (**36**), luteone (**55**), and cadlinaldehyde (**56**) in *C. luteomarginata*; and it suggested that the dialusterols were being made de novo but their biosynthesis was being down-regulated in California where the nudibranchs had access to a plentiful source of dietary chlorinated acetylenes. Feeding studies with [1,2-$^{13}C_2$] acetate confirmed that the polyketide fragment of diaulusterol A (**85**) was being made de novo by *D. sandiegensis* (Kubanek and Andersen 1999). However, there was no evidence for incorporation of [1,2-$^{13}C_2$] acetate or [2-^{14}C] mevalonate into the steroidal nucleus of **85**. Similar observations had been found with *C. luteomarginata*, where during most times of the year there was high incorporation of [1,2-$^{13}C_2$] acetate into the acetate residue of albicanyl acetate (**36**) but no incorporation into the terpenoid fragment. However, during *C. luteomarginata* egg-laying season, [1,2-$^{13}C_2$] acetate was effectively incorporated in the terpenoid fragment of **36** as well (Kubanek et al. 1997). Therefore, two possibilities were proposed for the observed feeding results with *D. sandiegensis*. In one scenario, the steroid core of **85** and **86** is being made by the nudibranch but either its biosynthesis was not active during the feeding experiments or the precursors did not get to the biosynthetic site. Alternatively, the nudibranch may acquire the steroid core from its diet and chemically modify it rather than producing it de novo. More work needs to be done to resolve this interesting question.

13.4
Conclusions

Chemical studies of skin extracts obtained from nudibranchs collected on the west coast of North America have resulted in the identification of approximately 100 chemically diverse secondary metabolites. The majority of these compounds are terpenoids but there are also a few examples of steroids and polyketides. There is a general assumption that nudibranch skin chemistry plays a role in defending the shell-less molluscs from predation. In support of this assumption, fish antifeedant studies have shown that albicanyl acetate (**36**), furodysin (**38**), furodysinin (**27**), pallescensin-A (**25**), idiadione (**29**), a mixture of isonitriles (**30**, **32**, **34**), and a mixture of isothiocyanates (**31**, **33**, **35**) from *C. luteomarginata*, the steroid **8** from *Aldisa sanguinea cooperi*, and the terpenoic acid glyceride **18** from *Archidoris montereyensis* are deterrent at ecologically relevant concentrations. The limited quantities of skin metabolites typically isolated as part of the chemical structural studies have prevented rigorous testing of most of the other compounds for ecologically significant biological activities.

The Pacific coastline of North America has served as an excellent natural laboratory to study the origins of nudibranch skin chemistry. Roughly 25% of the terpenoids isolated from nudibranchs collected along this coast have also been isolated from sponges, soft corals, sea pens, and bryozoans found at the same collecting sites, providing direct evidence for sequestration of skin chemicals from dietary organisms (vide supra). A second large group of terpenoids and polyketides (50%) have been isolated from sponges collected in other parts of the world or have structures that are very similar to known sponge metabolites. These compounds almost certainly have been sequestered from dietary sponges as well. The third major group of compounds (25%), which is comprised of terpenoids and polyketides, has been shown by stable isotope-feeding studies to be biosynthesized by the nudibranchs de novo. Diaulusterol A (**85**) and the cholic acid derivatives **8** and **9** represent examples of skin metabolites that might have an intermediate origin involving biosynthetic modification of dietary compounds.

The biosynthetic capabilities of northeastern Pacific nudibranchs are quite remarkable. Feeding studies have shown that they can make degraded monoterpenoids (i.e., **101**), sesquiterpenoids (i.e., acanthodoral (**4**)), diterpenoids (i.e., **15**), and degraded sesterterpenoids (i.e., luteone (**55**)). Included in the terpenoids made de novo by the nudibranchs are the first examples of the nanaimoane, acanthodorane, isoacanthodorane, luteane, and cadlinalane carbon skeletons. One northeastern Pacific nudibranch, *Triopha catalinae*, is capable of polyketide biosynthesis using both acetate and butyrate building blocks. Another nudibranch, *Diaulula sandiegensis*, makes a butyrate fragment attached to the side-chain of a steroid. Stable isotope-feeding experiments have shown that nine of the 28 species of nudibranchs studied from this coast are capable of de novo biosynthesis of their skin metabolites and three more (*D. kresii*, *D. albopunctata*, *D. janaina*) are assumed by analogy with *D. limbata* (Cimino et al. 1983) to also make their compounds de novo. Eleven of these are in the Suborder Doridina and one is in the Suborder Dendronotina. Only 13 of the 28 species chemically investigated are found along the more northerly extremes of the coast from BC to Alaska. Nine of these 13, a very large percentage, are capable of de novo biosynthesis. It is interesting to compare this to tropical nudibranchs, which have very rich skin chemistry; and it appears that most if not all of their skin chemistry is sequestered from their diets (Avila 1995; Cimino et al. 1999, 2001).

Tropical reef habitats have a high biodiversity of sponges, bryozoans, soft corals, and ascidicans; and these tropical invertebrates are a very rich source of bioactive secondary metabolites (Faulkner 2002). Therefore, tropical nudibranchs have ready access in their diets to sequesterable defensive chemicals and consequently they should have little need to make repugnant chemicals via de novo biosynthesis. In contrast, the cold waters along the northern coast of North America from BC to Alaska have a relatively low biodiversity of sponges, bryozoans, soft corals, and ascidicans and the organisms found there are much less rich in bioactive secondary metabolites. Under these latter conditions, the capability to make defensive chemicals de novo should enable nudibranchs to occupy habitats that lack a suitable dietary source of bioactive chemicals. These simple arguments might explain the high percentage of de novo biosynthesis in cold water nudibranchs.

Two northeastern Pacific nudibranchs, *C. luteomarginata* and *Diaulua sandiegensis*, simultaneously sequester dietary metabolites and carry out de novo biosynthesis. There is preliminary evidence that both might regulate the de novo production of skin metabolites according to the availability of dietary sources of defensive chemicals. These two molluscs possibly represent an intermediate step in the evolution of nudibranch chemical defenses (Faulkner and Ghiselin 1983; Cimino and Ghiselin 1999). They simultaneously use the presumably more primitive process of pure sequestration and the undoubtedly more highly evolved process of pure de novo biosynthesis to acquire their skin metabolites.

References

Andersen RJ, Sum FW (1980) Farnesic acid glycerides from the nudibranch *Archidoris odhneri*. Tetrahedron Lett 21:797–800

Avila C (1995) Natural products of opisthobranch molluscs: a biological review. Oceanogr Mar Biol Annu Rev 33:487–559

Ayer SW, Andersen RJ (1982) Steroidal antifeedants from the dorid nudibranch *Aldisa sanguinea cooperi*. Tetrahedron Lett 23:1039–1042

Ayer SW, Andersen RJ (1983) Degraded monoterpenes from the opistobranch mollusc *Melibe leonina*. Experientia 39:255–256

Ayer SW, Hellou J, Tischler M, Andersen RJ (1984a) Nanaimoal, a sesquiterpenoid aldehyde from the dorid nudibranch *Acanthodoris nanaimoensis*. Tetrahedron Lett 25:141–144

Ayer SW, Andersen RJ, Cun-heng H, Clardy J (1984b) Acanthodoral and isoacanthodoral, two sesquiterpenoids with new carbon skeletons from the dorid nudibranch *Acanthodoris nanaimoensis*. J Org Chem 49:2653–2654

Barsby T, Linington RG, Andersen RJ (2002) De novo terpenoid biosynthesis by the dendronotid nudibranch *Melibe leonina*. Chemoecology 12:199–202

Behrens DW (1980) Pacific coast nudibranchs. Sea Challengers, Los Osos

Behrens DW (2004) Pacific coast nudibranchs, supplement II. New species to the Pacific coast and new information on the oldies. Proc Calif Acad Sci 55:11–54

Burgoyne DL, Dumdei EJ, Andersen RJ (1993) Acanthenes A to C: a chloro, isothiocyanate, formamide sesquiterpene triad isolated from the northern pacific sponge *Acanthella* sp. and the dorid nudibranch *Cadlina luteomarginata*. Tetrahedron 49:4503–4510

Burreson BJ, Scheuer P, Finer J, Clardy J (1975) 9-Isocyanopupukeanane, a marine invertebrate allomone with a new sesquiterpene skeleton. J Am Chem Soc 97:4763–4764

Carte B, Faulkner DJ (1983) Defensive metabolites from three nembrothid nudibranchs. J Org Chem 48:2314–2318
Carte B, Faulkner DJ (1986) Role of secondary metabolites in feeding associations between a predatory nudibranch, two grazing nudibranchs, and a bryozoan. J Chem Ecol 12:795–804
Cimino G, Ghiselin MT (1999) Chemical defense and evolutionary trends in biosynthetic capacity among dorid nudibranchs (Mollusca: Gastropoda: Opisthobranchia). Chemoecology 9:187–207
Cimino G, De Rosa D, De Stephano S, Minale L (1974) Isoagatholactone, a diterpene of a new structural type from the sponge *Spongia officinalis.* Tetrahedron 30:645–649
Cimino G, De Rosa S, De Stephano S, Sodano G, Villani G (1983) Dorid nudibranch elaborates its own chemical defense. Science 219:1237–1238
Cimino G, Fontana A, Gavagnin M (1999) Marine opisthobranch molluscs: chemistry and ecology in sacoglossans and dorids. Curr Org Chem 3:327–372
Cimino G, Ciavatta ML, Fontana A, Gavagnin M (2001) Metabolites of marine opisthobranchs: chemistry and biological activity. In: Tingali C (ed) Bioactive compounds from natural sources. Taylor and Francis, London, pp 577–637
Dumdei EJ, Kubanek J, Coleman JE, Pika J, Andersen RJ, Steiner JR, Clardy J (1997) New terpenoid metabolites from the skin extracts, an egg mass, and dietary sponges of the northeastern pacific dorid nudibranch *Cadlina luteomarginata.* Can J Chem 75:773–789
Edmunds MJ (1966) Protective mechanisms in the *Eolidacea* (Mollusca: Nudibranchia). Zool J Linn Soc 46:27–71
Edmunds MJ (1968) Acid secretion in some species of *Doridacea* (Mollusca, Nudibranchia). Proc Malacol Soc London 38:121–133
van Engen D, Clardy J, Kho-Wiseman E, Crews P, Higgs MD, Faulkner DJ (1978) Violacene: a reassignment of structure. Tetrahedron Lett 19:29–32
Faulkner DJ (2002) Marine natural products. Nat Prod Rep 19:1–48
Faulkner DJ, Ghiselin MT (1983) Chemical defense and evolutionary ecology of dorid nudibranchs and some other opisthobranch gastropods. Mar Ecol Prog Ser 13:295–301
Faulkner DJ, Molinski TF, Andersen RJ, Dumdei EJ, de Silva ED (1990) Geographical variation in defensive chemicals from pacific coast dorid nudibranchs and some related marine molluscs. Comp Biochem Physiol 97C:233–240
Gavagnin M, Fontana A (2000) Diterpenes from marine opisthobranch molluscs. Curr Org Chem 4:1201–1248
Gavagnin M, Ungur N, Castellucio F, Munianin C, Cimino G (1999) New minor diterpenoid diacylglycerols from the skin of the nudibranch *Anisodoris fontaini.* J Nat Prod 62:269–274
Graziani EI, Andersen RJ (1996a) Investigations of sesquiterpenoid biosynthesis by the dorid nudibranch *Acanthodoris nanaimoensis.* J Am Chem Soc 118:4701–4702
Graziani EI, Andersen RJ (1996b) Stable isotope incorporation evidence for a polyacetate origin of the acyl residues in triophamine, a diacylguanidine metabolite obtained from the dorid nudibranch *Triopha catalinae.* Chem Commun 1996:2377–2378
Graziani EI, Andersen RJ, Krug PJ, Faulkner DJ (1996) Stable isotope incorporation evidence for the de novo biosynthesis of terpenoic acid glycerides by dorid nudibranchs. Tetrahedron 52:6869–6878
Gustafson K, Andersen RJ (1982) Triophamine, a unique diacylguanidine from the dorid nudibranch *Triopha catalinae* (Cooper). J Org Chem 47:2167–2169
Gustafson K, Andersen RJ (1985) Chemical studies of British Columbia nudibranchs. Tetrahedron 41:1101–1108
Gustafson K, Andersen RJ, Chen MHM, Clardy J, Hochlowski JE (1984) Terpenoic acid glycerides from the dorid nudibranch *Archidoris montereyensis.* Tetrahedron Lett 25:11–14

Gustafson K, Andersen RJ, Cun-heng H, Clardy J (1985) Marginatafuran, a furanoditerpene with a new carbon skeleton from the dorid nudibranch *Cadlina luteomarginata*. Tetrahedron Lett 26:2521–2524
Hellou J, Andersen RJ, Rafii S, Arnold E, Clardy J (1981) Luteone, a twenty-three carbon terpenoid from the dorid nudibranch *Cadlina luteomarginata*. Tetrahedron Lett 22:4173–4176
Hellou J, Andersen RJ, Thompson JE (1982) Terpenoids from the dorid nudibranch *Cadlina luteomarginata*. Tetrahedron 38:1875–1879
Hochlowski JE, Faulkner DJ (1981) Chemical constituents of the nudibranch *Chromodoris marislae*. Tetrahedron Lett 22:271–274
Hochlowski JE, Walker RP, Ireland C, Faulkner DJ (1982) Metabolites of four nudibranchs of the genus *Hypselodoris*. J Org Chem 47:88–91
Hochlowski JE, Faulkner DJ, Matsumoto GK, Clardy J (1983a) Norrisolide, a novel diterpene from the dorid nudibranch *Chromodoris norrisi*. J Org Chem 48:1142–1144
Hochlowski JE, Faulkner DJ, Bass LS, Clardy J (1983b) Metabolites of the dorid nudibranch *Chromodoris sedna*. J Org Chem 48:1738–1740
Johannes RE (1963) A poison-secreting nudibranch (Mollusca: Opisthobranchia). Veliger 5:104–105.
Krug PJ, Boyd KG, Faulkner DJ (1995) Isolation and synthesis of tanyolides A and B, metabolites of the nudibranch *Sclerodoris tanya*. Tetrahedron 51:11063–11074
Kubanek J (1998) Chemical studies on the origin of secondary metabolites in selected marine invertebrates. Ph.D. thesis, UBC, Vancouver
Kubanek J, Andersen RJ (1997) Evidence for the incorporation of intact butyrate units in the biosynthesis of triophamine. Tetrahedron Lett 38:6327–6330
Kubanek J, Andersen RJ (1999) Evidence for de novo biosynthesis of the polyketide fragment of diaulusterol A by the northeastern Pacific dorid nudibranch *Diaulula sandiegensis*. J Nat Prod 62:777–779
Kubanek J, Graziani EI, Andersen RJ (1997) Investigations of terpenoid biosynthesis by the dorid nudibranch *Cadlina luteomarginata*. J Org Chem 62:7239–7246
Kubanek J, Faulkner DJ, Andersen RJ (2000) Geographic variation and tissue distribution of endogenous terpenoids in the northeastern Pacific ocean dorid nudibranch *Cadlina luteomarginata*: implications for the regulation of de novo biosynthesis. J Chem Ecol 26:377–389.
Molinski TF, Faulkner DJ (1986) Aromatic norditerpenes from the nudibranch *Chromodoris macfarlandi*. J Org Chem 51:2601–2603
Molinski TF, Faulkner DJ, Cun-heng H, van Duyne GD, Clardy J (1986) Three new rearranged spongian diterpenes from *Chromodoris macfarlandi*: reappraisal of the structures of dendrillolides A and B. J Org Chem 51:4564–4567
Okuda RK, Scheuer PJ, Hochlowski JE, Walker RP, Faulkner DJ (1983) Sesquiterpenoid constituents of eight porostome nudibranchs. J Org Chem 48:1866–1869
Piers E, Chong M, Gustafson K, Andersen RJ (1984) Total synthesis of (±)-triophamine. Can J Chem 62:1–5
Ricketts EF, Calvin J, Hedgpeth JW (1968) Between Pacific tides. Stanford University, Stanford
Stonard RJ, Petrovich JC, Andersen RJ (1980) A new C26 sterol peroxide from the opisthobranch mollusk *Adlaria* sp. and the sea pen *Virgularia* sp. Steroids 36:81–86
Thompson TE (1960a) Defensive acid-secretions in marine gastropods. J Mar Biol Assoc 39:115–122
Thompson TE (1960b) Defensive adaptations in opisthobranchs. J Mar Biol Assoc 39:123–134
Thompson JE, Walker RP, Wratten SJ, Faulkner DJ (1982) A chemical defense mechanism for the nudibranch *Cadlina luteomarginata*. Tetrahedron 38:1865–1873
Tischler M, Andersen RJ (1989) Glaciolide, a degraded diterpenoid with a new carbon skeleton from the nudibranch *Cadlina luteomarginata* and the sponge *Aplysilla glacialis*. Tetrahedron Lett 30:5717–5720

Tischler M, Andersen RJ, Choudhary MI, Clardy J (1991) Terpenoids from the sponge *Aplysilla glacialis* and specimens of the nudibranch *Cadlina luteomarginata* found on the sponge. J Org Chem 56:42–47
Walker RP, Faulkner DJ (1981) Chlorinated acetylenes from the nudibranch *Diaulula sandiegensis*. J Org Chem 46:1475–1478
Williams DE, Andersen RJ (1987) Terpenoid metabolites from skin extracts of the dendronotid nudibranch *Tochuina tetraquetra*. Can J Chem 65:2244–2247
Williams DE, Ayer SW, Andersen RJ (1986) Diaulusterols A and B from the skin extracts of the dorid nudibranch *Diaulula sandiegensis*. Can J Chem 64:1527–1529
Williams DE, Andersen RJ, van Duyne G, Clardy J (1987) Cembrane and pseudopterane diterpenes from the soft coral *Gersemia rubiformis*. J Org Chem 52:332–335
Wratten SJ, Fenical W, Faulkner DJ, Wekell JC (1977) Ptilosarcone, the toxin from the sea pen *Ptilosarcus gurneyi*. Tetrahedron Lett 18:1559–1562

Biogenetic Proposals and Biosynthetic Studies on Secondary Metabolites of Opisthobranch Molluscs

A. Fontana

Abstract. Marine chemical diversity is generated by a large number of transformations often not noted in terrestrial counterparts. Life in the oceans differs in most respects from life on land and our knowledge of the genetics and biochemistry of marine organisms is still very limited to a small number of species. Biosynthetic studies and biogenetic speculations can therefore be crucial in predicting relevant enzymes and their encoding genes, with a view to setting the stage for rational engineering of marine natural products. A further useful outcome to the identification of biosynthetic pathways is the resulting classification of natural products, which can serve to correlate chemical diversity and biodiversity. This review summarizes the present knowledge on secondary metabolites biogenesis in marine opisthobranchs, a class of organisms that has been emerging as a prolific source of structurally diverse metabolites possessing a broad variety of biological activities.

14.1 Introduction

Historically, the screening of crude extracts has proven to be an effective method for identifying organisms that produce useful, sometimes surprising, therapeutic compounds. Early studies, which focused on terrestrial plants and microorganisms, proved extremely fruitful, yielding novel metabolites that were developed into commercial products. By the early 1960s, researchers began to view the oceans as a new and untapped source of potential pharmaceutical molecules. Until 2001, over 10,000 novel compounds had been described from a variety of marine organisms, proving that the marine environment is a rich source of bioactive compounds (Proksch et al. 2002). Considering that only a small fraction of these molecules is currently known to be produced by culturable organisms (Burja et al. 2001), compound availability represents a major obstacle in the development of marine-derived products. Biosynthetic studies and biogenetic speculation play, therefore, an essential role in the prediction of

Angelo Fontana
Instituto di Chimica Biomolecolare (ICB-CNR)
Via Campi Flegrei 34, 80078 Pozzuoli (Naples), Italy

Progress in Molecular and Subcellular Biology
Subseries Marine Molecular Biotechnology
G. Cimino, M. Gavagnin (Eds.): Molluscs

relevant enzymes and their encoding genes, with a view to setting the stage for rational engineering of marine natural products. A further useful outcome to the identification of biosynthetic pathways is the resulting classification of natural products, which can serve to correlate chemical diversity and biodiversity. In this view, characterization of the relevant biosynthetic pathway instead of the substance also represents an ultimate way of addressing the symbiotic origin. Regularly updated reviews of a broad spectrum of biosynthetic pathways are available in the literature (Moore 1999; Rawlings 1999, 2001a,b; Knaggs 2001; Staunton and Weissman 2001; Cox 2002; Moore and Hertweck 2002; Garson and Simpson 2004).

In this review, I have set out to highlight recent literature on biogenesis in marine molluscs of the subclass Opisthobranchia, which are a proven source of structurally diverse natural products, possessing a broad variety of biological activities (Cimino et al. 1999, 2001; Faulkner 2002; Blunt et al. 2005). Attention is focused on experimental investigations of polyketide and terpene biosynthesis, although biogenetic speculation will be taken into account to discuss general schemes for the production of specific classes of metabolites isolated from taxonomically or metabolically related sources. Large parts of these studies have been previously covered in articles (Garson 1989, 1993, 2001; Cimino and Sodano 1993; Davies-Coleman and Garson 1998; Fontana and Cimino 2001; Garson 2001; Cimino et al. 2004). The outstanding series of reviews dedicated to *Marine Natural Products* should be consulted for structural information (Faulkner 2002; Blunt et al. 2005). An appealing interpretation in relating dietary preference and de novo biogenesis within the subclass Opisthobranchia has been proposed by Cimino and Ghiselin (1998, 1999).

14.2 Polyketides

Complex polyketides are a large and structurally diverse class of natural products that includes many different compounds with valuable biological properties. The unifying element of all these compounds is the biosynthetic mechanism that relies on successive rounds of Claisen-type decarboxylative condensation between an activated malonic acid derivative and an acyl thioester. The β-keto thioester is then processed by a set of *optional* reactions, including ketoreduction (KR), enoyl reduction (ER) and dehydration (DH). Polyketides are produced by stepwise chain assembly on multi-catalytic enzymes, named polyketide synthases (PKS), which are roughly classified as type I, type II and type III on the basis of their enzyme architecture (Staunton et al. 1996). As in non-ribosomal peptide synthases or fatty acid synthase, the separate set or module of enzymes that together catalyze each successive cycle of chain extension are often encoded together in the order in which they are used. The processive nature of these systems has generated increasing interest in the possibility of using chimeric enzymes to make new, potentially valuable natural products. It has been proven that non-natural hybrids may be constructed by deleting, inserting or swapping one or more domains from natural enzymes (Rowe et al. 2001; Moss et al. 2004), providing support to the prospects for combinatorial deployment of individual modules from diverse sources.

Polyketides are produced by many different organisms, from protists and bacteria to plants and fungi. Knowledge of polyketide biogenesis in marine systems is still limited and the metabolic pathways operating in marine organisms are inevitably discussed in relation to terrestrial processes, especially those occurring in bacteria and plants. One outstanding example of such an approach is the comparison (Davies-Coleman and Garson 1998) of the so-called siphonariid model (Garson et al. 1994) with Celmer's model for macrolide, such as erythromycin (Celmer 1965), and the PAPA model for polyethers (Cane et al. 1983).

Although the first experiment targeting opisthobranch polyketides dates back to the 1970s (Ireland and Scheuer 1979), the number of biosynthetic studies in these molluscs is insignificant in comparison to the large number of cyclic and linear structures isolated (Davies-Coleman and Garson 1998; Cimino et al. 1999, 2001; Garson 2001). In fact, whereas many orders of the subclass include species containing bioactive polyketide compounds, some with promising therapeutic applications, the biogenesis of these molecules is unknown and in many cases, such as for metabolites isolated from molluscs of the order Anaspidea (e.g. *Aplysia*, *Dolabella*, *Stylocheilus*), even the origin is uncertain (Cimino et al. 1999, 2001).

A restricted group of opisthobranchs of the order Sacoglossa shows the unique ability to host viable chloroplasts in specific parts of their body, assimilated after ingestion of siphonous algae (Cimino et al. 1999, 2001). Trench and coworkers were the first to show the transfer of organic carbon fixed during photosynthesis from plastids to mollusc tissues (Trench et al. 1969). In some cases, the photosynthesis-derived carbon sustains the sacoglossans for several months in the absence of an algal food source (Trench et al. 1969; Green et al. 2000), giving rise to the issue of whether lateral gene transfer has occurred between eukaryotic alga and animal nucleus (Pierce et al. 2003). The fixed carbon appears in a wide variety of primary and secondary metabolites of sea slugs, including γ-pyrone polypropionates, as proved by the labelling of 9,10-deoxytridachione (1) and photodeoxytridachione (2) of *Placobranchus ocellatus* after the addition of radioactive sodium hydrogen carbonate to aquarium water (Ireland and Scheuer 1979). The absence of pyrone-containing compounds in siphonous algae was read as indirect proof of the origin of polyketides in animal tissues. A few years later, using the committed precursor [1-^{14}C]-propionate, feeding experiments with the Mediterranean

Elysia viridis demonstrated this assumption, showing that a related pyrone, namely elysione (3) (Dawe and Wright 1986), is truly a product of sacoglossan metabolism (Gavagnin et al. 1994a). To date Elysioidean metabolites, with a backbone from six to eight propionate units, have been independently characterized by several different research groups. These compounds can be grouped in four structural types, represented by the skeleton of 9,10-deoxytridachione, photodeoxytridachione[1], tridachyahydropyrone and elysiapyrone (Cimino et al. 1999; Blunt et al. 2005; Cueto et al. 2005; Manzo et al. 2005).

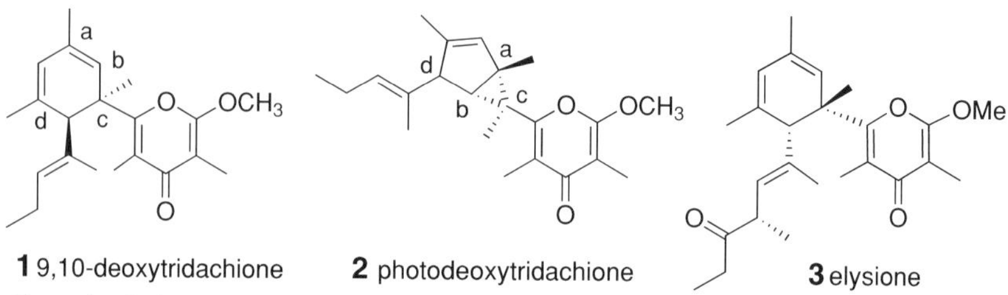

1 9,10-deoxytridachione **2** photodeoxytridachione **3** elysione

Formulae 1–3

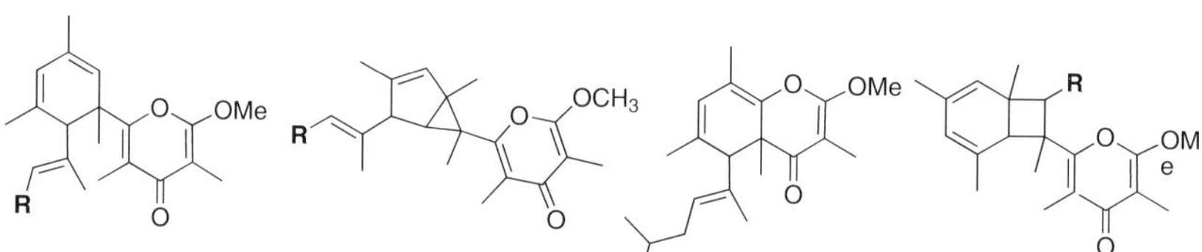

deoxytridachione skeleton type | photodeoxytridachione skeleton type | tridachiahydropyrone skeleton type | elysiapyrone skeleton type

polypropionate skeletons

As described in Fig. 14.1, stereochemical and structural considerations suggest that all these structural backbones may be the result of a unique process, with the only differences arising from the starter unit type (acetate, propionate or iso-valeriate) and the number of propionates added during the chain extension. Formation of the bicyclo[3,1,0]hexane of photodeoxytridachione has been proved to derive in vivo and in vitro by photoconversion of the deoxytridachione skeleton (Ireland and Scheuer 1979).

[1] Considering the recent synthesis of tridachiahydropyrone (Jeffrey et al. 2005), the structure of the sacoglossan metabolite should be revised. From a careful reading of the original paper (Gavagnin et al. 1990), it is likely that the stereochemistry of the substituents at C4 and C9 of the *Tridachia crispata* metabolite is *cis*, instead of *trans* as reported. This would be in agreement with the chemical shift differences occurring between the natural product and the synthetic compound.

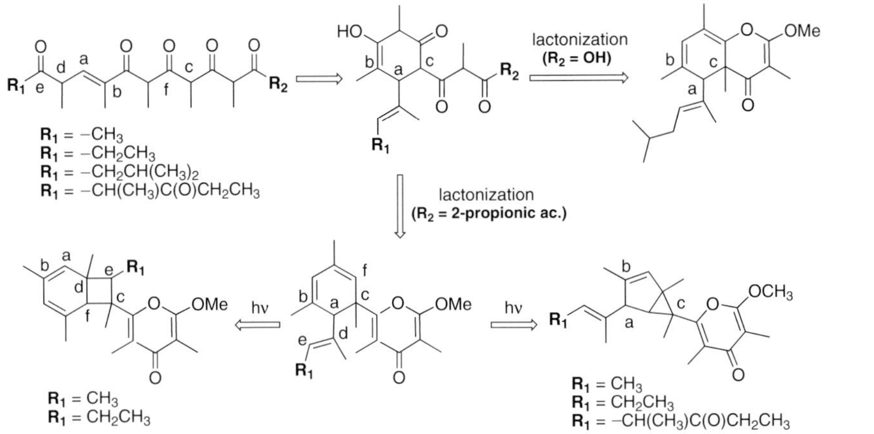

Fig. 14.1. Proposed biogenesis of polypropionate skeletons in opisthobranch molluscs of the family Elysioidea

Analogously, the bicyclo[4,2,0]octane of the elysiapyrones is suggested to involve a non-ezymatic electrocyclization through cleavage of the *a–c* bond and formation of the new linkages *d–f* and *e–c* (see Fig. 14.1; Cueto et al. 2005; Manzo et al. 2005). This mechanism is consistent with Black's proposal for the biogenesis of endriandic acids, a series of sponge compounds exhibiting a bicyclic core similar to that of elysiapyrones (Bandaranayake et al. 1980; Wickramasinge et al. 1980). It is, however, worth noting that attempts to repeat the incorporation with radioactive propionate in *Elysia timida*, another Mediterranean elysioidean, were unsuccessful in spite of the remarkable analogies with *P. ocellatus* (Gavagnin et al. 1994b). Negative results were also obtained by feeding experiments with deuterated propionate in *E. viridis* (Fontana, unpublished data).

De novo biosynthesis of polyketide-derived pyrones has also been documented in other two Mediterranean sacoglossans, namely *Cyerce cristallina* and *Ercolania funerea*, by feeding experiments with ^{14}C-labelled propionate (Di Marzo et al. 1991; Vardaro et al. 1992b). *Cyerce* contains a series of polypropionates (cyercenes, **4–10**) that are selectively compartmentalized in the mucous secretion, in the cerata and in the mantle (Di Marzo et al. 1991; Vardaro et al. 1991). Incorporation with $[2\text{-}^{14}C]$-propionate gave excellent incorporation in all cyercenes (Di Marzo et al. 1991). Although the authors are inclined to suggest a biogenetic origin of the metabolites from demethylation of regular polypropionate precursors, the cyercenes appear to be examples of compounds derived by mixed acetate/propionate biosynthesis, with occasional use of isobutyrate unit. Such a process is also consistent with the differences in the relative incorporations of labelled C_3 units that are effectively part of the polyketide skeletons of **4–10**.

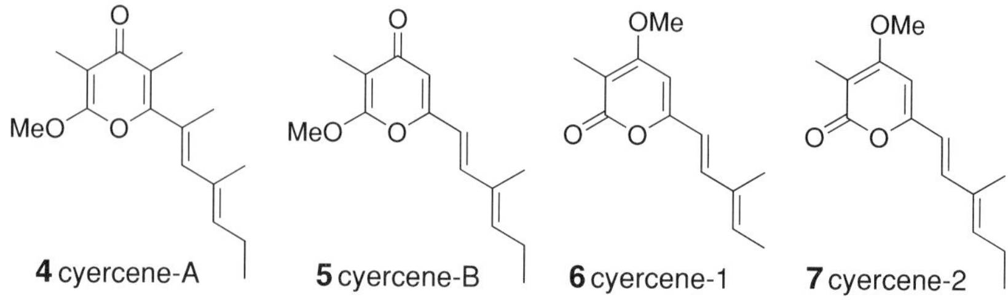

4 cyercene-A 57,200 cpm/mg **5** cyercene-B 18,416 cpm/mg **6** cyercene-1 14,656 cpm/mg **7** cyercene-2 22,847 cpm/mg

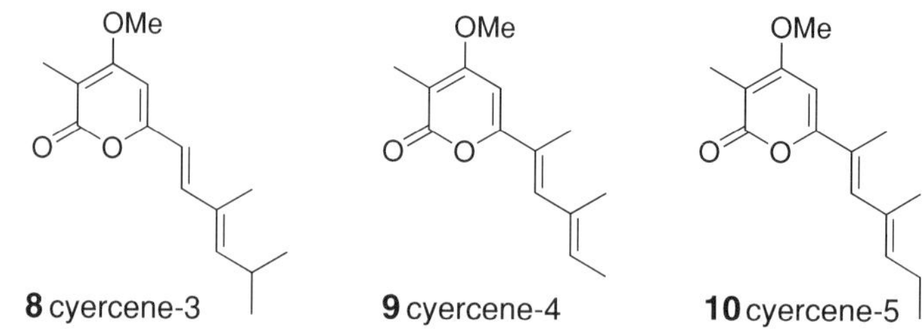

8 cyercene-3 18,947 cpm/mg **9** cyercene-4 28,783 cpm/mg **10** cyercene-5 18,305 cpm/mg

Formulae 4–10

Another Mediterranean opisthobranch, the sacoglossan *Placida dendritica* (Vardaro et al. 1992a; Cutignano et al. 2003a), possesses regular and irregular polypropionates (**11–19**) closely related to cyercenes. Although no feeding experiment has been carried out for testing the biosynthesis of placidenes, a de novo origin through a mixed acetate/propionate pathway has been suggested as the most plausible hypothesis (Cutignano et al. 2003a).

Interestingly, a recent paper sets out a possible photochemical link between cyercenes and placidenes, showing that in vitro placidene A (**11**) and isoplacidene A (**13**) are produced from cyercene A (**4**), when the latter is exposed to sunlight (Fig. 14.2; Zuidema and Jones 2005). In contrast, irradiation of **4** in the presence of Rose Bengal and oxygen leads to the hydroperoxide **19** through photochemical singlet oxygenation of the distal double bond (Fig. 14.2). The experiments represent a feasible explanation for the presence of isomeric forms of pyrone-containing polypropionates in these molluscs, such as the placidene/isoplacidene pairs (**11**/**13**, **12**/**14**) or cyercene 5 (**10**)/compound **21** and 7-methylcyercene B(**20**)/compound **22** isolated from *C. cristallina* (Vardaro et al. 1991) and *C. nigricans* (Hay et al. 1989; Roussis et al. 1990). However, the suggested hypothesis of the origin of placidenes from a photorearrangement in vivo of cyercene-like precursors appears not plausible in consideration of the fact that no organism is known to contain both series of compounds.

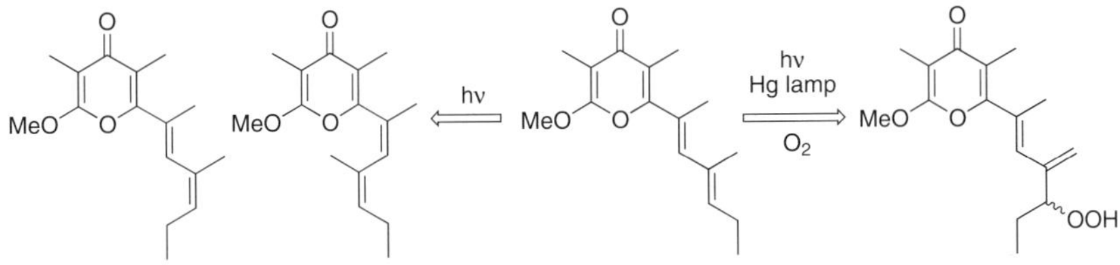

11 placidene A **13** isoplacidene A **4** cyercene A **19**

Fig. 14.2. Photoreactions of cyercene A

While incorporation of radioactive precursors has demonstrated at least the de novo origin of sacoglossan polyketides (*C. cristallina*, *E. funerea*), currently the true source of the carbon skeletons of pyrone-containing propionates remains uncertain, since similar molecules have been also reported from terrestrial fungi. Interestingly, feeding experiments in these latter organisms, namely the plant pathogens *Leptosphaeria maculans/Phoma lingam* and *Gliocladium vermoesenii* (Avent et al. 1992; Pedras and Chumala 2005), reveal that the polypropionate chains arise from methionine-mediated methylation of polyacetate precursors. This may represent an important divergence in the biogenesis of pyrone polyketides in the two groups of organisms (Davies-Coleman and Garson 1998), although further research on polyketide metabolism in marine molluscs should adequately consider these aspects.

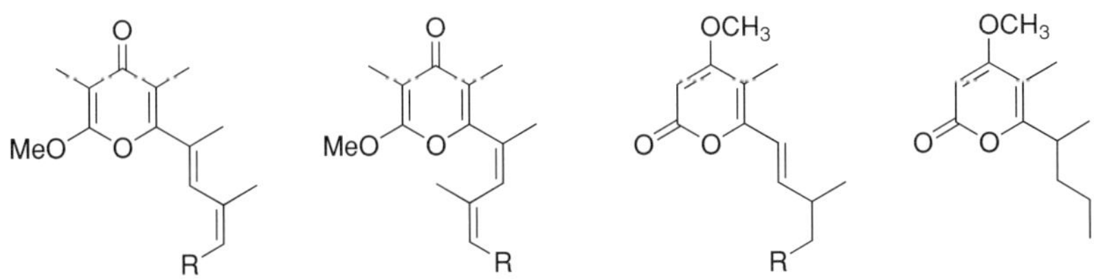

11 R = CH_2CH_3
12 R = CH_3

13 R = CH_2CH_3
14 R = CH_3

15 R = CH_3
16 R = CH_2CH_3

17

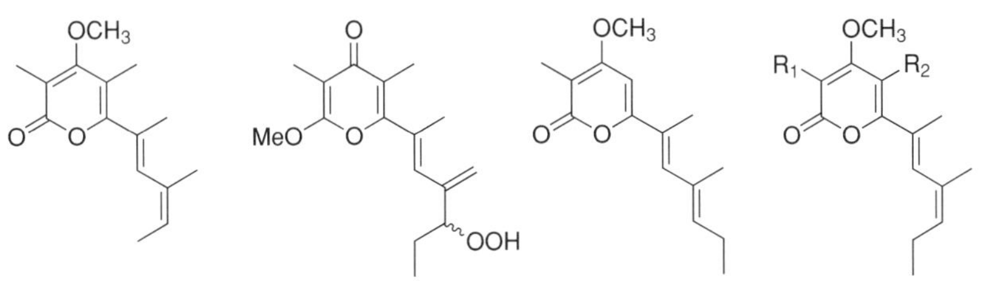

18

19

20

21 R_1 = H; R_2 = CH_3
22 R_1 = CH_3; R_2 = H

Formulae 11–22

In the 1980s, Cimino and coworkers suggested (Cimino et al. 1986, 1989) that the biosynthesis of the sponge alkaloids sarains [e.g. sarain 1 (**23**) and sarain A (**24**)] may involve a macrocyclic structure (**25**) derived by addition of two 3-alkylpyridines molecules, then known as bioactive constituents of the sea slug *Navanax inermis* (Sleeper and Fenical 1977; Sleeper et al. 1980; Fig. 14.3). The same proposal also put forth that head-to-tail addition of 3-alkylpyridine (or the biosynthetically equivalent 3-alkylpiperidine) units gave a reliable explanation of the origin of the cyclic and linear skeletons of petrosin (**26**), xestospongin (**27**) and halitoxins (**28**). The key elements of Cimino's hypothesis were subsequently used by several other authors to address the biogenesis of structurally complex marine alkaloids (for a more detailed discussion, see Andersen et al. 1996), until Baldwin and Whitehead (1992) published a milestone paper suggesting an *endo* Diels–Alder intramolecular cyclization of a *bis*-dihydropyridine intermediate for the biosynthesis of the complex structures of manzamines (e.g. manzamine B, **29**; Baldwin and Whitehead 1992). The basic principles of this proposal involve the direct formation of 3-alkyl-dihydropyridine units from ammonia, a C_{10} symmetrical dialdehyde and acrolein or an equivalent C_3 unit. As put in evidence by Kobayashi (Kobayashi et al. 1994) and Andersen (Kong et al. 1995), the biogenetic hypothesis also foresaw the presence of two intermediates that corresponded exactly with the skeletons of ircinals (e.g. ircinal B, **30**), isolated from *Amphimedon* sp. (Kondo et al. 1992), and ingenamine (**31**), isolated from *Xestospongia ingens* (Kong et al. 1995).

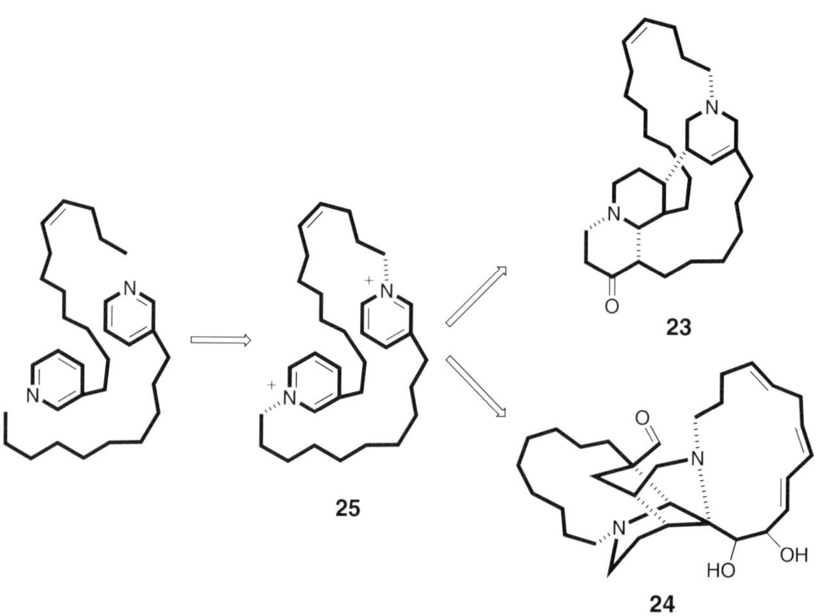

Fig. 14.3. Cimino's proposal for the biosynthesis of macrocyclic alkaloids in the sponge *Reniera sarai*

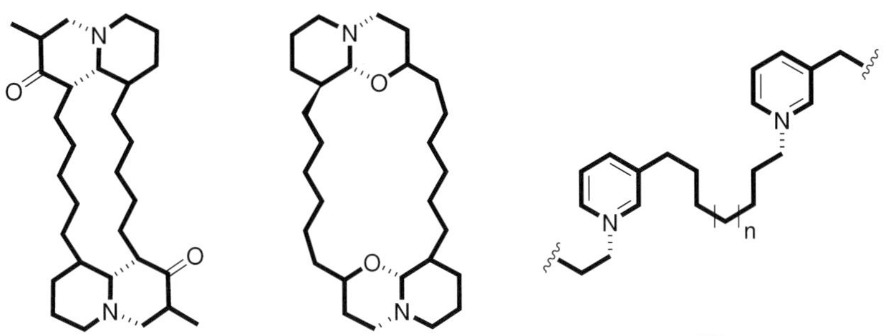

26 **27** **28**

Formulae 26–28

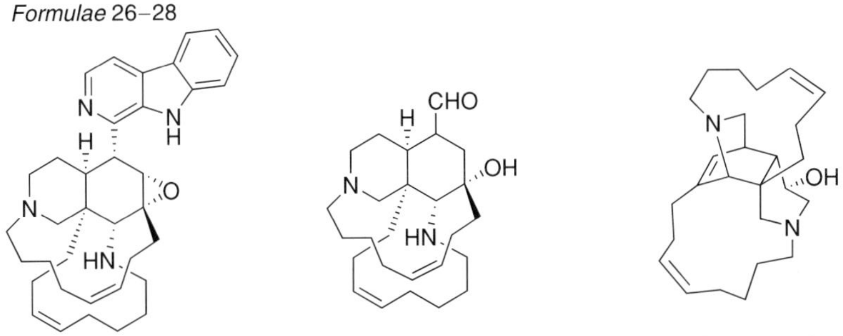

29 manzamine B **30** ircinal B **31** ingenamine

Formulae 29–31

In support to the Diels–Alder mechanism, a few years later Baldwin et al. (1998) published the biomimetic synthesis of keramaphidin B (**32**; Kobayashi et al. 1994; Tsuda et al. 1996), a dihydro derivative of the ingenamine-type cycloadduct suggested as the precursor of manzamines (Fig. 14.4). However, while the aptitude of alkylpyridine to polymerize is directly confirmed by the isolation of alkylpyridinium oligomers and polymers in different genera of sponges (Andersen et al. 1996 and references therein; Sepčić et al. 1997), the hypothetical condensation of C_{10}-dialdehyde, ammonia and acrolein to give alkylpyridine monomers is at present open to conjecture. In contrast, biosynthetic studies with marine molluscs of the genus *Haminoea* have recently demonstrated that the 3-alkylpyridine motif is polyketide-derived, arising from condensation of nicotinic acid and acetate units (Cutignano et al. 2003b, 2004).

Haminoea are shelled opisthobranchs containing a series of oxygenated 3-alkylyridine derivatives, generically named haminols [e.g. haminol-A (**33**) from *H. navicula*; haminol-1 (**34**), haminol-2 (**35**) from *H. orbignyana*], differing only in the position and number of the double bonds at C-2 of the linear chain (Spinella et al. 1993b; Marin et al. 1999). When molested, the molluscs release a white mucus containing haminols, which serve as alarm pheromones, inducing an escape reaction in conspecifics (Cimino et al. 1991; Spinella et al. 1993b). Feeding experiments were carried out with the Mediterranean *H. orbignyana* and the origin of the 3-alkylpyridine motif was elucidated by incorporation of deuterated nicotinic acid and ^{13}C-acetate into the molecular skeleton of haminol-1 (**34**) and haminol-2 (**35**; Cutignano et al. 2003b, 2004).

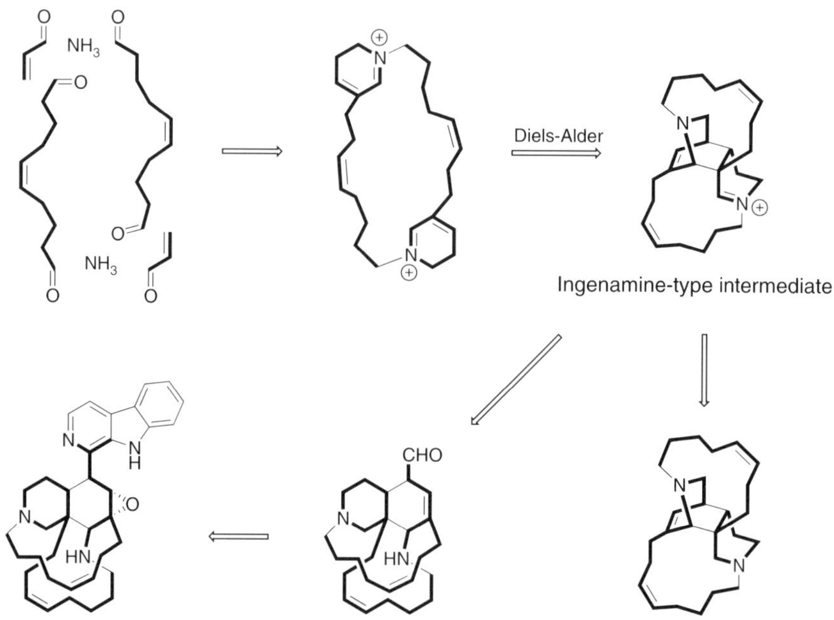

Fig. 14.4. Baldwin and Whitehead's proposal for the biosynthesis of manzamines from ammonia, C_3-unit and C_{10}-α,ω-dihaldehyde

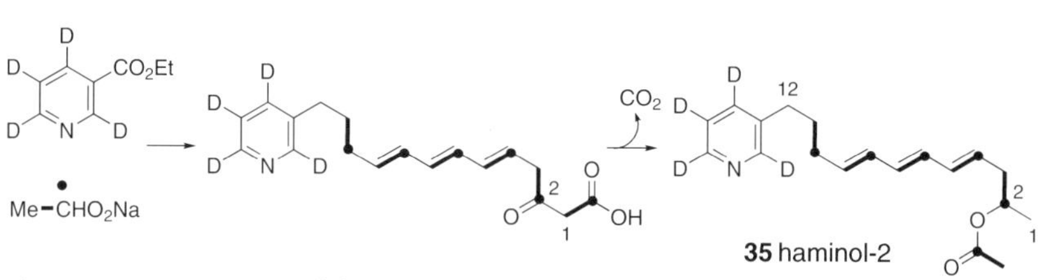

Fig. 14.5. Incorporation of d_4-nicotinic acid ethyl ester, [1-^{13}C]-acetate and [1,2-$^{13}C_2$]-acetate into haminol-2 (**35**)

As indicated in Fig. 14.5, MS and ^{2}H NMR analysis of haminol-2 (**35**) purified from animals fed with d_4-nicotinic acid showed that no deuterium atom was lost during the biosynthesis. Since dihydropyridine intermediates are required for the decarboxylation of the nicotinic acid (Mann 1994), the presence of all deuterium atoms indicated the preservation of the carboxylic carbon. Retention of deuterium proved, therefore, the incorporation of an intact molecule of the precursor, implying that the pyridine ring and the first carbon of haminol chain (C-12) derived from an intact unit of nicotinic acid. This entity serves as a starter unit of the emerging 3-alkylpyridine molecule that is assembled by progressive addition of acetate units. In particular, carbons C-2, C-4, C-6, C-8 and C-10 are all derived from C-1 of acetate, as proved by enrichment of the corresponding signals in the ^{13}C NMR spectrum. A subsequent study with $^{13}C_2$-acetate confirmed that five intact C_2 units were incorporated into the molecule. In agreement with the origin of C-12 from the carboxylic carbon of nicotinic acid, the related ^{13}C signal did not show coupling with other carbons. The signal of C-1 of haminol-2 (35) also appeared as a simple singlet due to the absence of coupling with vicinal carbons, thus revealing cleavage of the final C_2 unit and loss of the terminal carbon, probably through decarboxylation of the PKS-released intermediate (Fig. 14.5).

The ability of the *Haminoea* PKS to use nicotinic acid as a starter unit is unprecedented in nature. The double-bond positions in haminol-1 (**34**) and haminol-2 (**35**) suggest that the domain organization of the putative PKS does not correspond to the traditional modular pattern of this class of enzymes. Considering a processive assembly, Fig. 14.6 suggests a sequence of events consistent with the observed functionalization of the haminol chain. The probable sequence of operations starts with loading a unit of nicotinic acid. The maintenance of the carbonyl group at C-10 of the final structure directs the succeeding elimination of water to give the

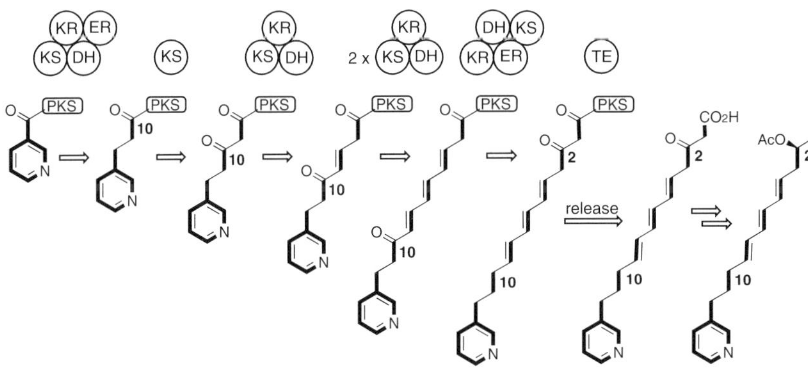

35 haminol-2

Fig. 14.6. Proposed processive assembly of haminol-2 in the marine mollusc *Haminoea orbygniana* through sequential addition of acetate unit. For simplicity, activities due to acyl transferase and acyl carrier protein have been omitted. *KS* β-ketoacylsynthase, *KR* β-ketoacylreductase, *DH* dehydratase, *ER* enoyl reductase, *TE* thioesterase

E double bond between C-8 and C-9. A repetitive process is consistent with the formation of the conjugated triene residue in the nascent chain. The final removal of the oxygenated function at C-10 anticipates the release of the β-hydroxy acid from which a haminol-1 skeleton is formed by decarboxylation. The sequence of events may not correspond in every detail to that predicted on the basis of the labelling results and product structure. There is also a chance that the released molecule could be more fully oxidized than that shown and, in addition to decarboxylation and acetylation, post-PKS modifications could include double-bond isomerization in a way that is consistent with the structure of haminols.

Considering that *Haminoea* metabolites are structurally similar to monomeric alkylpyridines isolated from sponges, e.g. theonelladins A (**36**) or ikimines (**37**), the PKS origin of haminols provides a mechanistic framework that can be generalized to formulate a biogenetic proposal for the formation of sarains, manzamines and the other aforementioned sponge alkaloids. As shown in Fig. 14.7, the hypothesis integrates the

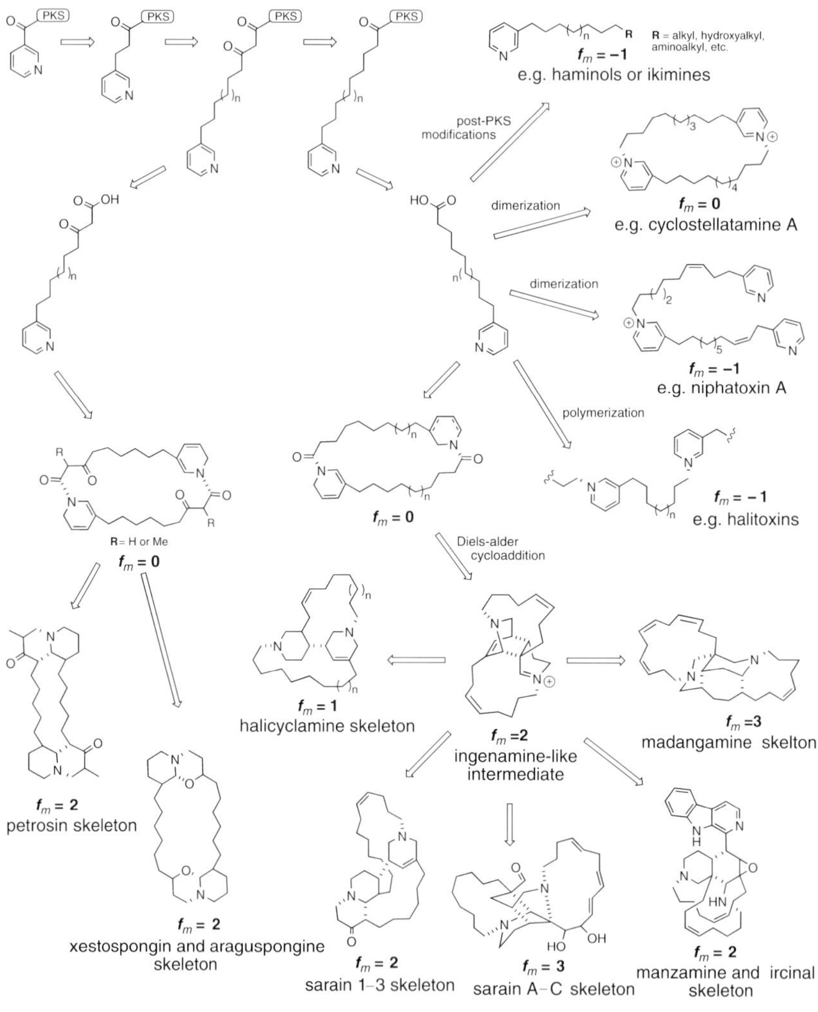

Fig. 14.7. An overall scheme for PKS biogenesis of 3-alkylpyridine alkaloids in marine molluscs and sponges. The hypothesis receives the key points previously suggested (see text). f_m, metabolic factor

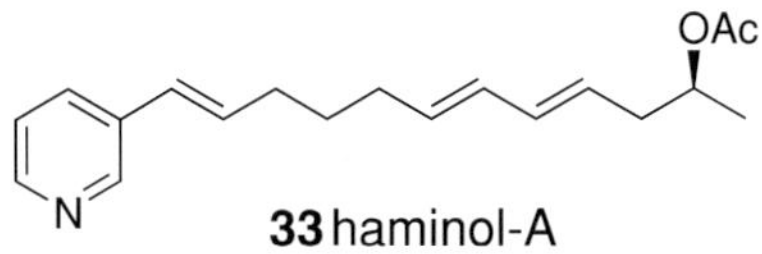

33 haminol-A

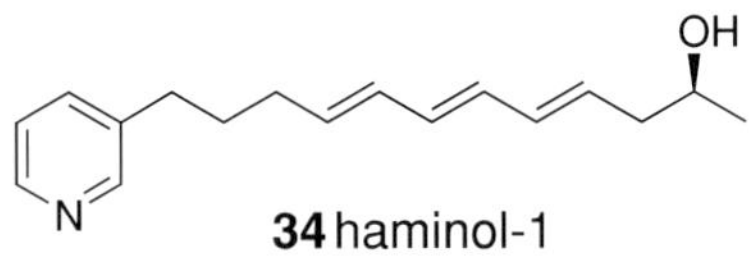

34 haminol-1

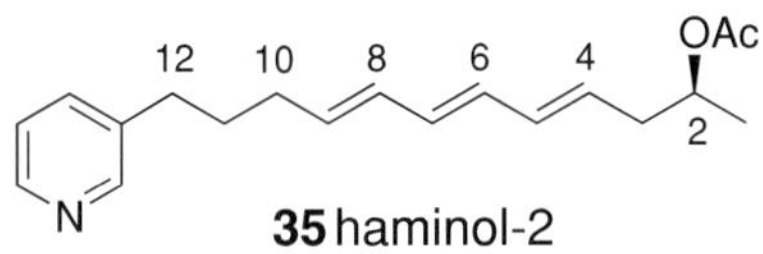

35 haminol-2

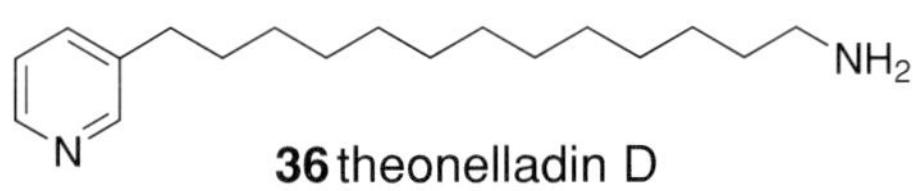

36 theonelladin D

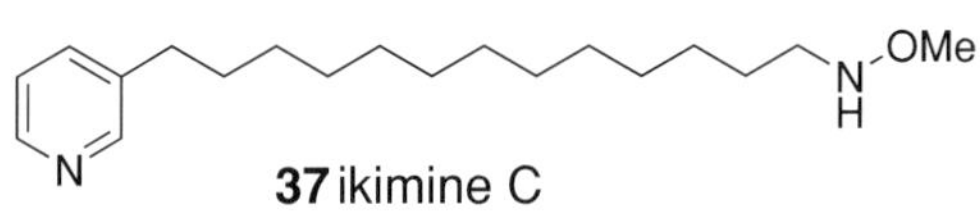

37 ikimine C

Formulae 33–37

suggestions put forth to date, including Cimino's idea of a dimeric *bis*-alkylpyridine intermediate and Baldwin and Whitehead's intuition of the intramolecular [4 + 2] cycloaddition. In consideration of the structural features of the alkaloids reported so far, it is possible to identify five subclasses of compounds. These are monomers, linear dimers, cyclic dimers, linear oligomers and polymers of the 3-alkylpyridine motif. In analogy with haminol biogenesis, it is conceivable that 3-(ω-carboxyalkyl)-pyridine or oxygenated analogues are the final product of the PKS sequence through a classic termination step catalyzed by thioesterase. After release, these entities undergo post-PKS modifications that include, for example, reduction, bond isomerization, decarboxylations, methylation and *trans*-amination (see structures of haminols, nyphatynes, niphatesines, ikimines). The remaining members of the family share the presence of at least one intramolecular bond from two or more 3-alkylpyridine units. The entire class can be organized by introducing a *metabolic factor* (f_m) that takes into account the biosynthetic complexity of each member on the basis of the number of linkages between two different units (interaction number, n_i) and the number of monomers (polymerization number, n_p):

$$f_m = (n_i - n_p).$$

The f_m index reflects the transformation of the alkylpyridine derivative besides the simple polymerization. In this view, monomers, linear oligomers, including niphatoxins (Talpir et al. 1992) or polymers, which derive from head-to-tail condensation of single units, are featured by a *metabolic factor* of −1. The factor f_m is 0 in cyclic dimers, such as cyclostellatamines (Fusetani et al. 1994), or the biosynthetic *bis*-dyhydropyridine intermediates (Cimino et al. 1986; Baldwin and Whitehead 1992) that should derive from two head-to-tail condensations of two 3-alkylpyridine units, followed by reduction of the amide groups.

Dimerization of polyketide products is not infrequent in nature and recent examples of such post-PKS transformations are the structures of the anthraquinone actinorhodin (**38**) from *Streptomyces coelicolor* (Bartel et al. 1990; Revill et al. 1995) and macrocyclic dilactone SCH 351448 (**39**) from *Micromonospora* (Hegde et al. 2000). Biosynthesis of dimeric polyketides in marine molluscs has been demonstrated for the diacylguanidine triophamine (**40**) that is produced by the nudibranch *Triopha catalinae* through an unusual acetate/butyrate pathway (Fig. 14.8; Kubanek and Andersen 1997).

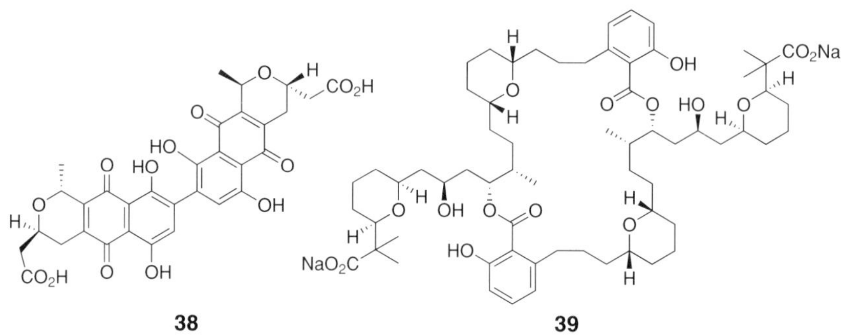

Formulae 38–39

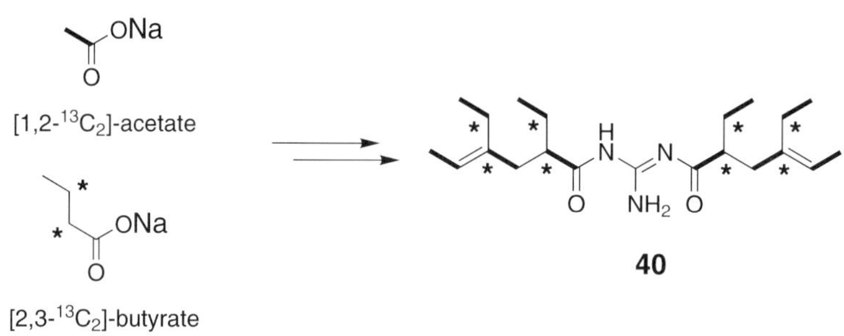

Fig. 14.8. Labelling pattern of triophamine (**40**) after injection of ^{13}C-labelled acetate and butyrate into specimens of the nudibranch *Triopha catalinae*

In agreement with the proposal of Cimino et al. (1986) and Kitagawa and Kobayashi (1993), further transformations of the macrocyclic *bis*-pyridine systems give the highly symmetric scaffold of petrosins ($f_m = 2$), xestospongins ($f_m = 2$), araguspongines ($f_m = 2$) and other related compounds. As in haminol biosynthesis, it is possible that the oxygen of the quinolazidine and oxaquinolizadine systems of these latter compounds may derive directly from PKS-inserted acetate units (Fig. 14.7). Biosynthesis of the most complex alkylpyridine alkaloids relies on Baldwin and Whitehead's suggestion and incorporates other elements already discussed by other authors (Andersen et al. 1996). The key aspect of this process is the intramolecular [4 + 2] cycloaddition to give the ingenamine-type intermediate ($f_m = 2$) that precedes the formation of the intriguing structures of sarains 1–3 ($f_m = 2$), manzamines ($f_m = 2$), sarains A–C ($f_m = 3$) and madangamines ($f_m = 3$).

Besides the genus *Haminoea*, the order Cephalaspidea embraces many other species possessing molecules with polyketide motifs, although the origin of the large majority of these compounds is still unexplored. Recently, labelling of the linear polyketides aglajne-1 (**41**) and aglajne-3 (**42**) was obtained in Mediterranean *Bulla striata* with sodium [1-^{14}C]-propionate (Fontana et al. 2004; aglajne-2 was not considered during this study). The radioactive precursor was supplied by injection into the digestive gland through the mollusc's shell. Animals were incubated for after three days, and labelled aglajnes were recovered by silica gel column. The authenticity of the incorporation was ascertained by radio-HPLC in both reversed and normal phase, confirming the specific labelling of the carbon skeletons of **41** and **42**. The data showed a coherent presence of radioactivity only in the fractions containing the polypropionates, with levels which, although low (incorporation yield less than 0.01%), were always much higher than those recorded for the other fractions and background. Since conversion of [1-^{14}C]-propionate to acetate would lead to loss of radioactivity, this experiment suggested propionate incorporation into the polyketide skeleton rather than the alternative addition of methyl groups to an acetate-derived backbone (Fig. 14.9).

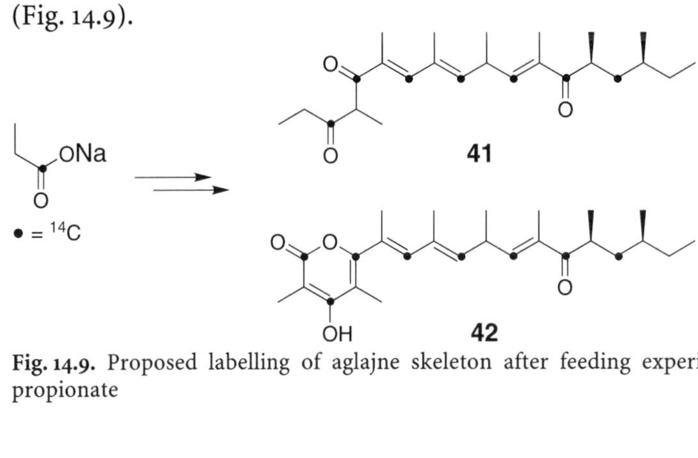

Fig. 14.9. Proposed labelling of aglajne skeleton after feeding experiment with [1-^{14}C]-propionate

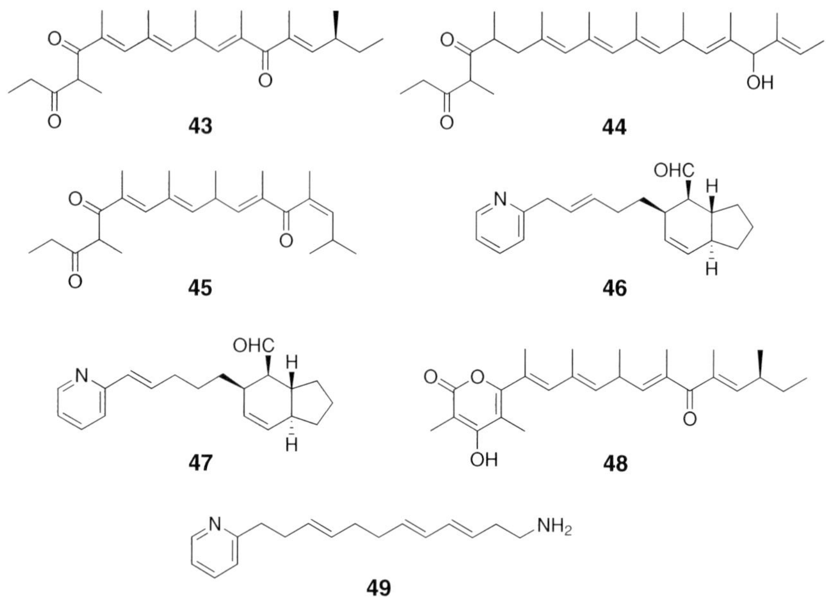

Formulae 43–49

Formation of aglajnes and haminols provides a suggestive tool for interpreting the polyketide biogenesis in cephalaspidean molluscs. The structure and stereochemistry of aglajne-1 are similar to those of niuhinone-B (**43**) and nalodionol (**44**). Compound **43** was reported from a Pacific *Philinopsis* (*Aglaja*) *speciosa*, together with niuhinone-A (**45**) and pulo'upone (**46**; Coval et al. 1985), as well as from a Pacific *B. gouldiana*, together with isopulo'upone (**47**) and dehydroaglajne 3 (**48**; Spinella et al. 1993a). Nalodionol (**44**), together with naloamine (**49**), is the major component of *Smaragdinella calyculata* (Cephalaspidea, Smaragdinellidae; Szabo et al. 1996). Considering a C_3-based pathway similar to that proposed for *B. striata*, the de novo biosynthesis of polypropionate pairs **41**/**43** and **42**/**48** should involve an acetate starter unit and eight propionates for the elongation steps. In agreement with Fig. 14.10, the processive reactions leading to the metabolites isolated from the Mediterranean and Pacific species of *Bulla* differ only in the enoyl reductase step following the addition of the second propionate unit. In contrast, polyketide synthesis in *Smaragdinella* appears to imply a different processing of the emerging chain, although the entire sequence is not different from that suggested for *Bulla* (Fig. 14.10). In both genera, it is presumably that release of the polypropionate product is followed by a decarboxylation step to give **41**, **43** and **45**. Remarkably, the parallelism between the two genera may also concern the biogenesis of the pyridine derivatives pulo'upone (**46**), isopulo'opone (**47**) (from *B. gouldiana*), and naloamine (**49**) (from *S. calyculata*).

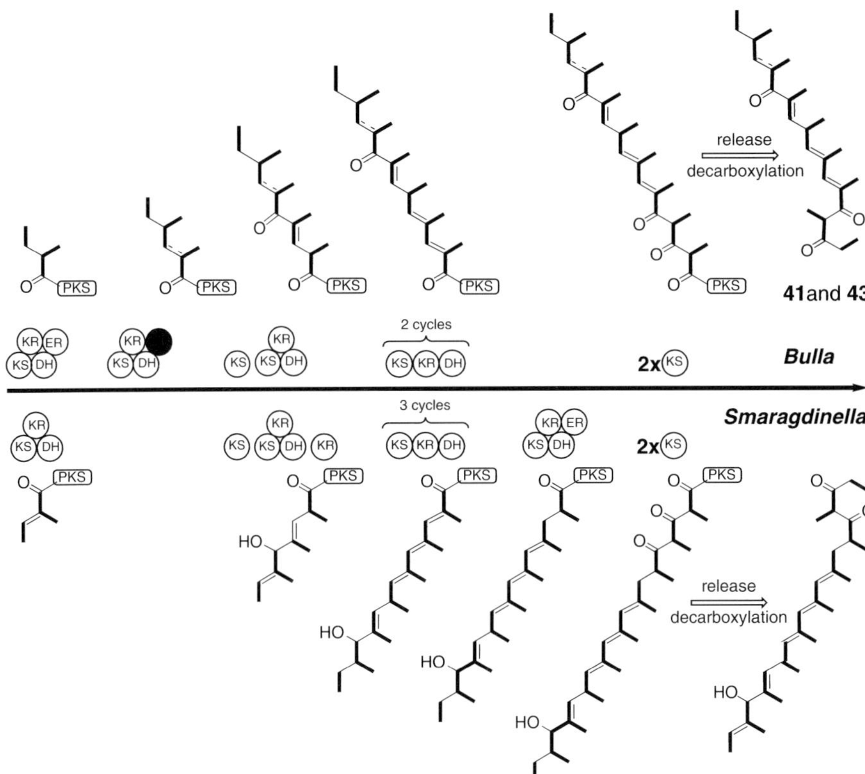

Fig. 14.10. Proposed processive assembly of polypropionates in cephalaspidean molluscs: **41** aglajine-1 from *Bulla striata*, **43** niuhinone-B from *B. gouldiana*, **44** nalodionol from *Smaragdinella calyculata*. For simplicity, the starter unit (acetate) and activities due to acyl transferase and acyl carrier protein have been omitted. *KS* •-ketoacylsynthase, *KR* •-ketoacylreductase, *ER* enoyl reductase, *DH* dehydratase, *TE* thioesterase. The *dark sphere* indicates the additional step that is supposed to generate the structural differences of the metabolites from *B. gouldiana* and *B. striata*

However, although the biosynthesis of the 3-alkylpyridine motif of haminols may suggest a similar PKS formation for **46**, **47** and **49**, there is little certainty about the origin of these compounds.

The demonstration of de novo biosynthesis in *B. striata* and *H. orbignyana* suggests that polypropionates, alkylpyridines, alkylaryls and alkylphenols (e.g. navenone-B (**50**), navenone-C (**51**) from *Navanax (Aglaja) inermis*; Sleeper and Fenical 1977; Spinella et al. 1993a) isolated from carnivorous Aglajdae molluscs, including *Philinopsis (Aglaja) depicta*, *N. inermis* and *P. (Aglaja) speciosa*, are of dietary origin. The ability of these latter molluscs to prey upon shelled cephalaspideans of the genera *Bulla* and *Haminoea* is well documented (Cimino et al. 2001; Fontana et al. 2004). This is analogous to that suggested for the origin of cytotoxic depsipeptides, e.g. kulolide (**52**; Reese et al. 1996) and kulokekahilide-2 (**53**; Nakao et al. 2002) of *P. speciosa*, which are thought to be sequestered by predation of other molluscs, such as *Stylocheilus longicauda* (Opisthobranchia, Anaspidea). These metabolites, which are also reminiscent to some extent of toxic components of other anaspidean opisthobranchs, e.g. aurilide (**54**; Suenaga et al. 1996) and dolastatin-16 (**55**; Pettit et al. 1997) of *Dolabella auricularia*, have ultimately been tracked to marine cyanobacteria, such as *Symploca kydnoides* or *Lyngbya majuscula* (*Microcoleus lyngbyaceus*; Burja et al. 2001). As suggested above for cyercene-like compounds in sacoglossans, the possible origin of cephalaspidean polypropionates from microbial symbionts warrants careful evaluation.

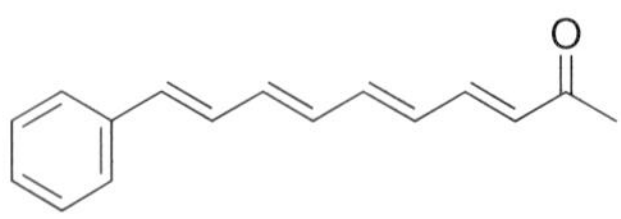

50 navenone-B

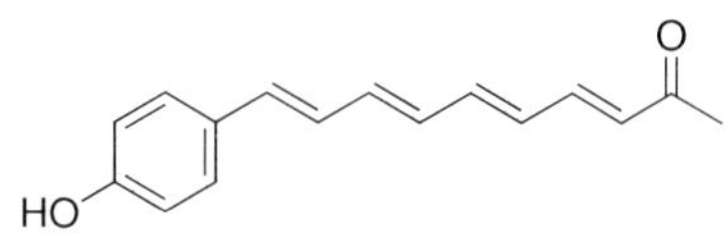

51 navenone-C

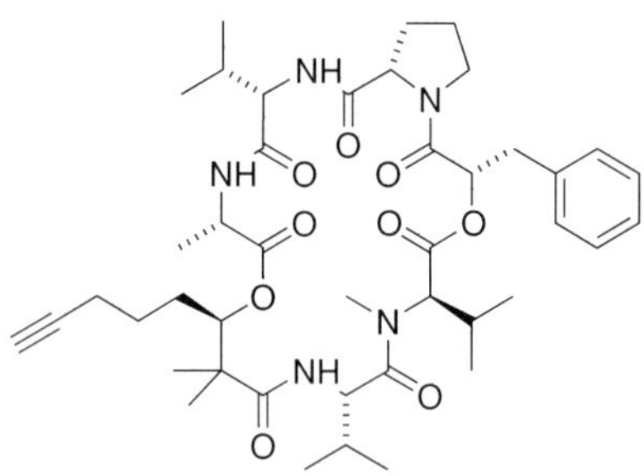

52 kulolide

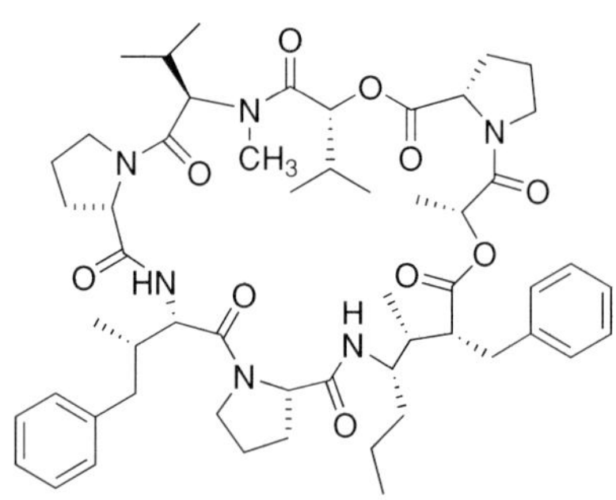

53 kulokehahilide-2

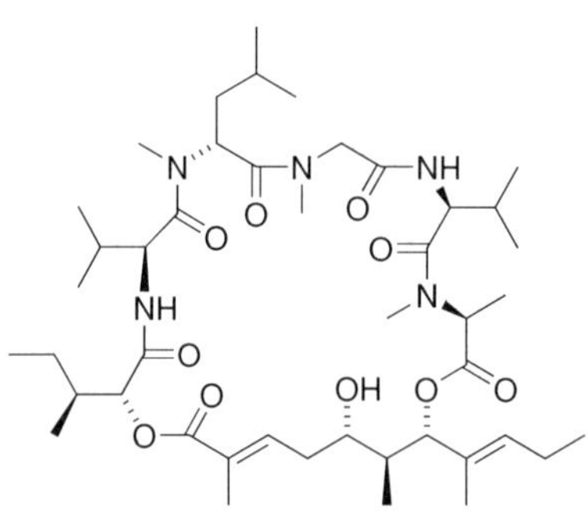

54 aurilide

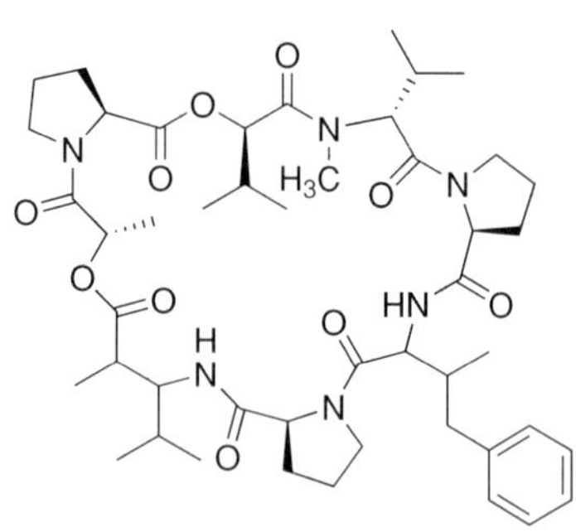

55 dolastatin 16

Formulae 50–55

14.3 Terpenes

Terpenes and steroids are classified on the basis of the number of C-5 isoprene units that are repetitively present in the molecule. Although the C-5 skeleton of isoprenoids is universally derived through condensations of the five-carbon metabolite isopentenyl diphosphate (IPP) and its isomer dimethylallylyl diphosphate (DMAP), two distinct and independent pathways to terpenes exist in prokaryotic and eukaryotic cells (Rohmer 1999). The initial source of isoprenoid units to be identified was mevalonic acid (MVA), which is synthesized from three molecules of acetyl-CoA. The acetate/mevalonate pathway was widely assumed to be responsible for terpene biosynthesis in all organisms for about 40 years, until it was found that IPP synthesis can also derive from a second pathway that requires 1-deoxy-D-xylulose 1-phosphate (DXP) as the earliest C-5 intermediate. Since formation of DXP does not involve either acetate or mevalonate, but proceeds via glyceraldehyde 3-phosphate/pyruvate, the corresponding route is also known as non-mevalonate or mevalonate-independent (Fig. 14.11).

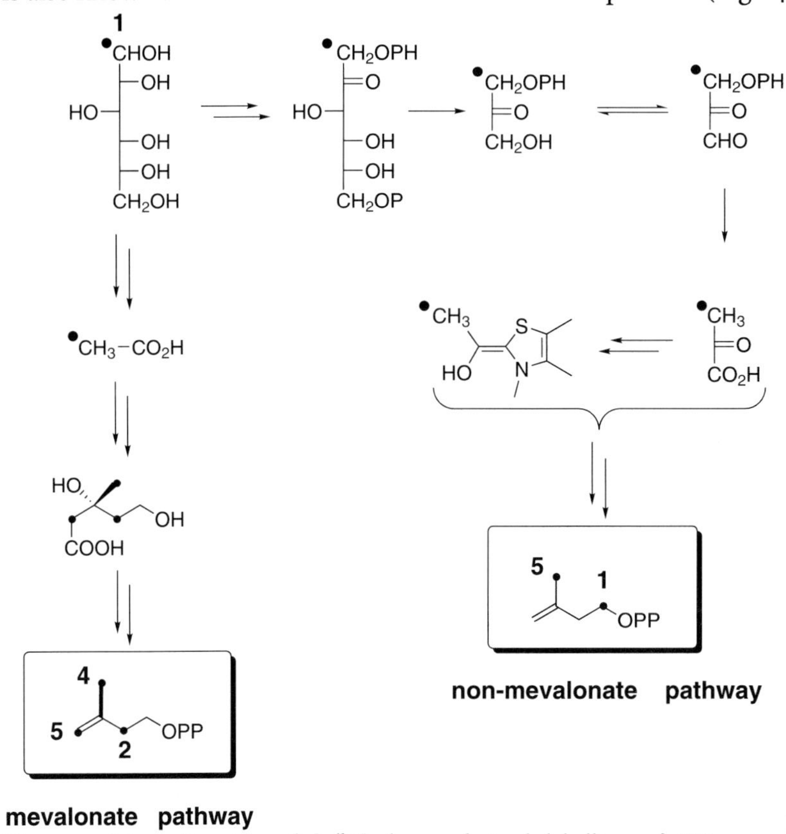

Fig. 14.11. Schematic view of [1-^{13}C]-glucose-derived labelling of IPP accordingly to mevalonate and non-mevalonate pathways

Genetic studies, consistent with labelling experiments, indicate that the occurrence of the two processes is differently distributed between bacteria, archeabacteria and eukaryotes (Langue et al. 2000). In prokaryotes, the synthesis of MVA appears to represent the ancestral pathway in archeabacteria, as the DXP pathway does in eubacteria. However, horizontal gene transfer seems to have contributed to the distribution across prokaryotic genomes, leading to the sporadic distribution of the MVA pathway among eubacteria. In contrast, with the important exception of plastids of plants and algae, eukaryotic organisms only produce terpenes from MVA. The complete absence of DXP genes in non-plastid-bearing eukaryotes suggests that plants acquired these genes from the cyanobacterial ancestor of chloroplasts.

Given their pronounced biological activity, it is not surprising that experimental attention to opisthobranch biosynthesis has focused primarily on terpenes in nudibranch molluscs. The mevalonate-derived origin of opisthobranch terpenes has been well documented in nudibranchs, especially in the families Dendrodoridae and Dorididae, by incorporation of both radioactive and ^{13}C-labelled precursors (Cimino et al. 1983; Fontana et al. 1999, 2000). The terpene biosynthesis in molluscs from the Pacific coast of North America and the experiments of his own research group in this area are discussed in greater detail by Ray Andersen in his chapter: *Skin Chemistry of Nudibranchs from the West Coast of North America*. Further information is available in a series of recent reviews (Cimino et al. 1999, 2001, 2004; Fontana and Cimino 2001).

The European nudibranch *Doriopsilla areolata*, Bergh 1980 is characterized by an array of defensive sesquiterpenoids showing enantiomeric stereochemistry of the decalin junctions of their skeletons (Gavagnin et al. 2001). The cooccurrence of enantiomeric sesquiterpenes, namely related to drimane (**56**–**58**) and *ent*-pallescensin A (**59**–**61**), is unique within opisthobranch molluscs, but molecules based on drimane and *ent*-pallescensin skeletons have been reported from sponges of the genus *Dysidea* (Butler and Capon 1993).

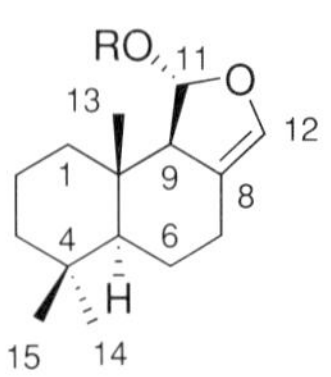

56 R = fatty acyl

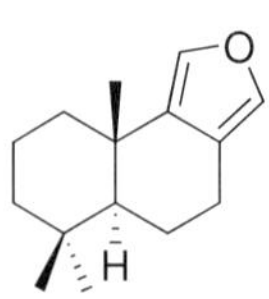

57

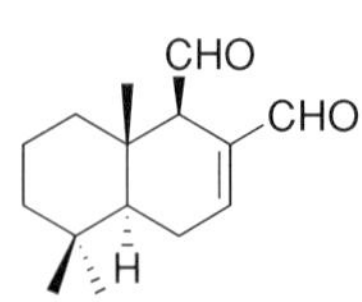

58

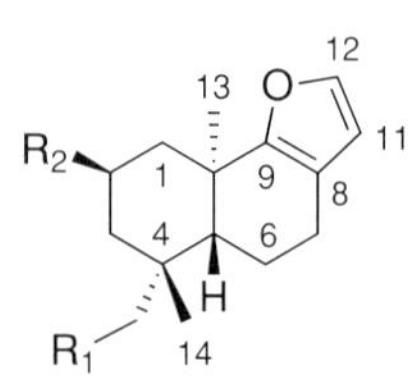

59 R_1 = H; R_2 = H
60 R_1 = OAc; R_2 = H
61 R_1 = OAc; R_2 = OAc

Formulae 56–61

The cooccurrence of similar compounds in different phyla is rather unusual and the case of *D. areolata* is even more singular, since the nudibranch and the sponge may represent a prey–predator pair. Biosynthetic studies of *D. areolata* compounds were addressed by feeding experiments with radioactive and stable isotopes. Drimane and *ent*-pallescensin-based compounds were found to incorporate radioactive labels after injection of [2-^{14}C]-mevalonate into specimens of *D. areolata* (Gavagnin et al. 2001). Subsequently, the biosynthesis of both skeletons was studied by injecting three distinct groups of molluscs every second day for two weeks with [1-^{13}C]-glucose, [1,2-$^{13}C_2$]-glucose or [1,2-$^{13}C_2$]-acetate (Fontana et al. 2003a). Injection of [1-^{13}C]-glucose led to an enrichment pattern of drimane esters (**56**) and 15-acetoxy-*ent*-pallescensin A (**60**), in total agreement with the origin via the acetate/mevalonate pathway (Fig. 14.11) of both skeletons (Rohmer 1999). Interestingly, while the acetate unit at C15 of **60** was as efficiently labelled as the terpene carbons, no incorporation was observed in the acetate-derived chains of the fatty acids of **56**, implying the origin of the fatty acyl substituents from a pre-existing exogenous reserve.

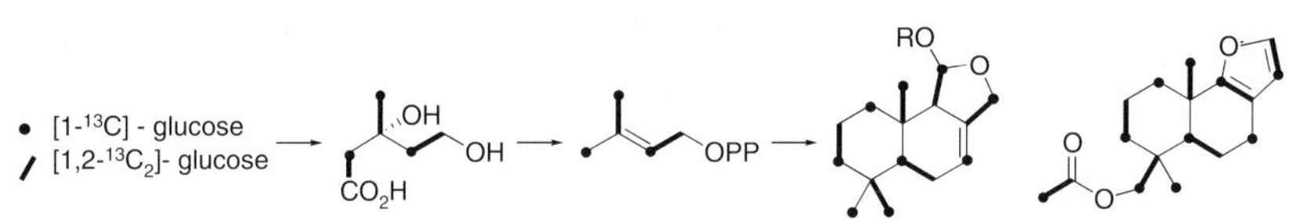

drimane esters
56 R = fatty acyl
60 15-acetoxy-*ent*-pallescensin A

Fig. 14.12. Incorporation of ^{13}C-labelled glucose in sesquiterpenoids of *Doriopsilla areolata*

In agreement with the scheme illustrated in Fig. 14.12, these results were largely corroborated by experiments with doubly ^{13}C-labelled precursors that showed incorporation into **56** and **60** of six intact acetate units. For both compounds, analysis of the coupling constants suggests a common biogenesis of the sesquiterpenoid skeletons, branching the pathway through the isomerization of the *E*,*E*,*E*-farnesol skeleton prior to the cyclization (Fig. 14.13). The biogenetic proposal for **60** predicts that isomerization may occur through two possible mechanisms, requiring either formation of β-farnesene or an oxygenated analogue (e.g. the non-cyclized furanosesquiterpene dendrolasin) after cleavage of the diphosphate group (path A) or 1,3-hydride shift with resulting formation of 4,15-isomer of farnesol pyrophosphate (path B). The cyclization of *Doriopsilla* sesquiterpenes is initiated by protonation of FPP at C-10 and proceeds through several discrete steps that may involve enzyme-bound intermediates. The role of the diphosphate in the formation of the oxygenated five-membered rings of **56** and **60** is another subject that warrants biosynthetic attention. Biochemical and genetic studies are

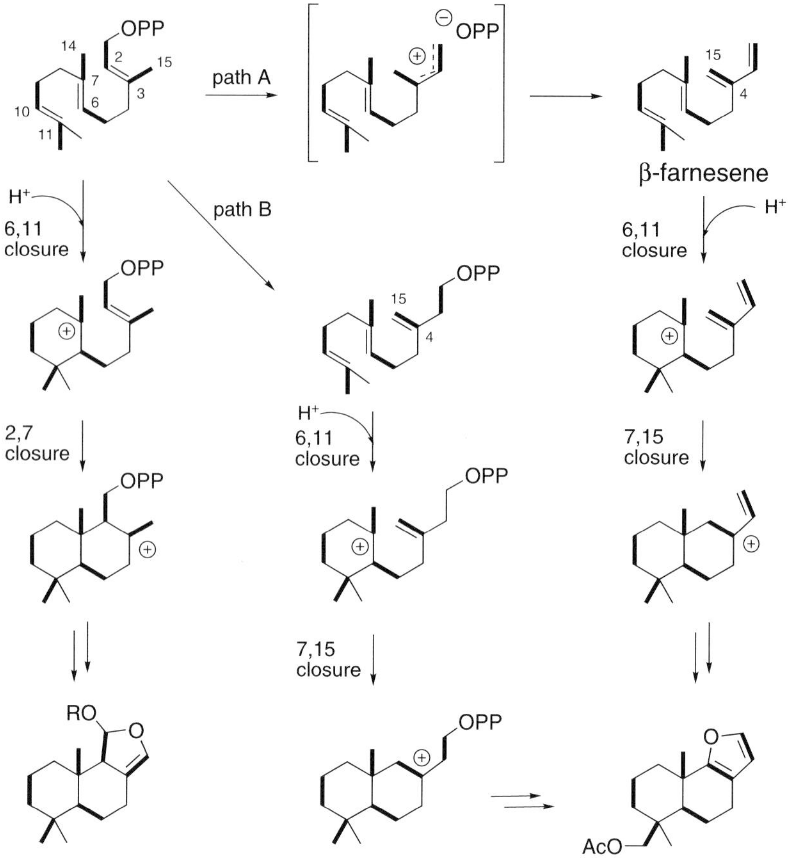

Fig. 14.13. Biogenesis of sesquiterpenoid skeletons in *Doriopsilla areolata*. The proposal is deduced on the basis of the biosynthetic experiments discussed in the text. In particular, the scheme takes into account the labelling pattern resulting from incorporation of the C_2-unit (*bold line*) from feeding experiments with [1,2-$^{13}C_2$]-acetate and [1,2-$^{13}C_2$]-glucose

required to sort out the detailed process leading to the enantiomeric ring closure of the drimane and *ent*-pallescensin skeletons. Although the aforementioned feeding experiments rule out acquisition of the sesquiterpenoids or their potential precursors from dietary sources, they do not exclude production by a symbiont, leaving unresolved the questions about the presence of identical chemical patterns in taxonomically distinct organisms, such as sponges and molluscs. Interestingly, drimane compounds **56–58** are also biosynthesized de novo by other dorid nudibranchs, *Dendrodoris grandiflora* and *D. limbata*, via the same general route proposed for *D. areolata* (Cimino et al. 1983; Fontana et al. 1999, 2000). The absence of *ent*-pallescensin-like compounds in *D. limbata* and *D. grandiflora* seems to indicate that, despite the close proximity of the two taxonomic genera, *Dendrodoris* and *Doriopsilla* may possess different terpene cyclases.

Another intriguing group of marine isoprenoids is represented by the glycerol esters of terpenoic acids from globally distributed molluscs of the family Dorididae (Gastropoda, Nudibranchia). Structural, biological and biosynthetic studies of this class of metabolites have been independently reported by several authors (for reviews, see Cimino et al. 1999; Gavagnin and Fontana 2000) and are also discussed elsewhere in this book.

Verrucosins (**62–74**) are toxic diterpenoyl glycerides isolated from the Mediterranean mollusc *Doris verrucosa* (Nudibranchia, Dorididae; Cimino et al. 1988; Avila et al. 1990; Gavagnin et al. 1990). The glycerides are composed of a mixture of eight different terpenoid skeletons that are thought to derive from a unique pathway through a number of methyl migrations and cyclizations. Recently, the de novo origin of these metabolites has been proved by feeding experiments with [^{14}C-U]-glucose (Fontana et al. 2003b). Injection of the radioactive precursor in six specimens of the mollusc led to the labelling of every verrucosin, although the specific incorporation was only a little higher than 1,000–1,500 DPM mg^{-1}. The specific labelling of the terpenoid skeleton was however corroborated by retention of the radioactivity into the rearranged isocopalanic derivative **75** prepared by refluxing labelled verrucosin B (**63**) with $LiAlH_4$ in THF. The low enrichment suffered during the experiments with ^{14}C-glucose also affected the studies with ^{13}C-containing precursors. In fact, several distinct incorporations with [6-^{13}C]-glucose, [5-^{13}C]-glucose and [2-^{13}C]-pyruvate failed to give a complete labelling of the terpenoid backbones of verrucosins, although, on the whole they were strongly indicative of a mevalonate origin for the mollusc metabolites (Fig. 14.14; Fontana et al. 2003b).

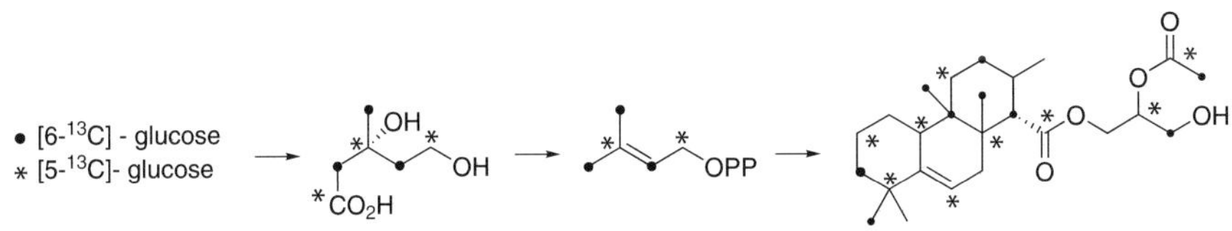

62 verrucosin A

Fig. 14.14. Proposed biogenesis of verrucosins via acetate/mevalonate pathway. The proposal is based on significant incorporations obtained by feeding experiments with ^{13}C-labelled glucose

It is, however, important to stress that only low levels of incorporation were obtained in feeding experiments with [2-^{14}C]-mevalonate (Avila et al. 1990), as well as no apparent conversion of putatively committed precursors, namely [3,20-$^{13}C_2$]-geranylgeranoic acid and the corresponding glyceride ([3,20-$^{13}C_2$]-verrucosin 4, **64**), was achieved after injection

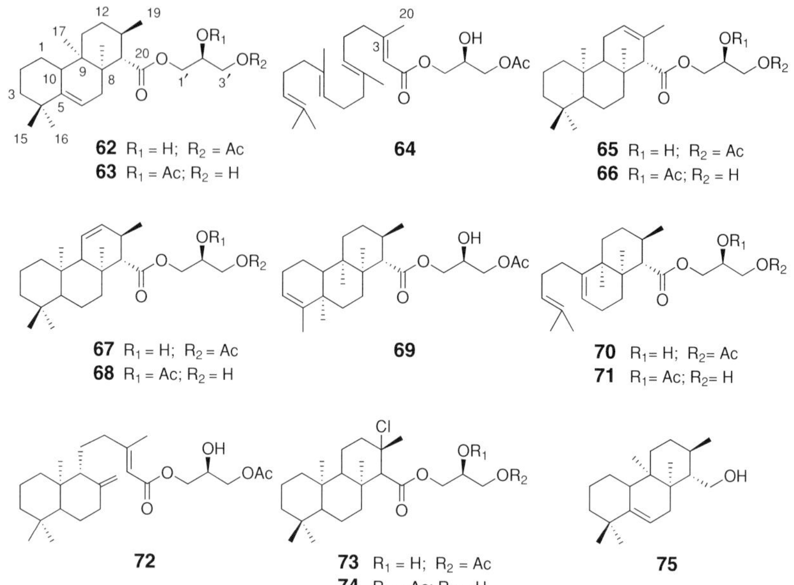

Formulae 62–75

into *D. verrucosa* (Gavagnin and Ungur, unpublished data). As outlined in Fig. 14.15, the specific labelling at C-3′ and C-2′ of glycerol in animals supplied, respectively, with [6-^{13}C]- and [5-^{13}C]-glucose indicated the origin of *sn*-glycerol from glycolysis-derived D-glyceraldehyde 3-phosphate (GAP). In analogy with the biosynthesis of triglycerides,

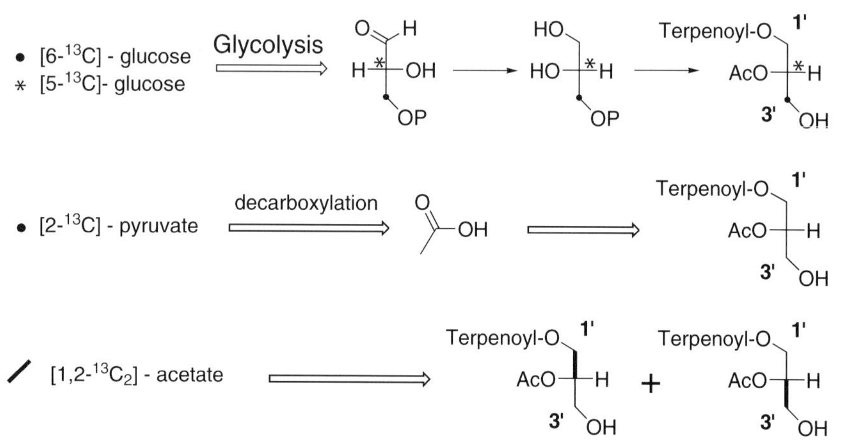

Fig. 14.15. Biogenesis of glycerol and mechanism for the stereospecific formation of the ester linkage between the diterpenoic acid and the *sn*-glycerol of verrucosins. The labelling pattern of glycerol from feeding experiments with [1,2-$^{13}C_2$]-acetate is in agreement with that reported by Graziani et al. (1996)

verrucosin synthesis involves formation of 1-terpenoyl-*sn*-glycerol 3-phosphate through stereoselective esterification of GAP-derived *sn*-glycerol 3-phosphate. The specific enrichment of C-1′ and C-3′ from the feeding experiment with [2-^{13}C]-pyruvate is analogous to the labelling pattern derived by injection of ^{13}C-labelled acetate into the Pacific nudibranchs *Archidoris montereyensis* and *Archidoris odhneri* (Gustafson and Andersen 1985; Graziani et al. 1996). It is obvious that pyruvate is converted to GAP via acetate but, ruling out the involvement of achiral intermediates, how the C_2-unit is incorporated in the glycerol backbone is not clear.

14.4 Conclusions

Opisthobranchs possess the ability to produce a large number of unique secondary metabolites, some of which (e.g. 3-alkylpyridine and polypropionates) are unlike those found in any terrestrial species. Our knowledge of the biosynthetic processes leading to these products is still in its infancy, although the studies described here and in other specialized articles provide a good starting point. The tremendous progress in genetic and genomic techniques is expected to bring a crucial contribution for the comprehension of secondary metabolic pathways in marine organisms, including opisthobranchs. However, the study of genomics alone will not provide all the answers, since the expression of a gene may have no definitive relationship to the ultimate production of the metabolites. This emphasizes the need for traditional biosynthetic work as a prerequisite for the full exploration of the biochemical aspects of the secondary metabolism in these organisms. The development of novel methodologies and many more studies are, however, needed to overcome the difficulties related to the slow rate of incorporations and the scarce knowledge of the basic biology of marine opisthobranchs.

Acknowledgements. Special thanks are due to Dr. Adele Cutignano and Dr. Giuliana d'Ippolito for their precious contribution to most of the scientific subjects described in this article. A.F. acknowledges financial assistance from the European Network of Excellence "Marine Biodiversity and Ecosystem Functioning".

References

Andersen RJ, van Soest RWM, Kong F (1996) 3-Alkylpyridine alkaloids isolated from marine sponges in the order Haplosclerida. In: Pelletier SW (ed) Alkaloids: chemical and biological perspectives, vol 10. Pergamon, London, pp 301–355

Avent AG, Hanson JR, Truneri A (1992) The biosynthesis of nectriapyrone and vermopyrone. Phytochemistry 31:3447–3449

Avila C, Ballesteros M, Cimino G, Crispino A, Gavagnin M, Sodano G (1990) Biosynthetic origin and anatomical distribution of the main secondary metabolites in the nudibranch mollusk *Doris verrucosa*. Comp Biochem Physiol B 97:363–368

Baldwin JE, Whitehead RC (1992) On the biosynthesis of manzamines. Tetrahedron Lett 33:2059–2062

Baldwin JE, Claridge TDW, Culshaw AJ, Heupel FA, Lee V, Spring DR, Whitehead RC, Boughtflower RJ, Mutton IM, Upton RJ (1998) Investigations into Manzamine alkaloid biosynthetic hypothesis. Angew Chem Int Ed Engl 37:2661–2663

Bandaranayake WM, Banfield JE, Black DStC (1980) Postulated electrocyclic reactions leading to endiandric acid and related natural products. J Chem Soc Chem Commun 1980:902–903

Bartel PL, Zhu CB, Lampel JS, Dosch DC, Connors NC, Strohl WR, Beale JM, Floss HG (1990) Biosynthesis of anthraquinones by interspecies cloning of actinorhodin biosynthesis genes in Streptomycetes: clarification of actinorhodin gene functions. J Bacteriol 172:4816–4826

Blunt WJ, Copp BR, Munro MHG, Northcote PT, Prinsep MR (2005) Marine natural products. Nat Prod Rep 22:15–61

Burja AM, Banaigs B, Abou-Mansour E, Burgess JG, Wright PC (2001) Marine cyanobacteria – a prolific source of natural products. Tetrahedron 57:9347–9377

Butler MS, Capon RJ (1993) Beyond polygodial – new drimane sesquiterpenes from a southern Australian marine sponge, *Dysidea* sp. Aust J Chem 46:1255–1267

Cane DE, Celmer WD, Westey JW (1983) Unified stereochemical model of polyether antibiotic structure and biogenesis. J Am Chem Soc 105:3594–3600

Celmer WD (1965) Macrolide stereochemistry. III. A configurational model for macrolide antibiotics. J Am Chem Soc 87:1801–1802

Cimino G, Ghiselin MT (1998) Chemical defense and evolution in the Sacoglossa (Mollusca: Gastropoda: Opisthobranchia). Chemoecology 8:51–60

Cimino G, Ghiselin MT (1999) Chemical defense and evolutionary trends in biosynthetic capacity among dorid nudibranchs (Mollusca: Gastropoda: Opisthobranchia). Chemoecology 9:187–207

Cimino G, Sodano G (1993) Biosynthesis of secondary metabolites in marine molluscs. Top Curr Chem 167:77–115

Cimino G, De Rosa S, De Stefano S, Sodano G, Villani G (1983) Dorid nudibranch elaborates its own chemical defense. Science 219:1237–1238

Cimino G, De Stefano S, Scognamiglio G, Sodano G, Trivellone E (1986) Sarains: a new class of alkaloids from the marine sponge *Reniera sarai*. Bull Soc Chem Belg 95:783–800

Cimino G, Gavagnin M, Sodano G, Puliti R, Mattia CA, Mazzarella L (1988) Verrucosin-A and -B, ichthyotoxic diterpenoic acid glycerides with a new carbon skeleton from the dorid nudibranch *Doris verrucosa*. Tetrahedron 44:2301–2310

Cimino G, Puliti R, Scognamiglio G, Spinella A, Trivellone E, Mattia CA, Mozzarella L (1989) Amazing new alkaloid skeletons from the marine sponge *Reniera sarai*. Pure Appl Chem 61:535–538

Cimino G, Passeggio A, Sodano G, Spinella A, Villani G (1991) Alarm pheromones from the Mediterranean opisthobranch *Haminoea navicula*. Experientia 47:61–63

Cimino G, Fontana A, Gavagnin M (1999) Marine opisthobranch molluscs: chemistry and ecology in sacoglossan and dorids. Curr Org Chem 3:327–372

Cimino G, Ciavatta ML, Fontana A, Gavagnin M (2001) Metabolites of marine opisthobranchs: chemistry and biological activity. In: Tringali C (ed) Bioactive compounds from natural sources – isolation, characterization and biological properties. Taylor & Francis, London, pp 579–637

Cimino G, Fontana A, Cutignano A, Gavagnin M (2004) Biosynthesis in opisthobranch molluscs: general outline in the light of recent use of stable isotopes. Phytochem Rev 3:285–307

Coval SJ, Schulte GR, Matsumoto GK, Roll DM, Scheuer PJ (1985) Two polypropionate metabolites from the cephalaspidean mollusk *Philinopsis speciosa*. Tetrahedron Lett 26:5359–5362

Cox RJ (2002) Biosynthesis. Annu Rep Prog Chem B 2002:223–251

Cueto M, D'Croz L, Maté JL, San-Martin A, Darias J (2005) Elysiapyrones from *Elysia diomedea*. Do such metabolites evidence an enzymatically assisted electrocyclization cascade for the biosynthesis of their bicyclo[4,2,0] core? Org Lett 7:415–418

Cutignano A, Fontana A, Renzulli L, Cimino G (2003a) Placidenes C–F, novel •-pyrone propionates from the mediterranean sacoglossan *Placida dendritica*. J Nat Prod 66:1399–1401

Cutignano A, Tramice A, De Caro S, Villani G, Cimino G, Fontana A (2003b) Biogensis of 3-alkylpyridine alkaloids in the marine mollusc *Haminoea orbignyana*. Angew Chem Int Ed Engl 42:2633–2636

Cutignano A, Cimino G, Giordano A, d'Ippolito G, Fontana A (2004) Polyketide origin of 3-alkylpyridines in the marine mollusc *Haminoea orbignyana*. Tetrahedron Lett 45:2627–2629

Davies-Coleman MT, Garson MJ (1998) Marine polypropionates. Nat Prod Rep 15:477–493

Dawe R, Wright JLC (1986) The major polypropionate metabolites from the sacoglossan mollusc *Elysia chlorotica*. Tetrahedron Lett 27:2559–2562

Di Marzo V, Vardaro RR, De Petrocellis L, Villani G, Minei R, Cimino G (1991) Cyercenes, novel pyrones from the ascoglossan mollusc *Cyerce cristallina*. Tissue distribution, biosynthesis and possible involvement in defense and regenerative processes. Experientia 47:1221–1227

Faulkner DJ (2002) Marine natural products. Nat Prod Rep 19:1–48

Fontana A, Cimino G (2001) Biosynthetic studies of secondary metabolites in marine opisthobranchs. Recent Res Dev Org Chem 5:89–105

Fontana A, Ciavatta ML, Miyamoto T, Spinella A, Cimino G (1999) Biosynthesis of drimane terpenoids in dorid molluscs: pivotal role of 7-deacetoxyolepupuane in two species of *Dendrodoris* nudibranchs. Tetrahedron 55:5937–5946

Fontana A, Villani G, Cimino G (2000) Terpene biosynthesis in marine molluscs: incorporation of glucose in drimane eaters of *Dendrodoris* nudibranchs via classical mevalonate pathway. Tetrahedron Lett 41:2429–2433

Fontana A, Tramice A, Cutignano A, d'Ippolito G, Gavagnin M, Cimino G (2003a) Terpene biosynthesis in the nudibranch *Doriopsilla aureolata*. J Org Chem 68:2405–2409

Fontana A, Tramice A, Cutignano A, d'Ippolito G, Renzulli L, Cimino G (2003b) Studies of the biogenesis of verrucosins, toxic diterpenoid glycerides of the Mediterranean mollusc *Doris verrucosa*. Eur J Org Chem 16:3104–3108

Fontana A, Cutignano A, Giordano A, Domènech Coll A, Cimino G (2004) Biosynthesis of aglajnes, polypropionate allomones of the opisthobranch mollusc *Bulla striata*. Tetrahedron Lett 45:6847–6850

Fusetani N, Asai N, Matsunaga S, Honda K, Yasumuro K (1994) Cyclostellettamines A–F pyridine alkaloids inhibit binding of methyl quinuclidinyl benzilate (QNB) muscarinic acetylcholine receptors from the sponge *Stelletta maxima*. Tetrahedron Lett 35:3967–3970

Garson MJ (1989) Biosynthetic studies on marine natural products. Nat Prod Rep 6:143–170

Garson MJ (1993) The biosynthesis of marine natural products. Chem Rev 99:1699–1733

Garson MJ (2001) Ecological perspectives on marine natural product biosynthesis. In: McClintock A, Baker MJ (eds) Marine chemical ecology. CRC Press, Boca Raton, pp. 71–114
Garson MJ, Simpson JS (2004) Marine isocyanides and related natural products structure, biosynthesis and ecology. Nat Prod Rep 21:164–179
Garson MJ, Goodman JM, Paterson I (1994) A configurational model for siphonariid polypropionates derived from structural and biosynthetic considerations. Tetrahedron Lett 35:6929–6932
Gavagnin M, Fontana A (2000) Diterpenes from marine opisthobranch molluscs. Curr Org Chem 4:1201–1248
Gavagnin M, Spinella A, Cimino G, Sodano G (1990) Stereochemistry of ichthyotoxic diacyglycerols from opisthobranch molluscs. Tetrahedron Lett 31:6093–6094
Gavagnin M, Marin A, Mollo E, Crispino A, Villani G, Cimino G (1994a) Secondary metabolites from Mediterranean Elysioidea: origin and biological role. Comp Biochem Physiol 108B:107–115
Gavagnin M, Spinella A, Castelluccio F, Cimino G, Marin A (1994b) Polypropionates from the Mediterranean mollusc *Elysia timida*. J Nat Prod 57:298–304
Gavagnin M, Mollo E, Castelluccio F, Ghiselin MT, Calado G, Cimino G (2001) Can molluscs biosynthesize typical sponge metabolites? The case of the nudibranch *Doriopsilla areolata*. Tetrahedron 57:8913–8916
Graziani EI, Andersen RJ, Krug PJ, Faulkner DJ (1996) Stable isotope incorporation evidence for the de novo biosynthesis of terpenoic acid glycerides by dorid nudibranchs. Tetrahedron 52:6869–6878
Green BJ, Li WY, Menhart JR, Fox TC, Summer EJ, Kennedy RA, Pierce SK, Rumpho ME (2000) Mollusc-algal chloroplast endosymbiosis. Photosynthesis, thylakoid protein maintenance, and chloroplast gene expression continue for many months in the absence of the algal nucleus. Plant Physiol 124:331–342
Gustafson K, Andersen RJ (1985) Chemical studies of British Columbia nudibranchs. Tetrahedron 41:1101–1108
Hay ME, Pawlik JR, Duffy JE, Fenical W (1989) Seaweed-herbivore-predator interactions: host-plant specialization reduces predation on small herbivores. Oecologia 81:418–427
Hegde VR, Puar MS, Dai P, Patel M, Gullo VP, Das PR, Bond RW, McPhail A (2000) A novel microbial metabolite, activator of low density lipoprotein receptor promoter. Tetrahedon Lett 41:1351–1354
Ireland C, Scheuer PJ (1979) Photosynthetic marine mollusks: in vivo ^{14}C incorporation into metabolites of the sacoglossan *Placobranchus ocellatus*. Science 205:922–923
Jeffrey DW, Perkins MV, White JM (2005) Synthesis of an analogue of the marine polypropionate tridachiahydropyrone. Org Lett 7:407–409
Kitagawa I, Kobayashi M (1993) Pharmacochemical investigations of marine sponge products. Gazz Chim Ital 123:321–327
Knaggs AR (2001) The biosynthesis of shikimate metabolites (1999). Nat Prod Rep 16:334–355
Kobayashi J, Tsuda M, Kawasaki N, Matsumoto K, Adachi T (1994) Keramaphidn B, a novel petacyclic alkaloid from a marine sponge *Amphimedon* sp.: plausible biogenetic precursor of manzamine alkaloids. Tetrahedon Lett 35:4383–4386
Kondo K, Shigemori H, Kikuchi Y, Ishibashi M, Sasaki T, Kobayashi J (1992) Ircinals A and B from the Okinawan marine sponge *Ircinia* sp.: plausible biogenetic precursors of manzamine alkaloids. J Org Chem 57:2480–2483
Kong FK, Andersen RJ, Allen TM (1995) Ingenamine alkaloids isolated from the sponge *Xestospongia ingens*: structures and absolute configurations. Tetrahedron 51:2895–2906
Kubanek J, Andersen R (1997) Evidence for the incorporation of intact butyrate units in the biosynthesis of triophamine. Tetrahedron Lett 38:6327–6330
Langue M, Rujan T, Martin W, Croteau R (2000) Isoprenoid biosynthesis: the evolution of two ancient and distinct pathways across genomes. Proc Natl Acad Sci USA 97:13172–13177

Mann J (1994) Alkaloids. In: Mann J, Davidson S, Hobbs, Banthorpe DV, Harborne JB (eds) Natural products: their chemistry and biological significance. Longman, London, pp 389–446
Manzo E, Ciavatta ML, Gavagnin M, Mollo E, Whidulla S, Cimino G (2005) New g-pyrone propionates from the Indian Ocean sacoglossan *Placobranchus ocellatus*. Tetrahedron Lett 46:465–468
Marin A, Alvarez LA, Cimino G, Spinella A (1999) Chemical defence in cephalaspidean gastropods: origin, anatomical location and ecological roles. J Molluscan Stud 65:121–123
Moore BS (1999) Biosynthesis of marine natural products: microorganisms and macroalgae. Nat Prod Rep 16:653–674
Moore BS, Hartweck C (2002) Biosynthesis and attachment of novel bacterial polyketide synthase starter units. Nat Prod Rep 19:70–99
Moss SJ, Martin CJ, Wilkinson B (2004) Loss of colinearity by modular polyketide synthases: a mechanism for the evolution of chemical diversity. Nat Prod Rep 21:575–593
Nakao Y, Yoshida WY, Takada Y, Kimura J, Yang L, Mooberry SL, Scheuer PJ (2002) Kulokekahilide-2, a cytotoxic depsipeptide from a cephalaspidean mollusk *Philinopsis speciosa*. J Nat Prod 67:1332–1340
Pedras MSC, Chumala PL (2005) Phomapyrones from blackleg causing phytopathogenic fungi: isolation, structure determination, biosynthesis and biological activity. Phytochemistry 66:81–87
Pierce SK, Massey SE, Hanten JJ, Curtis NE (2003) Horizontal transfer of functional nuclear genes between multicellular organisms. Biol Bull 204:237–240
Proksch P, Edrada RA, Ebel R (2002) Drugs from the sea – current status and microbiological implications. Appl Microbiol Biotechnol 59:125–134
Rawlings BJ (1999) Biosynthesis of polyketides (other than actinomycete macrolides). Nat Prod Rep 16:425–484
Rawlings BJ (2001a) Type I polyketide biosynthesis in bacteria (part B) 1995 to mid-2000. Nat Prod Rep 18:231–281
Rawlings BJ (2001b) Type I polyketide biosynthesis in bacteria (part A, erythromycin biosynthesis) 1994 to 2000. Nat Prod Rep 18:190–227
Reese MT, Gulavita NK, Nakao Y, Hamann MT, Yoshida WY, Coval SJ, Scheuer PJ (1996) Kulolide – a cytotoxic depsipeptide from a cephalaspidean mollusk, *Philinopsis speciosa*. J Am Chem Soc 118:11081–11084
Revill WP, Bibb MJ, Hopwood DA (1995) Purification of a malonyltransferase from *Streptomyces coelicolor* A3(2) and analysis of its genetic determinant. J Bacteriol 177:3946–3952
Rohmer M (1999) The discovery of a mevalonate-independent pathway for isoprenoid biosynthesis in bacteria, algae and higher plants. Nat Prod Rep 16:565–574
Roussis V, Pawlik JR, Hay ME, Fenical W (1990) Secondary metabolites of the chemically rich ascoglossan *Cyerce nigricans*. Experientia 46:327–329
Rowe CJ, Bohm IU, Thomas IP, Wilkinson B, Rudd BAM, Foster G, Blackaby AP, Sidebottom PJ, Roddis Y, Buss AD, Staunton J, Leadlay PF (2001) Engineering a polyketide with a longer chain by insertion of an extra module into the erythromycin-producing polyketide synthase. Chem Biol 8:475–485
Sep i K, Guella G, Mancini I, Pietra F, Dalla S, Minestrina G, Minestrina G, Tubb K, Ma ek P, Turk T (1997) Characterization of anticholisterase-active 3-alkylpyridium polymers from the marine sponge *Reniera sarai* in aqueous solutions. J Nat Prod 60:991–996
Sleeper HL, Fenical W (1977) Navenones A–C: trail-breaking alarm pheromones from the marine opisthobranch *Navanax inermis*. J Am Chem Soc 1977:2367–2368
Sleeper HL, Paul VJ, Fenical W (1980) Alarm pheromones from the marine opisthobranch *Navanax inermis*. J Chem Ecol 6:57–70
Spinella A, Alvarez LA, Cimino G (1993a) Predator–prey relationships between *Navanax inermis* and *Bulla gouldiana* – a chemical approach. Tetrahedron 49:3203–3210

Spinella A, Alvarez LA, Passeggio A, Cimino G (1993b) New 3-alkylpyridine from three Mediterranean cephalaspidean molluscs: structure, ecological role and taxonomic relevance. Tetrahedron 49:1307–1314

Staunton J, Weissman KJ (2001) Polyketide biosynthesis: a millennium review. Nat Prod Rep 18:380–416

Staunton J, Caffrey P, Aparicio JF, Roberts GA, Bethell SS, Leadlay PF (1996) Evidence for a double-helical structure for modular polyketide synthases. Nat Struct Biol 3:188–192

Suenaga K, Mutou T, Shibata T, Itoh T, Kigoshi H, Yamada K (1996) Isolation and stereostructure of aurilide, a novel cyclodepsipeptide from the Japanese sea hare *Dolabella auricularia*. Tetrahedron Lett 37:6771–6774

Szabo CM, Nakao Y, Yoshida WY, Scheuer PJ (1996) Two diverse constituents of the cephalaspidean mollusk *Smaragdinella calyculata*. Tetrahedron 52:9681–9686

Talpir R, Rudi A, Ilan M, Kashman Y (1992) Niphatoxin A and B: two new ichthyo- and cytotoxic tripyridine alkaloids from a marine sponge. Tetrahedon Lett 33:3033–3034

Trench RK, Greene R, Bystrom B (1969) Chloroplast as functional organelles in animal tissues. J Cell Biol 42:404–417

Tsuda M, Inaba K, Kawasaki N, Honma K, Kobayashi J (1996) Chiral resolution of (±)-keramaphidin B and isolation of manzamine L, a new •-carboline alkaloid from a sponge *Amphimedon* sp. Tetrahedron 52:2319–2324

Vardaro RR, Di Marzo V, Crispino A, Cimino G (1991) Cyercenes, novel polypropionate pyrones from the autotomizing Mediterranean mollusc *Cyerce cristallina*. Tetrahedron 47:5569–5576

Vardaro RR, Di Marzo V, Cimino G (1992a) Placidenes: cyercene-like polypropionate •-pyrones from the Mediterranean ascoglossan mollusc *Placida dendritica*. Tetrahedron Lett 33:2875–2878

Vardaro RR, Di Marzo V, Marin A, Cimino G (1992b) •- and •-pyrone-polypropionates from the Mediterranean ascoglossan mollusc *Ercolania funerea*. Tetrahedron 48:9561

Wickramasinge M, Bandaranayake WM, Banfield JE, Black DStC, Fallon GD, Gatehouse BMJ (1980) Endiandric acid, a novel carboxylic acid from *Endiandra introrsa* (Lauraceae): X-ray structure determination. J Chem Soc Chem Commun 1980:162–163

Zuidema DR, Jones PR (2005) Photochemical relationship in sacoglossan polypropionates. J Nat Prod 68:481–486

Total Synthesis of Bioactive Peptides and Depsipeptides from Marine Opisthobranch Molluscs

C. Della Monica, G. Della Sala, F. De Riccardis, I. Izzo, A. Spinella

Abstract. This chapter covers the synthetic aspects of both linear or cyclic peptides and depsipeptides isolated from opisthobranch molluscs. In many cases, synthetic effort not only determined the absolute stereostructure of these compounds but also made it possible to supply sufficient amounts for the evaluation of pharmacological activities. A summary of the synthetic work associated with each compound is reported after a short description of its natural source and biological properties. Discussion in the text concentrates on key reactions and synthetic efficiency.

15.1 Introduction

Research in the field of marine natural products has developed extensively during the past four decades and has shown that marine organisms are a rich source of bioactive compounds (Blunt et al. 2004). Many investigations have been focussed on opisthobranch molluscs because these slow-moving marine invertebrates, lacking shell protection, are able to use organic molecules for communication and protection against predators (Faulkner and Ghiselin 1983). In the course of these studies, a great variety of compounds from the mevalonate, polyketide and amino acid metabolic pathways have been isolated. Bioactive peptides from opisthobranch molluscs are of particular importance, not only for their ecological role but also because pharmacological evaluations have proved that some of these compounds are promising candidates for the supply of new medical resources. In fact, some peptides and depsipeptides from opisthobranch molluscs are among the most active compounds currently used in clinical evaluation (Kijjoa and Sawangwong 2004). The synthetic challenge presented, due to the incorporation of unusual amino acids, and the biological properties reported stimulated interest in their total synthesis. In this chapter, we describe the synthetic approaches employed in order to achieve the total synthesis of both linear or cyclic peptides and depsipeptides, isolated from opisthobranch molluscs. The literature is surveyed from the early 1990s

Aldo Spinella
Dipartimento di Chimica, Università di Salerno
Via S. Allende, 84081 Baronissi (Salerno), Italy

Progress in Molecular and Subcellular Biology
Subseries Marine Molecular Biotechnology
G. Cimino, M. Gavagnin (Eds.): Molluscs

until the close of 2004, coverage of the period 1960–1989 having been already reported (Albizati et al. 1992).

15.2 Linear Peptides

15.2.1 Janolusimide

Janolusimide (1) is a lipophilic tripeptide isolated from the nudibranch mollusc *Janolus cristatus* collected in the bay of Naples. This peptide is toxic to mice (LD 5 mg kg^{-1}) and, at lower concentrations, its neurotoxic action is antagonized by atropine, suggesting an interaction with acetylcholine receptors (Sodano and Spinella 1986). Structure **1** contains *N*-Me-L-Ala and two rather unusual constituents, 4-amino-3-hydroxy-2-methylpentanoic acid and the previously undescribed lactam, 5-isopropyl-3,3-dimethyl-pyrrolidine-2,4-dione of unknown stereochemistry.

The synthetic strategy developed to obtain all the 4-amino-3-hydroxy-2-methylpentanoic acid stereoisomers was based on Brown crotylboration of *N*-Boc-alaninal (**2**; Giordano et al. 1999). The stereochemistry of the single stereogenic center in the lactam ring was assigned by synthesis starting from Z-L-Val. Total synthesis of **1** (Scheme 15.1; see Appendix 15.1 for abbreviations used in the schemes) commenced with the deprotection of alkene **3**, prepared according to the methodology previously reported. Coupling of the obtained aminoalcohol and *N*-Boc-*N*-methyl-L-Ala afforded compound **4**, which was protected as *N*,*O*-acetonide and then oxidated to the protected dipeptide **5**. Compound **5** was transformed into the pentafluorophenyl ester **6** and coupled with the *N*-anion obtained from **8**. This reaction afforded two diastereomeric protected tripeptides **9** and **10** in a 2.8:1 ratio. Only mixed proton NMR spectra of the synthetic compounds with the natural one allowed the identification of **1** as the product, obtained after deprotection of **9** (Giordano et al. 2000).

15.2.2 Dolastatin 10

Dolastatin 10 (**11**) is the most active antineoplastic substance presently known (the average GI_{50} in 60 NCI panel tumor cell lines was in the 0.1-nM range). This peptide inhibits microtubule assembly by binding with tubulin β and is currently in clinical trials. It was originally isolated by Pettit et al. (1987a) from the sea hare *Dolabella auricularia* and by the Moore group (Luesch et al. 2001) from the marine cyanobacterium *Symploca* species VP642. In fact, although *D. auricularia* is the mollusc mostly implicated as source of bioactive peptides (Pettit 1997), recent isolation of identical or related peptides from blue-green algae suggests a dietary origin for some of them (Luesch et al. 2002).

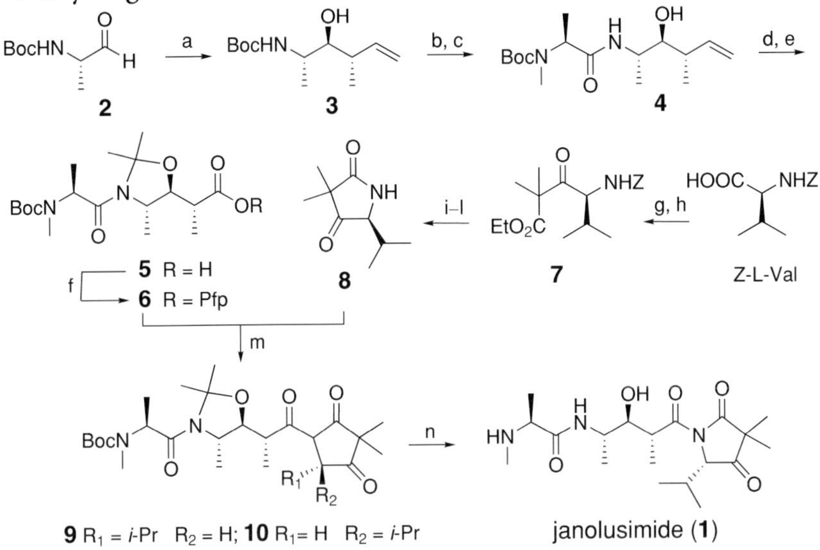

Scheme 15.1. Sodano synthesis of janolusimide. (a) (–)-Ipc_2B-(*E*)-crotyl, 72%. (b) HCl. (c) *N*-Boc-*N*-Me-L-Ala, EDC, DMAP, 38%, 2 steps. (d) $(CH_3)_2C(OCH_3)_2$, *p*-TsOH (cat.). (e) $NaIO_4$, $RuCl_3$, 36%, 2 steps. (f) C_6F_5OH, DCC, DMAP, 89%. (g) CDI, then $LiCH_2COOEt$. (h) K_2CO_3, MeI. (i) $NaBH_4$, 44%, 3 steps. (j) H_2, Pd/C. (k) NaOEt. (l) PDC, 86%, 3 steps. (m) *n*-BuLi, 63%. (n) *p*-TsOH, 59%

Up to now, four total syntheses of dolastatin 10 have been published. The synthesis reported by Pettit et al. (1989b, 1996) started with the reaction of L-Boc-prolinal (**12**), easily obtained from commercially available L-Boc-Pro, with the chiral lithium enolate of **13**, furnishing the (2*S*,3*R*,4*S*)-isomer **14**. O-Methylation, carboxy group deprotection and C-2 successful base-mediated epimerization, afforded the Boc-(2*R*,3*R*,4*S*)-dolaproine (Boc-Dap **16**). The elaboration of (3*R*,4*S*,5*S*)-dolaisoleuine fragment (Dil-*O*-*t*-Bu, **18**) was realized via manipulation of the Z-(*S*,*S*)-Ile. The preparation of (*S*)-dolaphenine·TFA (Doe·TFA, **22**) followed the probable biosynthetic route, via the known *N*-Boc-(*S*)-phenylalanal (**21**). Assembly of the five components, Boc-Dap (**16**), Dil-*O*-*t*-Bu (**18**), Z-Val, Doe·TFA (**22**), and *N*,*N*-dimethylvaline (Dov) pentafluorophenyl ester (**19**), in a convergent 3 + 2 fashion, gave the first synthetic dolastatin 10 (**11**) (Scheme 15.2).

The Pettit synthesis was not suitable for large scale preparation of peptide **11** (in view of the clinical investigation) because of its lack of stereoselectivity. The Shioiri–Hamada synthesis (Hamada et al. 1991; Shioiri et al. 1993) is based on more reliable and higher yielding stereoselective synthetic protocols. Preparation of the required Boc-

(3*R*,4*S*,5*S*)-Dil (**31**) was accomplished from L-Boc-Ile. The key step was a stereoselective low-temperature sodium borohydride-mediated reduction of a β-ketoester intermediate (9:1, d.r.). The synthesis of the Dap derivative **16** followed the Evans aldol methodology, as depicted in Scheme 15.3. The synthesis of the (*S*)-dolaphenine (Doe, **30**) followed a modified Hantzsch thiazol synthesis. Efficient assembling of all the components was carried out in a stepwise manner and gave pure dolastatin 10 (**11**).

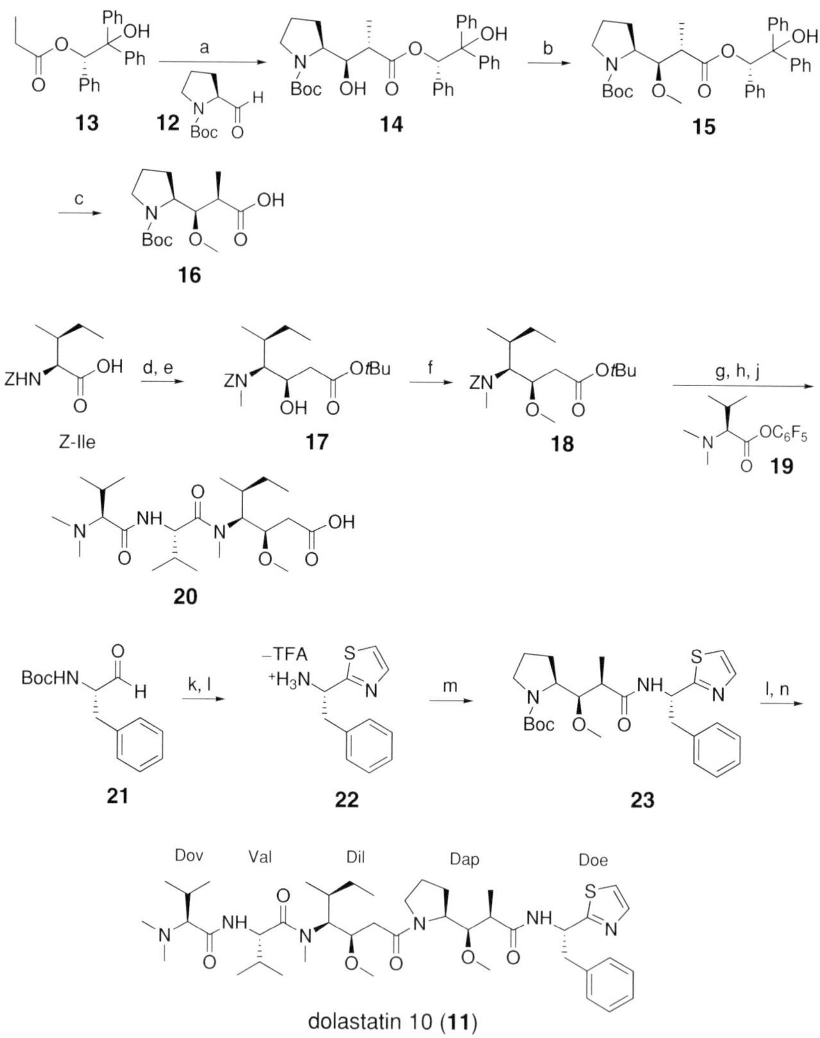

Scheme 15.2. Pettit synthesis of dolastatin 10. (a) LDA, **12**, $MgBr_2$, 47%. (b) CH_2N_2, $BF_3{\cdot}OEt_2$. (c) (1) *t*-BuOK, 57%, (2) H_2, Pd–C. (d) (1) NaH, CH_3I, (2) B_2H_6·THF, (3) DMSO-SO_3-Pyr, Et_3N, 74%, 3 steps. (e) $LiCH_2COO$*t*-Bu, 33%. (f) CH_2N_2, $BF_3{\cdot}OEt_2$, 67%. (g) (1) H_2, Pd–C, 63%, (2) Z-L-Val, $(CH_3)_2CHCOCl$, NMM. (h) **19**, H_2, Pd–C. (j) TFA. (k) (1) $H_2NCH_2CH_2SH$, 77%, (2) MnO_2. (l) TFA. (m) **16**, DEPC, TEA. (n) **20**, DEPC

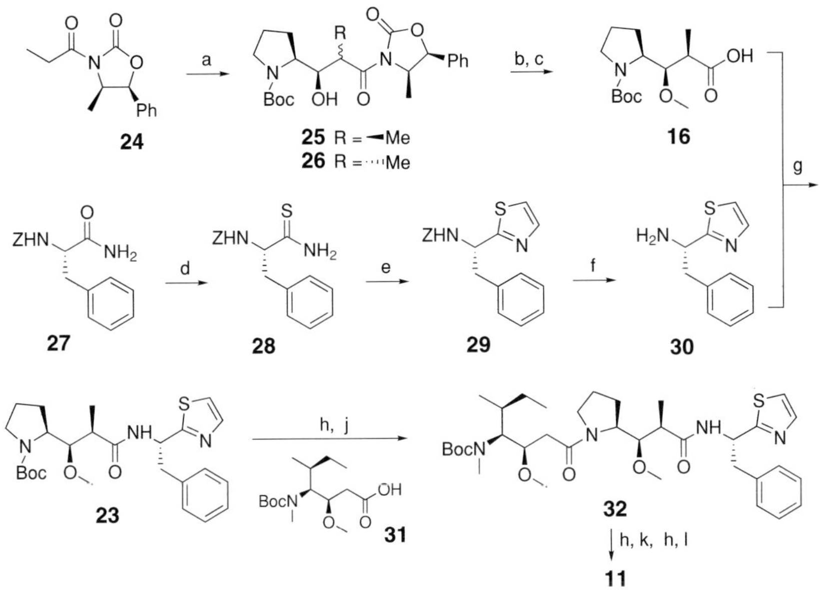

Scheme 15.3. Shioiri–Hamada synthesis of dolastatin 10. (a) Et_3N, n-Bu_2BOTf, **12**, H_2O_2, 99%. (b) **25**, LiOH, H_2O_2, 98%. (c) NaH, CH_3I, 89%. (d) Lawesson's reagent, 97%. (e) (1) $BrCH_2CHO$, K_2CO_3, (2) $(CF_3CO)_2O$, 66%, 2 steps. (f) (1) 25% HBr/AcOH, (2) aq. $NaHCO_3$, 99%. (g) DEPC, Et_3N, 97%. (h) TFA. (j) **31**, DEPC, Et_3N, 99%. (k) Boc-Val, Bop-Cl, Et_3N, 59%. (l) Dov, DEPC, NMM, 90%

The Koga synthesis (Tomioka et al. 1991) is worth mentioning mainly for its rapid construction of the Boc-Dap **16**, prepared in four steps (Scheme 15.4) from the L-Boc-prolinal (**12**). The preparation of the required Z-Dil and that of the Doe fragments followed the Shioiri–Hamada approach and Pettit's biomimetic synthesis, respectively. Also, in this case, the linear elongation strategy proceeded uneventfully to give target **11** in decent yields.

The final contribution came from the group guided by Poncet (Roux et al. 1994). In this synthesis (Scheme 15.5), useful derivatives of Dil and Dap were assembled in five and three steps, respectively, from readily available starting materials. The key step for the synthesis of Boc-Dil (**31**) was the stereoselective sodium borohydride reduction of an allyl ketone prepared from L-Boc-Ile. In contrast, the addition of an achiral (*Z*)-crotylboronate on the L-Boc-prolinal (**12**) gave the double bond oxidation to Boc-Dap (**16**). Sequential condensation of amino acids followed the Hamada–Shioiri synthesis.

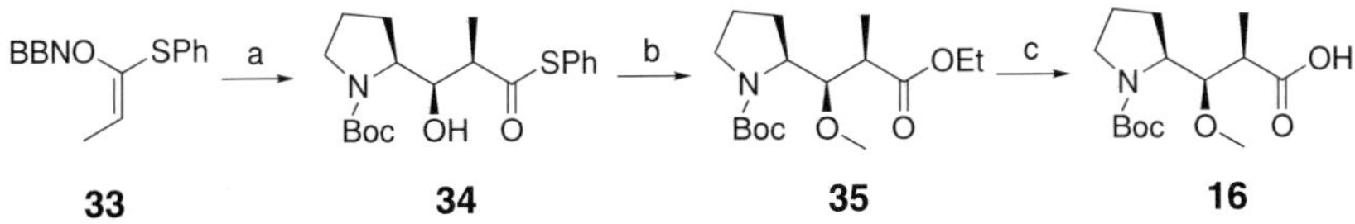

Scheme 15.4. Koga Boc-Dap preparation. (a) **12**, 64%. (b) (1) K_2CO_3 88%, (2) LHMDS, MeOTf, 83%. (c) LiOH, 91%

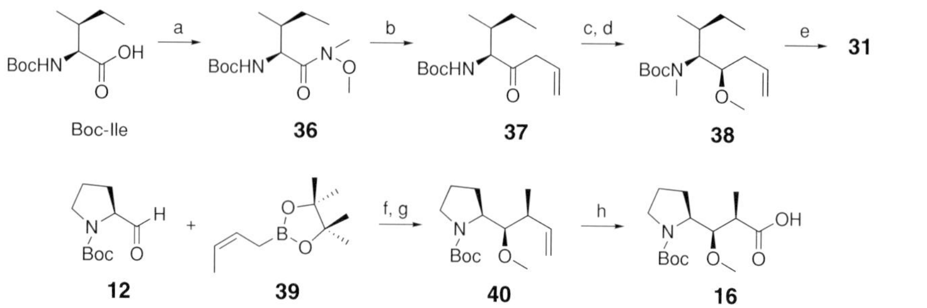

Scheme 15.5. Poncet Dil and Dap derivatives preparation. (a) MeNHOMe, PyBOP, DIEA, 90%. (b) AllylMgBr. (c) $NaBH_4$, 74%, 2 steps. (d) NaH, CH_3I, 85%. (e) RuO_2, 85%. (f) 54%. (g) NaH, CH_3I, 90%. (h) RuO_2, 81%, 3 steps

15.2.3 Dolastatin 18

Dolastatin 18 (**41**) was isolated from specimens of *D. auricularia* collected in Papua New Guinea by Pettit et al. (1997) during a bioassay-guided separation employing selected human cancer cell lines. The significant anticancer activity and the microscopic quantities obtained from the natural source stimulated the development of a synthetic access to this compound.

The synthesis of this peptide (Pettit et al. 2004) started with the coupling of Z-*N*-methyl-(*R*)-Phe and Doe (**30**), both prepared according to earlier described procedures. Deprotection was followed by coupling of the obtained dipeptide **42** to Fmoc-(*S*)-Leu, which afforded tripeptide **43**. Finally, coupling of the deprotected amine **44** with 2,2-dimethyl-3-oxohexanoic acid, prepared according to the Yonemitsu protocol, afforded dolastatin 18 (Scheme 15.6).

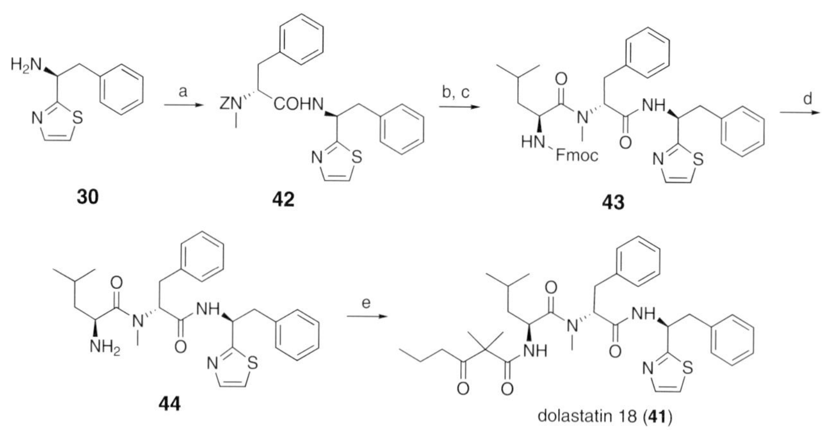

Scheme 15.6. Pettit synthesis of dolastatin 18. (a) Z-*N*-Me-(*R*)-Phe, DEPC, TEA, 98%. (b) HBr, AcOH. (c) Fmoc-L-Leu, PyBroP, DIEA, 58%, 2 steps. (d) TAEA. (e) 2,2-dimethyl-3-oxo-hexanoic acid, DEPC, TEA, 40%, 2 steps

15.2.4 Dolastatin H and Isodolastatin H

Dolastatin H (**45**) and its structural isomer isodolastatin H (**46**) are two potent cytotoxic peptides isolated by Yamada from the Japanese sea hare *D. auricularia* (Sone et al. 1996), which showed cytotoxicity against HeLa-S_3 cells (IC_{50} 0.0022, 0.0016 μg ml^{-1}). Peptide **45** exhibited in vivo antitumor activity with a T/C of 141% at a daily dose of 6 mg kg^{-1} against P388 leukemia.

The absolute stereostructure of both compounds was confirmed by Yamada by enantioselective synthesis of the possible stereoisomers (Sone et al. 1996). Since the absolute stereochemistry of the 3-phenylpropane-1,2-diol unit was unknown, (2*S*)-**45** and (2*S*)-**46** and their C-2-epimers (2*R*)-**45** and (2*R*)-**46** were all synthesized (Scheme 15.7). The 3-phenylpropane-1,2-diol unit **49** was prepared from the optically pure epoxy alcohol **47** (Scheme 15.7). The enantiomer was prepared from *ent*-**47**. The Dap and Dil units, Boc-Dap (**16**) and Z-Dil-*O*-*t*-Bu (**18**), were prepared by the methods of Shioiri and Koga, respectively, used for the synthesis of dolastatin 10 and congeners. Condensation of alcohol **49** with Boc-Dap (**16**) gave ester **50**. The coupling of deprotected Z-Dil-*O*-*t*-Bu (**52**) and Z-L-Ile afforded the dipeptide **53**, which was converted to tripeptide **54** by condensation with Dov and deprotection. The synthesis of (2*S*)-**45** was completed by the coupling of **54** and **51** followed by debenzylation. The synthesis of isodolastatin H (2*S*)-**46** was performed by a sequence similar to that used for the synthesis of (2*S*)-**45** starting from silyl ether **48**. The epimers (2*R*)-**45** and (2*R*)-**46** were also synthesized using *ent*-**49** and *ent*-**48**, respectively, in the same manner. Of the four stereoisomers, the spectral data for (2*S*)-**45** and (2*S*)-**46** resulted identical to those for natural compounds **45** and **46**, respectively.

15.3 Cyclic Peptides

15.3.1 Dolastatin 3

Dolastatin 3 (**56**), a very potent antineoplastic agent (ED $= 1 \times 10^{-4}$ to 1×10^{-7} μg ml^{-1} against P388 lymphocytic leukemia cells) was isolated for the first time by Pettit et al. (1982) from the Indian Ocean sea hare *D. auricularia*. On the grounds of NMR data, a cyclic pentapeptide structure, having the sequence *cyclo*[Pro-Leu-Val-(Gln)Thz-(Gly)Thz], was proposed.

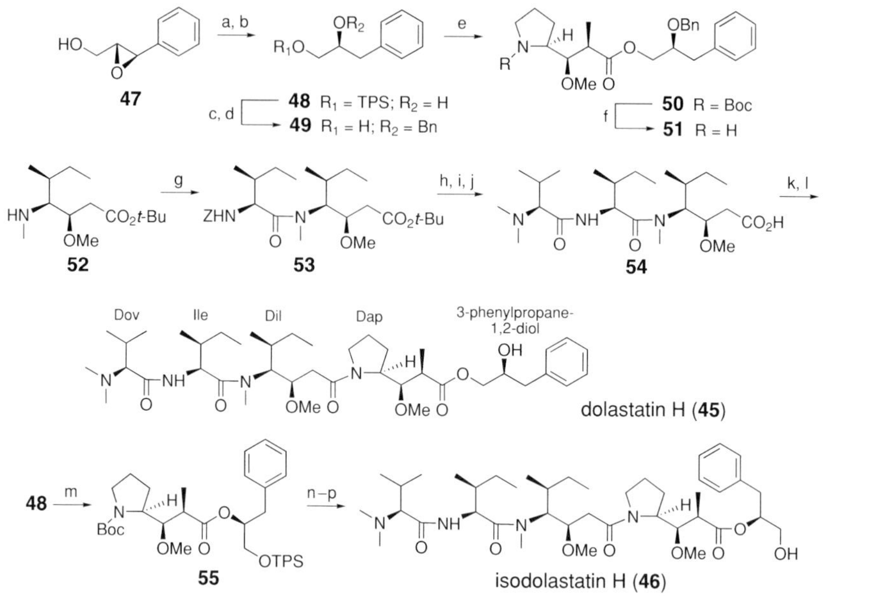

Scheme 15.7. Yamada synthesis of dolastatin H and isodolastatin H. (a) DIBAL, 62%. (b) TPSCl, imidazole, 78%. (c) BnBr, $LiN(SiMe_3)_2$, 75%. (d) HF, 97%. (e) **16**, DCC, DMAP, CSA, 86%. (f) TFA. (g) Z-Ile, PyBOP, DIPEA, 83%. (h) H_2, Pd/C. (i) Dov, DEPC, Et_3N, 95%, 2 steps. (j) TFA. (k) **51**, DEPC, Et_3N, 87%, 2 steps. (l) H_2, Pd/C, AcOH, 77%. (m) **16**, DCC, DMAP, CSA, 85%. (n) TFA. (o) **54**, DEPC, Et_3N, 52%, 2 steps. (p) HF, 87%

The synthesis of the proposed structure, however, afforded an inactive material displaying different physical properties. The definitive structure assignment of dolastatin 3 (**56**), with a different amino acidic sequence, was confirmed by a total synthesis reported by Pettit et al. (1987b). This synthetic approach was then followed by Holzapfel et al. (1990) in order to obtain a large amount of material for conformational behavior studies (Scheme 15.8). In this work, the linear *N*-Boc protected pentapeptide ester **62** was obtained by stepwise amino acid condensation. The unusual amino acid Boc-(Gln)Thz **59** was prepared from L-Glu in nine passages, employing a modified Hantzsch condensation as a key step. In the last cyclization step, the method devised for reagent addition avoided high dilution and long reaction time.

15.3.2 Dolastatin E

Dolastatin E (**63**) is a cyclic hexapeptide isolated from the Japanese sea hare *D. auricularia* (Ojika et al. 1995), showing a weak cytotoxicity against HeLa-S_3 cells (IC_{50} 22–40 μg ml^{-1}). This peptide consists of three modified amino acids containing oxazole, thiazole, and thiazoline rings.

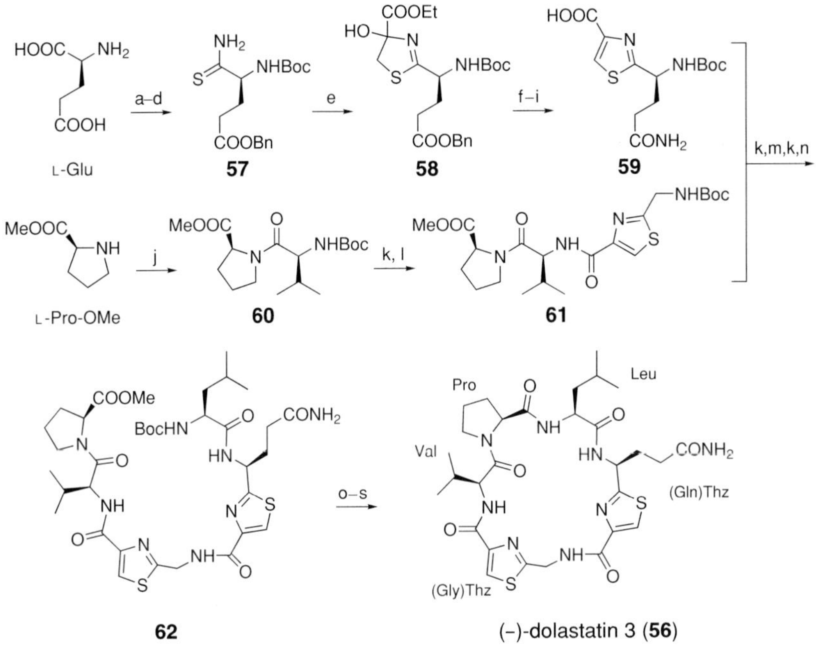

Scheme 15.8. Holzapfel synthesis of (–)-dolastatin 3. (a) Cu(II), BnBr. (b) (t-BuOCO)$_2$O, TEA. (c) NMM, ClCOO-i-Bu, NH_3. (d) Lawesson's reagent, 51% ,4 steps. (e) $BrCH_2COCO_2Et$, $KHCO_3$. (f) $(CF_3CO)_2O$, Py, 92%, 2 steps. (g) Pd–C, $HCOONH_4$, 96%. (h) NMM, ClCOO-i-Bu, NH_3, 86%. (i) NaOH, then HCl, 93%. (j) Boc-Val, DEPC, TEA, 92%. (k) TFA. (l) Boc-Gly(Thz), DEPC, TEA, 91%. (m) DEPC, TEA, 93%. (n) Boc-Leu, DEPC, TEA. (o) NaOH. (p) HCl. (q) DCCI, C_6F_5OH. (r) TFA. (s) 4-pyrrolidinopyridine, 53%, 6 steps

The complete stereostructure of dolastatin E (**63**) was confirmed by the Yamada group (Nakamura et al. 1995) by total synthesis of the eight possible stereoisomers. The synthesis of the correct one is shown in Scheme 15.9. Thiazole **65** was prepared in four steps (69% yield), starting from N-Boc-*allo*-D-Ile. As shown in Scheme 15.9, units **64** and **65** were connected to accomplish the synthesis of tetrapeptide **66**. Regioselective thionation of compound **66** provided, after Boc deprotection, thiopeptide **67**. Oxazole unit **68** was prepared starting from N-Boc-L-Ala and L-Ser-OMe·HCl. Cyclization to oxazoline was performed under Mitsunobu conditions and dehydrogenation to oxazole was realized through an ester α-selenylation and subsequent oxidation–elimination. Coupling of deprotected fragment **67** with unit **68** provided hexapeptide **69**. Macrolactamization was followed by cyclodehydration under Mitsunobu conditions, affording the thiazoline ring of cyclic hexapeptide **63**.

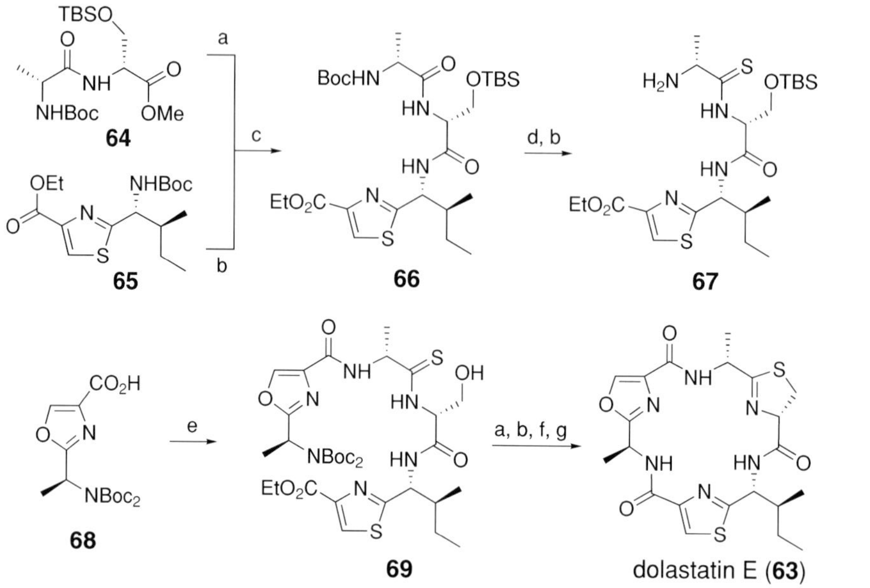

Scheme 15.9. Yamada synthesis of dolastatin E. (a) NaOH. (b) TFA. (c) DEPC, Et_3N, 78%, 2 steps from **65**. (d) 2,4-bis(4-phenoxyphenyl)-1,3-dithian-2,4-diphosphetane-2,4-disulfide, 44%. (e) **67**, DPPA, Et_3N, 65%. (f) DPPA, Et_3N, 22%, 3 steps. (g) DIAD, Ph_3P, 20%

15.3.3 Dolastatin I

Dolastatin I (**70**) is a cyclic hexapeptide, containing three kinds of five-membered heterocycles (oxazole, thiazole, oxazoline), isolated by the Yamada group (Sone et al. 1997) from the Japanese *D. auricularia,* which exhibited cytotoxicity against HeLa-S_3 cells (IC_{50} 12 μg ml^{-1}).

Kigoshi and Yamada (1999) confirmed the reported stereostructure by a convergent total synthesis (Scheme 15.10). Due to its lability, the oxazoline ring in compound **70** was constructed during the very last stage of the synthesis by cyclodehydration of a serine-containing cyclopeptide, which was prepared from the oxazole unit **72**, the thiazole unit **73**, *N*-Boc-L-Val, and *N*-Boc-L-Ser by condensation reactions and macrocyclization. The desired oxazole **72** was prepared starting from *N*-Boc-L-Ile and *N*-Boc-L-Thr-OMe·HCl through cyclization to oxazoline **71** and subsequent radical oxidation. The thiazole unit **73**, prepared by a modified Hantzsch thiazole synthesis, was transformed into the tetrapeptide **75** using classic peptide synthesis. The hexapeptide **76**, derived from a union of the free acid **72** and primary alcohol **75**, was cyclized under high-dilution conditions giving, after dehydration, the desired dolastatin I (**70**).

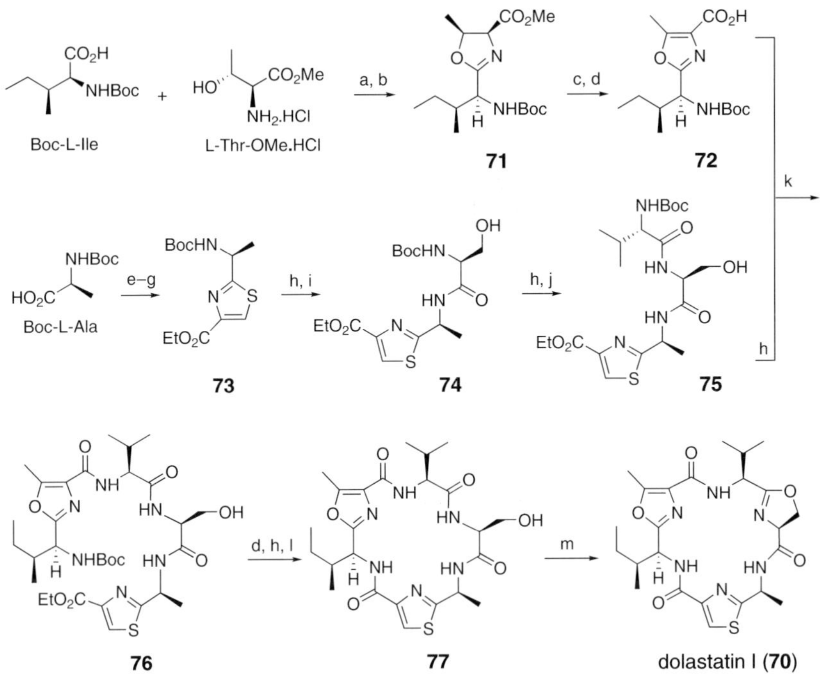

Scheme 15.10. Yamada synthesis of dolastatin I. (a) DPPA, Et_3N, 79%. (b) $MeO_2CNSO_2NEt_3$, 74%. (c) CuBr, PhCOOO-*t*-Bu, 71%. (d) NaOH. (e) DCC, HOBt, then NH_3(l). (f) $PhOC_6H_4PS_2PS_2C_6H_4OPh$. (g) $BrCH_2COCO_2Et$, $KHCO_3$, then $(CF_3CO)_2O$, 2,6-lutidine, 57%, 3 steps. (h) TFA. (i) Boc-L-Ser, DCC, HOBt, 92%, 2 steps. (j) Boc-L-Val, DCC, HOBt, DIPEA, 71%, 2 steps. (k) DPPA, Et_3N, 45%, 2 steps. (l) DPPA, Et_3N, 41%, 3 steps. (m) PPh_3, DIAD, 80%

15.4 Linear Depsipeptides

15.4.1 Dolastatin 15

Dolastatin 15 (**78**; $ED_{50} = 2.4 \times 10^{-3}$ μg ml^{-1} against P388 lymphocytic leukemia cells) was isolated in 1989 by Pettit et al. (1989c). The structure of this depsipeptide was found to be constituted of five amino acidic residues (Dov, Val, *N*-Me-Val, two Pro units) and two nonamino-acidic residues, dolapyrrolidone (Dpy) and 2-hydroxyisovaleric acid (Hiva, **79**); and it can be seen precisely as a pentapeptide (Dov-Val-*N*-Me-Val-Pro-Pro) esterified by the Dpy-Hiva carboxylic acid subunit. The stereochemical assignment was initially based on the assumption that amino acids possessed the L configuration and this was then confirmed by synthesis.

The first synthetic convergent approach, proposed by Pettit et al. (1991), was then improved (Pettit et al. 1994) for the preparation of dolastatin 15 (78) in a quantity useful for clinical trials (Scheme 15.11). It employed the coupling between tripeptide **82** and proline derivative **83**, obtained starting from Hiva (79). In the key step, mixed anhydride of carboxylic acid **80** was used to acylate Meldrum ester. Decarboxylative lactamization of the resulting adduct furnished the dolapyrrolidone ring. Subunit **82** was obtained starting from Z-*N*-Me-Val. Dov was introduced in the final step.

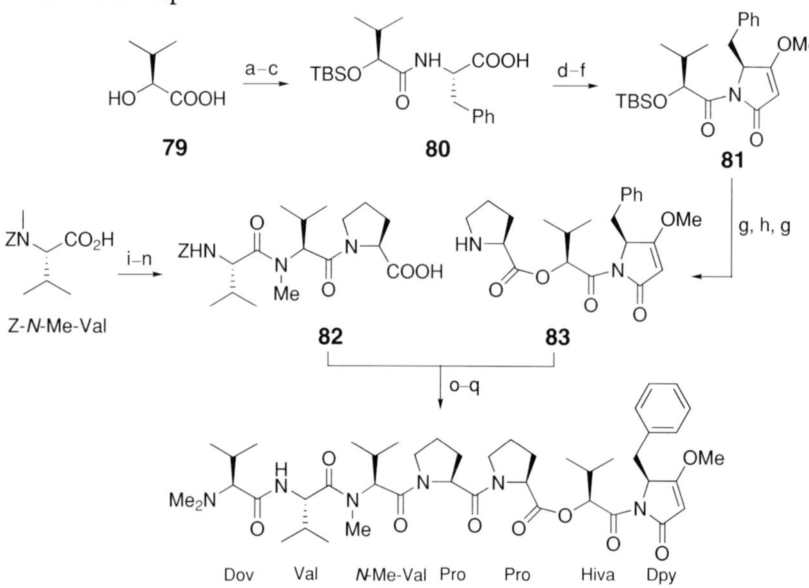

(–)-dolastatin 15 (**78**)

Scheme 15.11. Pettit synthesis of (–)-dolastatin 15. (a) Phe-OMe·HCl, DEPC, NMM, 61%. (b) TBSCl, imidazole, DMF, 87%. (c) NaOH, 90%. (d) Meldrum's ester, DMAP, $ClCO_2C(CH_3)=CH_2$. (e) toluene, Δ. (f) K_2CO_3, $(MeO)_2SO_2$, THF, 68%, 3 steps. (g) TFA, quant. (h) Boc-Pro, DCCI, 4-pyrrolidinopyridine, 92%. (i) Pro-OMe·HCl, DEPC, TEA, 77%. (l) H_2, Pd/C. (m) Z-Val, PivCl, NMM, 83%, 2 steps. (n) NaOH, quant. (o) DEPC, TEA, 89%. (p) H_2, Pd/C, 96%. (q) Dov, DEPC, TEA, 97%

In the slightly different synthetic plan by the Poncet group (Patino et al. 1992), the fragments condensed in the last step were pentapeptide **84** and α-hydroxyamide **85** (Scheme15.12). Structure **84** was constructed by one amino acid stepwise condensation employing PyCloP/DIPEA and following removal of the Z group at every step. Preparation of **85** was realized in a similar manner as described above. After coupling, the terminal Z-Val of the resulting product was deprotected and permethylated in situ leading to dolastatin 15 (78).

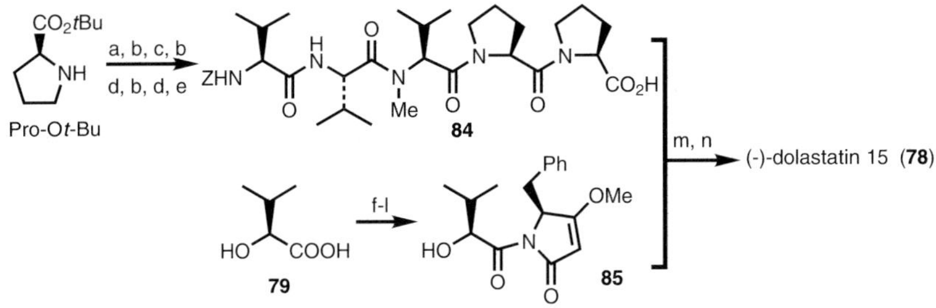

Scheme 15.12. Poncet synthesis of (–)-dolastatin 15. (a) Z-Pro, PyCloP, DIPEA. (b) H_2, Pd/C. (c) Z-*N*-Me-Val, PyCloP, DIPEA. (d) Z-Val, PyCloP, DIPEA, 32%, 7 steps. (e) TFA, 82%. (f) Phe-OBn, PyBOP, NMM, 98%. (g) TBSCl, imidazole, 80%. (h) H_2, Pd/C, quant. (i) Meldrum's ester, DMAP, $ClCOOC(CH_3)=CH_2$. (j) CH_3CN, Δ. (k) PPh_3, MeOH, DEAD, 60%. (l) TFA, 82%. (m) DMAP, IPCC, TEA, 76%. (n) H_2, Pd/C, 37% aq. HCHO, MeOH, 80%

15.4.2 Dolastatin C

The Yamada group isolated dolastatin C (**86**), a depsipeptide with weak cytotoxicity against HeLa-S_3 cells (IC_{50} 17 μg ml^{-1}), from the Japanese specimens of *D. auricularia*. The same researchers confirmed its structure by total synthesis (Sone et al. 1993b). L-Phenylalaninamide, Pro and Me-Ile were successively attached to afford tripeptide **88**. The deprotection product of compound **88** was coupled with the protected L-isoleucic acid **89**, prepared according to the Shiori method affording tetrapeptide **90**. After deprotection, esterification with L-Me_2-Ile afforded depsipeptide **86** (Scheme 15.13).

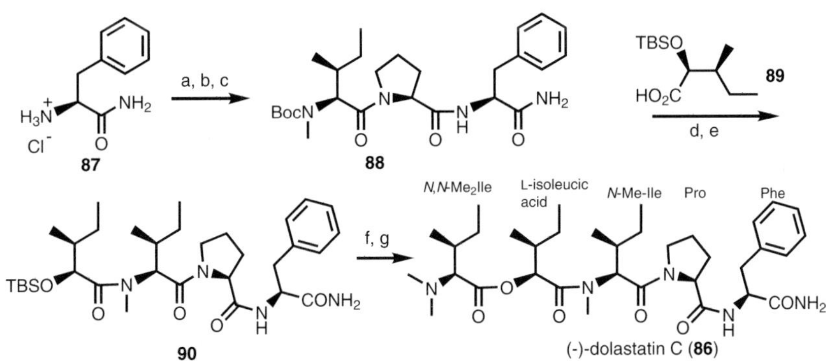

Scheme 15.13. Yamada synthesis of dolastatin C. (a) Boc-L-Pro, DEPC, Et_3N, 87%. (b) HCl, 78%. (c) Boc-L-Me-Ile, DEPC, Et_3N, 67%. (d) HCl, 85%. (e) **89**, BopCl, Et_3N, 40%. (f) TBAF, 98%. (g) L-Me_2-Ile, DCC, DMAP, 52%

15.5 Cyclic Depsipeptides

15.5.1 Dolastatin 11

Dolastatin 11 (**91**) ($ED_{50} = 2.7 \times 10^{-3}$ µg ml^{-1} against P388 mouse lymphocytic leukemia cells) is a cyclic depsipeptide isolated from *D. auricularia* (Pettit et al. 1989a). Its structure was found to contain six amino acids, Gly (two residues), *N*-Me-Leu, *N*-Me-Val, *O*,*N*-di-Me-Tyr, Ala, (2*S*,3*R*)-3-amino-2-methylpentanoic acid (Map) and two nonamino-acidic units, 4-amino-2,2-dimethyl-3-oxopentanoic acid (Ibu), and (2*S*,3*S*)-2-hydroxy-3-methylpentanoic acid (**95**, Hmp).

Five small subunits, namely **95**, *N*-Me-Val-*O*,*N*-di-Me-Tyr-*O*-Bn·HCl (**92**), Boc-Gly-*N*-Me-Leu-Gly (**93**), Boc-(2*S*,3*R*)-Map (**96**), and Boc-Ibu-Ala (**97**), were identified as building blocks to be assembled in order to synthesize dolastatin 11 (**91**; Bates et al. 1997). Amino acid **96** was obtained through two previously reported routes. The dipeptide **97**, containing the other new amino acid Ibu, was prepared starting from Boc-L-Ala, while fragment **92** was prepared by trimethylation of Boc-Val-Tyr and subsequent Boc cleavage. The synthesis of synthon **93** started from Z-L-Leu. The hydroxyacid Hmp was obtained from L-Ile by treatment with nitrous acid. The assembly of the units described above was realized as depicted in Scheme 15.14. Noteworthy are the valuable Map and Ibu units later introduced into the synthetic route, an expedient which helped to avoid epimerization of the acid/base-sensitive Ibu subunit. The cyclization of the linear depsipeptide **98** proved to be troublesome. HBTU gave the best results (20–24%).

15.5.2 Dolastatin D

Dolastatin D (**99**), a cyclic depsipeptide isolated by the Yamada group (Sone et al. 1993a) from the Japanese sea hare *D. auricularia*, showed cytotoxicity against HeLa-S_3 cells (IC_{50} 2.2 µg ml^{-1}). Two α-hydroxy acid units [(*S*)-2-hydroxyisovaleric acid and the (*S*)-phenyllactic acid], two α-amino acid units (L-Me-Val and L-Ile), were found as components, together with the novel β-amino acid (2*R*,3*R*)-3-amino-2-methylbutanoic acid.

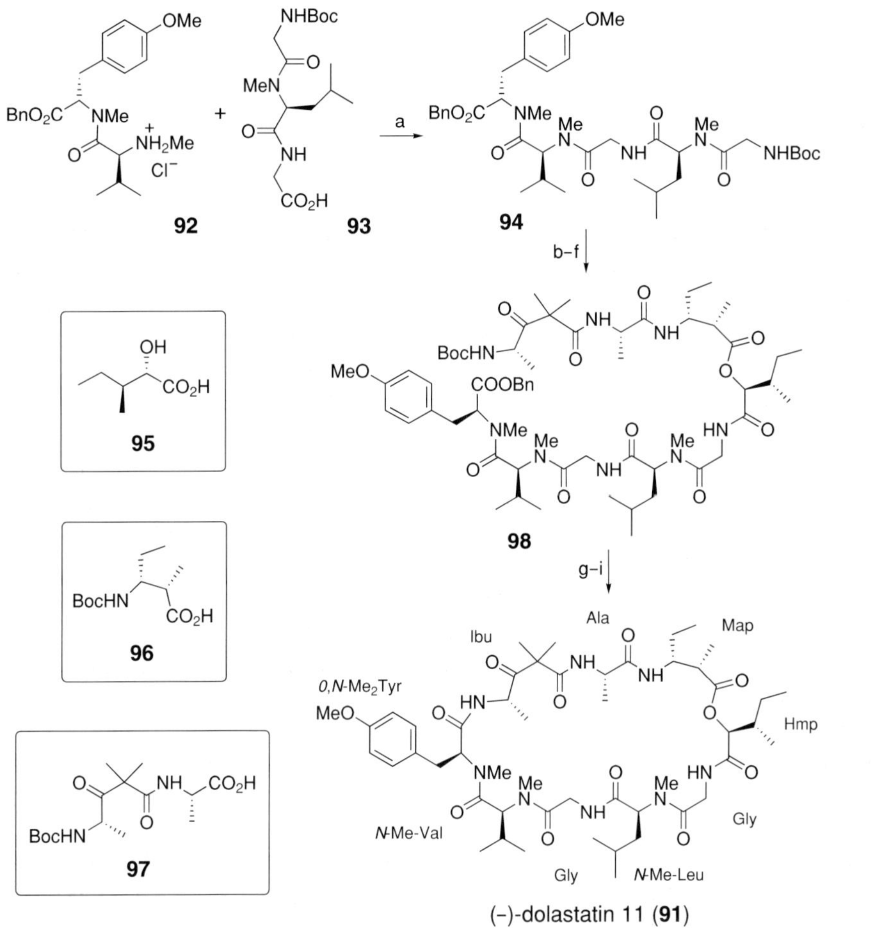

Scheme 15.14. Pettit synthesis of (–)-dolastatin 11. (a) $ClCO_2$-*i*-Bu, 61%. (b) TFA. (c) **95**, EDC. (d) **96**, EDC, DMAP. (e) TFA. (f) **97**, TBTU, 22%, 5 steps. (g) H_2, Pd/C. (h) TFA. (i) HBTU, 18%, 3 steps

The structure of dolastatin D (**99**) was confirmed by total synthesis (Sone et al. 1993a). The linear tridepsipeptide **104** was constructed starting from (*S*)-3-phenyllactic acid benzyl ester (**100**) by sequential amino acid coupling under classic conditions. Compound **104** was obtained by esterification of desilylated tridepsipeptide **102**, using the Keck method, with the enatiomerically pure Boc-(2*R*,3*R*)-3-amino-2-methylbutanoic acid (**103**). This Boc protected β-amino acid was prepared according to the reported procedure of Seebach with a slight modification. After introduction of Z-L-Ile and deprotection, the best macrocyclization conditions were achieved using BOP (Scheme 15.15).

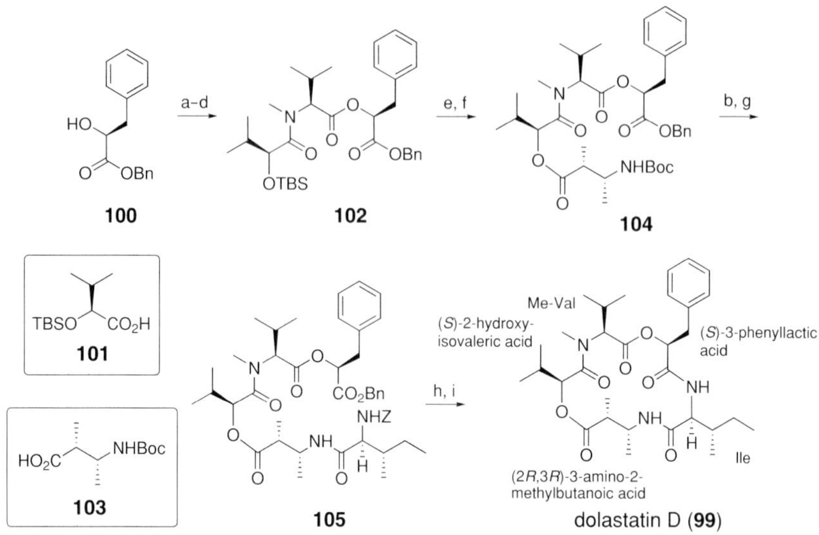

Scheme 15.15. Yamada synthesis of dolastatin D. (a) Boc-L-Val, DCC, DMAP, 88%. (b) TFA. (c) **101**, DEPC, Et_3N, 89%, 2 steps. (d) NaH, MeI, 75%. (e) 12% HF, 98%. (f) **103**, DCC, DMAP, CSA, 98%. (g) Z-L-Ile, DEPC, Et_3N, 82%, 2 steps. (h) H_2, Pd/C, 91%. (i) BOP, $NaHCO_3$, 66%

15.5.3 Dolastatin G and Nordolastatin G

Dolastatin G (**106**) and Nordolastatin G (**107**) are 35-membered cyclodepsipeptides isolated by the Yamada group (Mutou et al. 1996a,b) from the Japanese sea hare *D. auricularia,* which exhibited cytotoxicity against HeLa-S_3 cells (IC_{50} 1.0 μg ml^{-1}, IC_{50} 5.3 μg ml^{-1}, respectively).

The total synthesis realized by the Yamada group (Mutou et al. 1996a,b), was accomplished by a convergent approach, as depicted in Scheme 15.16. The stereochemistry of β-keto ester **108** was derived from commercially available levoglucosenone. Starting from (*R*)-2-hydroxy-3-methylbutanoate, the stereogenic centers into the dihydroxy acid **109** were introduced through an Evans aldol asymmetric reaction. The coupling of units **108** and **109** afforded the β-keto ester **110**, which was treated under acid conditions and immediately condensed with hexapeptide **111**, prepared in solution with a linear coupling of amino acids in the *C* to *N* direction, starting from L-Pro-*O*-Bn·HCl. Reduction with Zn provided the seco acid **112**. The desired macrolactone nordolastatin G (**107**) was obtained only under Yamaguchi conditions, although the yield was very low. Finally after extensive investigation, the formation of enol ether and synthesis of dolastatin G (**106**) were achieved using montmorillonite K 10 as catalyst.

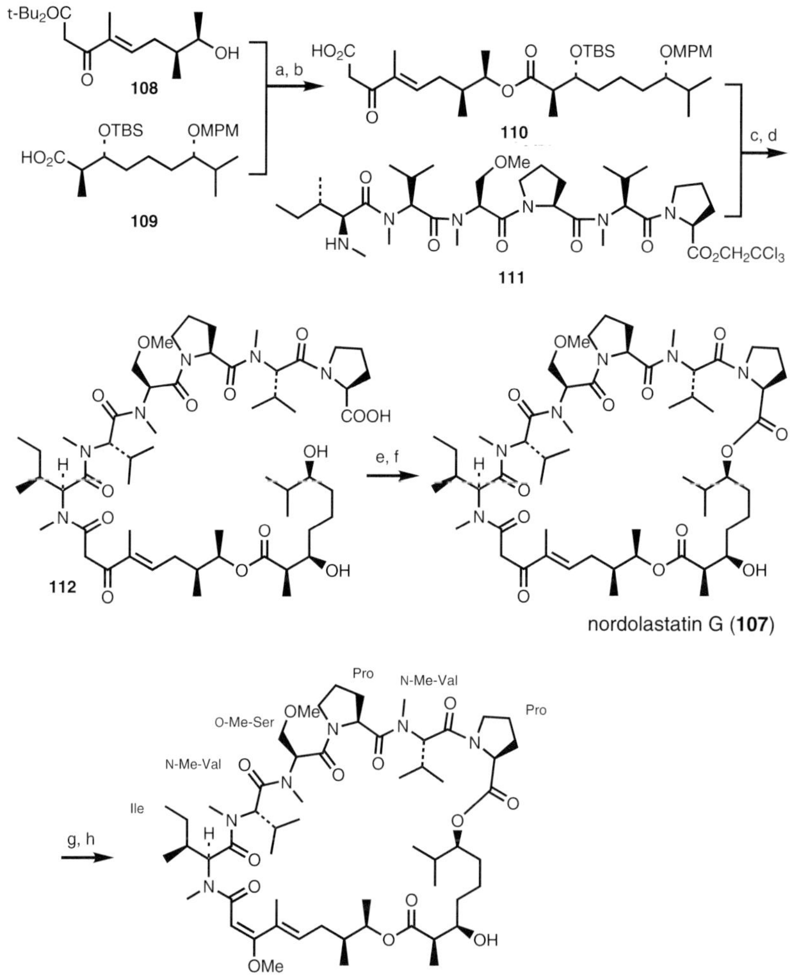

Scheme 15.16. Yamada synthesis of dolastatin G and nordolastatin G. (a) DCC, DMAP, CSA, 85%. (b) TFA. (c) PyBroP, DIPEA, 83%, 2 steps. (d) Zn, NH_4OAc, 96%. (e) 2,4,6-trichlorobenzoyl chloride, Et_3N. (f) DMAP, toluene, reflux, 3%, 2 steps. (g) montmorillonite K 10, $HC(OMe)_3$. (h) toluene, reflux, 29%, 2 steps

15.5.4 Doliculide

Doliculide (**113**) is a cyclic depsipeptide isolated from Japanese specimens of *D. auricularia* (Ishiwata et al. 1994c). This metabolite exhibits potent cytotoxic activity by interfering with normal actin assembly (Bai et al. 2002). It contains a unique 15-carbon deoxypropionate polyketide unit and a substituted D-tyrosine derivative.

The three synthetic approaches to doliculide (**113**), described up to now, differ from each other in the construction of the deoxypropionate unit. In the Yamada synthesis (Ishiwata et al. 1994a,b), the polyketide moiety was prepared using a strategy based on the Evans aldol reaction. Aldehyde **114**, prepared from (*S*)-4-methyl-1,3-pentanediol, gave compound **116** under Evans conditions. Simple manipulations produced aldehyde **117**, which underwent another Evans aldol reaction. The hydroxyl function generated in this process was then removed in two steps. Iteration of the same strategy was followed by selective removal of the MOM group and inversion of hydroxyl group by the Mitsunobu reaction. Union with Gly-*O*-*t*-Bu·HCl and debenzylation afforded alcohol **120**, which was coupled with *N*-Boc-3-iodo-*N*-methyl-*O*-TBS-D-Tyr (**121**). Linear depsipeptide **122** was converted into **113** (11% overall yield) by macrolactamization and removal of a silyl group (Scheme 15.17).

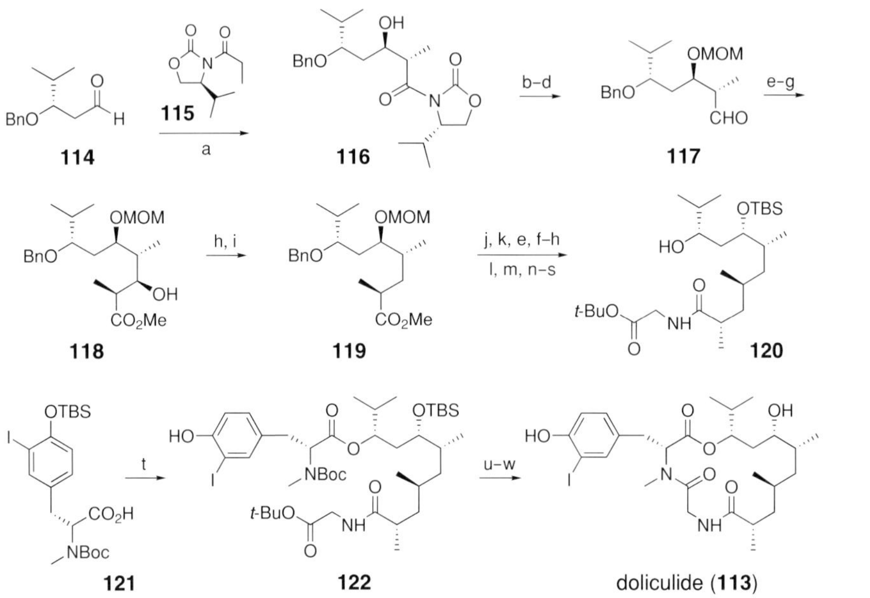

Scheme 15.17. Yamada synthesis of doliculide. (a) **115**, Bu_2BOTf, Et_3N, 95%. (b) Me(MeO)NH·HCl, Me_3Al. (c) $MeOCH_2Cl$, *i*-Pr_2NEt. (d) DIBAL, 80%, 3 steps. (e) **115**, Bu_2BOTf, Et_3N. (f) LiOH, H_2O_2. (g) CH_2N_2, 85%, 3 steps. (h) Im_2CS, THF. (i) Bu_3SnH, 73%, 2 steps. (j) $LiAlH_4$. (k) DMSO, $(COCl)_2$, Et_3N. (l) Bu_3SnH, 55%, 7 steps. (m) HCl. (n) Ph_3P, *p*-$NO_2C_6H_4COOH$, $(EtOOCN)_2$, 69%. (o) NaOH. (p) TBSOTf, Et_3N. (q) K_2CO_3, 99%, 3 steps. (r) Gly-*O*-*t*-Bu·HCl, DEPC, Et_3N, 93%. (s) H_2, $Pd(OH)_2/C$. (t) **120**, DCC, DMAP, 94%. (u) TFA. (v) BOP-Cl, Et_3N, 74%. (w) Bu_4NF, 99%

The strategy (Scheme 15.18) utilized by Ghosh and Liu (2001) involves an iteration of Charette asymmetric cyclopropanation for the construction of the deoxypropionate unit. The allylic alcohol **123**, easily prepared in three steps (66%) from (3*S*)-4-benzyloxy-3-methyl-butyronitrile, was subjected to highly diastereoselective (91% d.e.) asymmetric cyclopropanation producing alcohol **125**. Conversion of **125** into the corresponding iodide followed by metal-induced ring opening afforded the terminal alkene **126**. Alkene **128** was prepared following the iterative sequence described for alkene **126**, using as starting material the aldehyde **127**. Ozonolytic cleavage of alkene **128** furnished an aldehyde, which was converted into allyl alcohol **129**. Sharpless asymmetric epoxidation, two-carbon elongation and regioselective epoxide opening provided alcohol **131**. Epoxide **132**, obtained in three steps, was opened, affording the allylic alcohol **133**, which was hydrogenated with contemporary debenzylation and double bond saturation. Selective oxidation of the primary alcohol and esterification of the obtained acid afforded the *t*-butyl ester **134** ready to be coupled with a dipeptide, prepared from a D-Tyr derivative and *N*-Boc-Gly. Finally, deprotection, macrolactamization and the action of TBAF gave doliculide (**113**).

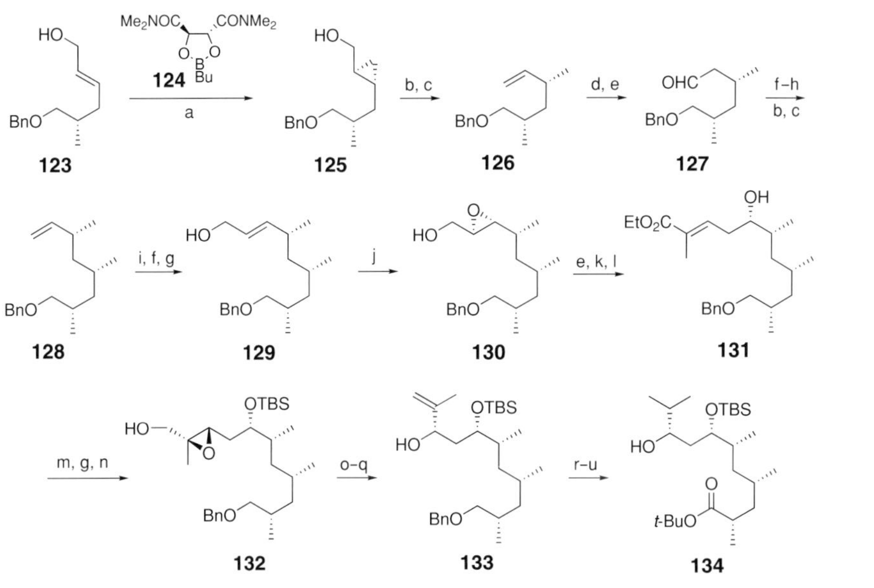

Scheme 15.18. Ghosh synthesis of doliculide deoxypropionate fragment. (a) **124** (cat.), $Zn(CH_2I)_2$·DME. (b) I_2, PPh_3, imidazole. (c) *n*-BuLi, TMEDA, 71%, 3 steps. (d) 9-BBN, H_2O_2, OH^-. (e) Swern oxidation. (f) NaH, $(EtO)_2P(O)CH_2CO_2Et$. (g) DIBAL. (h) **124** (cat.), $Zn(CH_2I)_2$·DME, 55%, 5 steps. (i) O_3, Ph_3P. (j) $Ti(O\textit{i}Pr)_4$, (–)DET, *t*-BuOOH, 53%, 4 steps. (k) $Ph_3P{=}C(Me)CO_2Et$, 81%, 2 steps. (l) $Pd_2(dba)_3$·$CHCl_3$, Bu_3P, HCO_2H, Et_3N, 90%. (m) TBSOTf, Et_3N. (n) $Ti(O\textit{i}Pr)_4$, (+)DET, *t*-BuOOH, 73%. (o) MsCl, Et_3N, DMAP. (p) NaI. (q) *n*-BuLi, TMEDA, 84%, 3 steps. (r) H_2, Pd/C, 84%. (s) TPAP, NMO. (t) $NaClO_2$. (u) Boc_2O, DMAP, 51%, 3 steps

The approach of the Hanessian group (Hanessian et al. 2004) to the deoxypropionate unit was based on an iterative stereocontrolled conjugate addition to α,β-unsaturated esters (Scheme 15.19). Enoate **135**, prepared in a few steps starting from L-ascorbic acid, was subjected to the addition of lithium dimethylcuprate in the presence of TMSCl. The single stereocenter of L-ascorbic acid allowed a stereocontrolled introduction of C-methyl group, leading to compound **136** as major isomer (14:1 ratio). This strategy was applied twofold, affording ester **138**, which was subjected to reduction and dehydration to give the olefin **139**. Primary alcohol, obtained by removal of TPS, was converted into an iodide useful for displacement with the lithium 2-isopropyl-[1,3]-dithiane. Conversion of compound **140** to the corresponding ketone and successive reduction afforded a 1:1 mixture of alcohol **141** and its epimer. Ghosh strategy was adopted for the completion of the synthesis, generating the needed acid for macrolactamization by ozonolysis of the double bond and oxidation of the obtained aldehyde.

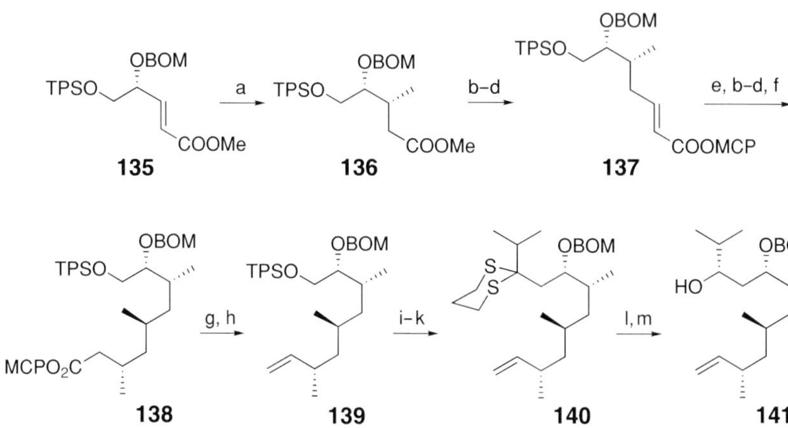

Scheme 15.19. Hanessian synthesis of doliculide deoxypropionate fragment. (a) MeLi·LiBr, CuI, TMSCl, 95% (*anti*/*syn* 14:1). (b) DIBAL. (c) Swern oxidation. (d) Ph_3P=C(H)CO_2MCP, 70%, 3 steps. (e) MeLi, LiBr, CuI, TMSCl, 87% (*anti*/*syn* 1:8). (f) MeLi·LiBr, CuI, TMSCl, 92% (*syn*/*anti*:*syn*/*syn* 1:12), 52%, 4 steps. (g) DIBAL, 92%. (h) $(n\text{-Bu})_3P$, *o*-nitrophenylselenocyanate, THF then H_2O_2, 68%. (i) Bu_4NF. (j) Ph_3P, I_2, imidazole, 93%, 2 steps. (k) 2-isopropyl-[1,3]-dithiane, *t*-BuLi, HMPA, 75%. (l) $AgNO_3$, NCS, 79%. (m) DIBAL, 96% (*syn*/*anti* 1:1)

15.5.5 Aurilide

Aurilide (**142**) is a 26-membered cyclodepsipeptide isolated from the Japanese sea hare *D. auricularia* (Suenaga et al. 1996) and has a potent cytotoxicity against HeLa-S_3 cells (IC_{50} = 0.011 μg ml^{-1}). It is constituted of five amino acid residues (two L-Val, Me-Gly, L-Me-Ala, D-Me-Leu), an *allo*-D-isoleucic acid and a new dihydroxy acid.

The first enantioselective synthesis of aurilide by Yamada (Mutou et al. 1997), unambiguously confirmed its stereostructure and was then much improved (Suenaga et al. 2004). Preparation of pentapeptide **143** (Scheme 15.20) was carried out starting from *N*-MeGly-*O*-*t*-Bu·HCl and using classic peptide synthesis techniques. The protected dihydroxy acid **146** was synthesized through an *anti* selective aldol condensation, under Heathcock conditions, followed by a vinylogous Mukaiyama aldol reaction. The coupling reaction between fragments **143** and **146**, followed by esterification of the deprotected alcohol with Fmoc-L-Me-Ala and removal of the trichloroethyl group, gave compound **147**.

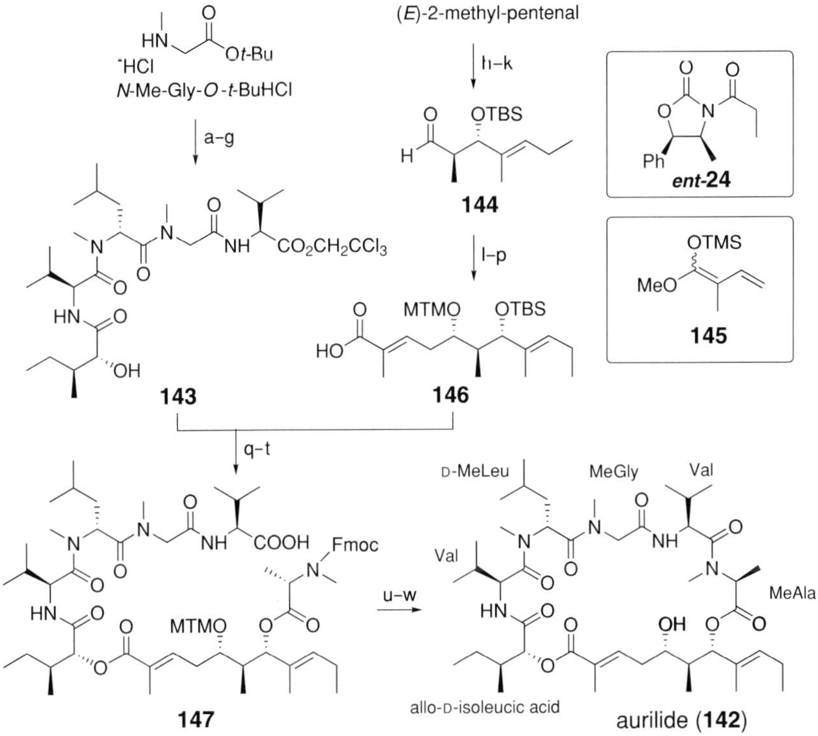

Scheme 15.20. Yamada synthesis of aurilide. (a) Z-D-Me-Leu, DEPC. (b) H_2, Pd/C. (c) Z-L-Val, PyBOP, *i*-Pr_2EtN. (d) H_2, Pd-/C. (e) *allo*-D-isoleucic acid Na salt, EDCI·HCl, HOBt. (f) TMSOTf, 2,6-lutidine. (g) L-Val-OCH_2CCl_3, EDCI·HCl, HOBt, 84%, 7 steps. (h) Bu_2BOTf, *i*-Pr_2EtN, *ent*-**24**. (i) Me_2AlN(Me)OMe. (j) TBSCl. (k) DIBAL, 78%, 4 steps. (l) **145**, BF_3·Et_2O. (m) Dess–Martin oxid. (n) $NaBH_4$. (o) DMSO, Ac_2O, AcOH. (p) LiOH, 44%, 5 steps. (q) EDCI·HCl, DMAP. (r) HF·Py, Py. (s) Fmoc-L-Me-Ala, EDCI·HCl, DMAP. (t) Zn, NH_4OAc, 83%, 3 steps. (u) Et_2NH. (v) EDCI·HCl, HOAt. (w) $AgNO_3$, 2,6-lutidine, 62%, 3 steps

Macrolactamization afforded a cyclic precursor in 66% along with an isomer (24%), which resulted from epimerization at C-6. Finally, deprotection yielded aurilide (12% overall yield, 16 steps).

In 2003, Takahashi et al. (2003; Scheme 15.21) described a solid-phase combinatorial approach toward aurilide (**142**) and related analogues. The tetrapeptide moiety **149** was assembled on trityl linker-functionalized SynPhase Crowns by extension of the N terminus, using an Fmoc strategy. The aliphatic moiety **148** was synthesized through a slightly modified Yamada pathway from the fragment **146**, obtained using an improved procedure. The coupling of the acid **148** with amine **149** was the key step of the synthesis. Cleavage from the solid support of deprotected linear precursor **150**, macrocyclization and removal of the MTM group gave aurilide (**142**) in 11% overall yield.

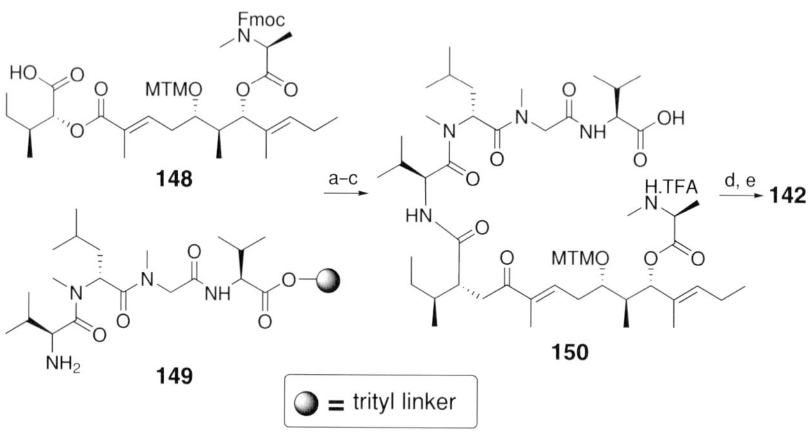

Scheme 15.21. Takahashi synthesis of aurilide. (a) DIC, HOBt. (b) piperidine. (c) TFA. (d) EDC, HOAt. (e) $AgNO_3$, 2,6-lutidine

15.5.6 Kahalalide B

The kahalalides are peptides isolated by Hamann and Scheuer (1993) from the sacoglossan mollusc *Elysia rufescens* and the green alga *Bryopsis* sp. on which it feeds. Kahalalide B (**151**; Hamann et al. 1996) is a cyclic depsipeptide formed by seven different amino acids (Gly, Thr, Pro, D-Leu, Phe, D-Ser, Tyr) and the fatty acid 5-methylhexanoic (5-MeHex), an aliphatic isoacid also present in the structure of other members of the series.

The synthesis of kahalalide B (**151**), achieved by the Albericio–Giralt group at the University of Barcelona (López-Macià et al. 2000), was based on elongating the peptide chain in the solid phase on a 2-ClTrt-Cl-resin with the Fmoc/*t*-Bu strategy and carrying out the cyclization in solution after cleavage of the peptide from the resin. First, 5-MeHex was prepared by a malonic synthesis in a two-step sequence from diethyl malonate and 1-bromo-3-methylbutane. The synthesis of peptide was developed through two different strategies (Scheme 15.22): in the first, the cyclization was carried out in solution through the ester bond formation between the carboxyl group of Gly and the side-chain hydroxyl of D-Ser (strategy A) and, in the second, the ester bond was formed in the solid phase and the cyclization was due to an amide bond between the carboxyl group of Thr and the amine group of Gly (strategy B).

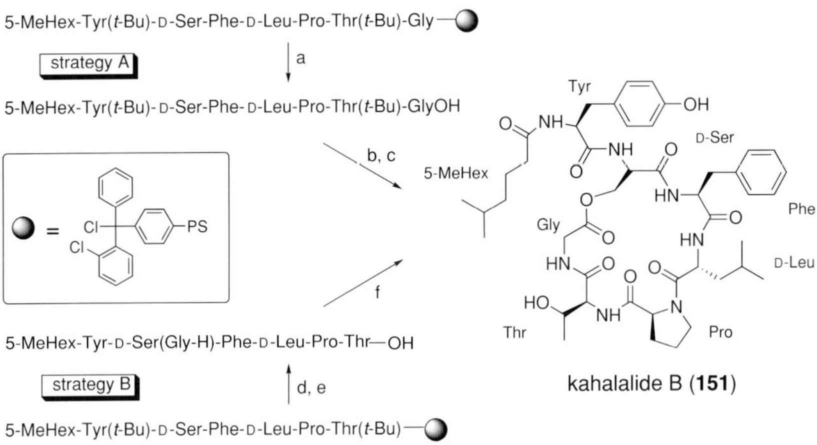

Scheme 15.22. Albericio–Giralt synthesis of kahalalide B. (a) TFA, 67%. (b) PyBOP-DIEA. (c) TFA, 28%, 2 steps. (d) Boc-Gly-OH, DIPCDI, DMAP. (e) TFA, 89%. (f) PyBOP-DIEA, 22%

15.5.7 Kahalalide F

Among kahalalides, the largest and the most active is kahalalide F (**152**; Hamann and Scheuer 1993), which exhibits a very interesting antitumor activity and is currently undergoing clinical trials. This cyclodepsipeptide, containing 13 amino acids and 5-MeHex at the N terminus, is responsible for the protection of *E. rufescens* from fish predation (Becerro et al. 2001).

The synthesis of kahalalide F (**152**) reported by the Albericio–Giralt group (López-Macià et al. 2001) allowed the assignment of the correct stereostructure. Two strategies (Scheme 15.23) based on the solid-phase approach were developed for this synthesis. Both involved the elongation of the synthetic chain in the solid phase, cleavage of the protected peptide from the resin, subsequent cyclization and final deprotection in solution.

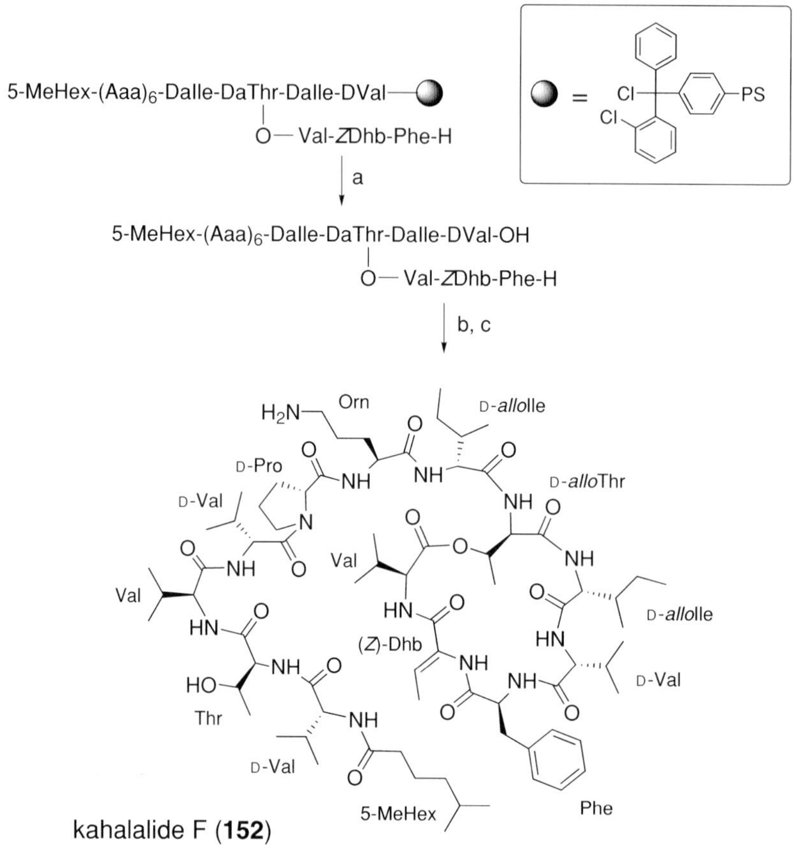

kahalalide F (**152**)

Scheme 15.23. Albericio–Giralt synthesis of kahalalide F. (a) TFA, 65–87%. (b) PyBOP-DIEA. (c) TFA

The difference is in the insertion of the double bond of didehydroamino acid in the linear sequence, which was synthesized on a 2-ClTrt-Cl-resin elongating the peptide chain with Fmoc/*t*-Bu strategy. Cleavage of the protected peptide, cyclization and final deprotection afforded depsipeptide **152**.

Appendix 15.1

The abbreviations used in the schemes are explained in Table 15.1

Table 15.1. Abbreviations used in the schemes

abbreviation	compound
Alloc	allyloxycarbonyl
BBN	9-borabiciclo[3.3.1]nonane
Boc	*t*-butoxycarbonyl
BOM	benzyloxymethyl
Bop, BOP	benzotriazol-1-yloxy-tris(dimethylamino) phosphonium hexafluorophosphate
BopCl	bis(2-oxo-3-oxazolidinyl) phosphinic chloride
CDI	carbonyldiimidazole
CSA	10-camphorsulfonic acid
DCC, DCCI	dicyclohexylcarbodiimide
DEAD	diethyl azodicarboxylate
DEPC	diethylphosphorocyanidate
DET	diethyl tartrate
DIAD	diisopropylazodicarboxylate
DIBAL	diisobutylaluminum hydride
DIC	diisopropylcarbodiimide
DIEA	diisopropylethylamine
DIPCDI	diisopropylcarbodiimide
DIPEA	diisopropylethylamine
DMAP	4-dimethylaminopyridine
DME	1,2-dimethoxyethane
DMF	*N,N*-dimethylformamide
DMSO	dimethylsulfoxide
DPPA	diphenylphosphoryl azide
EDC, EDCI	1-(3-dimethylaminopropyl)-3-ethylcarbodiimide (hydrochloride)
Fmoc	9-fluorenylmethoxycarbonyl
HBTU	*O*-(benzotriazol-1-yl)-1,1,3,3-tetramethyluronium hexafluorophosphate
HMPA	hexamethylphosphoramide
HOAt	7-aza-1-hydroxybenzotriazole
HOBt	hydroxybenzotriazole
Ipc	isopinocampheyl
IPCC	isopropenyl chloroformate
LDA	lithium diisopropylamide
LHMDS	lithium 1,1,1,3,3,3-hexamethyldisilamide
MCP	1-methyl-1-cyclopentyl
MOM	methoxymethyl
MPM	*p*-methoxybenzyl

abbreviation	compound
MTM	methylthiomethyl
NCS	*N*-chlorosuccinimide
NMM	*N*-methylmorpholine
PDC	pyridinium dichromate
Pfp	pentafluorophenyl
Piv	pivaloyl
Py, Pyr	pyridine
PyBOP	(benzotriazol-1-yloxy) tripyrrolidinophosphonium hexafluorophosphate
PyBroP	bromotripyrrolidino phosphonium hexafluorophosphate
PyCloP	chlorotripyrrolidino phosphonium hexafluorophosphate
TAEA	trisaminoethylamine
TBS	*t*-butyldimethylsilyl
TBTU	*O*-(benzotriazol-1-yl)-1,1,3,3-tetramethyluronium tetrafluoroborate
TEA	triethylamine
TFA	trifluoroacetic acid
THF	tetrahydrofuran
TIPS	triisopropylsilyl
TMEDA	*N,N,N′,N′* tetramethylethylenediamine
TMS	trimethylsilyl
TPAP	tetrapropylammonium perruthenate
TPS	*t*-butyldiphenylsilyl
Trt, Tr	triphenylmethyl, trityl
Z	benzyloxycarbonyl

References

Albizati KF, Martin VA, Agharahimi MR, Stolze DA (1992) Synthesis of marine natural products 2: nonterpenoids. In: Scheuer PJ (ed) Biorganic marine chemistry, vol 6. Springer, Berlin Heidelberg New York

Bai R, Covell DG, Liu C, Ghosh AK, Hamel E (2002) (–)-Doliculide, a new macrocyclic depsipeptide enhancer of actin assembly. J Biol Chem 277:32165–32171

Bates RB, Brusoe KG, Burns JJ, Caldera S, Cui W, Gangwar S, Gramme MR, McClure KJ, Rouen GP, Schadow H, Stessman CC, Taylor SR, Vu VH, Yarick GV, Zhang J, Pettit GR, Bontems R (1997) Dolastatins. 26 Synthesis and Stereochemistry of Dolastatin 11. J Am Chem Soc 119:2111–2113

Becerro MA, Goetz G, Paul VJ, Scheuer PJ (2001) Chemical defenses of the sacoglossan mollusk *Elysia rufescens* and its host alga *Bryopsis* sp. J Chem Ecol 27:2287–2291

Blunt JW, Copp BR, Munro MHG, Northcote PT, Prinsep MR (2004) Marine natural products. Nat Prod Rep 21:1–49

Faulkner DJ, Ghiselin MT (1983) Chemical defence and evolutionary ecology of dorid nudibranch and some other opisthobranch gastropods. Mar Ecol Prog Ser 13:295–301

Ghosh AK, Liu C (2001) Total synthesis of antitumor depsipeptide (–)-doliculide. Org Lett 3:635–638
Giordano A, Spinella A, Sodano G (1999) Stereoselective synthesis of 4-amino-3-hydroxy-2-methylpentanoic acids: stereochemistry of the amino acid occurring in the marine toxin janolusimide. Tetrahedron Asymm 10:1851–1854
Giordano A, Della Monica C, Landi F, Spinella A, Sodano G (2000) Stereochemistry and total synthesis of janolusimide, a tripeptide marine toxin. Tetrahedron Lett 41:3979–3982
Hamada Y, Hayashi K, Shioiri T (1991) New methods and reagents for organic synthesis. 95. Efficient stereoselective synthesis of dolastatin 10, an antineoplastic peptide from a sea hare. Tetrahedron Lett 32:931–934
Hamann MT, Scheuer PJ (1993) Kahalalide F: a bioactive depsipeptide from the sacoglossan mollusk *Elysia rufescens* and the green alga *Bryopsis* sp. J Am Chem Soc 115:5825–5826
Hamann MT, Otto CS, Scheuer PJ, Dunbar DC (1996) Kahalalides: bioactive peptides from a marine mollusk *Elysia rufescens* and its algal diet *Bryopsis* sp. J Org Chem 61:6594–6600
Hanessian S, Mascitti V, Giroux S (2004) Total synthesis of the cytotoxic cyclodepsipeptide (–)-doliculide: The "ester" effect in acyclic 1,3-induction of deoxypropionates. Proc Natl Acad Sci USA 101:11996–12001
Holzapfel CW, van Zyl WJ, Roos M (1990) The synthesis and conformation in solution of *cyclo*[L-Pro-L-Leu-L (Gln)Thz-(Gly)Thz-L-Val] (dolastatin 3). Tetrahedron 46:649–660
Ishiwata H, Sone H, Kigoshi H, Yamada K (1994a) Enantioselective total synthesis of doliculide, a potent cytotoxic cyclodepsipeptide of marine origin and structure-cytotoxicity relationships of synthetic doliculide congeners. Tetrahedron 50:12853–12882
Ishiwata H, Sone H, Kigoshi H, Yamada K (1994b) Total synthesis of doliculide, a potent cytotoxic cyclodepsipeptide from the Japanese sea hare *Dolabella auricularia*. J Org Chem 59:4712–4713
Ishiwata H, Nemoto T, Ojika M, Yamada K (1994c) Isolation and stereostructure of doliculide, a cytotoxic cyclodepsipeptide from the Japanese sea hare *Dolabella auricularia*. J Org Chem 59:4710–4711
Kigoshi H, Yamada S (1999) Synthesis of dolastatin I, a cytotoxic cyclic hexapeptide from the sea hare *Dolabella auricularia*. Tetrahedron 55:12301–12308
Kijjoa A, Sawangwong P (2004) Drugs and cosmetics from the sea. Mar Drugs 2:73–82
López-Macià A, Jiménez JC, Royo M, Giralt E, Albericio F (2000) Kahalalide B. Synthesis of a natural cyclodepsipeptide. Tetrahedron Lett 41:9765–9769
López-Macià A, Jiménez, JC, Royo M, Giralt E, Albericio F (2001) Synthesis and structure determination of kahalalide F. J Am Chem Soc 123:11398–11401
Luesch H, Moore RE, Paul VJ, Mooberry SL, Corbett TH (2001) Isolation of dolastatin 10 from the marine cyanobacterium *Symploca* species VP642 and total stereochemistry and biological evaluation of its analogue symplostatin 1. J Nat Prod 64:907–910
Luesch H, Harrigan GG, Goetz G, Horgen FD (2002) The cyanobacterial origin of potent anticancer agents originally isolated from sea hares. Curr Med Chem 9:1791–1806
Mutou T, Kondo T, Ojika M, Yamada K (1996a) Isolation and stereo structures of dolastatin G and nordolastatin G, cytotoxic 35-membered cyclodepsipeptides from the Japanese sea hare *Dolabella auricularia*. J Org Chem 61:6340–6345
Mutou T, Kondo T, Shibata T, Ojika M, Kigoshi H, Yamada K (1996b) Synthesis of dolastatin G and nordolastatin G, cytotoxic 35-membered cyclodepsipeptides of marine origin. Tetrahedron Lett 37:7299–7302
Mutou T, Suenaga K, Fujita T, Itoh T, Takada N, Hayamizu K, Kigoshi H, Yamada K (1997) Enantioselective synthesis of aurilide, a cytotoxic 26-membered cyclodepsipeptide of marine origin. Synlett 1997:199–201
Nakamura M, Shibata T, Nakane K, Nemoto T, Ojika M, Yamada K (1995) Stereochemistry and total synthesis of dolastatin E. Tetrahedron Lett 36:5059–5062
Ojika M, Nemoto T, Nakamura M, Yamada K (1995) Dolastatin E, a new cyclic hexapeptide isolated from the sea hare *Dolabella auricularia*. Tetrahedron Lett 36:5057–5058

Patino N, Frérot E, Galeotti N, Poncet J, Coste J, Dufour M-N, Jouin P (1992) Total synthesis of the proposed structure of dolastatin 15. Tetrahedron 48:4114–4122
Pettit GR (1997) The Dolastatins. Fortschr Chem Org Naturst 70:1–79
Pettit GR, Kamano Y, Brown P, Gust D, Inoue M, Herald CL (1982) Antineoplastic agents 3. Structure of the cyclic peptide dolastatin 3 from *Dolabella auricularia*. J Am Chem Soc 104:905–907
Pettit GR, Kamano Y, Herald CL, Tuinman AA, Boettner FE, Kizu K, Schmidt JM, Baczynskyj L, Tomer KB, Bontems RJ (1987a) The isolation and structure of a remarkable marine animal antineoplastic constituent: dolastatin 10. J Am Chem Soc 109:6883–6885
Pettit GR, Kamano Y, Holzapfel CW, van Zyl WJ, Tuinman AA, Herald CL, Baczynskyj L, Schmidt JM (1987b) Antineoplastic agents. 150. The structure and synthesis of dolastatin 3. J Am Chem Soc 109:7581–7582
Pettit GR, Kamano Y, Kizu H, Dufresne C, Herald DL, Bontems R, Schmidt JM, Boettner FE, Nieman RA (1989a) Antineoplastic agents. 173. Isolation and structure of the cell growth inhibitory depsipeptides dolastatins 11 and 12. Heterocycles 28:553–558
Pettit GR, Singh SB, Hogan F, Lloyd-Williams P, Herald DB, Burkett DD, Clewlow PJ (1989b) Antineoplastic agents. Part 189. The absolute configuration and synthesis of natural (–)-dolastatin 10. J Am Chem Soc 111:5463–5465
Pettit GR, Kamano Y, Dufresne C, Cerny RL, Herald DL, Schmidt JM (1989c) Isolation and structure of the cytostatic linear depsipeptide dolastatin 15. J Org Chem 54:6005–6006
Pettit GR, Herald DL, Singh SB, Thornton TJ, Mullaney JT (1991) Antineoplastic agents. 220. Synthesis of natural (–)-dolastatin 15. J Am Chem Soc 113:6692–6693
Pettit GR, Thornton TJ, Mullaney JT, Boyd MR, Herald DL, Singh SB, Flahive EJ (1994) Antineoplastic agents. 248. The Dolastatins 20. A convenient synthetic route to dolastatin 15. Tetrahedron 50:12097–12108
Pettit GR, Srirangam JK, Singh SB, Williams MD, Herald DB, Barkoczy J, Kantoci D, Hogan F (1996) Dolastatins. 24. Synthesis of (–)-dolastatin 10. X-ray molecular structure of *N,N*-dimethylvalyl-valyl-dolaisoleucine tert-butyl ester. J Chem Soc Perkin Trans 1:859–863
Pettit GR, Xu J-P, Williams MD, Hogan F, Schmidt JM, Cerny RL (1997) Antineoplastic agents. 370. Isolation and structure of dolastatin 18. Bioorg Med Chem Lett 7:827–832
Pettit GR, Hogan F, Herald DL (2004) Synthesis and X-ray crystal structure of the *Dolabella auricularia* peptide dolastatine 18. J Org Chem 69:4019–4022
Roux F, Maugras I, Poncet J, Niel G, Jouin P (1994) Synthesis of dolastatin 10 and [R-Doe]-dolastatin 10. Tetrahedron 50:5345–5360
Shioiri T, Hayashi K, Hamada Y (1993) Stereoselective synthesis of dolastatin 10 and its congeners. Tetrahedron 49:1913–1924
Sodano G, Spinella A (1986) Janolusimide, a lipophilic tripeptide toxin from the nudibranch mollusk *Janolus cristatus*. Tetrahedron Lett 27:2505–2508
Sone H, Nemoto T, Ishiwata H, Ojika M, Yamada K (1993a) Isolation, structure, and synthesis of dolastatin D, a cytotoxic cyclic depsipeptide from the sea hare *Dolabella auricularia*. Tetrahedron Lett 34:8449–8452
Sone H, Nemoto T, Ojika M, Yamada K (1993b) Isolation, structure, and synthesis of dolastatin C, a new depsipeptide from the sea hare *Dolabella auricularia*. Tetrahedron Lett 34:8445–8448
Sone H, Shibata T, Fujita T, Ojika M, Yamada K (1996) Dolastatin H and isodolastatin H, potent cytotoxic peptides from the sea hare *Dolabella auricularia*: isolation, stereo structures, and synthesis. J Am Chem Soc 118:1874–1880
Sone H, Kigoshi H, Yamada K (1997) Isolation and stereostructure of dolastatin I, a cytotoxic cyclic hexapeptide from the Japanese sea hare *Dolabella auricularia*. Tetrahedron 53:8149–8154
Suenaga K, Mutou T, Shibata T, Itoh T, Kigoshi H, Yamada K (1996) Isolation and stereostructure of aurilide, a novel cyclodepsipeptide from the Japanese sea hare *Dolabella auricularia*. Tetrahedron Lett 37:6771–6774

Suenaga K, Mutou T, Shibata T, Itoh T, Fujita T, Takada N, Hayamizu K, Takagi M, Irifune T, Kigoshi H, Yamada K (2004) Aurilide, a cytotoxic depsipeptide from the sea hare *Dolabella auricularia*: isolation, structure determination, synthesis, and biological activity. Tetrahedron 60:8509–8527

Takahashi T, Nagamiya H, Doi T, Griffiths PG, Bray AM (2003) Solid phase library synthesis of cyclic depsipeptides: aurilide and aurilide analogues. J Comb Chem 5:414–428

Tomioka K, Kanai M, Koga K (1991) An expeditious synthesis of dolastatin 10. Tetrahedron Lett 32:2395–2398

Kahalalide F and ES285: Potent Anticancer Agents from Marine Molluscs

G. Faircloth, C. Cuevas

Abstract. The marine environment is proving to be a very rich source of unique compounds with significant activities against cancer of several types. Finding the sources of these new chemical entities has made it necessary for marine and medical scientists to find enterprising ways to collaborate in order to sample the great variety of intertidal, shallow and deep-water sea life. Recently these efforts resulted in a first generation of drugs from the sea undergoing clinical trials. These include PharmaMar compounds: Yondelis, Aplidin, kahalalide F, ES285 and Zalypsis. Two of these compounds, kahalalide F and ES285, have been isolated from the Indopacific mollusc *Elysia rufescens* and the North Atlantic mollusc *Spisula polynyma*, respectively.

16.1 Introduction

Potent cytotoxic agents have a well-established role in the treatment of cancer. Many of the anticancer agents currently in use are of natural origin (i.e. vinca alkaloids, taxanes, anthracyclines), derived from terrestrial plants and microorganisms, or are natural product derivatives. The relevance of the sea as a source to discover novel anticancer compounds was validated by the discovery, development and marketing approval of 1-beta-D-arabinofuranosylcytosine (ARA-C). The available results clearly anticipated the potential of the marine ecosystem in cancer therapy. This chapter describes the progress made and the perspectives in clinical development of two innovative marine anticancer compounds: kahalalide F and ES285. The mechanistic data generated in parallel with the clinical program confirms the potential of the marine ecosystem in the discovery of new agents acting against new relevant cellular targets in cancer cell biology.

16.2 Kahalalide F

Kahalalide F is a partially cyclic depsipeptide with the unusual feature of having a short-chain fatty acid amide at its amino terminal residue

Maria del Carmen Cuevas Marchante
PharmaMar SA, Avda de los Reyes 1, P.1. La Mina–Norte
28770 Colmenar Vieja, Madrid, Spain

Progress in Molecular and Subcellular Biology
Subseries Marine Molecular Biotechnology
G. Cimino, M. Gavagnin (Eds.): Molluscs

(Fig. 16.1; Hamann and Scheuer 1993; López-Maciá et al. 2001). A number of other natural kahalalides (A, B, G) have also been described (Hamann et al. 1996); and many of them show activity against cancer and AIDS-related opportunistic infections. These compounds were discovered in the sea mollusc *Elysia rufescens* (a marine gastropod of the subclass Opisthobranchia) collected at Kahala Bay in Honolulu (Hawaii). Several species of animals in this class are known to acquire, process and accumulate chemicals produced by the algae on which they feed. Surprisingly, some species are able to retain chloroplasts from these algae that remain photosynthetically active within the animal for prolonged periods of time (Green et al. 2000). In the case of kahalalide F, the compound is found in the algae (*Bryopsis pennata*) on which the *Elysia* molluscs feed, albeit in a much reduced concentration (Becerro et al. 2001). In this natural setting, a biological role has been proposed for kahalalides as a deterrent to the feeding behaviour of predators of *Elysia* nudibranchs.

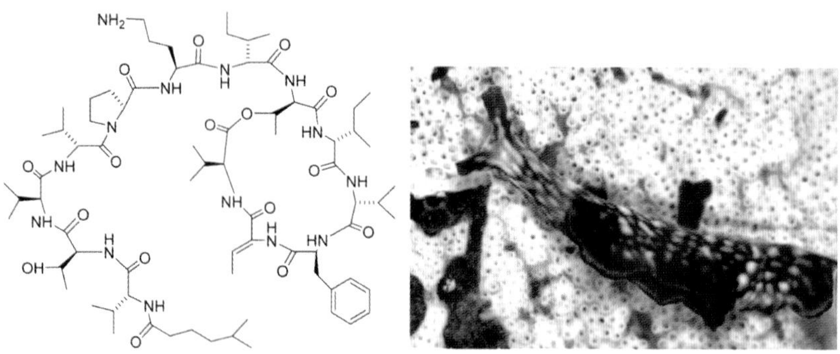

Fig. 16.1. Structure and source organism of kahalalide F

16.2.1 Mechanism of Action

The primary mechanism of kahalalide F action has not been identified; however, early experimental results have provided some insight into the physiological events that correlate with tumour cell killing by this compound. In the NCI COMPARE analysis, kahalalide F did not exhibit significant correlations to any other standard chemotherapeutic agent, suggesting that the compound may possess a unique way of achieving its biological effectiveness. In these in vitro studies, selectivity was noted to colon, central nervous system (CNS), melanoma, prostate and breast tumour cell lines, where PC-3 and DU-145 prostate tumour cell lines happened to be most sensitive with LC_{50} concentrations below 1 μM (NCI data).

Neither protein nor nucleic acid syntheses have been found to be inhibited in cultured cells by sublethal concentrations of kahalalide F. Topoisomerase enzymatic activities (I or II) are also not affected; and no damage to DNA has been specifically correlated with exposure to kahalalide F. A cell cycle block in G0–G1 has been identified in a variety of tumour cell lines that include prostate (DU-145), cervical (HeLa), colon (HT-29), head and neck (HN30) and non-small cell lung carcinoma (NSCLC; HOP62) all with IC_{50} values in the 1 μM range (Córdoba et al. 2003).

Kahalalide F is strongly cytotoxic to both wild-type p53 and mutated p53 tumour cells in the NCI panel. A number of cell lines overexpressing multi-drug resistance proteins (e.g. PC-3 prostate, CACO-2 colon, UO-31 renal, MCF7 breast) as well as cell lines resistant to topoisomerase II inhibitors are sensitive to kahalalide F. This suggests that the compound may act independently of the respective resistance mechanisms.

Cultured cells exposed to biologically relevant concentrations of kahalalide F detach from their substrata and become markedly swollen. This is associated with the formation of large intracellular vacuoles (García-Rocha et al. 1996). Within minutes, these engorged vesicles move from the periphery of the cell to a perinuclear location, as observed by confocal laser scanning microscopy. Confocal fluorescence microscopy studies using a fluorescent acidophilic probe (LysoTracker Green), specific to the inner organelle membrane and confirmed by antibodies to the lysosomal-located enzyme cathepsin D (Fig. 16.2), have identified major and immediate effects on lysosomes. Moreover, there is an increase in lysosomal pH. However, cytoskeletal structures and in particular the microtubule network appear intact; and the morphologies of the endoplasmic reticulum and Golgi apparatus appear to be unaffected by the action of kahalalide F. Thus, it appears that its actions are mostly of a lysosomal nature.

It has been suggested that the subcellular effect might be explained if kahalalide F is inserted as an ionophore in membranes favouring an increase in cation permeability, thus causing a passive water influx and resulting in cisternal dilation. The hydrophobic nature of the compound would not be incompatible with a model in which membrane-associated events trigger cell death. These effects would be similar to those of compounds like the carboxylic ionophore, monensin (Tartakoff 1983).

It has also been suggested that the compound (NCI655128) blocks the EGF receptor and inhibits TGF-β gene expression, receptors that initiate an important signal transduction pathway for proliferation (Wosikowski et al. 1997). These growth factors control the tyrosine kinase class I subfamily that normally mediates signal transduction in the signalling pathway mediated by the "“ras” oncogene. This is overexpressed in many cancer types, especially some breast and ovarian tumour cells. In vitro studies have shown that kahalalide F is selectively cytotoxic to neu$^+$ cells overexpressing Her2, suggesting it may interfere with ErbB2 transmembrane tyrosine kinase activity, a key effector mediating this intracellular pathway. However, it does not inhibit autophosphorylation of the receptors or MEK kinase activity (unpublished data). More recently, kahalalide F has been shown to cause rapid and potent cytotoxic effects in the ErbB2 (HER2/neu) overexpressing breast cancer cell lines (i.e. SKBR3, BT474). This was associated with induction of a hypodiploid cell population, dramatic cell swelling and permeabilization of the plasma and lysosomal membranes (Suárez et al. 2003).

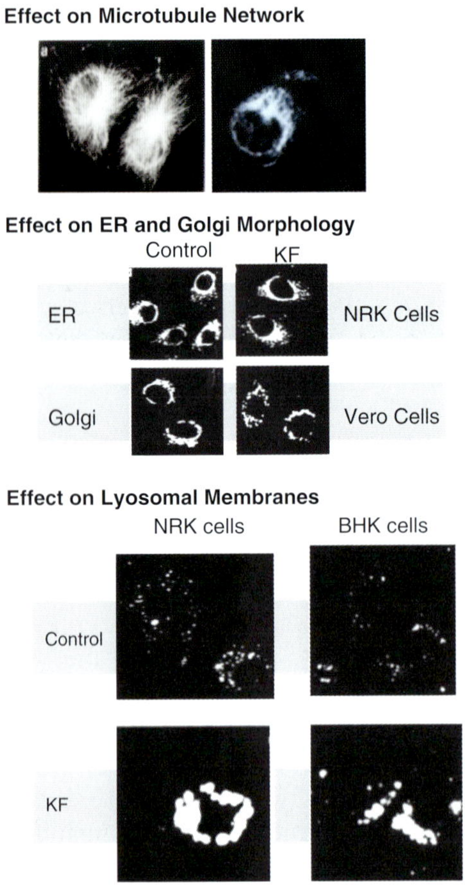

Fig 16.2. Effect on the micotuble network and membrane-bound organelles. (**a**) Effect on Microtubule Network. Typsinized COS-1 cells incubated in the absence or presence of 2μM KF. A reduction of cytoskeletal space is seen as well as the appearance of relatively intact microtubule network. (**b**) Effect on ER and Golgi Morphology Control and treated cells with 2μM Kf. Stained fo ER with anti-PDI (protein disulfide isomerase) and for Golgi apparatus with anti-Golgi alpha mannosidase and showing no difference in organelle morphologies. (**c**) Effect of Lyosomal Membranes Contol and treated cells with 2μM KF. Labeled with the fluorescent acidophilic probe LysoTracker Green and observed unde the laser scanning microscope showing an increase in lysosomal volume as a consequence of kahalalide F treated cells.

Several markers of caspase-dependent apoptosis were negative after kahalalide F exposure, including the externalization of phosphatidyl serine, the release of cytochrome *c* from mitochondria and cleavage of caspase-3 and PARP. Moreover, molecular or chemical inhibition of caspases by ectopic overexpression of Bcl-2 or a pan-caspase inhibitor (zVAD-fmk), respectively, failed to protect against kahalalide F cytotoxicity. Specific inhibitors of cathepsin B (CA-074 Me, zFA-fmk) or D (pepstatin A) also failed to protect against cell death induced by kahalalide F. Taken together, these results suggest that kahalalide F induced cytotoxicity is predominantly due to a process of necrotic cell death involving oncosis rather than apoptosis. This effect has also been reported in hepatoma cell lines (Sewell et al. 2004).

The sensitivity to kahalalide F in a panel of human tumour cell lines derived from breast (SKBR3, BT474, MCF7), vulval (A431), NSCLC (H460, A549, SW1573, H292) and hepatic carcinoma (SKHep1, HepG2, Hep3B) significantly correlated with protein expression levels of ErbB3 (HER3) but not other ErbB receptors. Exposure to kahalalide F for 4 h induced down-regulation of ErbB3 expression in sensitive cell lines, as well as inhibition of the PI3K-Akt/PKB signalling pathway, which is directly linked to ErbB3 (Janmaat et al. 2004). Moreover, ectopic expression of a constitutively active Akt mutant had a protective effect against kahalalide F cytotoxicity. This suggests ErbB3 and the Akt pathway are major determinants of kahalalide F action on these cell lines.

16.2.2 Preclinical Pharmacology

Early preclinical data identified kahalalide F as a potent new chemical entity showing significant cytotoxic activity below 10 μM (IC_{50}) against solid tumour cell lines. Further evaluation demonstrated that this activity was selective for, but not restricted to, prostate tumour cells. Subsequent studies have identified tumour cells that overexpress the Her2/neu and Her3/neu oncogenes as potentially sensitive targets for kahalalide F. Moreover, kahalalide F cytotoxicity was not schedule-dependent (unpublished data) and it was not a strong multi-drug resistance substrate, as it was effective against many multi-drug resistant tumour cell lines (NCI data).

Preliminary in vitro screening studies identified micromolar activity of kahalalide F against mouse leukaemia (P388) and two human solid tumours: non-small cell lung (A549) and colon (HT-29). Gastric (HS746T, 0.01 μM) and prostate (PC-3, 0.08 μM) tumours were shown to be very sensitive as well. In vitro studies in cell lines of human origin evaluated by the NCI confirmed these results and identified selective activity against colon, NSCLC, melanoma, prostate and breast cancer cells, with potencies ranging from 200 nM (prostate) to 10 μM (leukaemia). Extended in vitro selectivity studies reveals that kahalalide F was active against neu^+ (Her2-overexpressing) human breast tumour cells (Suárez et al. 2003), some primary sarcoma lines, but not hormone-sensitive LNCAP tumour cells (Table 16.1).

Table 16.1. In vitro cytotoxic activity of kahalalide F against solid tumours

tumour	line	IC_{50} (Molar)
chondrosarcoma	CHSA	1.58 μM
osteosarcoma	OSA-FH	1.65 μM
prostate	PC-3	1.02 μM
prostate	DU-145	1.78 μM
prostate	LNCAP	not active
breast	SK-BR-3	2.50 μM
breast	BT-474	2.00 μM

Kahalalide F has also been evaluated in a human tumour colony-forming unit (TCFU) assay from surgically derived tumours (Table 16.2; Córdoba et al. 2003). In particular, complete inhibition against breast, colon, kidney, NSCLC, ovary, prostate, stomach and uterine tumour specimens. To date, prostate and stomach tumour specimens are the most sensitive, with preliminary IC_{50} activities of less than 10 nM in a limited number of specimens that have been tested.

Table 16.2. In vitro activity of kahalalide F in the TCFU assay: long-term exposure

primary tumour	10 nM	100 nM	1.0 μM
breast	0/10	1/10	9/10
colon	3/6	2/6	5/6
head and neck	0/1	0/1	0/1
kidney	0/4	1/4	2/4
NSCL	1/8	1/8	7/8
melanoma	1/2	0/2	1/2
neuroblastoma	0/1	0/1	1/1
ovary	2/10	2/10	9/10
peritoneum	0/1	0/1	1/1
prostate	1/1	1/1	1/1
stomach	1/1	1/1	1/1
unknown primary	0/1	0/1	1/1
uterus	0/1	0/1	1/1
	9/47(19%)	9/47 (19%)	39/47(83%)

[a] Less than 50% survival of TCFU's

In vivo antitumour activity was observed against human breast, colon, prostate and lung tumour cells xenografted into athymic mice. Interestingly, chemotherapy-resistant DU-145 (hormone refractory prostate) tumours, initially responded to kahalalide F at the maximum tolerated dose (MTD) and half MTD levels in the first cycle. One week after the first cycle was completed, a second cycle was started and tumour growth was inhibited further (Table 16.3). In similar fashion, studies with PC-3 xenografted tumours confirmed this tumour growth inhibition (Table 16.4).

Table 16.3. In vivo antitumour activity of kahalalide F to DU-145 human prostate tumours

group	injection dose ($\mu g\ kg^{-1}$)	regimen	cycle 1 volume (mm^3)	day 8 %T/C	cycle 2 volume (mm^3)	day 26 %T/C
vehicle	–	Q2D × 5, iv	598	100	1,398	100
MTD	490	Q2D × 5, iv	240	40	704	50
1/2 MTD	245	Q2D × 5, iv	347	58	697	50
1/4 MTD	123	Q2D × 5, iv	633	106	1,510	108

Table 16.4. In vivo antitumour activity of kahalalide F to PC-3 human prostate tumours

group	injection dose ($\mu g\ kg^{-1}$)	regimen	cycle 1 volume (mm^3)	day 10 %T/C	cycle 2 volume (mm^3)	day 35 %T/C
Vehicle	–	Q2D × 5, iv	734	–	3,512	–
MTD 4	490	Q2D × 5, iv	229	32	2,210	63
1/2 MTD	245	Q2D × 5, iv	489	67	2,466	70
1/4 MTD	123	Q2D × 5, iv	372	51	2,294	65

16.2.3 Preclinical Toxicology

The pharmacokinetic (PK) behaviour of kahalalide F was characterized in mice and confirmed in rats. In conjunction with in vitro and in vivo antitumour activity studies, a pattern of systemic exposure to the drug associated with efficacious dosing regiments was developed (Nuijen et al. 2001; Brown et al. 2002; Gómez et al. 2003).

- A dose of 278 μg kg^{-1} given as a rapid intravenous (i.v.) bolus injection afforded an initial plasma concentration of 1.55 μM. The plasma concentration time profile was distinctly bi-exponential with half-lives of 15.8 min and 4.4 h for the initial and terminal disposition phases, respectively.
- The apparent volumes of distribution of the drug were very large, more than 100 times body weight, suggesting that the compound distributed extensively into peripheral tissues. The total body clearance, 14.5 ml min^{-1} kg^{-1}, was only 23% of hepatic blood flow.
- The C_{max} of the MTD i.v. dose was comparable to the in vitro IC_{50} values for the most sensitive human tumour cell lines.

No drug was found in mouse plasma 24 h after i.v. injection. There was no accumulation upon repeated i.v. injection at an interval of 24 h. Surprisingly, a slightly greater than MTD was too toxic when given as an i.v. single bolus. Daily serial i.v. injections of the MTD dose were not toxic. This preliminary data suggested that kahalalide F was rapidly eliminated from plasma with limited binding to extra vascular tissues. In addition, increased tolerance to repeat dosing without acute cumulative adverse effects may be favourable to clinical development of the drug, if confirmed in the human setting.

In rodents, a single 250 μg kg^{-1} dose bolus injection of kahalalide F was established as the MTD, although there were gender differences. The dose-limiting toxicity was predominantly renal, albeit reversibility was reached by the final necropsy day (day 29). Furthermore, minimal histologic evidence of nephrotoxicity was apparent at the half MTD. Signs of liver function alteration were seen on day 4 at the MTD dose, with recovery by day 29 necropsy. If the MTD was exceeded, there were lethalities with signs of neurotoxicity.

In fractionated dose studies in beagle dogs, slight non-regenerative anaemia was seen in males and females at all dose levels. The effect was typically most severe on day 8, with general recovery by days 15 and 22. No other haematological changes were observed. No correlative histologic changes, i.e. in the bone marrow, spleen or liver, were seen. No other indices of toxicity were seen and drug-related clinical signs or histologic lesions were not apparent.

16.2.4 Synthesis of Kahalalide F

Thus far, only one synthesis of kahalalide F has been published, as described by López-Maciá et al. (2001). They used a solid-phase strategy to obtain kahalalide F in a multi-gram scale. The methodology involves elongation of the synthetic chain in the solid phase. With the linear peptide in hand, cyclization in solution follows and finally deprotection allows preparation of the natural compound in a straightforward manner. Moreover, the solid-phase methodology used is easy to scale up and could be applied to generate a wide variety of new analogues.

After synthesis of the linear peptide, the Fmoc/tBu strategy and 2-chlorotrityl chloride resin allowed cleavage of the peptide under mild acid conditions. Next, aminoacids *d-allo*-Thr and the Thr precursor of the *Z*-Dhb were both introduced without protection of the hydroxyl function. For the formation of all the amide bonds, HATU/DIEA was used.

Before deprotecting the peptide from the resin, the alloc group was removed under standard conditions. The cyclization reaction was then performed with PyBOP/DIEA using DMF as a solvent. Finally, the deprotection of the Boc group afforded the natural compound (Scheme 16.1).

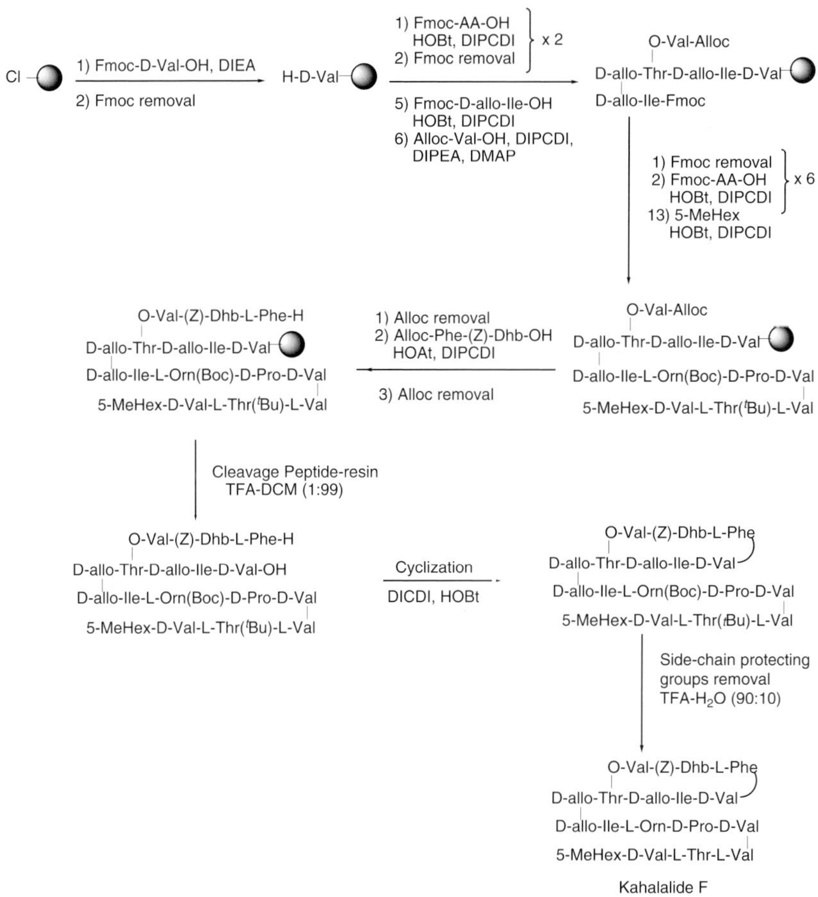

Scheme 16.1. Synthesis of kahalalide F

16.2.5 Clinical Trials

Results from a dose-escalating Phase I study in patients with advanced androgen-resistant prostate cancer were presented at the 2002 Annual Meeting of the American Society for Clinical Oncology (Schellens et al. 2002). As predicted, kahalalide F was found to have rapid plasma clearance in this study. Moreover, the dose of kahalalide F could be safely escalated up to 930 μg m^{-2} day^{-1}. Kahalalide F also demonstrated a favourable safety profile and treatment-related side-effects were non-cumulative and rapidly reversible.

Data from a Phase I study in patients with advanced solid tumours that had failed to respond to previous chemotherapy was presented at the 2002 EORTC-NCI-AACR annual meeting (Ciruelos et al. 2002). In this study, kahalalide F was administered as a weekly 1-h i.v. infusion and the dose could be escalated up to 1,200 μg m^{-2} $week^{-1}$. Signs of activity in a variety of cancers were observed at 400–1,200 μg m^{-2} $week^{-1}$. Overall, this data suggested a favourable safety profile for kahalalide F, with no reports from bone marrow or renal toxicities, mucositis, alopecia or general cumulative toxicity.

16.3 ES285

ES285 (Fig. 16.3) is a marine compound found in the mollusc *Mactromeris* (formerly *Spisula polynyma*) by Rinehart et al. (1998). ES285·HCl consists of a linear 18-carbon chain bearing amine and alcohol groups at positions 2 and 3, respectively. Each chiral centre is a single configuration (2*S*,3*R*). Drug substance is synthesized as the hydrochloride salt from commercially available raw materials. The molecular formula of the synthesized material is $C_{18}H_{39}NO$·HCl, with the molecular weight 321.97. ES285 shows antitumour selectivity for certain slow-growing solid tumours, such as those of the liver, prostate and kidney, and it is currently in Phase I clinical trials in Europe.

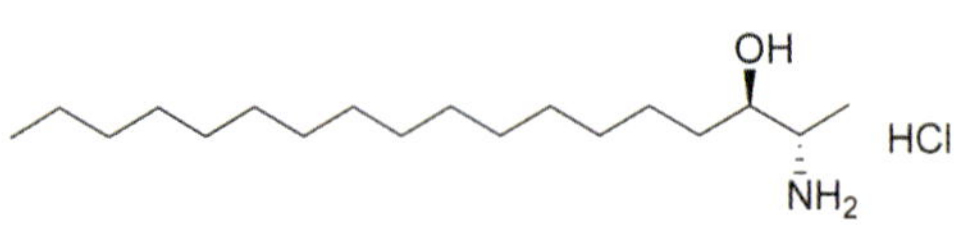

Fig. 16.3. Structure and source organism of ES285

16.3.1 Mechanism of Action

The mechanism of the action of ES285 is under investigation. Available data suggests that the antitumour activity of ES285 may be associated with disruption of the cytoskeleton in cancer cells. The in vitro cytotoxicity of ES285 is schedule-dependent from 1 to 24 h at 100 nM, 1 μM and 10 μM. Cell cycle analysis shows a delayed G2/M transition and an accumulation of cells in G1 after variable drug-washout experiments, regardless of pre-treatment duration of exposure (1 or 24 h; Salcedo et al. 2003).

Cultured tumour cells change their morphology in the presence of ES285, acquiring first a fusiform shape and later becoming rounded without focal adhesions (Fig. 16.4; Cuadros et al. 2000). The transition to bipolar, spindle-shaped cells is also associated with apoptosis that eventually leads to immediate cell death in most, but not all, cells. The selective induction of apoptosis in tumour cells is an area of active study for this compound.

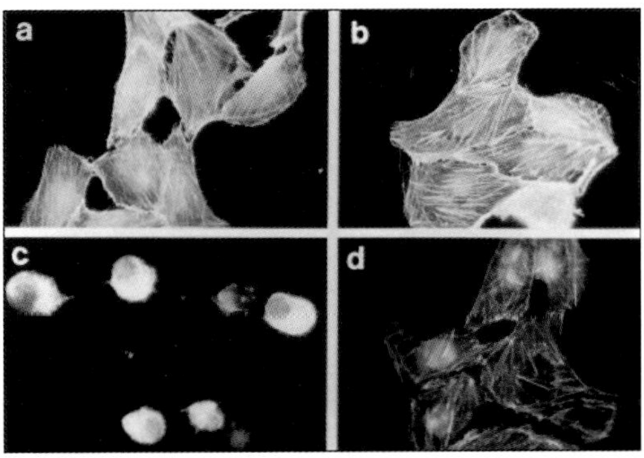

Fig 16.4. Change in cell morphology in the presence of ES285. a, appearance of normal stress fibers; **b**, increase in stress fibers from LPA; **c**, rounding of cells by ES-285; **d**, reduction of LPA-included stress fibers by ES-285.

Microscopic analysis of the cytoskeleton of treated cells indicates that there is an absence of actin stress fibres that are typically regulated by Rho, a small GTP-binding protein. Moreover, stimulation of Rho by lysophosphatidic acid (LPA) is blocked by ES285 (Fig. 16.4). These indirect findings led us to speculate that Rho may be a tentative target for ES285. In contrast, studies from other investigators using both overexpression and siRNA-mediated knockout of RhoA do not support a direct RhoA function or that of signalling pathways under the control of RhoA (Lacal et al. 2004).

ES285 may act through G protein-coupled endothelium differentiation gene (EDG) receptors, considering its structural homology with bioactive lipids such as sphingosine-1-phosphate. In particular, ES285 could utilize EDG receptors coupled to several G proteins and thus activate RhoA through a signalling pathway originated from one of the former proteins. Recently, Salcedo (2005) showed that ES285 treatment effects were due to interaction with EDG receptors. However, this data also showed that these receptors were not essential for ES285-induced cell death.

ES285 produces cell vacuolation that precedes apoptosis. The resulting multi-nucleated, dividing cells are unable to separate (note G2/M arrest preceding apoptosis). Some vacuolated cells undergo blebbing and die quickly, while other cells take longer for this to occur. Further details are beginning to emerge of the molecular targets involved in the early and late events induced by ES285. Using either HeLa or Jurkat cell lines, ES285 has a specific effect on cell cycle distribution (Fig. 16.5). Following prolonged exposure, however, cells in early G1 become progressively more sensitive to the drug and, within 24 h, sub-G1 cells represent 70–80% of all apoptotic cells when exposed to 10 μM ES285.

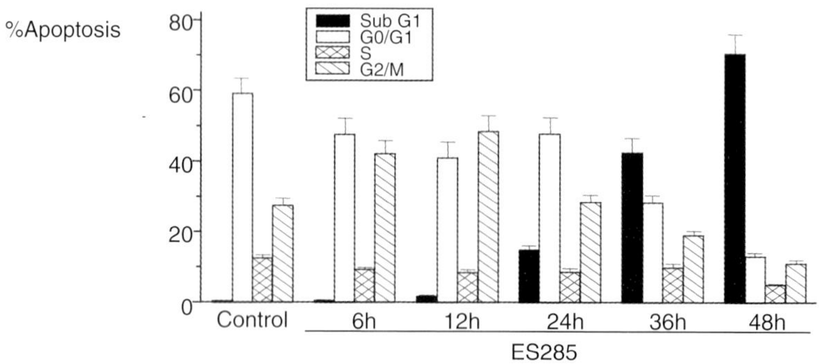

Fig. 16.5. Time-dependent induction of apoptosis in sub-G1 HeLa cells

After 24 h exposure, 10 μM ES285 produces internucleosomal DNA breakdown or apoptotic cell death. The Fas/FasL systems are not involved in this process. However, caspase 3 and its substrate, PARP, are activated by ES285 and subsequent markers of apoptosis evolve within 24 h. Potent and persistent activation of JNK is not affected by ES285. There is only a transient activation in 12 h. ERK activation, a survival signal, is clearly involved within 24 h in a strong time-dependent induction. Blocking ERK leads to an increased apoptotic response to ES285; and the use of cells transfected with antiapoptotic genes renders them unresponsive to ES285. Finally, ES285 induces mitochondrial release of cytochrome *c*.

Other tumour cell types exhibit a consistent pattern of delayed apoptosis as described earlier; and overall ES285 has shown that it can profoundly influence several targets in the induction pathway of apoptosis. For instance, this compound activates caspase 3 and 12 and modifies the phosphorylation level of p53, thus suggesting that ES285 triggers an atypical cell death program (study UIC/TRL 391: single i.v. dose in rats).

16.3.2 Non-Clinical Studies

Preliminary in vitro studies indicate that ES285 has potent activity against cell line subpanels containing solid tumours, lymphomas and leukaemias, with selectivity for certain solid tumours (i.e. colon, gastric, pancreas, pharynx, renal) at IC_{50} potencies in the nanomolar range. SK-HEP-1 hepatoma tumour cells deserve special mention because they showed an IC_{50} of 0.562 pM. The activities against solid tumour cell lines were generally tenfold more potent than those for leukaemias and lymphomas and, more specifically, the slow-growing adherent tumour cells seemed to be more sensitive to ES285. In vitro studies by the NCI have confirmed that the antitumour activity of ES285 ranges from 0.1 to 10 μM.

The ES285 effects appear to be long-lasting. In particular, human HCT-116 N7 colon tumour cells were pre-treated with various concentrations of ES285 for either 1 or 24 h (Fig. 16.6). After 1 h of pre-treatment and subsequent drug removal, the cytotoxic effects of 10 μM ES285 continued for up to 72 h. After 24 h of pre-treatment, in turn, cytotoxic activity continued to be present for both 1 and 10 μM concentrations of the drug. Moreover, following treatment with 0.1 μM ES285, cytotoxic activity remained for at least 48 h after 24 h exposure.

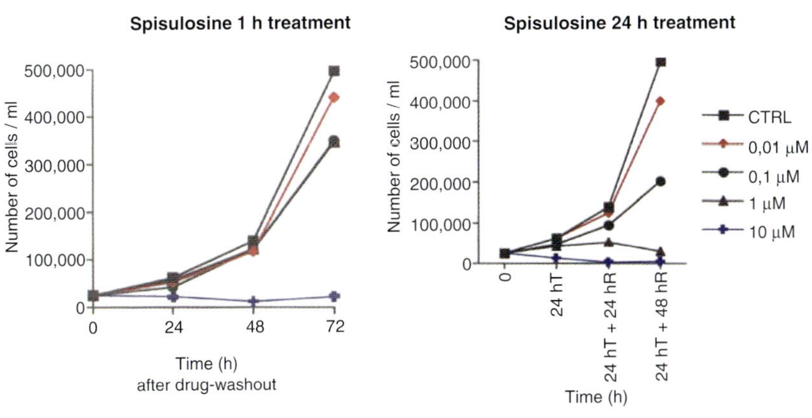

Fig. 16.6. Exposure-dependent in vitro cytotoxicity

ES285 is active in vivo against certain slow-growing solid human tumours (hepatoma, renal, prostate; Table 16.5).

The efficacy of ES285 has been demonstrated when administered as a continuous infusion in male athymic rats bearing a human liver adenocarcinoma tumour. Analysis of net tumour growth of the corresponding treated (T) groups relative to the vehicle control (C) group indicated that the optimal value of %T/C occurred on day 6 after group randomization, i.e. –80 and –73% for the high-dose and low-dose ES285 infusion groups, respectively.

Table 16.5. In vivo activity in mice of ES285 against human tumours

tumour type/line	MTD ($mg\ kg^{-1}$)	regimen	%T/C	score	model
leukaemia					
ip P388	10	QD × 5, *ip*	104	–	survival
melanoma					
iv B16	10	QD × 9, *ip*	111	–	survival
Hepatoma					
iv SK-HEP-1	10	Q2D × 5, *iv*	122	+	survival
	10	QD × 9, *ip*	14	+++	hollow fibre
sc SK-HEP-1	5		19	+++	
	2.5		22	++	
colon cancer					
sc HT-29	10	QD × 9, *ip*	184	–	hollow fibre
pancreatic cancer					
sc PANC-1	10	QD × 9, *ip*	105	–	hollow fibre
Melanoma					
sc MRI-H-187	10	Q2D × 5, *iv*	48	+/–	xenograft
renal cancer					
sc MRI-H-121	25	Q4D × 3, *ip*	28	++	xenograft
prostate cancer					
sc PC-3	25	Q4D × 3–5, *ip*	<1	++++	xenograft
sc DU-145	25	Q4D × 3–5, *ip*	34	+	xenograft

Non-clinical toxicity has been tested in mice, rats, dogs and monkeys via the i.v. route. Across all species at high doses, the heart was the common target organ of toxicity and the histopathologic change noted was myocardial degeneration. However, the severity decreased inversely with the higher order of animal. Serum troponin I levels were predictive of toxicity in the monkey. Elevations in liver enzyme function tests were frequently observed in the dog, monkey and rat, with cellular changes noted in the rat liver (degenerative). The kidney indicative of nephropathy was also observed to be a target organ of toxicity in the mouse, rat and monkey. The tests and epididimides were affected in both rat and dog, in that degenerative cellular changes were noted. Although injection site changes were noted, predominantly in the rodents (mouse, rat), no such inflammatory, edematous or degenerative changes were noted in the monkey.

16.3.3 Synthesis of ES285

A synthetic scheme that allows preparation of sufficient quantities of ES285·HCl for non-clinical and clinical evaluation has been developed at PharmaMar (Scheme 16.2). ES285 is prepared in four steps from the commercially available chiral starting material (*S*)-2 (*N*,*N*-dibenzylamino)-1-propanol, **1**.

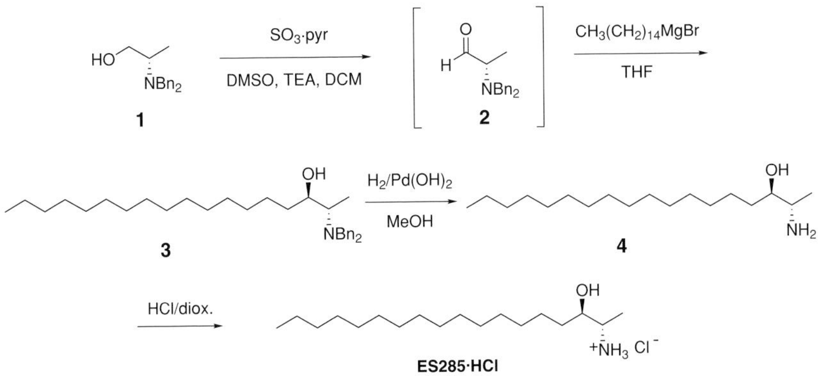

Scheme 16.2. Synthesis of ES285

The chiral alcohol **1** is oxidized to the corresponding aldehyde, which is then coupled with the Grignard reagent derived from 1-bromopentadecane. The coupling reaction proceeds with high *anti:syn* selectivity with the desired isomer representing about 90% of the crude reaction mixture. Compound **3** is purified by preparative HPLC. Next, this intermediate is deprotected by hydrogenation to produce the free amine. Last, the hydrochloride salt of ES285 is generated under anhydrous conditions to produce the final drug substance. Using this approach, ES285 has been prepared on thc multi-gram scale.

16.3.4 Clinical Trials

ES285 is currently under investigation in patients with advanced malignant solid tumours as a single agent. Four Phase I clinical trials are ongoing. All these trials are still at the dose-escalation stage and no reliable data is available at present.

16.4 Conclusions

The continuing PharmaMar exploration program has shown the potential use of marine ecosystems as a source of new anticancer compounds. As result of this program, we are developing five new chemical entities as anticancer agents. Further research and development of these marine-derived compounds requires the collaboration of a diverse range of disciplines, including marine biology, biochemistry, toxicology, pharmacology and oncology.

References

Becerro MA, Paul VJ, Goetz G, Scheuer PJ (2001) Chemical defences of the sacoglossan mollusk *Elysia rufescens* and its host alga *Briopsis* sp. J Chem Ecol 27:2287–2299

Brown AP, Morrissey RL, Faircloth GT, Levine BS (2002) Preclinical toxicity of Kahalalide F, a new anticancer agent: single and multiple dosing regimens in the rat. Cancer Chemother Pharmacol 50:333–340

Ciruelos C, Trigo J, Pardo J, Paz-Ares L, Estaun N, Cuadra C, Domínguez M, Marín A, Jimeno JM, Izquierdo M (2002) A phase I clinical and pharmacokinetic study with Kahalalide F in patients with advanced solid tumours with a continuous weekly 1-hour iv infusion schedule. Eur J Cancer 38:33

Córdoba S, Zapata I, Romero J, de la Torre A, López-Martin JA, Vargas JA, Sánchez-Prieto R, Corbacho C, Regueiro CA, Magallón R (2003) In vitro chemosensitivity, cell cycle redistribution and apoptosis induction by Kahalalide F, a new marine compound in a panel of human tumoural cells. Eur J Cancer Suppl 1:S173

Cuadros R, Montejo de Garcini E, Wandosell F, Faircloth GT, Fernández-Sousa JM, Avila J (2000) The marine compound spisulosine, an inhibitor of cell proliferation, promotes the disassembly of actin stress fibers. Cancer Lett 152:23–29

Gajate C, An F, Nieto-Miguel T, Mollinedo F (2005) Induction of apoptosis by the novel marine-derived antitumour agent ES-285 (spisulosine): cell cycle arrest, and mitochondrial and caspase involvement (in press)

<AQ. Please update Gajate et al. (2005).>

García-Rocha M, Bonay P, Avila J (1996) The antitumoural compound Kahalalide F acts on cell lysosomes. Cancer Lett 99:43–50

Gómez SG, Bueren JA, Faircloth GT, Jimeno J, Albella B (2003) In vitro toxicity of three new antitumoural drugs (trabectedin, aplidin and kahalalide F) on hematopoietic progenitors and stem cells. Exp Hematol 31:1104–1111

Green BJ, Li W, Manhart JR, Fox TC, Summer EJ, Kennedy RA, Pierce SK, Rumpho ME (2000) Mollusc–algal chloroplast endosymbiosis. Photosynthesis, thylakoid protein maintenance, and chloroplast gene expression continue for many months in the absence of the algal nucleus. Plant Physiol 124:331–342

Hamann MT, Scheuer PJ (1993) Kahalalide F: a bioactive depsipeptide from the sacoglossan mollusk *Elysia rufescens* and the green alga *Bryopsis* sp. J Am Chem Soc 115:5825–5826

Hamann MT, Otto CS, Scheuer PJ, Dunbar DC (1996) Kahalalides, bioactive depsipeptides from the sacoblassan mollusk *Elysia rufescens* and the green alga *Bryopsis* sp. J Org Chem 61:6594–6600

Janmaat ML, Kruyt FAE, Jimeno J, Rodríguez JA, Giaccone G (2004) Kahalalide F induces caspase-independent cytotoxicity that correlates with HER2 and/or HER3 expression levels and is accompanied by down-regulation of AKT signaling. Proc AACR 95:5328

Lacal JC, Rodríguez-González A, Alvarez-Miranda M (2004) Study of the mechanism of action of ES285. PharmaMar, Madrid

López-Maciá A, Jiménez JC, Royo M, Giralt E, Albericio F (2001) Synthesis and structural determination of Kahalalide F. J Am Chem Soc 123:11398–11401

Nuijen B, Bouma M, Talsma H, Manada C, Jimeno JM, López-Lazaro L, Bult A, Beijnen JH (2001) Development of a lyophilized, parenteral pharmaceutical formulation of the investigational polypeptide marine anticancer agent Kahalalide F. Drug Dev Ind Pharm 27:767–780

Rinehart KL, Fregeau NL, Warwick RA (1998) US patent 6107520c

Salcedo M (2005) The marine antitumour compound ES 285 activates EGD receptors and PKC, and in parallel triggers an atypical cell death pathway (in press)
Salcedo M, Cuevas C, Sánchez-Puelles JM, Otero G, Sousa JM, Avila J, Wandosell F (2003) Proc Am Assoc Cancer Res 44:3649
Schellens JH, Rademaker JL, Horenblas S, Meinhardt W, Stukvis E, de Reijke TM, Jimeno JM, López-Lazaro L, López-Martín JA, Beijnen JH (2002) Phase I and pharmacokinetic study of Kahalalide F in patients with advanced androgen refractory prostate cancer. Proc ASCO 38:451
Sewell JM, Langdon SP, Smyth JF, Jodrell I, Guichard S (2004) Kahalalide F appears to promote necrotic cell death in hepatoma cell lines. Proc AACR 95:1509
Suárez J, Gonzalez L, Cuadrado A, Berciano M, Lafarga M, Muñoz A (2003) Kahalalide F, a new marine-derived compound induces oncosis in human prostate and breast cancer cells. Mol Cancer Ther 2:863–872
Tartakoff AM (1983) Perturbations of vesicular traffic with the carboxylic ionophore monensin. Cell 32:1026–1028
Wosikowski K, Schuurhuis D, Johnson K (1997) Identification of Epidermal growth factor receptor and c-ErbB2 pathway inhibitors by correlation with gene expression patterns. J Nat Cancer Inst 89:1505–1515

Subject Index

List of Cited Species

Printing: Krips bv, Meppel
Binding: Stürtz, Würzburg